Collins
dictionary *of*
Geology

Dorothy Farris Lapidus
Consultant Editor: Isobel Winstanley

Revised and updated by
Dr James MacDonald
Dr Christopher Burton

Collins

HarperCollins,
77-85 Fulham Palace Road,
London, W6 8JB

www.collins.co.uk

First published as *The Facts on File Dictionary of Geology and Geophysics*
by Facts on File Publications, New York and Oxford, 1987

First published by HarperCollins Publishers 1990
This revised edition first published by HarperCollins
Publishers 2003

Reprint 10 9 8 7 6 5 4 3 2

ISBN-13 978-0-00-723226-0

A catalogue record for this book is available from the British Library

Printed and bound in Great Britain by Clays Ltd, St Ives plc

Mixed Sources
Product group from well-managed
forests and other controlled sources
www.fsc.org Cert no. SW-COC-1806
© 1996 Forest Stewardship Council

FSC is a non-profit international organisation established to promote the
responsible management of the world's forests. Products carrying the FSC
label are independently certified to assure consumers that they come
from forests that are managed to meet the social, economic and
ecological needs of present and future generations.

Find out more about HarperCollins and the environment at
www.harpercollins.co.uk/green

The Authors

James G. MacDonald, MBE, BSc, PhD, FGS, FMinSoc, FRSA
Honorary Research Fellow, Faculty of Education, University of Glasgow

Dr Jim MacDonald, until recently Director of Lifelong Learning of the Faculty of Education at the University of Glasgow, has over 35 years of experience in the education of adults. A graduate of the University of Glasgow, much of his career has been dedicated to the public understanding of geology. His wide-ranging interests include igneous petrology and volcanic activity, environmental geology and the geology of the Balkans, and he has led numerous geological study tours to Brittany, Bulgaria, the Canary islands, Greece, the Hawaiian Islands, Iceland, Madeira and Norway. He has carried out geological research on the Carboniferous igneous rocks of the Midland Valley of Scotland and on the age and structural relationships of the Rhodope Complex of Southern Bulgaria. He is a member of Council of the Mineralogical Society, joint principal editor of *Mineralogical Abstracts* and a past president of the Geological Society of Glasgow.

Christopher J. Burton, BSc, PhD, FGS
Head of the Division of Earth Sciences, University of Glasgow

Dr Chris Burton, a graduate of the University of Exeter, is highly experienced in the teaching of geology, especially in the fields of, Earth history, palaeontology, economic geology, geological mapping, stratigraphy, environmental geology, hydrogeology and field skills. His research interests cover Ordovician Chitinozoa with particular reference to the Highland Border Complex, Carboniferous paleaoecology, Scyphozoa and Porifera, and the Tertiary and Quaternary faunas and stratigraphy of Macaronesia (Central Atlantic Islands). During his career he has carried out particular studies of the geology of southwest England, Brittany, Montagne Noir (south of France), western Germany, central Scotland and Madeira. In addition to his duties as Head of Division he is active in quality assurance, school/university liaison, is a member of the Steering Committee of the Scottish Earth Science Education Forum and a council member of the Geological Society of Glasgow.

Isobel Winstanley graduated in geology from the University of Edinburgh, where she also completed a PhD. Since 1975 she has been a member of the part-time staff of the Open University, teaching the science foundation course.

Dorothy Farris Lapidus holds a degree in mathematics and physics from the City College of New York. She is a writer, lecturer and translator, and has worked in a number of scientific and technical fields with companies such as NASA and Bell Telephone. She has travelled extensively throughout the United States, Canada and Iceland on geological studies.

Preface

Since publication of the previous revision the breadth of geology has expanded to include a wider coverage of Earth science. There have been a number of important developments in the subject as well as changes in nomenclature and classification. The latter are reflected in particular in the entries relating to divisions of the stratigraphical column, sedimentology and some groups of fossils, particularly the brachiopods. New entries include a number relating to extraterrestrial impacts, the snowball Earth hypothesis and the Earth's deep interior. The scope of a geological dictionary covers a wide range of specialisms within Earth science. We are particularly grateful to our colleagues in the Division of Earth Sciences at the University of Glasgow and the Scottish Universities' Environmental Research Centre, East Kilbride, for their patience and ready provision of advice and comments. We also wish to thank Clare Crawford for her invaluable help with editing and checking of cross-references.

James G. MacDonald, Christopher J. Burton, 2003

Preface to the Collins' 1990 Edition

This edition has been revised to appeal particularly to undergraduates studying outside North America, but its hoped that it will also prove useful to advanced school pupils and to anyone with a general or professional interest in geology who requires an up-to-date source of reference. The coverage of most subjects has been expanded. I have tried to include mainly those terms that may be encountered in the major geological textbooks. Technical terms used in palaeontology have been added, in addition to stratigraphical stage names in general use, although some specifically British and European stratigraphical divisions have also been given. The main North American stratigraphical divisions have been retained, as they are often encountered in the geological literature.

Where cross-references are necessary, the keyword is given in small capital letters. Italics are used to indicate subsidiary terms, which are defined in the entry in which they appear.

I would like to thank Drs Euan Clarkson, Stewart Monroe, John Craven, John Dixon and Roy Gill, and Professor Brian Upton for their advice; I was also grateful for the helpful comments of Dr Douglas

Weedon, who read the proofs on Collins' behalf. I would like to thank Monica Thorp and Jim Carney of Collins for their assistance and my husband and family for their patience and support.

Isobel Winstanley

Preface to Original Edition

During the past few decades, the results of particular studies have had a profound impact on developments in many areas of geology and geophysics. Certain concepts have had to be abandoned and others modified or revised.

The concept of sea-floor spreading, the constant generation of new sea floor, offered a reason for why oceanic crust is younger than continental crust and not older, as had been believed. The same concept was one of the ideas that was basic to an acceptance of the notion of continental drift. Both these hypotheses were later joined by the theory of plate tectonics, a singular contribution of which was the derivation of some sense of order from the distribution patterns of earthquakes and volcanoes.

Many geological terms have now been redefined in the context of the components of modern geology. Where the association is not direct, additional references are given.

I am deeply grateful to Professor Donald Coates for all his help, particularly in the areas of geomorphology and glaciology. My thanks are also due to Dr Reed Craig for his generous assistance and to Milton Kerr for all the drawings and figures in the Facts On File addition.

Dorothy Farris Lapidus

Contents

Å see ANGSTRÖM.

a'a' or **aa** *n*. (*pronounced* ah ah) the Hawaiian term for LAVA FLOWS characterised by a rough, jagged or spiny surface. Compare PAHOEHOE.

Aalenian *n*. the lowest STAGE of the Middle JURASSIC.

a-axis *n*. (*crystallography*) one of the referential axes used in the description of a CRYSTAL; it is directed horizontally, from front to back. See B-AXIS, C-AXIS (see Fig. 23).

Abbé refractometer *n*. an instrument that determines the refractive index of liquids, minerals and gemstones. A sample is enclosed between two similar glass prisms, and the total refraction at the interface is noted.

ablation *n*. **1.** a reduction in the volume of snow, ice or FIRN from glaciers, snow fields or sea ice. It results from evaporation, sublimation, deflation and melting caused by rain and warm air, and by CALVING.
2. removal by vaporisation of the molten surface layers of meteorites during passage through the atmosphere.

ablation cone *n*. a debris-covered cone of ice, snow or FIRN formed by differential ABLATION.

ablation form *n*. a FORM produced by the reduction in the volume of snow, ice or FIRN from glaciers, snowfields or ice.

aboral *adj*. describing a point or area opposite the mouth of an invertebrate.

aboral cup see CALYX.

aboral surface *n*. the upper surface of the TEST (skeleton) of an ECHINOID (see Fig. 27).

abrasion *n*. **1.** the mechanical wearing down or grinding away of rock surfaces by the friction of rock particles carried by wind, running water, ice or waves. Abrasion polishes, smoothes, scratches or pits exposed rock faces. See CORRASION.
2. the effect of abrading, e.g. abrasion left by glacial action.

abrasion pH *n*. acidity resulting from the absorption of OH⁻ or H⁺ ions at the surfaces of finely ground minerals suspended in water.

abrasion platform see WAVE-CUT PLATFORM.

absarokite see SHOSHONITE.

absolute age *n*. the geological age of a rock, FOSSIL or FORMATION expressed in units of time, usually years. It commonly refers to ages determined by radiometric dating, but other methods such as FISSION TRACK DATING and DENDROCHRONOLOGY may also be used. The term 'absolute' is somewhat misleading, since it implies an exactness that is not always readily achieved. See DATING METHODS; compare RELATIVE AGE.

absolute humidity *n*. the content of water in the atmosphere, expressed as the mass of water vapour per unit volume of air. Compare RELATIVE HUMIDITY.

absolute permeability *n*. the ability of a rock to transmit a fluid when the saturation with that fluid is 100 per cent. See also EFFECTIVE PERMEABILITY, RELATIVE PERMEABILITY.

absolute temperature *n*. temperature measured with respect to absolute zero (–273.15°C) and expressed in kelvins (K).

absorptance see ABSORPTION COEFFICIENT.

absorption *n.* **1.** assimilation or incorporation by molecular or chemical action, as of liquids in solids or of gases in liquids. Compare ADSORPTION.
2. reduction in light intensity during transmission through a medium.
3. the process by which one energy form may be converted into another, e.g. seismic wave energy to heat.
4. penetration of surface water into the LITHOSPHERE.

absorption coefficient or **absorptance** *n.* the ratio of the amount of energy absorbed by a given material to the total amount incident on that material.

abstraction *n.* **1.** the simplest type of STREAM CAPTURE.
2. that part of precipitation that is stored, transpired, evaporated or absorbed, i.e. does not become direct RUNOFF.

abyss see DEEP.

abyssal *adj.* pertaining to ocean depths of about 2000 to 6000 metres and to the organisms of that environment. This includes all areas below the base of the continental slope, with the exception of oceanic trenches. The upper region of the abyssal zone, at about 2000 metres, is referred to as the BATHYAL zone. The *abyssal zone* may also be defined as the zone, excluding trenches, where water temperature never exceeds 4ºC. See CONTINENTAL SHELF. See also HADAL.

abyssal cones see SUBMARINE FANS.

abyssal deposits see PELAGIC DEPOSITS.

abyssal fans see SUBMARINE FANS.

abyssal hills *n.* submarine geomorphic features found only in the deep sea (3000 to 6000 metres). They are typically dome-shaped, rising as much as one kilometre above the surrounding abyssal plain and may be up to several kilometres wide at the base. Abyssal hills occur in all oceanic basins but are most prevalent in the Pacific. They are common along the outer margins of the MID-ATLANTIC RIDGE. Their origin is uncertain. Marine geologists and geophysicists tend towards the theory of a volcanic origin for these structures, but it is also possible that some hills are lithified and compacted sedimentary material. See also ABYSSAL PLAIN.

abyssal plain *n.* a flat area of the ocean-basin floor with a slope of less than 10º. Topographic and sedimentary studies of abysmal plains indicate they were formed by deposits from TURBIDITES that obscured the pre-existing geomorphic features. Long sediment cores taken from these plains all contain silts and sand apparently issuing from shallow continental areas. *Trench abyssal plains* are those that lie on the bottom of deep-sea trenches.

Acadian *n.* an obsolete name for the Middle Cambrian in North America. ALBERTAN is the modern equivalent.

Acadian orogeny *n.* a series of deformational events in the early Devonian that affected the northern part of the APPALACHIAN FOLD BELT in North America and the area south of the IAPETUS suture in Britain. The orogeny took its name from Acadia, in the Canadian Maritime Provinces, where it was first recognised. The Acadian orogeny is recognised as a phase of the CALEDONIAN OROGENY of northwest Europe and is ascribed to the collision of the crustal blocks of Eastern and Western AVALONIA with LAURENTIA.

acceleration *n.* (*physics*) the ratio of the increment in velocity to the increment in time:

$$a = \frac{\Delta v}{\Delta t}$$

Velocity has both magnitude and direction and is thus a vector quantity.

acceleration due to gravity *n*. ACCELERATION because of the Earth's gravitational attraction of a freely falling body in a vacuum. The value adopted by the International Committee of Weights and Measures is 9·80665 m s^{-2}; however, the actual value varies with altitude, latitude and composition of the underlying rocks.

accelerometer *n*. a seismograph that records ground ACCELERATION.

accessory mineral *n*. any mineral in a rock not essential to classification of the rock. When present in small amounts it is a *minor accessory*. If the amount is greater or has particular significance, the mineral is called a *varietal mineral* and may be added to the name of the rock (e.g. biotite in biotite granite). The typical formation of accessory minerals occurs during solidification of rocks from magma. Typical minor accessory minerals include topaz, zircon, sphene, rutile and tourmaline. Varietal accessories include biotite, amphibole, olivine and pyroxene although in some rocks these are ESSENTIAL MINERALS. In sedimentary rocks, accessory minerals are mostly HEAVY MINERALS. Compare SECONDARY MINERAL.

accessory plates *n*. thin slices of minerals, cut in particular crystallographic directions and mounted for use with a PETROLOGICAL MICROSCOPE. They are used for determining various optical properties of minerals, including the maximum POLARISATION COLOURS, the fast and slow VIBRATION DIRECTIONS and whether a mineral is optically positive or negative (see INDICATRIX).

The *quartz wedge* is ground to produce polarisation colours from the beginning of the first to the end of the third or fourth order. The *sensitive tint* (first order red) is cut in gypsum so that the interference colour it shows under crossed polars is red of the first order spectrum. The *mica plate* (quarter-wave plate) gives a standard retardation of a quarter of the wavelength of sodium light. All the plates are marked with their fast or slow vibration direction. See also BIREFRINGENCE.

accidental inclusion *n*. a XENOLITH.

accordant *adj*. matching or in agreement. It describes two streams the surfaces of which are at the same level at the junction location. Compare DISCORDANT.

accordant summit level *n*. a hypothetical level that intersects the summits or hilltops of certain regions. In an area of high topographic relief it may imply that the summits are remnants of a plain formed during an earlier erosion cycle. See also SUMMIT ACCORDANCE.

accreting plate boundary, constructive plate boundary or **divergent plate boundary** *n*. a borderline between two crustal plates that are separating and along the seam of which new oceanic LITHOSPHERE is being created. See also MID-OCEANIC RIDGE.

accretion *n*. **1.** gradual enlargement of a land area through the accumulation of sediment carried by a river or stream. See AGGRADATION, PROGRADATION.
2. the increase in size of inorganic bodies by the addition of new material to the exterior.
3. the theory that the continents have increased their surface area during geological history. See CONTINENTAL ACCRETION.
4. or **accretionary hypothesis**, an increase in mass of a celestial body by the incorporation of smaller bodies that collide with it. Gravitational attraction accelerates the accretion process when the body becomes sufficiently large. Accretion is now thought to have been

a significant factor in the formation of the planets from dust grains. See also NEBULAR HYPOTHESIS, CATASTROPHISM.

accretionary lapilli *n*. spheroidal, concentrically layered pellets of VOLCANIC ASH, usually between 2 and 10 millimetres in diameter. They are believed to form through the accretion of ash and dust by condensed water in a moisture-rich ERUPTION COLUMN. The nucleus may be a solid particle or a condensing water droplet. Accretionary lapilli may be produced by rain flushing through the eruption column or the ash cloud accompanying a PYROCLASTIC FLOW, or may result from a phreatomagmatic or PHREATIC EXPLOSION (see also PHREATO-MAGMATIC ERUPTION). They are also found segregated in pipes in pyroclastic flow deposits, where they appear to have been formed by gas streaming up through the flow.

accretionary prism *n*. a complex of deformed wedges of sediment scraped off the descending plate at a SUBDUCTION ZONE and incorporated in the inner trench wall. See ISLAND ARC, Fig. 53.

accretion vein *n*. a vein formed by the repeated filling of channelways with mineral deposits and by the reopening of these channels by fractures that develop within the zone of mineralisation.

accumulation *n*. **1.** the sum of all processes that contribute mass to a GLACIER or to floating ice or snow cover; these include snowfall, avalanching and the transport of snow by wind. Compare ABLATION.
2. the amount of precipitation or hoarfrost added to a glacier or snowfield.

accumulation area *n*. the part of a glacier or snowfield in which, during the period of a year, the mass balance is positive, i.e. ACCUMULATION exceeds ABLATION.

ACF diagram *n*. a triangular diagram used to represent the MINERAL ASSEMBLAGES that occur in rocks of a variety of chemical compositions under a particular range of metamorphic conditions. ACF diagrams are used for rocks of BASIC igneous composition and also for impure LIMESTONES. Most chemical components can be represented by an equilateral triangle with corners A, C and F. Corner A represents Al_2O_3–Na_2O–K_2O; C is CaO–$^{10}/_3(P_2O_5)$–CO_2 (this allows for the presence of apatite and calcite); F is $FeO + MgO$ (sometimes with corrections made for the presence of ILMENITE and MAGNETITE). All are in molecular proportions.

One requirement for use of an ACF diagram is that the rocks contain silica in amounts sufficient to form not only all the SILICATE minerals present but quartz as well. Compare AFM DIAGRAM, A'KF DIAGRAM.

achneliths *n*. solidified fragments of LAVA spray formed during fountaining of very fluid basaltic lava. They have smooth glassy surfaces moulded by surface tension, e.g. the drop-shaped particles called PELE'S TEARS.

achondrites *n*. STONY METEORITES in which CHONDRULES are absent. They are categorised on the basis of calcium content and are similar in texture and composition to basaltic igneous rocks on Earth and to some lunar rocks. The principal minerals are PYROXENES and PLAGIOCLASE, with little or no nickel/iron phase. Evidence indicates they were formed at or near the surface of a differentiated planet.

acicular *adj*. needle-shaped, as in certain CRYSTALS.

acid or **acidic** *adj*. **1.** (of igneous rocks) containing more than 63 per cent SiO_2, as distinct from INTERMEDIATE and BASIC rocks. The term 'acidic' is sometimes

incorrectly used as equivalent to FELSIC, but the latter includes some rock types not usually considered acidic. The term SILICIC is synonymous with acid. See SILICA SATURATION, SILICA CONCENTRATION.
2. (*metallurgy*, of a slag) in which the proportion of silica exceeds the amount required to form a 'neutral' slag with the earthy bases present.
3. (of hydrothermal, pegmatitic or other aqueous fluids) having a high hydrogen ion concentration (a low pH).

It should be noted that the first two definitions of the term 'acidic' have no reference to the hydrogen ion content, or pH, of a substance, as used in chemistry.

acidisation or **acid treatment** *n.* the forcing of acid into limestone, dolomite or sandstone in order to increase porosity and permeability by removing some of the rock constituents. The process is also used to remove mud introduced during drilling.

acid mine drainage (AMD) *n.* drainage with a pH of 2.0 to 4.5, issuing from mines and their wastes. The process is initiated with the oxidation of sulphides exposed during mining, which produces sulphuric acid and sulphate salts. The quality of the drainage water continues to be lowered as the acid dissolves minerals in the rocks.

acid treatment see ACIDISATION.

aclinic line see MAGNETIC EQUATOR.

acme zone, epibole or **peak zone** *n.* a BIOZONE consisting of a body of strata in which a particular species or genus of an organism occurs at maximum abundance. HEMERA is the corresponding unit of geological time. Compare ASSEMBLAGE ZONE, RANGE ZONE.

acmite see AEGERINE.

acoustic log *n.* a generic term for a well log that shows any of various measurements of acoustic waves travelling in rocks exposed by the sinking of a borehole. To obtain such a log, certain devices are used to generate an excitation through the adjacent materials in a borehole. The excitation PROFILES are recorded on a measuring device and the differences in the inherent acoustical properties of the various materials are determined from this record. See WELL LOGGING.

acoustics *n.* the study of the production, transmission, reception and utilisation of sound. Because sound waves are not strongly absorbed by water, sound is used to probe the ocean's depths, to study the nature of sediments and to locate objects in the oceans. Reflection and refraction of sound are used by marine geologists and geophysicists for seismic profiling and ECHO SOUNDING. The velocity of sound in the ocean depends on temperature, salinity and pressure (depth). The empirical formula for velocity in terms of these factors is:

$$C = 1449 + 4.6t - 0.055t^2 + 0.0003t^3 + (1.39 - 0.012t)(S - 35) + 0.017d$$

where C is the velocity in metres per second, t is temperature in degrees Celsius, S is salinity in parts per thousand and d is below surface depth in metres.

acoustic wave *n.* a longitudinal wave. Its use is commonly restricted to fluids but often includes P waves travelling in the solid part of the Earth.

acoustic-velocity log see WELL LOGGING.

acritarch *n.* a group of unicellular organic-walled microfossils of unknown biological affinity, probably polyphyletic, variously characterised by a smooth or spiny texture. The presence of the remains of these organisms in sedimentary rocks of marine origin indicates they were

planktonic (see PELAGIC, definition 2).
Acritarchs range from the Precambrian
to the present but are most abundant in
the Precambrian and early Palaeozoic.

actinolite *n*. a mineral, $Ca_2(Mg,Fe)_5Si_8$
$O_{22}(OH)_2$, composed of hydrous
calcium, magnesium and iron silicate.
It is a green fibrous AMPHIBOLE, generally found in crystalline schists, and
similar to tremolite in chemical
composition.

activation *n*. **1.** the treatment of bentonitic clay with acid to further its bleaching action or to improve its adsorptive
properties.
2. the process of rendering a substance
radioactive by bombarding it with
nuclear particles.

activation analysis *n*. a method for
identifying stable isotopes of elements.
If a sample is irradiated with neutrons,
charged particles or gamma rays, the
elements in the sample are made
radioactive and can then be identified
by their characteristic radiations.

active continental margin see CONTINEN-
TAL MARGIN.

active layer *n* **1.** the surface layer above
the permafrost, alternately frozen in
the winter and thawed in the summer.
Its thickness ranges from several
centimetres to a few metres.
2. (*engineering geology*) surficial
material that undergoes seasonal
changes of volume, expanding when
wet or frozen and shrinking when dry
or thawing.

acute bisectrix see BISECTRIX.

adamantine *adj*. (of a mineral) having a
brilliant lustre resembling that of a
diamond.

adamellite see QUARTZ MONZONITE.

adaptation *n*. the adjustment or modification by natural selection of an
organism, or of its parts or functions,
so that it becomes better suited to cope

with its environmental conditions. The
development of protective coloration
among fish is an example.

adaptive radiation *n*. a rapid evolutionary process whereby lineages separate
physically and genetically within a
short interval of geological time to
occupy a wide range of habitats.
Compare CLADOGENESIS.

adcumulate see CUMULATE.

adductor muscles *n*. the muscles that
pull closed the two VALVES of a shell, for
example in BRACHIOPODS and BIVALVES.

adhesion *n*. the molecular attraction
between closely contiguous surfaces of
unlike substances, e.g. liquid in contact
with a solid. Compare COHESION.

adiabatic process *n*. a process in which
a change of state takes place without
the transfer of heat between a system
and its environment. For example,
when a confined gas or other fluid is
compressed without loss or gain of
heat, the gas is warmed adiabatically. If
a gas is allowed to expand without
gain or loss of heat in the system, it is
cooled adiabatically. In the lower
atmosphere, the cooling of rising air
and the warming of descending air are
largely the result of adiabatic expansion and adiabatic compression of air,
respectively.

adit *n*. a gently sloping passage made
from the surface of the Earth, usually
to intersect a mineral vein or coal seam.
Although frequently called a tunnel, an
adit, unlike a tunnel, is closed at one
end. Adits are commonly horseshoe-
shaped but may also be square, round
or elliptical in cross-section. They can
be driven only in hilly country, where
the lower elevation of the entranceway
will provide an adequate slope for the
outflow of water and will facilitate the
removal of mineral material.

adobe *n*. a mixture of CLAY and SILT found

in the southwestern United States and Mexico. It dries to a hard, uniform mass, and its use for brickmaking dates back thousands of years; see LOESS.

adoral surface *n.* the lower, oral surface of the TEST (skeleton) of an ECHINOID.

adsorption *n.* the attraction of molecules of gases or molecules in solution to the surfaces of solid bodies with which they are in contact. Solids that can adsorb gases or dissolved substances are called *adsorbents*; the adsorbed molecules are referred to collectively as the *adsorbate*. Adsorption can be either physical or chemical. *Physical adsorption* resembles the condensation of gases to liquids. In *chemical adsorption*, gases are bound to a solid surface by chemical forces defined for each surface and each gas. Compare ABSORPTION.

adularia *n.* a colourless variety of low-temperature orthoclase feldspar, ranging from transparent to translucent. See MOONSTONE.

advance *n.* **1.** the gradual seaward progression of a shoreline as a result of ACCUMULATION or EMERGENCE.
2. the net seaward progression during a specific time interval.
3. the forward movement of a glacier front. Compare RECESSION.

advection *n* **1.** horizontal transport of air or atmospheric properties within the Earth's atmosphere.
2. the horizontal or vertical movement of sea water as a current.
3. lateral mass movement of material in the Earth's mantle. Compare CONVECTION.

aegerine, aegirite or **acmite** *n.* a brown or green mineral of the pyroxene group, $NaFe^{3+}Si_2O_6$. Aegerine augite is $Na,Ca(Fe^{2+}Fe^{3+}Mg)Si_2O_6$. Both can accept some Al^{3+} in the structure. They commonly occur in ALKALI IGNEOUS ROCKS, such as SYENITES and ALKALI GRANITE, and also in GLAUCOPHANE SCHISTS.

aeolian *adj.* pertaining to or caused by wind. LOESS and DUNE sands are *aeolian deposits*, and RIPPLE MARKS in sand are formed by wind. VENTIFACTS, YARDANGS and ZEUGEN are products of aeolian erosion.

aeolian placer see PLACER.

aerial photograph *n.* a photograph of the Earth's surface taken from the air by the use of cameras mounted on aircraft. They are usually taken in overlapping series from an aircraft flying in a systematic pattern at a given altitude. Aerial photographs are used for mapping land divisions and to provide information on geology, vegetation, soils, hydrology, etc. See PHOTOGEOLOGY.

aerobic *adj.* (of organisms) able to exist only in the presence of free oxygen; (of processes and activities) requiring free oxygen. Compare ANAEROBIC.

aerolites *n.* STONY METEORITES composed mainly of silicate minerals.

aeromagnetic *adj.* pertaining to observations and studies made with an AIRBORNE MAGNETOMETER.

aerosol *n.* a colloidal system in which the dispersion medium is a gas (usually air) and the dispersed phase consists of liquid droplets or solid particles, e.g. haze, mist, most smoke, some fog. Dust and liquid droplets from volcanic eruptions form aerosols in the upper layers of the Earth's ATMOSPHERE that may remain for a few years. The dust veil decreases the amount of incoming solar radiation and promotes surface cooling. An increase in the frequency of volcanic activity could have considerable effects on the Earth's climate.

affine *n.* **1.** a homogeneous DEFORMATION, i.e. one in which initially straight lines remain straight after deformation or initial lineations remain linear after deformation.

2. a transformation that maps parallel lines to parallel lines and finite points to finite points.

affinity *n*. a term used in biology to indicate relationship but not specific identity.

affinity of elements *n*. a classification of elements according to their preference, or affinity, for different types of chemical environments. Such preference is the consequence of differing bonding characteristics among the atoms of different elements:
(a) *siderophile elements*, e.g. cobalt and nickel, have an affinity for iron and are readily soluble in it. They are believed to be most concentrated in the CORE of the Earth and are found in the metal phases of METEORITES.
(b) *chalcophile elements* have a strong affinity for sulphur. They tend to concentrate in sulphide minerals and so occur in sulphide ORE deposits. Examples include copper and zinc.
(c) *lithophile elements* (e.g. lithium, oxygen, sodium, silicon) prefer the silicate phase and are most concentrated in the Earth's crust.

See also ATMOPHILE ELEMENTS, BIOPHILE. Elements are not necessarily restricted to one group.

AFM diagram *n*. a triangular diagram representing the simplified compositional character of a metamorphosed pelitic rock; the diagram is constructed by plotting the molecular properties of the three components: A=Al_2O_3, F=FeO and M=MgO. Compare ACF DIAGRAM, A'KF DIAGRAM.

aftershock *n*. an EARTHQUAKE following a larger earthquake and originating at or near the focus of the larger one. Generally, major shallow earthquakes are succeeded by several aftershocks. Although these decrease in number as time goes on, they may continue for days or even months. See also SEISMIC FOCUS, TSUNAMI.

Aftonian *n*. the first interglacial stage of the Pleistocene Epoch (Lower Pleistocene) in North America. It followed the Nebraskan and preceded the Kansan glacial stages.

agate *n*. a CRYPTOCRYSTALLINE variety of silica composed of varicoloured bands of CHALCEDONY. Agate usually occurs within rock cavities; its natural colour generally ranges from white through brown, red and grey to black. Compare MOSS AGATE.

age *n*. **1.** a unit of geological time, longer than a chron and shorter than an EPOCH, during which rocks of a particular stage were formed (see GEOLOGICAL TIME UNIT).
2. a frequently used term for a geological time interval within which were formed the rocks of any STRATIGRAPHICAL UNIT.
3. the time during which a particular event occurred or a time characterised by unusual physical conditions, e.g. the Ice Age.
4. the position of any feature relative to the geological time scale, e.g. 'rocks of Eocene age'.

age equation *n*. the relationship between geological time and radioactive decay, expressed mathematically as:

$$t = \frac{1}{\gamma} \times ln\left(1 + \frac{d}{\rho}\right)$$

where *t* is the age of the sample rock or mineral, γ is the decay constant of the particular radioactive series used for the calculation, *ln* is the logarithm to base *e*, and d/ρ is the present ratio of radiogenic daughter atoms to the parent isotope. See also DATING METHODS.

age of the Earth *n.* approximately 4530 Ma. This is based on analysis of the abundances of uranium and lead isotopes in meteorites and lunar rocks. The oldest meteorites give ages of 4560 Ma, and chemical similarities between meteorites and the Earth suggest that the Earth and the other planetary bodies were formed approximately 4530 Ma ago. The oldest lunar rocks give ages of 4460 Ma, and if the impact origin of the Moon from the Earth is correct then the age of lunar rocks gives a minimum age for the Earth. Zircon crystals from Australia have given ages of 4200 Ma, the oldest so far recorded from rocks on Earth.

age ratio *n.* the ratio of daughter element to parent isotope with which radiometric age is determined. Its validity requires a system that has remained closed since the time of its solidification, sedimentation or metamorphism; in addition, the sample must be representative of the rock from which it came and the decay constant must be known. See DATING METHODS.

age spectrum *n.* a group of ages for a rock specimen obtained from the proportion of argon driven off at increasingly higher temperatures in the $^{40}Ar/^{39}Ar$ spectrum dating method. See DATING METHODS.

agglomerate *n.* a coarse PYROCLASTIC deposit, or its lithified equivalent, containing a large proportion of rounded, fluidally shaped VOLCANIC BOMBS. The average grain size is greater than 64 millimetres. Agglomerate is a PYROCLASTIC FALL deposit that is a very good indicator of proximity to a vent. In the past the name 'agglomerate' has been used for almost any VOLCANIC BRECCIA regardless of how it formed, so care must be taken in interpreting accounts of such deposits, especially in ancient deposits. See PYROCLASTIC DEPOSITS.

agglutinate or **agglutinated spatter** *n.* a near-vent deposit of welded SPATTER fragments, usually of basaltic composition, formed during fountaining of lava that remains sufficiently fluid so that when the lava particles fall to the ground they coalesce to form a solid mass. See SPATTER RAMPART.

aggradation *n.* a gradational process that contributes to the general levelling of the Earth's surface by means of deposition that builds up. Some possible agents of this process are running water, wind, waves and glaciers. Compare DEGRADATION.

aggrading stream *n.* a stream that progressively builds up its channel or floodplain by receiving more load than it can transport. A stream can be caused to aggrade rather than deepen its valley by any change in the adjacent land resulting in ALLUVIATION that causes the stream's gradient to become too low for transport of its load, e.g. sinking of the surrounding land.

aggregate *n.* **1.** a mass of rock particles or mineral grains, or a combination of both.
2. construction material consisting of sand, gravel, crushed rock or other materials, such as recycled concrete.

agmatite *n.* a MIGMATITE in which angular blocks of GNEISS are separated by a mesh of veins of solidified melt-rock generated by extreme conditions of REGIONAL METAMORPHISM.

Agnatha *n.* a paraphyletic group of primitive vertebrates, the jawless fish. Included are the lampreys, hagfish and extinct groups. Range, Upper Cambrian to present.

agonic line *n.* the line through all points on the Earth's surface at which the magnetic declination is zero; it is the locus of all points at which true north and magnetic north coincide. See ISOGONIC LINE.

ahermatypic corals *n*. CORALS that have no symbiotic algae, so can live in deeper water than can the HERMATYPIC CORALS. Most ahermatypic corals are solitary and do not form reefs.

A horizon see SOIL PROFILE.

aiguilles *n*. **1.** pointed granitic rocks. **2.** (*physical geography*) peaks of mountains. **3.** needle-shaped peaks, especially certain peaks or clusters of needle-like rocks near Mont Blanc in the French Alps. **4.** another name for VOLCANIC SPINES, particularly in the West Indies.

airborne magnetometer or **flying magnetometer** *n*. an aircraft-transported instrument that measures variations in the Earth's magnetic field.

air drilling or **air-flush drilling** *n*. ROTARY DRILLING that uses high-velocity air instead of conventional DRILLING MUD; it is unsuitable where appreciable quantities of water may be encountered.

air fall see PYROCLASTIC FALL DEPOSITS.

air-flush drilling see AIR DRILLING.

air gun *n*. an energy source frequently used in marine seismic studies; the explosive release of highly compressed air generates a shockwave. It can also be modified for use in borehole velocity surveys.

air shooting *n*. detonation of an explosive charge above the surface of the Earth so as to produce a seismic pulse. It is also used in geophysical exploration.

air wave *n*. the acoustic energy pulse transmitted through the air as a result of a seismic shot. See SEISMIC SHOOTING.

Airy hypothesis *n* a concept of balance for the Earth's solid outer crust, which assumes that the crust has uniform density throughout and floats on a more liquid substratum of greater density. It is one explanation of the principle of *crustal flotation*. Since the thickness of the crustal layer is not uniform, this theory supposes that the thicker parts of the crust, such as mountains, sink deeper into the substratum. This is analogous to an iceberg, the greater part of which is beneath the water. See also PRATT HYPOTHESIS, ISOSTASY.

A'KF diagram *n*. a triangular diagram used to show the simplified componential character of a metamorphic rock. The molecular quantities of the following rock components are plotted:

$A' = Al_2O_3 + Fe_2O_3 - (Na_2O + K_2O + CaO)$

$K = K_2O$

$F = FeO + MgO + MnO$.

$A' + K + F$ (in mols) are recalculated to 100 per cent. When it is necessary to represent K minerals, this diagram is used in addition to the ACF DIAGRAM. Compare AFM DIAGRAM.

alabaster *n* fine-grained, massive gypsum, normally white and often translucent. It is used for statuary, vessels, etc.

alar septa *n*. two of the six first-formed (primary) septa (walls) in a CORALLITE. See RUGOSE CORAL, Fig. 77.

alaskite *n*. a plutonic rock consisting chiefly of ALKALI FELDSPAR and quartz; MAFIC constituents are few or absent. It is a LEUCOCRATIC variety of alkali GRANITE and a commercial source of feldspar.

albedo *n*. the ratio of the amount of solar radiation reflected from an object to the total amount incident upon it. An albedo of 1.0 indicates a perfectly reflecting surface, while a value of 0.0 indicates a totally black surface that absorbs all incident light. The albedo of

Earth is calculated as 0.39, more than half of which is caused by reflection from clouds. The greater reflection from new snow, relative to that of old snow, is an important factor in accounting for the melting of glaciers and snow.

Albertan *n.* the term used for the Middle Cambrian of North America, taken from the Canadian Rockies of Alberta where it is well represented. *Acadian* is the obsolete name. Above Waucoban, below Croixian.

Albian *n.* the top STAGE of the Lower CRETACEOUS.

albite *n.* **1.** a white or colourless feldspar mineral, $NaAlSi_3O_8$, that forms triclinic crystals. It is a variety of plagioclase that occurs commonly in metamorphic and igneous rocks.
2. in the plagioclase feldspar series, albite represents the pure sodium end member. Compare POTASSIUM FELDSPAR.

albite-epidote-hornfels facies *n.* a metamorphic mineral assemblage (typically developed in MAFIC rocks) that sometimes appears in the outer fringes of contact AUREOLES. It is a low-pressure FACIES characterised by imperfect recrystallisation and the presence of unstable relict phases from the premetamorphic condition.

albitite *n.* a PLUTONIC rock composed almost entirely of ALBITE, a variety of alkali feldspar SYENITE, produced by soda fenitisation (see FENITE).

Alexandrian or **Medinan** *n.* the Lower Silurian of North America.

alexandrite *n.* a transparent variety of CHRYSOBERYL used as a gemstone. Its colour is green in daylight and deep red by artificial light.

algae *n.* (*sing.* **alga**) a widely diverse group of photosynthetic plants, almost exclusively aqueous, that includes the seaweeds and freshwater forms. They range in size from unicellular forms to the giant kelp, tens of metres in length. Although considered primitive plants, since they lack true leaves, roots, stems and vascular systems, they manifest extremely varied and complex life cycles and life processes. Algae range from the Precambrian to the present.

algal bloom *n.* a relatively sudden increase in algal growth on the surface of a lake, pond or stream. Phosphate or other nutrient enrichment of the waters stimulates the growth of algal bloom.

algal limestone see LIMESTONE.

algal structure *n.* calcareous sedimentary structure formed in oceans or lakes by carbonate-depositing ALGAE. Forms include crusts, cabbage-head shapes and laminated structures. Most organically produced carbonate structures are not algal (*sensu stricto*) in origin but are produced by CYANOBACTERIA. Such structures include STROMATOLITES.

Algoman orogeny *n.* the OROGENY and granitic emplacement shown in Precambrian rocks of northern Minnesota and the Algoma district of western Ontario, dated about 2400 Ma. It is synonymous with the KENORAN OROGENY of the Canadian SHIELD.

alidade *n.* an instrument used in topographical surveying and mapping by the PLANE TABLE method; also, any sighting device or indicator used for angular measurement. A *surveying alidade* has a telescope with an attached vertical circle mounted on a flat, moveable base that carries a straight edge. An alidade is also a component of STADIA surveying, the method usually used to measure the distance and vertical height differences of distant points. See also GALE ALIDADE.

alkali 1. *n.* carbonate of sodium or potassium, or any bitter-tasting salt, occurring at or near the surface in arid and semi-arid regions.

2. *n.* a strong base (hydroxide), e.g. NaOH or KOH.

3. *adj.* (of minerals) having a high content of sodium or potassium, e.g. ALKALI FELDSPAR.

4. *adj.* (of IGNEOUS ROCKS) (a) containing FELDSPATHOIDS and/or alkali AMPHIBOLES or PYROXENES in the MODE, or (b) containing normative feldspathoids or acmite (see NORM, AEGERINE, compare SUB-ALKALI).

alkali basalt *n.* a major group of basaltic rocks defined chemically as containing normative NEPHELINE (see NORMATIVE MINERAL, BASALT). In addition to essential PYROXENE and PLAGIOCLASE, alkali basalts usually contain OLIVINE, both as PHENO-CRYSTS and in the GROUNDMASS. The plagioclases are richer in Na_2O and K_2O than those in the THOLEIITIC BASALTS, and ALKALI FELDSPARS and/or FELDSPATH-OIDS are often present in the ground-mass.

alkali feldspar *n.* FELDSPAR rich in sodium or potassium, e.g. albite, microcline, orthoclase or sanidine.

alkali flat *n.* a plain or a level area encrusted with alkali salts as a result of evaporation and poor drainage, found in arid or semi-arid regions; a salt (saline) flat. See also PLAYA.

alkali granite *n.* a variety of GRANITE essentially composed of ALKALI FELDSPAR and QUARTZ. The subsidiary minerals may be any of BIOTITE, MUSCOVITE, HORNBLENDE, alkali AMPHIBOLE or alkali PYROXENE. Less than 10 per cent of the total feldspar is PLAGIOCLASE.

alkali lake *n.* a SALT LAKE, often found in arid regions, the waters of which contain large amounts of sodium and potassium carbonates in solution as well as NaCl and other alkaline compounds; for example, Lake Magadi in the Eastern Rift Valley of Kenya. See also SODA LAKE.

alkali-lime index *n.* the percentage of silica (by weight) in a sequence of igneous rocks as shown on a variation diagram, where the weight percentages of CaO and those of $K_2O + Na_2O$ are equal, i.e. the intersection point of the curves for CaO and $(K_2O + Na_2O)$.

alkali metal *n.* any metal of the alkali group; Group 1a of the periodic classification: lithium, sodium, potassium, rubidium, caesium or francium.

alkaline *adj.* **1.** showing the qualities of a base (chemical).

2. pertaining to igneous rocks, as in ALKALI, definition 4(a).

alkalinity *n.* **1.** in a lake, the quantity and kinds of compounds that collectively shift the pH to the alkaline end of its range.

2. in sea water, the number of mill-equivalents of hydrogen ion neutralised by one litre of seawater at 20ºC. It has no relation to the hydroxyl content of sea water.

Alkemade line *n.* a straight line in a ternary PHASE DIAGRAM that joins the composition points of two phases the primary phase fields of which share a boundary. The phases can exist stably together. The Alkemade theorem states that the intersection of an Alkemade line with the boundary curve to which it is related represents a temperature maximum for that boundary. It is thus possible to show the direction of temperature change along the boundary.

Alleghenian orogeny *n.* a late Palaeozoic collisional episode in eastern North America, coeval with the VARISCAN (Hercynian) OROGENY, caused by the collision of Africa with North America. See APPALACHIAN FOLD BELT.

Alling grade scale *n.* a metric scale of

grain size for thin and polished sections of sedimentary rocks on which measurements can be taken only in two dimensions. Major divisions (boulder, cobble, gravel, sand, silt, clay, colloid) are in the constant ratio of 10; minor divisions are in the ratio of the fourth root of 10.

allivalite *n.* a pyroxene-poor olivine gabbro in which the feldspar is calcic bytownite or anorthite, with accessory augite, apatite and iron oxides. See TROCTOLITE.

allochem *n.* the carbonate particles that form the framework in most mechanically deposited LIMESTONES, as distinguished from MICRITE matrix. Silt-, sand- and gravel-sized intraclasts are some important allochems. Other examples are ooliths, FOSSILS and PELLETS. See OOLITE.

allochemical metamorphism *n.* METAMORPHISM accompanied by the removal or addition of material, such that the chemical composition of the greater mass of the rock is altered. See METASOMATISM, compare ISOCHEMICAL METAMORPHISM.

allochthon *n.* **1.** a term used for rock masses displaced considerable distances by tectonic processes such as overthrusting. Compare AUTOCHTHON. **2.** formerly, sedimentary rocks the constituents of which have been transported to and deposited in locations some distance from their origin.

allochthonous *adj.* describing a TERRANE displaced from its parent geological unit and forming part of an unrelated geological unit. Compare AUTOCH-THONOUS.

allocyclicity *n.* cyclical deposition within a sedimentary system resulting from mechanisms external to that system. Such mechanisms include EUSTATIC sea-

level change and climatic variations driven by MILANKOVITCH cycles.

allogene *n.* an ALLOGENIC mineral or rock constituent such as a pebble in a conglomerate or a xenolith in an igneous rock.

allogenic *adj.* **1.** (of minerals and rock constituents) deriving from pre-existing rocks transported from their original location. Compare AUTHIGENIC. **2.** (of a stream) fed by water from distant terrain.

allometry *n.* anisometric growth, a differential ontogenetic growth rate in one morphological attribute compared to growth in another morphological attribute or the whole body.

allopatric speciation *n.* speciation by geographical isolation of a population from a parent group or by the fragmentation of the parent group. Compare SYMPATRIC SPECIATION.

allotriomorphic see XENOMORPHIC.

allotrope *n.* a chemical element that may exist in two or more forms that have differing atomic arrangements as crystalline solids. For example, both graphite and diamond are allotropes of the element carbon.

alluvial *adj.* **1.** composed of or pertaining to ALLUVIUM, or deposited by running water. **2.** formed by running water, applied to a PLACER or to the mineral associated with it.

alluvial cone *n.* a sharply inclined alluvial deposit formed where a stream emerges on to a lowland after its descent from a steep upland area. Alluvial cones are steeper than ALLUVIAL FANS and show a greater average particle size. Fans are characterised by a much greater degree of stratification and by a higher incidence of mudflows. Compare TALUS CONE; see also COLLUVIUM, DEBRIS CONE.

alluvial fan *n*. an outspread mass of ALLUVIUM deposited by flowing water where it debouches from a steep, narrow canyon on to a plain or valley floor. The abrupt change of gradient eventually reduces the transport of sediment by the issuing stream. Viewed from above, the deposits are generally fan-shaped; they are especially prominent in arid regions but may occur anywhere. Compare BAJADA, ALLUVIAL CONE.

alluvial placer see PLACER.

alluvial plain *n*. the general name for a plain produced by the deposition of ALLUVIUM by rivers, e.g. floodplain, delta plain, alluvial flat.

alluviation *n*. **1.** the deposition of ALLUVIUM along streamways; AGGRADATION.
2. the filling of a depression or covering of a surface with ALLUVIUM.

alluvium *n*. the general term for detrital deposits made by rivers or streams or found on ALLUVIAL FANS, floodplains, etc. Alluvium consists of gravel, sand, silt and clay, and often contains organic matter that makes it a fertile soil. It does not include the subaqueous sediments of lakes and seas.

almandine *n*. the iron-aluminium end member of the GARNET group, $Fe_3Al_2(SiO_4)_3$. It characteristically occurs in SCHISTS and GNEISSES, with colours ranging from light red to brown-red. Its presence in rocks indicates the grade of METAMORPHISM. It is occasionally found in GRANITES. Deep-red crystals are valued as gemstones.

ANFO *acronym for* an explosive used in rock extraction. It is a mixture of ammonium nitrate and fuel oil, providing a high-gas, low-energy explosion.

alnöite see LAMPROPHYRE.

alpha decay *n*. radioactive decay caused by the loss of an ALPHA PARTICLE from the nucleus of an atom. The mass of the atom decreases by 4 and the atomic number by 2. Both radioactive isotopes of uranium ($^{235}_{92}U$ and $^{238}_{92}U$), which are used in the radiometric dating of rocks, decay in this way. See also DATING METHODS.

alpha particle *n*. the nucleus of a helium atom. It consists of two protons and two neutrons; no electrons are present, hence the particle charge is +2. Alpha particles are emitted during ALPHA DECAY of radioactive elements, they are highly ionising and have a short range through matter. Their energy is always uniform for a particular reaction.

alpha quartz, α-quartz or **low quartz** *n*. the polymorph of quartz stable below 573°C. It has a higher refractive index and BIREFRINGENCE than BETA QUARTZ, the form into which it passes at temperatures above 573°C. It is common in igneous, metamorphic and sedimentary rocks, and in veins and geodes. See also CRISTOBALITE, TRIDYMITE.

Alpides *n*. the long east-west orogenic belt that includes the European Alps and the Himalayas. The name specifically refers to the northern part of the belt, the southern part being referred to as the DINARIDES.

alpine *adj*. **1.** pertaining to the European Alps.
2. (of topographical features) resembling the form of the European Alps, e.g. the alpine topography of parts of New Zealand.
3. relating to the time of formation of the Alps (see following entry).

Alpine-Himalayan orogenic belt *n*. a complex series of arcuate fold mountain chains stretching from the Mediterranean region of southern Europe through the Aegean, Turkey, the Zagros Mountains in Iran to the Himalayan ranges. The formation of these mountain belts is ascribed to collision of the African, Arabian and Indian plates with the Eurasian plate during the closure of the TETHYS Ocean.

Deformation of the European sector appears to have started in the Mid to Late CRETACEOUS, reaching a climax in the MIOCENE. The collision between India and Eurasia that produced the Himalayan sector is thought, from palaeomagnetic evidence, to have begun about 38 Ma ago (the beginning of the OLIGOCENE).

Alportian *n.* a STAGE of the NAMURIAN (CARBONIFEROUS) in Britain and western Europe.

alteration *n.* changes in the chemical or mineralogical composition of a rock, usually produced by hydrothermal solutions or WEATHERING.

alternation of generations *n.* alternation of sexual and asexual phases in the life cycle of an organism, as seen in some CNIDARIANS. The phases are often morphologically dissimilar and sometimes chromosomally distinct.

altimeter *n.* an instrument that measures elevation or altitude. The two main types are the *pressure altimeter* (an ANEROID BAROMETER) and the *radio altimeter*, which measures the time required for a radio pulse to traverse the distance from an object in the atmosphere to the ground and back.

altiplanation *n.* the development of terrace-like surfaces or flattened summits of SOLIFLUCTION and related mass movements. Altiplanation is most common at high elevations and at latitudes in which periglacial conditions or processes predominate. Compare CRYOPLANATION, EQUIPLANATION.

altiplano *n.* a tableland plateau at high elevation, specifically the Altiplano of western Bolivia and southwestern Peru, which is a series of INTERMONTANE basins.

altithermal 1. *n.* a period of elevated temperature, especially the postglacial thermal optimum.
2. *adj.* pertaining to a climate characterised by high or rising temperatures. See GLACIERS.

altitude *n.* **1.** the vertical distance measured between a point and some reference surface, usually mean sea level.
2. the vertical angle measured between the plane of the horizon and a reference line at some higher point such as a summit peak. See also ELEVATION.

alum *n.* **1.** aluminium sulphate, $Al_2(SO_4)_3$, extracted from BAUXITE and clay minerals. It has a number of industrial uses.
2. a name applied to a group of hydrous alkali aluminium sulphate minerals. See ALUNITE.

alumina *n.* aluminium oxide, Al_2O_3.

alum shale *n.* an argillaceous, and often carbonaceous, rock impregnated with ALUM. It derives from a shale originally containing iron sulphide, which, upon decomposition, forms sulphuric acid that reacts with the SERICITE of the rock to produce aluminium sulphates.

alunite or **alumstone** *n.* a rhombohedral mineral, $KAl_3(SO_4)_2(OH)_6$, used in the production of ALUM. It usually occurs as grey, white or pink masses in hydrothermally altered feldspathic rocks. The introduction of alunite as a replacement mineral is called *alunitisation*.

alunitisation see ALUNITE.

alveolus see BELEMNITE.

amalgam *n.* **1.** a naturally occurring alloy of silver. Gold and palladium amalgams are also known.
2. an alloy of mercury and one or more metals. Amalgams are generally

crystalline in structure, but those with a high mercury content are liquid.

amazonite or **amazonstone** *n*. a green or blue-green variety of microcline feldspar, used as ornamental and gem material.

amber *n*. fossil tree resin that has achieved a stable state after ground burial, through chemical change and the loss of volatile constituents. Amber has been found throughout the world, most famously around the Baltic Sea where extensive deposits represent extinct flora of the Tertiary Era (2 to 65 million years ago); widespread deposits also occur in Ukraine. Some specimens of amber are opaque white, but it most often occurs in shades from yellow to brown. Fossil insects and plants may be found as INCLUSIONS. The hardness of amber is 2 to 3 on the Mohs scale, and its specific gravity is between 1.00 and 1.09. Optically, amber is mainly isotropic; its average index of refraction is 1.54 to 1.55. Isolation and identification of many of the resin components are still under investigation. Amber was believed to be completely amorphous, but X-ray diffraction studies have revealed crystalline components in certain fossil resins.

ambitus *n*. the circumference of the TEST (skeleton) of an ECHINOID, i.e. the edge of the specimen as seen from above or below.

amblygonite *n*. a triclinic mineral, $(Li,Na)Al(PO_4)(F,OH)$. It is found in pegmatites as white or greenish masses and used as a source of lithium.

ambulacra *n*. (*sing*. **ambulacrum**) the five double columns of plates in an ECHINODERM skeleton, which are perforated with *pore pairs* to carry the tube feet and overlie the WATER-VASCULAR SYSTEM. They are radially arranged and alternate with the *interambulacra*, which do not carry tube feet. See ECHINOID, Fig. 27.

AMD see ACID MINE DRAINAGE.

amethyst *n*. a transparent purple or blue-violet variety of quartz, SiO_2, valued as a semiprecious gem.

ammonite see AMMONOID.

ammonoid *n*. any of a group of CEPHALOPODS of the subclass Ammonoidea that appeared in the Lower DEVONIAN and were extinct by the end of the CRETACEOUS period. They typically have an external chambered shell, planispirally coiled and often ornamented with ribs and knobs (Fig. 1(a) opposite). *Heteromorph* ammonoids, which appeared at various times during the MESOZOIC, have partly uncoiled, curved or even helically coiled shells (Fig. 2(a)). Planispiral coiling may be described as *evolute* (all WHORLS visible) or *involute* (the last whorl covers the earlier ones). Terms used to describe shell shape include:
(a) *planulate*, an evolute shell with a nearly oval whorl section;
(b) *oxycone*, laterally flattened with a sharp venter;
(c) *cadicone*, evolute with a very broad whorl section;
(d) *sphaerocone*, an almost spherical, very involute shell;
(e) *serpenticone*, an evolute shell with slender coils that overlap very little (see Fig. 2(b)).

In most ammonoids the SIPHUNCLE lies near the outer margin of the shell, and the septal necks are prochoanitic (point forwards) in all but the earliest forms, (Fig. 1(a)).

Ammonoids are characterised by a complex SUTURE LINE compared with the simple NAUTILOID suture. Generally, suture lines increased in complexity from the Devonian to the Mesozoic. On

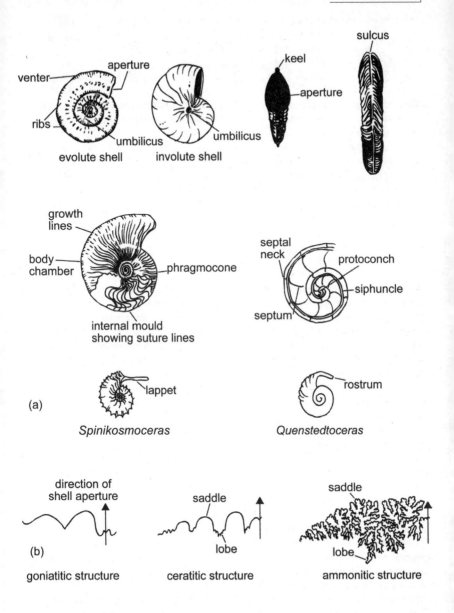

Fig. 1 **Ammonoid**. (a) external and internal structures of
ammonoids. (b) ammonoid sutures

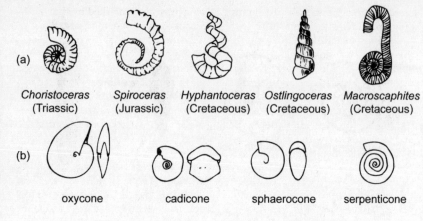

Fig. 2 **Ammonoid**. (a) some heteromorph ammonoids.
(b) some planispiral shell shapes.

the basis of the suture pattern, ammonoids can be classified into three broad groups: the *goniatites*, characterised by angular, zig-zag sutures, the *ceratites*, in which the sutures have filled lobes, and the *ammonites*, with extremely complex frills (Fig. 1(b)).

Ammonoids are important as INDEX FOSSILS because of their rapid evolution and wide distribution in shallow marine waters. They became almost extinct at the end of the PERMIAN and again at the end of the TRIASSIC, then declined slowly to extinction during the Upper Cretaceous.

amorphous *adj.* (of rocks and minerals) showing no definite crystalline structure, e.g. obsidian. Compare CRYSTALLINE.

amphibian *n.* a poikilothermic (cold-blooded) vertebrate, belonging to a grade that is midway in evolutionary development between fishes and reptiles. Amphibians breathe by means of gills in the early stages of life (a larval tadpole stage) and by means of lungs in the adult stage. Examples are frogs, toads, salamanders. Range, Upper Devonian to Recent.

amphiboles *n.* an important group of INOSILICATE rock-forming minerals that occur in a wide range of igneous and metamorphic rocks. Their molecular structure is based on SILICA TETRAHEDRA linked to give double chains, $(Si_4O_{11})n$, in a lattice that gives scope for very extensive ionic substitution. As a result there are a great many varieties of amphibole, this being reflected in the different SOLID SOLUTION series and varied CRYSTAL HABITS that they display. Amphibole crystals typically have good prismatic cleavage in two directions, intersecting at angles of 56° and 124°. The chemical variability of the amphiboles is typified by the widely distributed monoclinic mineral HORNBLENDE, with a general formula $(Na,K)_{0-1}Ca_2(Mg, Fe^{2+}, Fe^{3+}, Al)_5[Si_{6-7} Al_{2-1}O_{22}](OH, F)_2$, very common in AMPHIBOLITES. Four end members, *hastingsite* (Na,Ca,Fe-rich), *tschermakite* (Ca,Mg-rich), *edenite* (Na,Ca,Mg-rich) and *pargasite* (Na,Ca,Mg-rich), can be used to define the compositional range of hornblende. Other monoclinic amphiboles include the Ca-poor *cummingtonite* series, $(Mg,Fe^{2+})_7(Si_8O_{22})(OH)_2$, extending to the iron-rich end member grunerite; the Ca-rich amphiboles of the TREMOLITE $Ca_2Mg_5Si_8O_{22}(OH)_2$ series extending to ACTINOLITE $Ca_2(Mg,Fe^{2+})_5Si_8O_{22}(OH)_2$ with Mg>Fe and *ferroactinolite*, with Fe>Mg; and the *sodium amphiboles* including GLAUCOPHANE, $Na_2(Mg_3Al_2)[Si_8O_{22}](OH,F)_2$, RIEBECKITE, $Na_2(Fe_3^{2+}Fe_2^{3+})[Si_8O_{22}](OH,F)_2$, and *eckermannite-arfvedsonite*, $NaNa_2(Mg,Fe^{2+})_4Al[Si_8O_{22}](OH)_2$: eckermannite has Mg>Fe, arfvedsonite has Fe>Mg. *Oxyhornblende* and *kaersutite* are Ti-bearing amphiboles; *barkevikite* is similar to kaersutite but contains more Fe and less Ti. Fibrous forms of amphibole include varieties of ASBESTOS. NEPHRITE is a form of JADE. The Ca-poor orthorhombic ANTHOPHYLLITE series, $(Mg,Fe^{2+})_7Si_8O_{22}(OH)_2$, includes the Al- and Na-bearing variety *gedrite*, $(Mg,Fe^{2+})Al_2(Si_6Al_2O_{22})(OH)_2$, with Fe>Mg.

amphibolite *n.* a METAMORPHIC ROCK consisting mainly of AMPHIBOLE and PLAGIOCLASE, little or no QUARTZ and having CRYSTALLOBLASTIC texture. Amphibolite grades into hornblende-plagioclase gneiss as the content of quartz increases.

amphibolite facies *n.* one of the major divisions of metamorphic mineral assemblages, the rocks of which form under conditions of moderate to high temperature and pressure. Less intense temperature and pressure conditions rocks of the *epidote-amphibolite facies*, and more intense conditions form rocks of the GRANULITE FACIES. The epidote-amphibolite facies is typically represented by HORNBLENDE + EPIDOTE + ALBITE (in rocks of BASIC igneous composition). An increase in metamorphic intensity is marked by the disappearance of epidote and the formation of more calcic plagioclase to give the assemblage that is diagnostic of the amphibolite facies. Amphibolite facies rocks are widely distributed in Precambrian gneiss; this is construed as an indication that the rocks formed in the deeper parts of fold mountain belts. See also METAMORPHIC FACIES.

amphibolisation see URALITISATION.

amphidectic see LIGAMENT.

amphineuran *n.* a bilaterally symmetrical marine mollusc of the class Amphineura. Examples are usually oval and have a segmented univalve shell in the form of eight overlapping plates. A common form in the subclass Polyplacophora is the CHITON. Fossil forms date back to the Cambrian Period, these differing from Mesozoic to Recent forms in that the shell plates are not articulated. Amphineurans have little geological importance.

amphoteric *adj.* (of a substance) acting as an acid in the presence of a base and as a base in the presence of an acid. Water is an example.

amplitude *n.* 1. (*physics*) the maximum displacement from rest position of a point on a vibrating body or wave. As measured, it is equal to one half the length of the vibration path, i.e. the distance between zero position and trough, or zero position and crest. 2. (*geology*) (a) in a symmetrical fold, half the orthogonal distance between antiformal crest and synformal trough; (b) in meandering channels, half the wavelength of a cycle, i.e. one crest and one trough.

amygdale or **amygdule** *n.* SECONDARY MINERAL deposits filling elongated, rounded or almond-shaped VESICLES in IGNEOUS ROCKS, especially BASALT. They range in size from one millimetre to 30 centimetres. Among the secondary minerals that may fill these vesicles are QUARTZ, CARBONATES and ZEOLITES, all of which are precipitates from percolating ground water. See also DRUSE, VUG.

amygdaloidal *adj.* 1. (of rocks) containing AMYGDALES.
2. (of minerals) forming almond-shaped aggregates in VESICLES in LAVA flows.

amygdule see AMYGDALE.

anaerobic *adj.* 1. (of organisms) not requiring free atmospheric or dissolved oxygen.
2. (of biofacies) permanently anoxic benthic conditions. Compare DYSAEROBIC.

anaerobic sediment *n.* sediment that occurs where the limited circulation of water results in the absence or near absence of oxygen at and below the sediment surface. Such conditions can occur within basins and in shallow water. Water in contact with such sediments is rich in hydrogen sulphide, H_2S.

analcime *n.* a mineral, $NaAlSi_2O_6 \cdot H_2O$, of the FELDSPATHOID group, often found as a secondary infilling of vesicles in association with CALCITE, prehnite and ZEOLITES in basalts. It also occurs as a primary constituent of alkali DOLERITES and SYENITES.

analcite *n.* redundant name for ANALCIME.

analyser *n.* a polarising filter or NICOL PRISM mounted in the barrel of a PETROLOGICAL MICROSCOPE. Only light vibrating in a direction at right angles to that of the polariser will pass through. The analyser is removable, so THIN SECTIONS can be examined either in PLANE-POLARISED light (with the analyser out) or between crossed polars (analyser in). This arrangement is also known as CROSSED NICOLS. See also POLARISATION COLOURS.

anaptychus *n.* a flattened, butterfly-shaped plate of conchiolin that is sometimes found in the BODY CHAMBER of AMMONOIDS and thought to be part of the jaw. Compare APTYCHI.

anastomosing *adj.* 1. branching and recombining in a reticulated pattern, as in a BRAIDED STREAM.
2. interveined, as in leaves showing netlike vein patterns.

anatase or **octahedrite** *n.* a mineral, TiO_2, trimorphous with rutile and brookite.

anatexis *n.* the partial melting of pre-existing rock, which generates variable magmas that can subsequently be further diversified by mixing, contamination and magmatic differentiation. Compare SYNTEXIS.

anatexite or **anatectite** *n.* a rock formed by ANATEXIS. Compare ARTERITE. See also SYNTECTITE.

anauxite *n.* a clay mineral, $Al_2(SiO_7)OH_4$, similar in composition to KAOLINITE but with a higher $SiO_2:Al_2O_3$ ratio.

andalusite *n.* a mineral, Al_2SiO_5, that occurs as orthorhombic prisms in schists and gneisses. It is trimorphous with KYANITE and SILLIMANITE.

Andean-type mountain belt see CORDILLERA.

andesine *n.* a mineral of the plagioclase feldspar group, composed of 50 to 70 per cent albite and 30 to 50 per cent anorthite (i.e. An_{30} to An_{50}, see FELDSPAR, Fig. 33). It is a primary constituent of IGNEOUS ROCKS such as ANDESITES and DIORITES and of high-grade METAMORPHIC ROCKS.

andesite *n.* a fine-grained, dark-coloured VOLCANIC ROCK, frequently PORPHYRITIC, the extrusive equivalent of DIORITE. The PHENOCRYSTS consist chiefly of PLAGIOCLASE (often zoned from LABRADORITE to OLIGOCLASE) and one or more ferromagnesian minerals such as

PYROXENES, HORNBLENDE and/or BIOTITE; the GROUNDMASS is composed generally of the same minerals as the pheno-crysts (except hornblende). Chemically, andesites have a composition interme-diate between subalkaline BASALT and DACITE, with approximately 58 per cent SiO_2, a relatively high content of Al_2O_3 (about 17 per cent) and $Na_2O + K_2O$ (about 5 per cent). Andesites are commonly found in association with high-Al basalts, dacites and RHYOLITES in DESTRUCTIVE PLATE BOUNDARY settings, such as the ISLAND ARCS and active continental margins that ring the Pacific Ocean (see CONTINENTAL MARGIN. See also CALC-ALKALI SUITE). Andesitic rocks that are found in anorogenic areas (such as Iceland or the Tertiary volcanic district of Scotland) contain less alumina and more iron than orogenic andesites. Andesitic eruptions tend to be explosive, producing pyroclastic material and building conical stratovolcanoes; sometimes the eruptions are extremely violent and produce NUÉES ARDENTES or IGNIMBRITES, the latter often accompanied by CALDERA collapse.

Models for the production of andesi-tic MAGMAS suggest that hydrous magmas with compositions ranging from Mg-rich THOLEIITIC BASALT to BASALTIC ANDESITE could be produced by partial melting of the wet subducted oceanic crust/overlying MANTLE wedge at depth below a destructive plate boundary. MAGMATIC DIFFERENTIATION could produce a variety of compositions, from andesitic to dacitic. Some ASSIMILATION of crustal rocks in continental areas is possible. Andesite takes its name from the Andes mountains, where it was first applied to a series of lavas.

andesite line *n.* a boundary surrounding the Pacific Ocean, separating the ISLAND ARC and continental regions of andesitic VOLCANIC ROCK from the typically basaltic volcanic rock of the oceanic crust.

andradite *n.* the calcium-iron END MEMBER of the GARNET group, $Ca_3Fe_2(SiO_4)_3$, which characteristically occurs in contact-metamorphosed limestones. Its colour may be in various shades of yellow, green or brown.

anelastic behaviour *n.* a time-dependent response to STRESS. The STRAIN or change in shape is recoverable but not immedi-ately so. Rocks commonly display this type of behaviour, and it is an impor-tant consideration in mining, tunnel-ling and quarrying.

aneroid barometer *n.* a barometer consisting basically of a partially evacuated short, hollow cylinder with a flexible diaphragm at one end and a recording point at the other end. Atmospheric pressure changes are registered by movements of the diaphragm.

Angaraland or **Angara craton** *n.* a small SHIELD of exposed Precambrian rocks in north-central Siberia, once believed to have formed the nucleus about which all other tectonic structures of Asia were assembled. Current palaeo-geographies refer to it as the continen-tal unit of 'Siberia'.

angiosperm *n.* any member of the Angiospermae, the entire range of plants with true flowers having seeds enclosed in an ovary. Such plants originated in the Early Cretaceous or perhaps earlier. Compare GYMNOSPERM.

angle of emergence *n.* the angle formed between a ray of energy – acoustic, optic or electromagnetic – and the horizontal; the complement of the ANGLE OF INCIDENCE.

angle of incidence *n.* the angle that a ray of energy – acoustic, optic or electro-magnetic – makes with the normal to the surface on which it impinges; the complement of the ANGLE OF EMERGENCE. See also CRITICAL ANGLE.

angle of repose *n*. the maximum slope or angle at which loose earth materials, such as soil or sand, remain stable. It varies greatly with the composition, grain size and shape of the material as well as with its water content. For dry sand, the angle of repose is between 32° and 34°. TALUS slopes and the slopes of cinder cones (see VOLCANIC CONE) are commonly at the angle of repose.

anglesite *n*. a white orthorhombic lead sulphate mineral, $PbSO_4$, a common secondary mineral that is a minor ore of lead. It is usually formed by the oxidation of GALENA.

Anglian *n*. cold STAGE of the Middle Pleistocene, *c*. 0.45 Ma. See CENOZOIC ERA, Fig, 13.

Anglian glaciation *n*. a glacial episode in the Middle Pleistocene, named after East Anglia. It corresponds to the Elsterian glaciation of northwest Europe.

Angström *n*. a unit of length equal to 10^{-8} cm, used in measuring wavelengths of light and inter-atomic distances in mineral structures. It is indicated by the symbol Å.

angular *adj*. having sharp angles or edges, such as a particle or fragment showing little or no abrasion.

angular momentum *n*. the product of the linear momentum of a point on the surface of a rotating planet or globe and the perpendicular distance of that point from the axis of rotation. It is a vector quantity,

$$M = r \times mv$$

where *M* is the angular momentum about a point, *r* the position vector from the axis to the particle, *m* the mass of the particle and *v* the linear velocity of the particle. The angular momentum of the Earth is greatest at the Equator, where the position vector is at a maximum, and decreases with increasing north and south latitudes until its value becomes zero at the Poles.

angular unconformity see UNCONFORMITY.

angular velocity *n*. the rate at which a rotating body turns, expressed in revolutions per unit of time, radians or degrees. The Earth's angular velocity is the speed of its rotation about the geographic pole and not with reference to a point in space. The angular velocity of the earth is approximately 15.0411 seconds of arc per second of time.

anhedral *adj*. (of mineral crystals) not bounded by their typical crystal faces (*rational faces*). Compare EUHEDRAL, SUBHEDRAL.

anhydrite *n*. an orthorhombic mineral, anhydrous calcium sulphate, $CaSO_4$. It is common in evaporite deposits and alters to gypsum in humid conditions.

anhydrous *adj*. totally or essentially without water, e.g. an anhydrous mineral.

anion *n*. a negatively charged ion that moves toward the positive electrode (anode) during electrolysis. See ION.

Anisian *n*. the lowest STAGE of the Middle TRIASSIC.

anisometric *adj*. (of crystals) having unequal dimensions. Compare EQUANT.

anisomyarian see BIVALVE.

anisotropic *adj*. (of material) having some physical property that varies with direction. In the case of crystals, these variations are relative to the CRYSTAL AXES. All crystals except those of the isometric system are anisotropic with regard to some property, e.g. elasticity, conductivity, refractive index. The term commonly refers to optical properties. Compare ISOTROPIC. See PLEOCHROISM, BIREFRINGENCE, INDICATRIX.

anisotropy *n*. the condition or state of having different properties in different directions. In geological strata, sound waves may travel at one velocity in the vertical direction and at a different velocity in the horizontal.

ankerite *n*. a variety of iron-containing dolomite, $Ca(Fe,Mg,Mn)(CO_3)_2$, in which magnesium is replaced by iron in a SOLID SOLUTION series. Solid solution replacement by manganese leads to *kutnahorite*, $CaMn(CO_3)_2$.

annelid *n*. any wormlike invertebrate belonging to the phylum *Annelida*, characterised by a segmented body with a distinct head and appendages. Since they lack skeletal structures (hard parts) except for their jaws (scolecodonts), their fossil evidence is usually only in the form of trails and burrows. Their fossil record extends from the late Proterozoic to the present.

annual layer *n*. any sedimentary or precipitated layer of material of variable composition and thickness, presumed to have been deposited during the course of one year. Annual layers are most frequently utilised in the study of varved deposits laid down in glacially fed lakes. Certain non-glacial deposits also exhibit annual layers, e.g. the alternating laminae of anhydrite and calcite in a salt intrusion. It is thought that the anhydrite and calcite represent deposits of dry and humid seasons respectively. See RHYTHMITE.

annual ring *n*. the increment of wood added by a year's growth. Annual rings can be seen in cross-sections of the stems of woody plants. The number of annual rings in a tree trunk at its base indicates the age of the tree. See also DATING METHODS.

annular drainage pattern see DRAINAGE PATTERN.

anomalous lead *n*. lead deposits that give model ages older or younger than the age of the enclosing rock. Anomalous, or *nonconformable*, leads contain either too much or too little radiogenic lead. The first type, *B*- or *Bleiberg-type*, gives older model ages, and the second, *J*- or *Joplin-type*, younger ages than the age of the enclosing rock. Bleiberg, in Austria, and Joplin, Missouri, USA, are locations in which notable deposits of such leads are found. The existence of anomalous lead ores can be recognised from a model showing the evolution of common lead isotopes. Although both ordinary and anomalous lead evolved from primordial lead through the addition of radiogenic lead, the anomalous variety, unlike ordinary lead, does not follow the requirement that the radiogenic fraction be produced in a region having but a single source of uranium and thorium. See also DATING METHODS.

anomaly see GRAVITY ANOMALY, MAGNETIC ANOMALY.

anorogenic *adj*. 1. (of features) not related to tectonic disturbance. 2. (of a region) crustally inactive.

anorthite *n*. a feldspar mineral, $CaAl_2Si_2O_8$, found as white or greyish triclinic crystals in ultrabasic igneous rocks or in the cores of zoned phenocrysts in some basalts. It is the most calcic member of the plagioclase series. See FELDSPAR and Fig. 33.

anorthoclase *n*. a high-temperature mineral of the alkali feldspar group, $(Na,K)AlSi_3O_8$. It is a sodium-rich feldspar with constituents $Or_{37}Ab_{63}$ to $Or_{10}Ab_{90}$ (Or = orthoclase, Ab = albite). See FELDSPAR and Fig. 33.

anorthosite *n*. a plutonic igneous rock composed of more than 90 per cent PLAGIOCLASE FELDSPAR, usually LABRADORITE or BYTOWNITE, with minor amounts of PYROXENES, OLIVINE and/or Fe-Ti oxides. Anorthosites are found in layers within differentiated intrusions (STRATIFORM type), as large plutonic masses (massif type, often associated with plagioclase-rich GABBROS and a variety of syenitic and granitic rocks), and as layers in ARCHAEAN GNEISSES. The Archaean anorthosites are mostly of the stratiform type, which have been subsequently deformed and recrystallised. Anorthosites are also abundant in the lunar highlands. The majority of anorthosites on Earth are PRE-CAMBRIAN in age, mostly PROTERO-ZOIC. PHANEROZOIC anorthosites are mainly of the stratiform type. Most models for the origin of anorthosites involve MAGMATIC DIFFERENTIATION processes operating on a variety of andesitic, basaltic or picritic magmas, with the formation of plagioclase CUMULATES.

antecedent stream *n.* a stream estab-lished before the onset of some remoulding event, such as uplift, and maintaining its course unaffected by that event.

anteclese *n.* a positive structure of regional extent produced by crustal upwarp.

anthophyllite *n.* an amphibole mineral, $(Mg,Fe)_7Si_8O_{22}(OH)_2$, a variety of asbestos commonly produced by regional metamorphism of ultrabasic rocks. See AMPHIBOLE.

anthozoan *n.* any marine benthonic cnidarian of the class Anthozoa, characterised by an external skeleton with a stonelike or leathery texture. Sea anemones and the reef-building corals are both included in Anthozoa, but only the corals have geological importance. Range, Ordovician to present.

anthracite or **hard coal** *n.* the highest RANK of COAL and the only one that can

be considered a metamorphic rock. It is jet black with a somewhat metallic lustre and shows semi-conchoidal fracture (see CONCHOIDAL). Anthracite has the lowest percentage of volatile matter (less than 8 per cent), the highest percentage of fixed carbon (over 90 per cent on a dry mineral-matter-free basis) and a calorific value of about 15 MJ kg^{-1} (megajoule = 10^6 J). It is a clean coal that burns with a long-lasting, short blue flame without smoke. See also META-ANTHRACITE.

anthraxolite *n.* a hard, black ASPHALTITE found in sedimentary rocks, especially in association with oil shales but not restricted to such locations.

anthraxylon *n.* one of the components of the Thiessen-Bureau of Mines system of coal classification developed in the USA. The other components are ATTRITUS and FUSAIN. Anthraxylon is a composite term for a COAL component (MACERAL) that has a vitreous lustre, and partly equivalent to VITRAIN. In banded coal (see LITHOTYPE) it forms bands interlayered with opaque or dull attritus. Anthraxylon is derived from woody tissues of plants, the structures of which remain filled with humic matter or resin following the coalification process.

anticlinal axis see ANTICLINE.
anticlinal valley see ANTICLINE.
anticline *n.* a FOLD, of upwardly convex shape, that contains stratigraphically older rocks in its core. An *anticlinorium* is a large composite anticline com-posed of lesser folds. The *anticlinal axis* is that line which, if displaced parallel to itself, generates the form of an anticline. Valleys coinciding with this axis are referred to as *anticlinal valleys*. Compare SYNCLINE. See also ANTIFORM, ROLLOVER ANTIFORM.
anticlinorium see ANTICLINE.

antidune *n.* a transient accumulation of sediment on a stream bed, comparable to a sand DUNE, that moves upstream because of deposition of sediment on the upstream slope. Antidunes are an indication of high current velocities but are rarely preserved.

antiferromagnetism *n.* a phenomenon displayed by substances that are characterised by MAGNETIC DOMAINS that are aligned in opposite directions. The FERROMAGNETIC effects cancel each other out and there is no net ferromagnetism. This effect is destroyed on heating above a critical temperature, called the *Neel temperature*, when the substance behaves as a PARAMAGNETIC.

antiform *n.* a FOLD of upwardly convex shape in strata of unknown stratigraphical sequence. Compare ANTICLINE, SYNFORM.

antigorite *n.* a variety of SERPENTINE, $(Mg,Fe)_3Si_2O_5(OH)_4$. It occurs either in corrugated plates or in fibres.

antimony *n.* a silvery-white metal with a crystalline hexagonal structure occurring as the native element Sb. The most important antimony mineral is the sulphide STIBNITE, Sb_2S_3.

antiperthite *n.* an intergrowth of ALKALI FELDSPAR in a PLAGIOCLASE host caused by unmixing of a formerly homogeneous crystal during slow cooling. Compare PERTHITE; see also PERISTERITE, FELDSPAR.

antistress mineral *n.* a mineral the formation of which in metamorphosed rocks is aided by conditions not influenced by shearing stress but rather by thermal action and hydrostatic pressure. Feldspar and pyroxene are examples.

antithetic fault *n.* a minor normal FAULT the orientation of which is opposite to that of the major fault with which it is associated. Compare SYNTHETIC FAULT, LISTRIC FAULT.

apalhraun *n.* an Icelandic term for an AA lava flow.

apatite *n.* a group of hexagonal phosphate minerals composed of calcium phosphate together with chlorine, fluorine, hydroxyl or carbonate in varying amounts; general formula $Ca_5(PO_4)_3(F,OH,Cl)$; FLUORAPATITE is the commonest form; *chlorapatite* and *hydroxylapatite* are rarer. Apatite occurs widely as an ACCESSORY MINERAL in igneous and metamorphic rocks and less commonly is concentrated in ore deposits associated with alkalic igneous rocks. See also COLLOPHANE.

apex *n.* **1.** the highest point of a mountain or other land form.
2. the crest of an ANTICLINE.
3. in mining, the highest point of a vein relative to the surface, regardless of whether or not it appears as an outcrop.
4. the first-formed part of the shell in GASTROPODS, lying at the tip of the spire; see Fig. 37.

aphanite *n.* a rock with APHANITIC texture.

aphanitic *adj.* (of igneous rock) having crystalline aggregates of such fine grain size that constituent minerals are not visible to the naked eye. These aggregates form by rapid crystallisation or, secondarily, by the devitrification of glasses. A rock with such a fine texture is designated as aphanitic, regardless of the presence of phenocrysts.

Aphebian *n.* the lowest division of the Canadian PROTEROZOIC, from 2500 Ma to 1800 Ma.

aphotic *adj.* relating to ocean regions that do not allow the penetration of sunlight. These regions are normally found at depths of 200 metres and greater.

API gravity scale *n.* a standard scale

33

adopted by the American Petroleum Institute for designating the specific weight of oils. API gravity is expressed as:

API° (degrees API) = (141.5/sp. gr. at 60°C) – 131.5.

The API scale allows the linear calibration of hydrometers, which are used to measure specific gravity.

The alternative *Baumé gravity scale* is the specific weight of a liquid measured on a scale based on the weight of water. For petroleum, degrees Baumé are equal to 14°/SPGpet at 130°C.

apical disc *n.* a ring of small plates (genital and ocular) surrounding the PERIPROCT on the upper surface of the TEST (skeleton) of an ECHINOID (see Fig. 27). The genital plates are larger and have pores that are the outlets of the gonads. The ocular plates carry pores that are part of the WATER-VASCULAR SYSTEM.

aplite *n.* a fine-grained LEOCOCRATIC granitic rock, most commonly containing oligoclase, potassium-rich alkali feldspar and quartz with very small amounts of biotite and muscovite. It occurs as veins cutting GRANITE intrusions and often extending into the adjacent country rock. See also PEGMATITE.

apodemes *n.* small paired knobs on the inner surface of the axial ring of the dorsal exoskeleton of a TRILOBITE, which may have been points of attachment for ventral appendage muscles (Fig. 89).

apophysis *n.* in geology, a TONGUE or other direct offshoot of a larger intrusive body, such as a VEIN or DYKE.

Appalachian fold belt *n.* a MOUNTAIN BELT extending along the east coast of North America from Newfoundland to Alabama. It was affected by several orogenic episodes during the PALAEO-

ZOIC. In the northern Appalachians the TACONIAN OROGENY (ORDOVICIAN-SILURIAN) and ACADIAN OROGENY (DEVONIAN), which correspond to the CALEDONIAN OROGENY of Britain and Scandinavia, are related to the closure of the IAPETUS Ocean and collision between the North American Laurentian continent and Baltica (the European/Scandinavian continent).

In the central and southern part of the Appalachian belt the *Alleghenian orogeny* corresponds to the VARISCAN OROGENY of Europe. The main Alleghenian deformation occurred in late CARBONIFEROUS-PERMIAN times and is attributed to the collision of Africa with North America.

apparent density see BULK DENSITY.

apparent dip *n.* the angle of inclination of a fault plane or bedding surface exposed on the face of an outcrop. It is an angular measurement made on a dipping surface other than on a plane perpendicular to the strike. The result is always less than the TRUE DIP of the planar feature.

apparent movement of a fault *n.* the movement observed on any randomly viewed section across a FAULT. It is a function of the disrupted strata, attitude of the fault, particular site of observation and actual slip of the fault.

apparent resistivity *n.* the electrical RESISTIVITY of rocks, as measured by a series of voltage and current electrodes positioned on the surface of the Earth or in a borehole. If the Earth were homogeneous, this value would be equivalent to the actual resistivity; in practice, however, it is the weighted average of the resistivities that is measured. See RESISTIVITY METHOD.

apparent surface velocity *n.* the velocity at which the phase of a SEISMIC WAVE train appears to travel along the surface of the Earth. If the wave train is not travelling parallel to the surface, then the apparent velocity is greater than the actual value.

applanation *n*. all the processes that tend to reduce the relief of a region and cause it to become progressively plainlike. High parts are diminished by denudation and low parts are filled in or raised by the addition of material.

applied geology *n*. the application of GEOLOGY to human activities, including economic, environmental, engineering and hydrogeological. It also includes geological hazards and other fields.

apron *n*. a broadly extended deposit of unconsolidated material at the base of a mountain or in front of a glacier.

Aptian *n*. a stratigraphical STAGE of the Lower CRETACEOUS.

aptychi *n*. paired calcite plates often found associated with AMMONOIDS. It is suggested they may have formed an OPERCULUM, i.e. closed the aperture when the animal was drawn up inside.

aquamarine *n*. a transparent, pale bluegreen gem variety of BERYL. It occurs in pegmatite, in which it may form crystals of tremendous size.

aqueous *adj*. **1.** pertaining to water. **2.** prepared with water. **3.** (of rocks) formed of matter deposited by or in water.

aqueous ripple mark *n*. a RIPPLE MARK in sand, soil, etc, made by waves or water currents, as distinct from one made by wind.

aquiclude *n*. a body of rock that contains water but within which water cannot move under normal hydraulic gradients, e.g. a fine-grained permeable silt or clay layer that will absorb water but not release it in any significant quantity. An aquiclude will act as a CONFINING BED. Compare AQUIFUGE, AQUITARD, AQUIFER.

aquifer or **groundwater reservoir** *n*. a body of rock that contains water and releases it in significant quantities for use. The rock contains water-filled pore spaces that are sufficiently connected to allow the water to flow through the rock matrix to wells and springs.

aquifuge *n*. a rock body or rock layer that has no interconnected openings or interstices and which, thus, neither absorbs nor conducts water. Compare AQUICLUDE, AQUITARD.

Aquitanian *n*. the lowest STAGE of the MIOCENE.

aquitard *n*. a CONFINING BED that retards but does not completely stop the flow of water to or from an adjacent AQUIFER. Although it does not readily release water to wells or springs, it may function as a storage chamber for ground-water. Compare AQUIFUGE, AQUICLUDE.

aragonite *n*. an orthorhombic mineral, a form of calcium carbonate, $CaCO_3$, polymorphous with calcite. Aragonite is normally found as a low-temperature near-surface deposit, e.g. in caves as stalactites, near hot springs as sinter deposits and occasionally in vesicles in basaltic lava. It is the mineral normally found in pearls and is an important constituent in the shells and TESTS of many marine invertebrates that secrete this mineral from waters that would ordinarily yield only calcite.

arborescent see DENDRITIC.

arch see NATURAL BRIDGES AND ARCHES.

Archaea *n*. the domain of life that contains bacteria with archaeobacterial rRNA and the membrane lipids of which have ether-linked fatty acids. Compare EUBACTERIA, EUCARYA.

Archaean or **Archean** *n*. **1.** the eon of the PRECAMBRIAN between 3900 Ma and 2500 Ma. It follows the HADEAN eon. **2.** the rocks that formed during that time.

archaeocyathid *n*. any extinct marine organism of the phylum Archaeocyatha, characterised by a twin-walled tubular or conical skeleton of calcium carbonate, perforated throughout. They were often associated with calcareous ALGAE in REEF structures, and they lived in shallow waters. These organisms have been variously classified as sponges, corals and calcareous algae but seem most closely to resemble the calcareous sponges. They occur as fossils in marine limestones of the Lower and Middle Cambrian throughout the world, for example, in Russia.

Archaeozoic n. a name that is sometimes used for the earlier part of the PRECAMBRIAN, corresponding to the ARCHAEAN eon.

Archean see ARCHAEAN.

arc-trench gap n. the 50–400 km wide region between the volcanic arc and the oceanic trench at a DESTRUCTIVE PLATE BOUNDARY. (See ISLAND ARC, Fig. 52). The width of the gap appears to depend upon the dip of the subduction zone, this being related to rate of convergence, high convergence rates being associated with a large arc-trench gap. See SUBDUCTION ZONE.

arcuate adj. curved or bowed.

areal adj. pertaining to an area or the nature of an area, e.g. its surface features; the word should not be confused with *aerial*.

arenaceous adj. 1. (of a sediment or sedimentary rock) composed wholly or partly of sand-sized fragments, or having a sandy appearance or texture. 2. relating to sand or arenite, or to such a texture. The term connotes no particular composition and should not be used synonymously with SILICEOUS. 3. (of organisms) occupying sandy habitats.

Arenig n. a SERIES of the Lower Ordovician.

arenite n. 1. according to the classification of Grabau, arenites are consolidated or lithified sedimentary rocks composed of sand-sized particles (0.06 to 2 millimetres in diameter) irrespective of composition, e.g. ARKOSE, SANDSTONE, QUARTZITE, GREYWACKE. Composition may be indicated by a prefix, e.g. *litharenite*. See also LUTITE, RUDITE, PSAMMITE. 2. one of two main groups of sandstone recognised by the Dott classification on the basis of their matrix (particles less than 30 microns) content. Arenites have less than 15 per cent matrix, whereas WACKES have more than 15 per cent matrix but less than 75 per cent matrix. Arenites are subdivided on mineral and rock particle content into quartz arenite (see ORTHOQUARTZITE), ARKOSIC SANDSTONE and LITHIC ARENITE. See also SEDIMENTARY ROCK.

areola see TUBERCLE.

arête n. a sharp-crested, serrated or knife-edged ridge separating the heads of abutting valleys (CIRQUES) that previously contained glaciers. It may also be a divide between the glaciation of two parallel valley glaciers. An arête sometimes culminates in a high triangular horn, or peak, formed by three or more cirques eroding toward the same point. See also GLACIER.

argentiferous adj. containing silver.

argentite n. a silver sulphide mineral, Ag_2S, the most important ore of silver. Argentite is isometric and stable at temperatures above 180°C; below this point it inverts to monoclinic acanthite, which is stable at low temperatures.

argillaceous adj. 1. pertaining to rocks or SEDIMENTS composed of CLAY MINERALS or having a significant amount of clay in their composition. See also LUTACEOUS, PELITIC. 2. pertaining to rocks or sediments composed of SILT or clay-sized DETRITAL particles. Silt-sized particles range from

0.06 to 0.004 millimetres in diameter and clay-sized particles from 0.004 millimetres in diameter downwards.

argillic I. *adj.* pertaining to clay.

2. *n.* also called **argillic alteration**, the process in which certain minerals are converted to minerals of the clay series.

argillite *n.* a rock derived from mudstone or shale that has been altered and indurated by pressure and cementation. Argillites are midway in metamorphism between shale and slate, lacking the fissility of the former and the cleavage of the latter. See also SEDIMENTARY ROCK.

arid *adj.* (of a climate or area) characterised by extreme dryness, which is defined in various ways, including an annual rainfall of less than 25 centimetres, an evaporation rate that significantly exceeds the precipitation rate and insufficient rainfall for most plant life or crops to grow without irrigation.

Aristotle's lantern *n.* a structure of calcareous plates supporting five teeth, which forms the jaw of an ECHINOID.

arkose *n.* a coarse pink, red or grey SANDSTONE formed by the disintegration of crystalline rocks (for example, GRANITES and GNEISSES) without appreciable chemical decomposition. Arkose is composed chiefly of angular to subangular QUARTZ and FELDSPAR grains, with more feldspar than rock fragments. Arkoses are sometimes cemented by CALCITE or SILICA, and some have a clay matrix (more than 15 per cent clay), commonly coloured red by iron oxides. Arkoses are distinguished from quartzose sandstones by a high feldspar content (greater than 25 per cent) and from greywackes (which may also contain considerable amounts of feldspar) by their low matrix content. Since arkoses are often derived from the disintegration of granitic rocks,

they are rich in K-feldspar while greywackes generally contain Na-feldspar. When arkose consists of granitic detritus it is referred to as *granite wash*. The high feldspar content indicates that arkose deposits are a product of conditions that would retard the normal decomposition of feldspar, e.g. arid or glacial climate, high relief. See also SANDSTONE, GREYWACKE.

arkosic sandstone or **arkosic arenite** *n.* a sandstone with a high feldspar content (at least 25 per cent of the mineral grains) together with fragments of various other minerals (mica, quartz) in a matrix containing less than 15 per cent clay.

Armorican see HERCYNIAN.

armoured mud ball *n.* a spheroidal mass of clay or silt (5 to 10 centimetres in diameter) in which gravel and sand become embedded as it rolls downstream.

Arnsbergian *n.* a STAGE of the NAMURIAN (CARBONIFEROUS) in Britain and western Europe.

arrival time *n.* the initial appearance of SEISMIC energy as shown on a seismic record. At locations close to an EARTHQUAKE source, P waves and S waves will arrive fairly close together, even though P waves travel significantly faster than S waves. At locations more removed from the source, however, S waves will lag behind the P waves. The difference in arrival time between the two types of waves at any one station is used to calculate the distance of the earthquake from the station.

arroyo see EPHEMERAL STREAM.

arsenate *n.* a mineral compound formed from the arsenate group, AsO_4, e.g. scorodite, $FeAsO_4 \cdot 2H_2O$. Many arsenate minerals form SOLID SOLUTIONS with both phosphates and vanadates.

arsenic *n.* an element (As) existing in

both grey and yellow crystalline forms and having both non-metallic and metallic (mineral) forms. The metallic form, which is the more common and more stable, is grey and very brittle.

arsenopyrite or **mispickel** *n*. an iron sulpharsenide mineral, FeAsS, that forms metallic-white to steel-grey orthorhombic crystals. It is the principal ore of ARSENIC and a common mineral with lead and tin ores in ore veins and in pegmatites, probably having been deposited by the action of both hydrothermal solutions and fluids.

arterite *n*. a MIGMATITE with artery-like granite intrusions in the metamorphic host rock. See also INJECTION GNEISS.

artesian *adj*. pertaining to artesian conditions or their utilisation and application. Artesian conditions obtain when the hydrostatic pressure exerted on a confined AQUIFER is great enough to cause the water to rise above the WATER TABLE of the aquifer or the POTENTIOMETRIC SURFACE in the overlying CONFINING BED. Artesian conditions depend upon certain requirements:
(a) an inclined aquifer, the lower end of which is buried and the upper end of which is exposed;
(b) impervious layers (confining beds) above and below the aquifer that prevent leakage and allow hydrostatic pressure to develop (an aquifer so specified is an *artesian aquifer*);
(c) precipitation that infiltrates the aquifer at its exposed end;
(d) a spring, well or borehole that allows the water of the aquifer to discharge.

An *artesian spring* occurs where *artesian water* (confined GROUNDWATER) flows through faults or joints in the overlying impervious rock layer. An *artesian well* or *borehole* is one that is made below the exposed end of the aquifer, with the result that the water rises above the potentiometric surface in the upper confining bed at that position. If the potentiometric surface is above ground level, water will flow from the well or borehole, this being a *flowing artesian well* or *borehole*. An *artesian basin* is an area, commonly basin-shaped, containing an artesian aquifer the structure and orientation of which allow easy and wide-ranging access to its water. The Great Artesian Basin of Australia, underlying about one-fifth of the continent, is the world's largest area of artesian water.

arthrodiran *n*. one of an extinct order of armoured, jawed fishes of the class Placodermi, found in fresh water and marine deposits of the Devonian. They reached their peak during the Late Devonian, when some genera attained lengths of 9 metres.

arthrophycus *n*. a TRACE FOSSIL, originally described as a fossil seaweed, assigned to the genus *Arthrophycus*. It is an annulated, curving, branching sand-filled form, variously regarded as a feeding burrow or trail produced by an annelid, mollusc or arthropod.

arthropod *n*. any of a group of invertebrates of the phylum *Arthropoda*, characterised by segmented bodies, jointed appendages and an exoskeleton or carapace of chitin, reinforced in some groups by calcite. Phylum *Arthropoda* is the largest phylum in the animal kingdom, with species inhabiting all environments. Some typical members are TRILOBITES, CRUSTACEANS, chelicerates (for example EURYPTERIDS) and insects. Range, Lower Cambrian to present.

Articulata *n*. former taxonomic subdivision of phylum *Brachiopoda*; see BRACHIOPOD.

artificial recharge *n*. the engineered transfer of surface water into the

GROUNDWATER system. It most commonly involves infiltration from an artificial basin above the AQUIFER.

Artinskian *n*. a stratigraphical STAGE of the Lower PERMIAN.

Arundian *n*. a STAGE of the VISEAN (CARBONIFEROUS) in Britain and western Europe.

asbestos *n*. a general term applied to varieties of several fibrous minerals. Although different in chemical composition, their heat-resistant properties led to their use for fireproofing and heat-insulating materials; their link with lung cancer, however, has largely resulted in their abandonment for industrial use. Chrysotile, a fibrous variety of serpentine, $Mg_3Si_2O_5(OH)_4$, comprised about 95 per cent of commercial asbestos. All other types are amphiboles. All varieties occur in metamorphic rock. See also AMPHIBOLE, CROCIDOLITE.

Asbian *n*. a STAGE of the VISEAN (CARBONIFEROUS) in Britain and western Europe.

aseismic ridge *n*. a submarine ridge characterised by a lack of seismic activity, unlike the seismically active MID-OCEANIC RIDGE. Many aseismic ridges are linear volcanic chains and may have been generated by MANTLE PLUMES rising under a moving plate, or volcanic activity along a major fracture zone. Some, like the Lomonosov Ridge in the Arctic Ocean, are fragments of CONTINENTAL CRUST.

ash *n*. **1.** VOLCANIC ASH. See PYROCLASTIC MATERIAL.
2. the inorganic residue remaining after COAL is burned.

ash-cloud surge *n*. the turbulent low-density cloud of ash and gas above a PYROCLASTIC FLOW. It forms a fine-grained ash-fall deposit. Surges may become detached from rapidly moving pyroclastic flows, as in the case of one that devastated St Pierre, Martinique, in 1909.

A shell see SEISMIC REGIONS.

Ashgill *n*. the top SERIES of the ORDOVICIAN.

asphalt *n*. a dark brown to black bitumen with a consistency varying from viscous liquid to solid. Natural asphalt is found in oil-bearing rocks and is believed to be an early stage in the breakdown of organic marine materials into petroleum.

asphalt-base crude *n*. crude oil that contains a high percentage of asphaltic and naphthenic hydrocarbons. Compare PARAFFIN-BASE CRUDE.

asphaltic sand *n*. a naturally occurring mixture of asphalt and sand.

asphaltite *n*. any naturally occurring solid black bitumen that is insoluble in carbon disulphide, for example, glance pitch, GRAHAMITE.

assay 1. *v*. to determine the proportions of metal in an ore, or to test an ore or a mineral for composition, weight, purity, etc.
2. *n*. the analysis itself of such an ore or mineral.

assay grade *n*. the percentage of valuable or useful constituents in an ore, as determined from ASSAY.

assay limit or **cut-off limit** *n*. the limits of an ore body as defined by assay rather than by structural, stratigraphical or other geological parameters.

Asselian *n*. the lowest STAGE of the PERMIAN.

assemblage *n*. **1.** a grouping of similar organisms; in particular, a group of fossils found at the same stratigraphical level. Compare ASSOCIATION, BIOCOENOSIS, COMMUNITY.
2. the minerals that make up a rock, particularly an IGNEOUS or METAMORPHIC ROCK.

assemblage zone *n*. a biostratigraphical unit defined by a grouping of associated fossils rather than by any single guide fossil. Compare BIOZONE.

assimilation *n*. the incorporation of solid

or fluid matter from wall rocks, or from physically ingested inclusions, into magma to form a contaminated or HYBRID igneous rock. The degree of assimilation of cool wall rock into a magma is determined by factors such as the structure and fabric of the wall rock and the thermal energy of the magma itself.

association *n.* **1.** a group of organisms, living or fossil, that occur together because of similar environmental tolerances or requirements.
2. ROCK ASSOCIATION.

asterism *n.* a starlike optical phenomenon displayed by some crystals when viewed in reflected light (e.g. star sapphire) or in transmitted light (e.g. some phlogopite mica). It is the result of minute, oriented acicular inclusions.

asteroid *n.* **1.** any member of the Asteroidea, a class of phylum Echinodermata, that includes the starfish. The asteroids have broad, hollow arms that are radially arranged and not separable from the central disc, unlike the ophiuroids (see Fig. 3 below). The number of arms is typically five but may be in multiples of five. The tube feet, which are on the lower surface of the arms, are used for opening the MOLLUSCS, particularly BIVALVES, which form the usual prey of these animals. They are not common fossils, except in a few horizons ('starfish' beds) where they appear to have been buried very rapidly. Range. Lower ORDOVICIAN to present. See ECHINODERM.
2. one of several small celestial bodies in orbit around the Sun. Most of these have orbits between those of Jupiter and Mars. It is thought they are the parent bodies of METEORITES.

asthenosphere *n.* the layer or shell of the Earth directly below the LITHOSPHERE that behaves as a ductile solid. Its upper boundary begins about 75 kilometres below the surface (lying deeper beneath the continents than beneath the ocean) and extends to a depth of about 250 kilometres. The upper limit is usually defined on the basis of an abrupt decrease in the velocity of SEISMIC WAVES as they enter this region. The temperature and pressure in this zone reduce the strength of rocks, so that they flow plastically. For this reason, the asthenosphere is referred to as the 'weak zone'. See also GLOBAL TECTONICS, PLATE TECTONICS, SEA-FLOOR SPREADING, LOW VELOCITY ZONE.

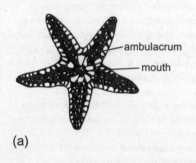

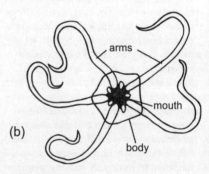

(a)   (b)

Fig. 3 **Asteroids**. Oral view of (a) an asteroid and (b) an ophiuroid or brittle star

astraeoid see COLONIAL CORAL.

astrobleme *n*. the eroded remains on the Earth's surface of an ancient impact structure produced by a cosmic body, in particular a meteor or comet. Astroblemes are usually characterised by a circular to elliptical outline, highly disturbed rocks and evidence of melting. Since features of crater walls are altered by erosion, astroblemes are identified by their outlines, the presence of SHATTER CONES, which are impact structures that form under the point of impact, high-pressure minerals and evidence of brecciation associated with melting. See IMPACTITE.

astrogeology or **extraterrestrial geology** *n*. a science that applies the knowledge and principles of all terrestrial branches of the geological sciences to the study of the origin and history of the condensed matter and gases of other bodies in the solar system. See also PLANETOLOGY.

asymmetrical *adj*. without symmetry or having no axis, centre or plane of symmetry.

asymmetrical fold see FOLD.

Atdabanian *n*. the middle STAGE of the Lower CAMBRIAN.

Atlantic-type coastline *n*. a coastline that develops where the structural features of the land, such as mountain ranges, are transverse to the boundary of the ocean basin. Such coastlines are usually irregular, with many inlets. Coastlines of this type are seen ideally along Ireland, Brittany and Newfoundland. See also PACIFIC-TYPE COASTLINE.

atmophile elements *n*. **1.** the most representative elements of the atmosphere: H, N, He, C, O, I and INERT GASES. See also AFFINITY OF ELEMENTS.
2. elements that were concentrated in the gaseous primordial atmosphere or that occur in the uncombined state.

atmosphere *n*. **1.** the gaseous envelope surrounding the Earth; it consists (by volume) of 78 per cent nitrogen, 21 per cent oxygen, 0.9 per cent argon, 0.03 per cent carbon dioxide, and small amounts of helium, krypton, neon and xenon. The mobility of the Earth's atmosphere is one of the determining factors of the Earth's climate and wind systems. The atmosphere is a poor conductor of heat but a rather good medium for the transmission of vibrations. Atmospheric density at sea level is 1.225 kg m^{-3}. This value decreases markedly with elevation, because half the total mass of the atmosphere is less than 5.5 kilometres from the Earth's surface. See also STRATOSPHERE, TROPOSPHERE.
2. a unit of pressure equal to the pressure exerted by a vertical column of mercury 760 millimetres in height at 0ºC, with gravity taken at 9.80665 ms^{-1}; equivalent to about 101.3 kPa.

atmospheric pressure *n*. the force per unit area exerted by an atmospheric column; it is most commonly measured with a mercury or aneroid barometer. Some expressions for standard sea-level pressure are: 760 millimetres or 29.92 inches of mercury; 1033.3 centimetres or 33.9 feet of water; 101.3 kPa; 14.66 lb/in^2 and 1013.25 millibars or 1 atmosphere.

Atokan *n*. a SERIES of the PENNSYLVANIAN of North America. It designates the Lower Middle Pennsylvanian in North America.

atoll *n*. a ring-shaped CORAL REEF composed of closely spaced coral islets that enclose a shallow lagoon; it may be circular to elliptical in plan view. Atolls are most abundant in the Pacific Ocean where the surrounding sea floor may be thousands of metres in depth; they may be as great as 130 kilometres in diameter, but most measure only a few kilometres. The theory of atoll origin that is most compatible with modem data is based on the subsidence of volcanic islands and the simultaneous growth of a coral reef. The reef develops initially as a fringe around a volcanic island; subsidence of the extinct volcano and upward growth of the reef produce a BARRIER REEF; progressive subsidence and reef growth eventually leave a lagoon surrounded by an atoll. See also SEAMOUNT.

atomic absorbtion spectroscopy see Appendix 4.

atomic percentage *n*. the percentage by atoms of a given element, as distinct from reference to the weight or number of molecules, in a substance consisting of two or more elements.

attapulgite *synonym for* PALYGORSKITE.

attenuation *n*. **1.** a reduction in the magnitude or energy of a signal.
2. a decrease in the amplitude of SEISMIC WAVES because of DIVERGENCE, reflection and scattering, and ABSORPTION.
3. that portion of the reduction in seismic or sonar signal strength that depends on the physical characteristics of the transmitting medium and not on geometrical divergence.

attenuation constant *n*. a mathematical parameter in an equation that expresses the relationship between the initial value x_0 of some physical quantity and x_1, the value it assumes by virtue of travelling or by virtue of a unit of elapsed time. An example in geophysics is the relation between the initial amplitude I_0 of a seismic disturbance and its amplitude I at a distance r.

$$r = I = \frac{I_0 e^{-qr}}{r}$$

where q is the attenuation constant.

Atterberg limits system *n*. the effect of water content on clay mineral proper-ties considered in terms of the plastic properties of soils, including LIQUID and PLASTIC LIMITS and PLASTICITY INDEX; it is based on a series of tests that delineate the transition between states of consistency. The system is important in civil engineering where it is used to determine rock and soil properties prior to construction work.

Atterberg scale *n*. a decimal scale of grain size, based on the unit value of 2 millimetres. Each grade stands in a fixed ratio of 10 to each successive grade. Subdivisions are the geometric means of the grade limits.

attitude *n*. the relation between some directional feature in a rock and a horizontal plane. The attitude of planar features, e.g. joints, bedding, foliation, is defined by the STRIKE and DIP. The attitude of a linear feature, e.g. fold axis, lineation, etc, is defined by the strike of the horizontal projection of the linear feature and its plunge.

attrital coal see ATTRITUS.

attrition *n*. wearing away by friction, in particular the deterioration that rock fragments in transit, such as pebbles in a stream, undergo through mutual grinding, bumping and scraping, with a resulting decrease in size and angularity.

attritus *n*. one of the components of the Thiessen-Bureau of Mines system of COAL classification, which was developed in the USA but is now less commonly used. Grey to black coal constituents of varying MACERAL content range from translucent to opaque. Attritus approximates to DURAIN and may form the bulk of some coals or may be interlayered with ANTHRAXYLON bands in others. *Attrital coal* is that coal in which the ratio of anthraxylon to attritus ranges from 1:1 to 1:3 or to the matrix of banded coal (see LITHOTYPE) in which vitrain and frequently fusain are embedded.

augen *n.* large lenticular mineral grains or aggregates in foliated metamorphic rock. Feldspar, quartz and garnet are common as augen, which are generally associated with the schistose and gneissic varieties of metamorphic rock. Such rocks are usually referred to as *augen schist* and *augen gneiss*, and are said to show *augen structure*.

augite *n.* $(Ca,Na)(Mg,Fe,Al)(Si,Al)_2O_6$, a dark-coloured clinopyroxene mineral that occurs in igneous rocks and is an essential mineral in BASALTS and GABBROS. Titanium-rich varieties (*titanaugite*) are typical of ALKALI BASALTS. See PYROXENE.

aulacogen *n.* an elongate fault-bounded depression, the 'failed arm' of a three-fold rift system, that did not develop into an active spreading rift during the break-up of continental LITHOSPHERE to form an ocean. See Fig. 4 below. The Benue Trough in West Africa is a good example. See also RIFT VALLEY.

aureole or **contact zone** *n.* a zone surrounding an igneous intrusion where contact METAMORPHISM of the host rock has occurred. Aureoles are products of mineralogical and chemical changes produced by heat and the migration of chemically active fluids from the magma into the earlier formed adjacent solid rock. The alteration is greatest along the solid rock-magma contact and decreases with distance from the contact to a point where no alteration has occurred. See also HORNFELS.

auric *adj.* **1.** pertaining to or containing gold.
2. pertaining to trivalent gold, e.g. auric chloride, $AuCl_3$.

auriferous *adj.* containing gold.

authigenic *adj.* (of mineral constituents) formed in a rock at the site where the rock is found. It is applied chiefly to sedimentary material. Compare ALLOGENIC.

autoclastic *n.* a term applied to rocks that have been fragmented *in situ* by mechanical processes such as faulting

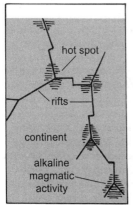
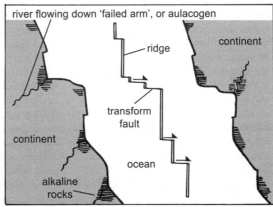

Fig. 4 **Aulacogen**. (a) three-fold rift systems form in the continental crust over hot spots. (b) continental break-up through linking of the rifts leaves aulacogens, or 'failed arms', on the continental margin

or by shrinkage following desiccation. Compare CRUSH BRECCIA. Autoclastic processes create essentially monolithologic, poorly sorted deposits of angular juvenile clasts.

autochthon *n.* a massive rock body that has remained essentially at its place of origin although it may show moderate to extensive DEFORMATION. Although CONTINENTAL DRIFT could have moved the entire framework containing the rock body into a different geographical location, the autochthon has not been displaced by the 'local' tectonic processes, such as overthrust faulting, that transport the ALLOCHTHON.

autochthonous *adj.* **1.** produced or formed in the location where it is found. For example, COAL and EVAPORITES are autochthonous SEDIMENTARY ROCKS. The term refers to whole formations rather than constituents, (see AUTHIGENIC). See also ALLOCHTHON. **2.** a TERRANE displaced from, but later returned to, its continent of origin.

autocyclic *adj.* cyclical deposition within a sedimentary system resulting from mechanisms internal to that system.

autogenetic or **autogenic** *adj.* **1.** (of land forms) having developed within the boundaries of local conditions, without the influence of orogenic movement. **2.** (of a drainage system) determined solely by the land surface over which streams flow.

autointrusion or **autoinjection** *n.* the injection of the residual liquid fraction of a differentiating magma into veins or irregular patches formed in the crystallised fraction at a late stage in the solidification of an intrusion. See also PEGMATITE, SCHLIEREN.

autolith *n.* **1.** also called **cognate inclusion**, an INCLUSION in an igneous rock to which it is genetically related. It

is an aggregate of early-formed crystals precipitated from the magma itself. Compare XENOLITH. **2.** Fe-Mg minerals of uncertain origin that are found in a granitoid rock and which may have the form of round, oval or elongate clots.

autometamorphism *n.* **1.** pervasive mineralogical and fabric changes that occur in an igneous rock body under conditions of decreasing temperature, and which are attributed to the action of the volatiles of the rock body itself. **2.** the DEUTERIC alteration of an igneous rock body that proceeds more or less automatically during cooling of the body in the presence of its own aqueous fluids.

autometasomatism *n.* a DEUTERIC effect in which newly crystallised igneous rock is altered by its own residual fluids that are held within the rock. See also METASOMATISM.

automorphic see EUHEDRAL.

autotheca see GRAPTOLITE.

auxiliary fault or **branch fault** *n.* a minor fault branching from a major one or abutting it.

auxiliary minerals *n.* as defined in the Johannsen classification of igneous rocks, those light-coloured, relatively rare minerals such as APATITE, CORUNDUM, FLUORITE, MUSCOVITE and TOPAZ.

available relief *n.* the vertical distance between an upland divide and the mouth of a valley floor. It determines the degree of headwater EROSION and the depth to which valleys can be cut.

Avalonia *n.* a continental crustal block of Precambrian age rifted away from GONDWANA in the Early Ordovician. Avalonia includes Eastern Avalonia (England, Wales, southern Ireland) and Western Avalonia (Maritime Canada).

aven *n.* a vertical shaft that leads upwards from a cave passage and sometimes connects with passages higher up.

aventurine *n.* a gem mineral that is a translucent variety of quartz or plagioclase feldspar. Its spangled appearance is caused by tiny inclusions of such minerals as mica and hematite.

Avonian see DINANTIAN.

avulsion *n.* the sudden diversion of a river, with formation of a new CHANNEL, that occurs in meandering and braided rivers. See also DELTA.

axial angle see OPTIC AXIAL ANGLE.

axial elements see CRYSTAL AXES.

axial plane *n.* **1.** a crystallographic plane in which two of the CRYSTAL AXES are included.
2. the plane in which lie the optic axes of an optically biaxial crystal.
3. planar surface connecting the HINGE LINES of the strata in a fold.

axial-plane cleavage *n.* cleavage that is generally parallel with the axial planes of folds in rock. Although some axial-plane cleavage parallels the regional fold axes, most is related to the minor folds occurring in individual outcrops. See also BEDDING-PLANE CLEAVAGE, STRAIN-SLIP CLEAVAGE, SCHISTOSITY.

axial-plane folding *n.* axial planes are folded during extensive secondary folding of pre-existing folds in rock. Such secondary changes are the result of movements that were considerably different from those effecting the original folding. See FOLD.

axial ratio see CRYSTAL AXES.

axial ring see THORAX.

axial structure *n.* any structure in the central (*axial*) region of a RUGOSE CORAL. There are three main types: the *columella*, which is a vertical rod formed from the dilated end of a counter SEPTUM (Fig. 77, rugose coral); the *axial vortex*, in which

the joined ends of the major septa are twisted; and the *axial column*, a zone of incomplete TABULAE.

axial surface *n.* a surface joining the HINGE LINES of strata in a fold.

axial trace *n.* the intersection of the axial plane of a FOLD with a specified surface, e.g. the surface of the Earth.

axis *n.* **1.** a line passing through a body on which the body rotates or may be imagined to rotate; a line centrally bisecting a body or system around which the parts are arranged.
2. see CRYSTAL AXES.
3. see ANTICLINE, SYNCLINE.

axis of symmetry see CRYSTAL SYMMETRY.

azimuth *n.* **1.** (*surveying*) the angle of horizontal difference, measured clockwise, of a bearing from a standard direction, as from north or south.
2. a horizontal angle, measured clockwise, between the north meridian and the arc of the great circle connecting an earthquake epicentre and the receiver.

azimuth compass *n.* a magnetic compass fitted with sights for determining the angle between a line on the Earth's surface (or the vertical circle through a heavenly body) and the magnetic meridian.

azonal soil or **immature soil** *n.* a soil that lacks a definite profile and resembles its parent material. Stream alluvium and dry sands are examples.

azurite *n.* a deep-blue monoclinic mineral, $Cu_3(CO_3)_2(OH)_2$. It is an ore of copper and occurs in the upper (oxidised) zones of copper deposits, where it is usually found with MALACHITE. Compact azurite is the source of semiprecious stones used in jewellery and decorative objects.

B*b*

back *n*. the roof or ceiling of an underground mine.

back-arc basin or **marginal basin** *n*. a basin located on the overriding plate behind the volcanic ISLAND ARC at a SUBDUCTION ZONE. The basin may be divided into an active and an inactive region containing submarine ridges (the REMNANT ARCS). The active part of the basin (sometimes called an *inter-arc basin*) is a zone of extension that splits the active arc and trench complex from the remnant arc and continent. It is suggested that such basins are formed by SEA-FLOOR SPREADING caused by the diapiric rise of hot mantle material released by the subduction process.

back bearing see BACKSIGHT.

backfill *n*. **1.** any material used to refill a quarry or other excavation.
2. waste rock used to support the roof after removal of ore from a mined space (stope or gallery).

background *n*. **1.** the normal radioactivity of the environment, attributed to the Earth's naturally radioactive substances and cosmic rays.
2. in geochemical prospecting, the normal concentration of a given element in a material under investigation.
3. the amount of pollutants in the surrounding air caused by natural sources and processes.

back reef *n*. **1.** the landward side of a reef. It may include lagoon, back reef and any terrestrial deposits entering the area.
2. the side of a reef away from the open sea even though no land is immediately near. Compare FORE REEF.

backset bed *n*. a cross-bed that inclines against the flow direction of a depository current.

backset eddy *n*. a small current swirling in the direction opposite those of the major oceanic eddies; it is often observed between the main current and the coastline.

backshore *n*. **1.** the upper zone of a BEACH or SHORELINE, bounded by the high-water line of mean spring tides and the upper limit of shore-zone processes. Only unusually large tides or severe storms affect this area. Compare FORESHORE.
2. the area at the base of a sea cliff.
3. a BERM or *beach berm*, a horizontal surface at the rear of the BEACH crest.

backsight *n*. a sight or bearing calculated on an already established survey point. A comparison between backsighting and foresighting is made for the purposes of calibrating and verifying the precision of surveying instruments. Compare FORESIGHT.

back slope *n*. the gentler slope of a FAULT BLOCK ridge or CUESTA; it is not necessarily related to the DIP of the underlying rocks.

back thrusting *n*. thrust faulting in which the direction of displacement is opposite that of the general direction of tectonic transport.

backwash see BREAKER.

bacteria *n*. a general term for prokaryotic, unicellular micro-organisms, or organisms that lack a nucleus, and which may be AEROBIC or ANAEROBIC.

They have been present on Earth since the Precambrian. Bacteria or bacteria-like FOSSILS have been found in ARCHAEAN rocks such as the Gunflint Chert (2000 Ma) in Canada and the Warrawoona Group (3500 Ma) of western Australia. See ARCHAEA, EUBACTERIA.

bacteriogenic *adj.* 1. (of mineral deposits) formed by the action of particular bacteria. These include bog iron ore and iron pyrites. See IRON BACTERIA, SULPHUR BACTERIA.
2. (of sediments) associated with bacteria, which, by their activity, are responsible for the deposition of carbonates. Many structureless limestones of geological periods dating back as far as the early Palaeozoic are bacteriogenic.

badlands *n.* a complex, erosional topography produced by intense periodical RUNOFF in the American West. Badlands (e.g. those of South Dakota) are characterised by a very fine drainage network and short steep slopes with little or no vegetative cover. They develop in regions of erodible sediments where vegetation has been destroyed, e.g. by cattle grazing, or where vegetation is lacking.

bajada, bahada, compound alluvial fan or **piedmont alluvial plain** *n.* a broad, gently sloping detrital surface formed in a basin, most commonly under arid or semi-arid conditions. A bajada is formed by the lateral coalescence of several ALLUVIAL FANS or is built up by material washed down from the fans. It is a depositional feature, as distinct from a PEDIMENT, but distinguishing between the two is often difficult. Bajadas are common in the Basin and Range Province of the southwestern United States.

Bajocian *n.* a stratigraphical STAGE of the Middle JURASSIC.

balance or **regimen** *n.* the difference between ACCUMULATION and ABLATION of a GLACIER over a specific time interval, determined as a value at a point, averaged over an area, or the total mass change of the glacier. Changes in the balance of a glacier determine whether it is in a state of advance, recession or stagnation. Compare NET BALANCE.

balanced section *n.* a section showing the pre-deformational state of folded and/or faulted rocks.

ball and pillow structure *n.* a sedimentary structure produced in unconsolidated sediments. It is caused by differential settling of coarser sediments into finer sediments, producing irregularly spheroidal and pillow-shaped masses of the coarser sediment.

ballas see INDUSTRIAL DIAMOND.

ball clay or **pipe clay** *n.* an extremely plastic kaolinitic clay, usually of lacustrine origin, commonly containing organic matter, used as a bonding substance of ceramic wares. Its unfired colours range from light buff to various shades of grey.

Baltica *n.* a continental crustal block consisting of Scandinavia plus the Baltic.

band *n.* 1. an informal stratigraphical term for a STRATUM or LAMINA distinguishable from adjacent layers by colour or lithological difference.
2. a range of frequencies between designated limits, e.g. the ultraviolet band of electromagnetic radiation ranging from 4×10^{-7} to 5.0×10^{-9} metres.
3. a glacier band, such as an OGIVE.

banded *adj.* (of a vein, sediment or other deposit) showing alternating layers differing in texture or colour, and possibly in mineral composition, for example BANDED IRON FORMATION.

banded agate *n.* an agate with colours arranged in alternating bands or stripes that are often wavy or zigzag and sometimes concentric; the bands may be clearly demarcated or may blend into one another. Compare ONYX.

banded coal see LITHOTYPE.

banded iron formation (BIF) *n.* sedimentary rocks that are typically bedded or laminated and composed of at least 25 per cent iron, mostly as oxides (HEMATITE, MAGNETITE) and microcrystalline quartz (chert, chalcedony, jasper). Banded iron formations occur in the Precambrian crust of all continents, from 3800 Ma onwards, with a peak at 2500 Ma, and a scatter of occurrences in the late Neoproterozoic. They are commonly used as low-grade iron ore. Because banded iron formations have not been formed since Precambrian time, it is thought that special conditions must have existed contemporaneously with their formation. Their origin has been ascribed primarily to a marine biochemical process in which oxygen from photoautotrophic bacteria produced seasonally rhythmic deposition of oxidised iron from ferrous iron in solution. Other factors, including volcanic activity and the chemical oxidation of ferrous iron, may be involved. See TACONITE.

banded ore *n.* ore that consists of bands or, strictly speaking, layers of different minerals or of the same minerals differing in texture or colour.

banded structure see LAYERING.

banding *n.* **1.** the appearance of banded structure, because of LAYERING, in an outcrop of igneous or metamorphic rock. Compare FLOW BANDING. See also FOLIATION.
2. in sedimentary rocks, thin bedding, conspicuous in cross-section, produced by different materials in alternating layers.
3. in GLACIER ice, layered structures caused by alternating layers of fine- and coarse-grained ice, or of clear and clouded ice.

bank *n.* **1.** a sandy or rocky submerged elevation of the sea floor. It may be a local prominence on continental or island shelves.
2. along the Atlantic coast of the USA, a long, narrow island composed of sand forming a barrier between a lagoon and the ocean.
3. a shoal.
4. the rising ground edging a stream.
5. a mound or ridge-like deposit of shells formed *in situ* by organisms such as brachiopods and crinoids. Compare REEF.
6. the surface of a coal deposit that is being worked.

banket *n.* a gold-bearing CONGLOMERATE from the PRECAMBRIAN Witwatersrand SYSTEM, South Africa. It consists mainly of pebbles of vein quartz in a siliceous matrix. It is a FOSSIL PLACER deposit.

bankfull stage *n.* the position of the water surface of a stream when it is at CHANNEL CAPACITY. Discharge at this stage is *bankfull discharge*. The bankfull discharge is commonly believed to be the most effective agent in forming the shape of the channel.

bar *n.* **1.** any of various types of submerged or emergent elongated embankments of sand and gravel built by waves and currents. See BARRIER ISLAND, also SPIT, TOMBOLO.
2. a RIVER BAR, point bar (see FLOODPLAIN) or CHANNEL-MOUTH BAR.
3. a unit of pressure equal to 105 Pa, equivalent to 750.076 millimetres at 0ºC (mercury barometer). It is equal to the mean ATMOSPHERIC PRESSURE at about 100 metres above sea level.

barbed drainage pattern see DRAINAGE PATTERN.

barchan see DUNE.

bar finger *n.* a long narrow sand mass of lenticular cross-section underlying a DISTRIBUTARY in a bird-foot DELTA.

barite, barytes or **heavy spar** *n.* a mineral, $BaSO_4$, specific gravity 4.5, with orthorhombic crystals ranging from colourless to white, yellow, grey and brown. Its major uses are as drilling mud, in paint manufacture and as a filler for paper and textiles. It is the principal ore of barium.

barite rose see DESERT ROSE.

barometric elevation *n.* an elevation located above mean sea level and established by the use of a barometer.

barred basin *n.* any sedimentary basin in which water movement is restricted by a natural barrier, such as rock or sediment. In oceanic basins the presence of a barrier may lead to oxygen depletion.

Barremian *n.* a STAGE of the Lower CRETACEOUS.

barrier *n.* **1.** a BARRIER ISLAND.
2. an ICE SHELF.
3. a GROUNDWATER BARRIER.

barrier flat *n.* the relatively flat area of a BARRIER ISLAND, often occupied by small persistent pools of water. A barrier flat lies between the seaward edge of a barrier island and the lagoon.

barrier ice see SHELF ICE.

barrier island *n.* an elongate coastal island representing a sand ridge rising above high-tide level and lying roughly parallel to the coast but separated from it by a marsh or a lagoon. One explanation for the existence of barrier islands is that they originated as submerged sand bars that subsequently became emergent after being moved by wave action towards the shoreline. Another possibility is that most barrier islands

are remnants of sand SPITS built from the mainland, then severed by inlets cut through them during storms. Many geologists subscribe to a third theory, which is that they result from the partial drowning and reworking of mainland beach sediments by rising seas at the conclusion of the last glacial period. Barrier islands are sometimes imprecisely called *offshore bars*.

barrier reef *n.* a long narrow REEF running roughly parallel to a shore and separated from it by a body of water, e.g. a lagoon, usually too deep to permit coral growth. The Great Barrier Reef off the coast of northeast Australia extends for more than 1600 kilometres with only minor breaks in it. A barrier reef is usually pierced by channels (*passes*), giving access to the lagoon and the coastline beyond it. Barrier reefs may change laterally into FRINGING REEFS by joining a coast. See also ATOLL, CORAL REEF.

Barrovian-type metamorphic sequence *n.* a succession of METAMORPHIC ZONES characterised by the appearance of certain INDEX MINERALS, observed in PELITIC rocks in the southeastern Scottish Highlands. First CHLORITE, then BIOTITE, almandine GARNET, STAUROLITE, KYANITE and finally SILLIMANITE appear with increasing METAMORPHIC GRADE. This sequence is thought to have developed under higher pressures than those of the BUCHAN-TYPE METAMORPHIC SEQUENCE.

bar theory *n.* a theory to account for the origin of large marine deposits of salt, gypsum and other EVAPORITES. It assumes a standing body of water, e.g. a lagoon, isolated from the ocean by a bar in an arid or semi-arid climate. As water in the lagoon evaporates, water of normal salinity continues to flow in from the ocean. If the rate of evaporation exceeds the rate of inflow over a prolonged period, the concentration of dissolved salts in the water increases and an evaporite is deposited by precipitation. An alternative theory for the origin of some evaporite sequences suggests they formed in a SABKHA-like environment.

Bartonian *n.* a stratigraphical STAGE of the EOCENE.

barytes see BARITE.

basal cleavage see CLEAVAGE.

basal conglomerate *n.* a well-sorted, homogeneous conglomerate at the base of a sedimentary sequence resting on a surface of erosion. It forms the initial stratigraphical unit in a marine series and marks an UNCONFORMITY. Basal conglomerate originates as a thin layer of coarse material deposited by an encroaching sea.

basalt *n.* a dark-coloured, fine-grained basic VOLCANIC ROCK essentially composed of calcic PLAGIOCLASE (labradorite or bytownite) and PYROXENE, usually AUGITE, with or without OLIVINE. MAGNETITE is commonly an important accessory. Basalts are often PORPHYRITIC, with PHENOCRYSTS of any of the ESSENTIAL MINERALS. They are the extrusive equivalents of GABBRO. Normally, basalts contain between 44 and 52 per cent SiO_2, $Na_2O + K_2O$ < 5 per cent and MgO, CaO and $FeO + Fe_2O_3$ generally in the range between 5 and 15 per cent. With increasing $Na_2O + K_2O$ they grade into BASALTIC ANDESITES or HAWAAITES. With increasing MgO content basalts grade into picro-basalts and basaltic komatiites (see also KOMATIITE, PICRITE).

A number of different varieties of basalt may be distinguished. The major division is based on the NORMATIVE MINERAL content and comprises ALKALI BASALTS, which contain normative NEPHELINE, and SUBALKALINE BASALTS, which do not. See also THOLEIITIC BASALT, CALC-ALKALI SERIES, LUNAR BASALT. Basalt is the most abundant lava type; it forms the upper layer of the OCEANIC CRUST and is the chief constituent of intraplate oceanic islands. Varieties of basalt are found in ISLAND ARCS and at active CONTINENTAL MARGINS, and vast amounts of continental flood basalts have been erupted, associated with tension and rifting of the continents (for example, the Deccan flood basalts of India).

basaltic andesite *n.* a VOLCANIC rock intermediate between BASALT and ANDESITE in composition, containing FELDSPARS characteristic of andesite (OLIGOCLASE-ANDESINE) and ferromagnesian minerals such as PYROXENES.

basal till see TILL.

basanite *n.* an UNDERSATURATED olivine basalt composed of calcic plagioclase, augite, olivine, and a FELDSPATHOID (nepheline, analcime, or leucite). See TEPHRITE.

base *n.* **1.** a substance that, when dissolved in water, forms hydroxyl ions.
2. a nontechnical term for the dominant hydrocarbon series in a crude oil, e.g. asphalt-base crude.

base correction *n.* an adjustment of geophysical measurements to express them relative to the values of a referential BASE STATION.

base exchange or **cation exchange** *n.* in geology, a property that enables certain minerals, notably the ZEOLITES, to exchange cations adsorbed on the surface for cations in the surrounding solution, e.g. the exchange of Ca for Na, when in a suitable environment.

baseflow *n.* the GROUNDWATER drainage component of stream flow, consisting of spring flows and direct groundwater discharges into surface watercourses. See INTERFLOW, OVERLAND FLOW.

base level *n*. the theoretical limit towards which the vertical erosion of the Earth's surface progresses but rarely, if ever, reaches. The general base level (*ultimate base level*) is the lowest possible base level; for the land surface, it is sea level; for a stream, it is the plane surface of the sea projected inland as an imaginary surface beneath the stream. A region is at the *temporary base level* whenever it is graded towards some level, other than sea level, below which the land area cannot be reduced for the time being by erosion, for example a level locally controlled by a resistant stratum in a stream bed.

base line *n*. any line established with precision instruments for the purpose of making additional azimuthal measurements.

base map *n*. any map that shows the basic outlines required for adequate geological reference, and on which supplementary or specialised information is plotted for some particular purpose.

basement *n*. **1.** also called **basement complex**, the *geological basement*, the surface beneath which sedimentary rocks are not found, the igneous, metamorphic, granitised or highly deformed rock underlying sedimentary rocks.
2. the *petroleum economic basement*, the surface below which there is no current interest in exploration, even though some sedimentary units may lie deeper.
3. the *magnetic basement*, the upper surface of igneous or metamorphic rocks that are magnetised more than the overlying sedimentary rocks.
4. the *electrical basement*, the surface below which RESISTIVITY is so high that its variations, as noted in electrical survey results, are not significant.

base metal *n*. **1.** any of the more common chemically active metals, e.g. copper, lead, tin, zinc, etc. Compare NOBLE METAL.
2. the principal metal of an alloy, e.g. the copper in bronze.

base of weathering *n*. in seismic studies, the boundary between the surface layer through which SEISMIC WAVES travel with low velocity and an underlying layer through which such waves travel with appreciably higher velocity. The base of weathering, which may correspond to geological WEATHERING or to the WATER TABLE, is used to derive time corrections for seismic records.

base station *n*. in geophysical surveys, a referential observation point to which measurements at additional points can be compared. See also BASE CORRECTION.

base surge *n*. a low cloud of PYROCLASTIC MATERIAL expanding radially outwards in all directions from the source of a PHREATIC or phreatomagmatic explosion or from the collapse of the eruption column produced by such an eruption. They are the result of the interaction of magma with water and in some cases are probably cold, wet and sticky although in others, in which the amount of water is restricted, they may be hot enough to char wood. Base surge deposits are often stratified, cross bedded, contain erosional channels and may contain BOMB SAGS. The term 'base surge' has sometimes been used less precisely to cover a wider range of pyroclastic surges including GROUND SURGE and ASH-CLOUD SURGE.

basic *adj*. (of igneous rocks) having a relatively low silica content, between 45 and 52 per cent, e.g. BASALT, GABBRO. Basic rocks contain relatively high amounts of iron, magnesium and/or calcium, and so include most MAFIC rocks as well as other types. Basic is one of four subdivisions (acid, intermediate, basic, ultrabasic) of a commonly used system for classifying igneous rocks by their silica content. See ACID, INTERMEDIATE, ULTRABASIC, SILICA CONCENTRATION. Compare FEMIC.

basin and range landscape (basin and range province) *n*. the Basin and Range physiographic province in the USA extends southwards from the Columbia Plateau, including most of Nevada and parts of Oregon, Idaho, Utah, California, Arizona and New Mexico. The topography is characterised by tilted fault blocks forming isolated, nearly parallel mountain ranges with intervening valleys or basins containing sediments derived from the mountains. In cross-profile the ranges are generally asymmetrical, the steeper slope commonly being fairly straight. Both sides of a range may be bounded by steep slopes, but generally only one side is so bounded. The origin of basin-and-range topography is related to tension in the continental crust and associated normal faulting.

bastnaesite *n*. a mineral, (Ce,La)CO$_3$F, wax-yellow to reddish-brown in colour. It occurs in alkaline igneous rocks, especially carbonatite, and is a source of RARE-EARTH ELEMENTS.

batholith *n*. a large, discordant PLUTONIC mass exceeding 100 square kilometres in area made up of multiple intrusions of granite and related rocks that individually rarely exceed 30 kilometres in lateral extent. Deep-earth temperature and density measurements suggest the floors of some batholiths are as much as 30 kilometres beneath the Earth's present surface. Batholiths are associated with CORDILLERA-type OROGENIC BELTS and may be intruded over a great range of time,

e.g. the Coastal Batholith of Peru in the Andes where the individual INTRUSIONS were emplaced over a time span of 60 to 70 million years. Batholiths appear to result from the repeated generation of large volumes of silica-rich MAGMA associated with SUBDUCTION at continent/ocean PLATE margins. Where this magma reaches the surface of the Earth it is erupted in major explosive volcanic centres often with formation of collapse CALDERAS, as in the Andes mountain chain. The GRANITE batholith underlying much of southwest England from Dartmoor to the Scilly Isles was emplaced during the Hercynian (VARISCAN OROGENY), some 280 Ma ago. Most batholiths are granitic or granodi-oritic in composition (see GRANODIORITE). Contacts with the COUNTRY ROCK may be sharp or diffuse, and veins of PEGMATITE, APLITE and various economic ORE minerals often invade the surrounding rocks. Contact metamorphic AUREOLES are common. Compare STOCK.

Bathonian *n*. a STAGE of the Middle JURASSIC.

bathyal zone *n*. the marine ecological zone that is deeper than the CONTINENTAL SHELF but shallower than the deep ocean floor, between the NERITIC ZONE and ABYSSAL zone. This range is usually given as 200 to 2000 metres. Fluctuations in temperature, oxygen concentration and sedimentation in the bathyal zone result in conditions that are much more variable than in the abyssal region. The bathyal zone is almost entirely APHOTIC, with general temperatures between 50 and 15ºC. Its salinity may range from 34 to 36‰, and current movement in this region is essentially geostrophic.

bathymetric chart *n*. a topographic record of the ocean floor or bottom of some other body of water; contour lines, called ISOBATHS, indicate points of equal elevation.

bathymetry *n*. the measurement and charting of ocean depths. A series of ECHO SOUNDINGS are made and the depths are plotted on a chart called an ECHOGRAM. Echograms are then used to establish contour maps and physiographic conditions for the area under investigation.

bathypelagic *adj*. pertaining to the part of open sea or ocean at BATHYAL depth.

Baumé gravity scale see API GRAVITY SCALE.

bauxite *n*. a grey, yellow or reddish-brown rock largely composed of a mixture of various aluminium oxides and hydroxides, along with free silica, silt, iron hydroxides and clay minerals. It is the principal commercial source of aluminium. Bauxite is formed by the weathering of many different rocks and varies physically according to the origin and geological history of deposits; its forms may be concretionary, earthy, pisolitic or oölitic. See BOEHMITE, DIASPORE, GIBBSITE.

bauxitisation *n*. the formation of bauxite from either primary aluminium silicates or secondary clay minerals under conditions of thorough tropical or subtropical weathering.

b-axis *n*. the crystallographic axis that is oriented horizontally, right to left, and perpendicular to the c-axis. See CRYSTAL SYSTEM, Fig. 23.

bay bar see BAYMOUTH BAR.

bayhead bar *n*. a bar formed at the head of a bay a short distance out from the shore. Wave refraction on bay shores produces currents that tend to move detritus toward the bayhead, where it accumulates.

baymouth bar or **bay bar** *n*. a ridge, commonly composed of sand, extending partially or completely across the mouth of a bay from one headland to another, usually formed by a SPIT growing from one or both of the headlands. As the bar develops, the bay becomes a lagoon. Complete closure is made only if shore currents are stronger than currents entering and leaving the lagoon.

bayou *n*. **1.** marshy offshoots and overflows of rivers and lakes, chiefly in the southern states of the USA.
2. the general term for a stagnant inlet or outlet of a bay or lake.
3. a stagnant or abandoned course of a meandering river.
4. an OXBOW LAKE.

beach *n*. the sloping shore of a body of water, especially those parts on which sand, gravel, pebbles and shells or other hard parts of marine organisms are deposited by waves or tides. The configuration of a beach and the nature of the deposits covering it are related to coastal profile, types of debris available, wave pattern and the rate of deposition. On mountainous coasts, beaches are narrow and broken, and often composed of poorly SORTED debris; along plain coasts, beaches may extend laterally for hundreds of kilometres and are usually composed of well-sorted sand. Beach sands on coasts of high relief may be rich in minerals such as feldspar; beach sands on plain coasts consist of quartz and minerals such as RUTILE and GARNET. Some tropical beaches are composed almost exclusively of shell fragments.

beach berm see BACKSHORE, BERM.

beach cusp *n*. an arcuate seaward projection of coarse sand or pebbles formed on the foreshore of a beach by the action of waves. The areas in between the cusps are composed of finer sand. The distance between cusps generally increases with an increase in wave height.

beach face *n.* **1.** the section of a beach that is normally subjected to the action of the wave surge.

2. the FORESHORE of a beach.

beach placer *n.* a PLACER deposit of valuable minerals on a present-day or ancient beach or along a coastline. The minerals are formed as a lag deposit (see LAG GRAVEL) by virtue of their greater hardness or density.

beach ridge *n.* near-parallel ridges of sand, shell or pebbles varying in height from a few centimetres to several metres. Depressions between them are called *swales*. Ridges are found behind the present shore, and each ridge marks the position of a previous BEACH FACE. See also CHENIER.

beaded drainage *n.* an arrangement of short streams joining small pools, characteristic of areas underlain by PERMAFROST.

Becke line see BECKE TEST.

Becke test *n.* a comparison of the refractive indices of two contiguous minerals (or of a mineral and a mounting medium or immersion liquid) in a thin section, using a microscope of moderate to high magnification. The difference in refractive index of the two substances is indicated by the *Becke line*, a bright line that separates the substances. When the distance between the microscope stage and the objective lens is increased, the line moves into the substance with the higher refractive index. See also SHADOW TEST.

bed *n.* **1.** the smallest formal lithostrati-graphical unit (see STRATIGRAPHICAL UNIT) that is lithologically distinguishable from beds beneath and above it. The term is usually applied to sedimentary

strata but may be used for other types. See LAMINA, STRATUM.

2. the bottom of a body of water, e.g. a stream bed.

bedded *adj.* **1.** (of a formation, especially sedimentary) arranged in layers.

2. (of any mineral deposit) aligned with the BEDDING in a SEDIMENTARY ROCK.

bedding *n.* a layered feature of SEDIMEN-TARY ROCKS characterised by differences in composition, texture or structure. The usually planar top and bottom surfaces that define the units (BEDDING PLANES or *bedding surfaces*) are parallel to the surface of deposition. Layers more than one centimetre thick are called BEDS (LAMINAE are less than one centimetre thick). Within an individual bed there may be internal layering approximately parallel to the bedding surfaces (*planar bedding* or *lamination*) or inclined at an angle (see CROSS-BEDDING). Some beds may show GRADED BEDDING; beds without any internal layering or other features are described as MASSIVE or structureless. See also CONVOLUTE LAMINATION, FLASER BEDDING.

bedding cleavage see BEDDING-PLANE CLEAVAGE.

bedding fissility see FISSILITY.

bedding plane *n.* the dividing plane that separates each layer in a sedimentary or stratified rock from layers below and above it. It ordinarily delineates a noticeable change in colour, texture or composition.

bedding-plane cleavage or **bedding cleavage** *n.* CLEAVAGE that is parallel with the BEDDING PLANE. Compare AXIAL-PLANE CLEAVAGE.

bed form *n.* any structure produced on the surface of a sediment by the action of a flowing medium (wind or water). The nature of the bed form depends on the flow velocity, for example RIPPLE

MARKS form at lower velocities than dunes. See also FROUDE NUMBER.

bed load, traction load or **bottom load** *n*. the material carried by moving water or by wind along the surface (stream bed, sea bed or ground) because it consists of particles too large or heavy to be carried in SUSPENSION. Movement of the bed load material is by rolling and sliding or by bouncing (SALTATION) and is a function of the fluid velocity, friction between grains and between grains and bed surface.

bedrock *n*. the solid rock lying beneath superficial material such as gravel or soil.

beds *n*. **I.** An informal term for stratigraphical units that are lithologically similar or that form a cohesive whole, for example beds of Devonian age.
2. STRATIGRAPHICAL UNITS that contain substances of economic importance, for example coal BEDS.

beef *n*. a quarryman's term for a fibrous variety of CALCITE.

beheaded stream see STREAM CAPTURE.

beheading *n*. **I.** the cutting off of the upper section of a stream and diversion of the headwaters into a new course by STREAM CAPTURE.
2. truncation of spurs by glacial action. See STREAM CAPTURE.

beidellite *n*. a dioctahedral member of the smectite group of CLAY MINERALS commonly found in soils and in certain clay deposits, e.g. metabentonite. See BENTONITE.

belemnite *n*. any member of the Belemnitida, an extinct order of CEPHALOPODS related to the modern squid and octopus. They have an internal skeleton that consists of two parts: a *phragmocone*, which fits into a cavity (the *alveolus*) at one end of the solid calcitic *guard*. Only the guard is normally preserved. Part of the phragmocone is drawn out at one end to form the *pro-ostracum* (see Fig. 5 below).

Belemnites were abundant during the MESOZOIC and are useful as INDEX FOSSILS. Range, Lower CARBONIFEROUS to early TERTIARY.

bench or **berm** *n*. an artificially formed elongated horizontal ledge used in embankments, road cuttings and also in quarrying.

bench mark *n*. a marker that is firmly implanted in stable ground from which local subsidiary measurements originate and which is used as a reference in topographic surveys and tidal readings.

beneficiation *n*. the upgrading of an ore by some process such as flotation,

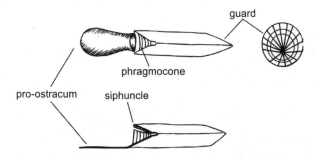

Fig. 5 **Belemnite**. General structure of a belemnite

milling, gravity concentration or sintering.

Benioff zone or **Wadati-Benioff zone** *n.* an inclined zone of earthquake foci that dips beneath an ISLAND ARC or active CONTINENTAL MARGIN, intersecting the sea floor at an OCEANIC TRENCH. It marks the site where one lithospheric plate plunges down beneath another, at a DESTRUCTIVE PLATE BOUNDARY. The zones are named after an American (Hugo Benioff) and a Japanese seismologist (K. Wadati) who located them. See also PLATE TECTONICS, SUBDUCTION ZONE.

benmoreite *n.* an alkaline VOLCANIC ROCK intermediate in composition between MUGEARITE and TRACHYTE, characterised by ALKALI FELDSPAR (ANORTHOCLASE or sodic SANIDINE), iron-rich OLIVINE and ferroaugite (see PYROXENE). See IGNEOUS ROCK, Fig. 47.

benthic or **benthonic** *adj.* descriptive of the ocean bottom at any depth.

benthos *n.* those organisms that live on the sea floor. *Benthic* organisms include mobile forms, such as crabs and starfish, and nonmobile forms, such as barnacles and hydroids.

bentonite *n.* a variety of MUDROCK composed almost totally of montmorillonite and colloidal silica, produced as the alteration product of glassy volcanic debris, usually VITRIC TUFF or ash. Bentonites contain evidence of their volcanic origin in the presence of RELICT SHARD structures and EUHEDRAL mineral grains that were originally PHENOCRYSTS.

Although bentonite is usually of the calcium variety, sodium and potassium bentonites are also known; these contain sanidine and quartz grains in addition to the montmorillonite. Palaeozoic bentonites are generally composed of ILLITE-SMECTITE layers. A bed of bentonite is the result of a single volcanic eruption, which may extend over a wide area, and thus is useful as a MARKER HORIZON.

Because of certain properties, bentonites have some economic importance. Calcium bentonites are highly absorbent and also useful as bleaching and decolorising agents. The fact that sodium bentonites will absorb large quantities of water makes them useful as DRILLING MUDS but also causes them to form very unstable slopes in outcrop.

bergschrund *n.* a crevasse that may form at the head of a valley GLACIER and which separates the mobile snow and ice of the glacier from the relatively stationary snow and ice adhering to the rocky headwall. It develops when a glacier moves downvalley, away from bedrock or a stationary snowfield. See also CIRQUE.

berm *n.* **1.** or **beach berm** a low shelf or terrace-like feature on the backshore of a beach, one side of which is generally bounded by a BEACH RIDGE or beach scarp. A beach berm is formed of material deposited by storm waves. The seaward limit and usually highest point of a berm on a beach is called the *berm crest*.
2. an artificial BENCH.

berm crest see BERM.

Bernoulli's theorem *n.* for fluids (liquid or gas) whose compressibility and viscosity are negligible and whose flow is laminar, Bernoulli's theorem states that the total mechanical energy of the flowing fluid, which comprises the kinetic energy of the fluid motion, the potential energy of elevation and the energy associated with fluid pressure, remains constant. In equation form:

$$\frac{v^2}{2} + gz + \frac{p}{d} = \text{constant}$$

where v is the velocity of the fluid, g

the gravitational acceleration, z the total vertical distance moved, p the pressure and d the density. Bernoulli's theorem has been used to describe flows in rivers and in the Gulf Stream, and in meteorology to describe air flow over mountains. See also LAMINAR FLOW.

Berriasian n. the lowest STAGE of the CRETACEOUS.

Bertrand lens n. a removable lens in a petrological microscope, used with convergent light to form INTERFERENCE FIGURES.

beryl n. a hexagonal mineral, $Be_3Al_2Si_6O_{18}$, from which beryllium is extracted; its principal occurrence is in granite pegmatites. Beryl is a major gemstone; when green, it is emerald, when blue or bluish-green, aquamarine, and when pink, morganite. Beryl is the principal source of the metal beryllium.

beta decay n. radioactive decay because of the loss of a beta particle (an electron) from one of the neutrons in the nucleus, thus changing the neutron into a proton. The atomic mass remains the same, but the atomic number increases by one. For example, $^{87}_{37}Rb$ decays to $^{87}_{38}Sr$ with the emission of a beta particle (β). See also DATING METHODS.

beta particle or **beta ray** n. an electron that has been ejected from an atom. Beta particles carry a negative unit charge and are moderately ionising. They have a longer range than ALPHA PARTICLES and can penetrate light metals. See BETA DECAY.

beta quartz, β-quartz or **high quartz** n. the polymorph of quartz that is stable from 573 to 867°C; it has a lower refractive index and birefringence than ALPHA QUARTZ. It occurs as phenocrysts in quartz porphyries and granite pegmatites. See also CRISTOBALITE, TRIDYMITE.

beta ray see BETA PARTICLE.

B horizon see SOIL PROFILE.

biaxial adj. (of crystals) having two optic axes, i.e. directions showing no double refraction of light. Crystals of the orthorhombic, monoclinic and triclinic systems can exhibit this property. Compare UNIAXIAL. See INDICATRIX, INTERFERENCE FIGURE.

bicarbonate n. a salt containing the HCO_3 radical, e.g. $NaHCO_3$.

BIF see BANDED IRON FORMATION.

bilateral symmetry n. a pattern of symmetry of an organism in which the individual parts are arranged along opposite sides of a median axis, so that one and only one plane can divide the whole into essentially identical halves. Compare RADIAL SYMMETRY.

binary system see PHASE DIAGRAM.

biochron n. the length of time represented by a BIOZONE.

biochronology n. that branch of GEOCHRONOLOGY based on the relative dating of geological occurrence by biostratigraphic or palaeontological evidence.

bioclastic rock n. a biochemical sedimentary rock consisting of fragmental remains of organisms, such as limestone composed of shell fragments. Compare BIOGENIC ROCK.

biocoenosis or **life assemblage** n. 1. an ASSEMBLAGE of FOSSIL remains found where the animals had lived. Compare THANATOCOENOSIS.
2. a group of organisms forming a natural ecological unit. Compare COMMUNITY.

biofacies n.. 1. the biological features or fossil character of a stratigraphical FACIES.
2. a rock or sediment formation differentiated from surrounding bodies purely on the basis of its FOSSIL content.
3. an ecological association of fossils.

biogenesis n. 1. formation by the activity

or processes of living organisms.
2. the tenet that all life must derive from previously living organisms.

biogenic rock *n*. an ORGANIC ROCK that is produced directly by the life activities or processes of organisms, e.g. coral reefs, shelly limestone, pelagic ooze, coal. Compare BIOCLASTIC ROCK.

biogeochemistry *n*. a branch of GEO-CHEMISTRY that is concerned with the interrelationships between organisms and the distribution and fixation of chemical elements in the BIOSPHERE, especially the soil. (Compare HYDROGEO-CHEMISTRY, LITHOGEOCHEMISTRY.) A buried deposit may impart to the soil an abnormal amount of the metal it contains, and the soil, in turn, may provide some amount of the same metal to the plants above. Analysis of such plants for the purpose of detecting an ore body is called *biogeochemical prospecting*. Compare GEOBOTANICAL PROSPECTING.

bioherm *n*. a mound-like mass of rock built by sedentary organisms such as COLONIAL CORALS, CYANOBACTERIA or CALCAREOUS ALGAE. Voids in the super-structure built by these organisms are filled with skeletons of other organisms as well as organic and inorganic detritus. Compare BIOSTROME.

biohorizon *n*. a surface of a particular biostratigraphical character or showing some biostratigraphical change. It is usually a BIOZONE boundary. Theoretically, it is a surface or interface, but in practice it may be a very thin bed with clearly distinguishable biostratigraphic features. Compare CHRONOHORIZON, LITHOHORIZON.

biolithite *n*. a LIMESTONE composed principally of the remains of reef-building organisms and associated debris. It is an AUTOCHTHONOUS REEF ROCK. See FOLK CLASSIFICATION.

biomicrite *n*. a LIMESTONE composed of FOSSIL skeletal remains (*bioclasts*) and CARBONATE MUD (micrite) in varying proportions. When the term is used, the predominant organism should be specified, e.g. 'gastropod biomicrite.' Compare MICRITE. See FOLK CLASSIFICA-TION.

biophile *n*. an element that typically occurs in organisms and organic material or is concentrated in and by living plants and animals.

biosome *n*. a mass of sediment deposited under uniform biological conditions. It is the biostratigraphic equivalent of LITHOSOME, not to be confused with BIOSTROME.

biosparite *n*. a LIMESTONE composed of skeletal remains and clear calcite cement (SPAR) in variable proportions. When the term is used, the predominant organism is usually specified, e.g. 'crinoidal biosparite'. Compare SPARITE. See FOLK CLASSIFICATION.

biosphere *n*. **I.** the zone at or near the Earth's surface occupied by living organisms. It includes parts of the LITHOSPHERE, HYDROSPHERE and ATMOS-PHERE. Compare ECOSPHERE.
2. all living organisms of the Earth, collectively.

biostratigraphical unit see STRATIGRAPHI-CAL UNIT.

biostratigraphical zone see BIOZONE.

biostratigraphy *n*. **I.** STRATIGRAPHY based on the palaeontological features of rocks.
2. the differentiation, dating and correlation of rock units by means of the fossils they contain.

biostrome *n*. a clearly bedded, sheetlike mass of rock built by sedentary organisms and composed chiefly of their remains, e.g. a bed of shells. Not to be confused with BIOSOME. Compare BIOHERM.

biotite *n*. a mineral of the MICA group, $K(Mg,Fe^{2+})_3(Al,Fe^{3+})Si_3O_{10}(OH)_2$, widely distributed in a variety of rocks: ACID and INTERMEDIATE PLUTONIC igneous rocks, some BASIC plutonic rocks, metamorphic rocks and some sedimentary rocks. Biotite is black in hand-specimen and has perfect basal CLEAVAGE.

biotope *n*. **1.** an area characterised by uniform ecology and organic adaptation. **2.** an environment in which an assemblage of flora and fauna lives or has lived.

bioturbation *n*. the disturbance of a sediment by the activities of organisms, often destroying its structure. An example is the feeding by burrowing of many sea-floor organisms through freshly deposited sediment, which disturbs bedding and lamination. Bioturbation often results in recognisable tracks, trails, and burrows. See TRACE FOSSIL.

biozone or **biostratigraphical zone** *n*. **1.** a general term for any type of biostratigraphical unit. See STRATIGRAPHICAL UNIT. **2.** the stratigraphical interval between two specified biohorizons (interval zone). See BIOHORIZON. Biozones form the fundamental units in biostratigraphical classification. Compare ACME ZONE, ASSEMBLAGE ZONE, RANGE ZONE.

3. in oceanography, a depth division within the ocean characterised by particular conditions and organisms.

bird-foot delta see DELTA.

birefringence *n*. the separation of an ordinary light beam by certain crystals into two unequally refracted beams. The measure of the birefringence of a crystal is the difference between the greatest and smallest refractive indices for light passing through it. Crystals having this property are said to be *birefringent*, since they have more than one INDEX OF REFRACTION. The property is characteristic of crystals in which the velocity of light rays is not the same in all directions (ANISOTROPIC crystals), i.e. all crystals except those of the isometric system. The effect can be demonstrated using a cleavage rhomb of CALCITE (Iceland spar). If the rhomb is placed over a dot marked on a piece of paper, two dots will be seen (see Fig. 6 below).

biscuit-board topography *n*. a glaciated landscape in which initial or incomplete erosion in CIRQUES may leave gently rolling sections of the proglacial upland surface. The cirques on the side of the upland resemble bits removed by a biscuit cutter at the edge of the dough. An example may be found in the Mackenzie Mountains of Canada.

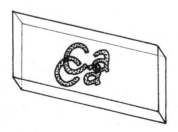

Fig. 6 **Birefringence**. A cleavage rhomb of calcite ($CaCO_3$) placed over printed text transmits a double image

bisector *n*. a line or plane of symmetry.

bisectrix *n*. an imaginary line bisecting either of the angles between the two OPTIC AXES of a BIAXIAL crystal. The *acute bisectrix* bisects the acute angle, and the *obtuse bisectrix* bisects the obtuse angle. See also INDICATRIX.

biserial *adj*. (of graptolites) having two rows of THECAE (cups) back-to-back along a STIPE; see GRAPTOLITE, Fig. 42.

bitheca see GRAPTOLITE.

bittern *n*. **1.** the residual brine from which the more soluble EVAPORITE minerals such as polyhalite, kieserite and carnallite (known as the bitterns) precipitate.
2. a bitter liquid remaining in salt-making after sea water or brine has evaporated and the salt has crystallised out, used as a source of bromides, iodides and certain other salts.

bitumen *n*. any of various solid and semisolid substances consisting mainly of hydrocarbons, e.g. petroleum, asphalts.

bituminous coal or **soft coal** *n*. a dark brown to black COAL, usually banded (see LITHOTYPE). It is the most abundant RANK of coal, much of it dating from the CARBONIFEROUS PERIOD, and is formed through the further COALIFICATION of LIGNITE. Bituminous coal, which burns with a smoky, yellowish flame, has more than 14 per cent volatiles on a dry, ash-free basis and a calorific value of about 14–16 MJ kg^{-1} (megajoule = 10^{6}J) (moist, mineral-matter free). Coals that are of gradations between lignite and bituminous are called SUB-BITUMI-NOUS.

bivalve, lamellibranch or **pelecypod** *n*. an aquatic MOLLUSC of the class Bivalvia, characterised by a calcareous bivalved shell that completely encloses the animal. Oysters, cockles and mussels are living examples. The VALVES are usually mirror images of each other (*equivalve*) and are right and left with respect to the body (Fig. 7(a) opposite). Normally the interior of each valve is marked by two adductor MUSCLE SCARS joined by the PALLIAL LINE, see Fig. 7(a). *Isomyarian* shells have approximately equal-sized scars; in *anisomyarian* shells the anterior scar is smaller, while *monomyarian* shells have no anterior scar and an enlarged posterior scar.

Bivalves have adapted to different modes of life: fixed by BYSSUS (fibrous threads) or cementation; free-swimming; free-lying; shallow or deep-burrowing; boring into rock; this is reflected in shell shape. The habitat may be marine, brackish or freshwater.

Classification is based on a number of features, including DENTITION (Fig. 7(b)), the nature of the muscle scar, LIGAMENT, pallial line and shell shape. Bivalves have been useful as zonal indices. Range, Lower CAMBRIAN to present.

blackband ironstone *n*. a ferruginous SEDIMENTARY ROCK, associated with the COAL MEASURES. The iron may be in the form of the carbonate SIDERITE (blackband), or as a clay ironstone (blackband ironstone) with as much as 20 per cent carbonaceous matter (COAL) present, so blackband ironstones are easy to smelt.

blackdamp or **chokedamp** *n*. a non-explosive coal-mine gas consisting of about 85 per cent nitrogen and 15 per cent carbon dioxide. Compare FIREDAMP.

black earth see CHERNOZEM SOIL.

blackjack *n*. a dark-coloured, iron-rich variety of SPHALERITE.

black light *n*. an ultraviolet or infrared lamp used to detect mineral FLUORES-CENCE.

black mud *n*. a marine-derived mud that acquires its colour from black sulphides of iron and from organic matter.

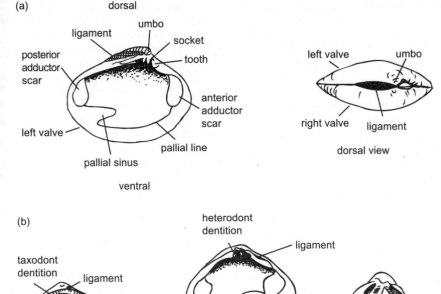

Fig. 7 **Bivalve**. (a) internal and external structures of a bivalve.
(b) the principal types of dentition

Black mud forms in bodies of water characterised by limited circulation and anaerobic conditions, e.g. lagoons, bays of water with weak tides, deep holes in neritic regions. The bottom waters in such locations are often rich in hydrogen sulphide and can be tolerated by few organisms. See also EUXINIC.

black sand *n*. an alluvial or beach sand containing heavy, dark minerals (magnetite, ilmenite, rutile) that have been transported by rivers and concentrated by wave and wind action.

black smoker *n*. a deep-sea FUMAROLE, found associated with submarine volcanic activity on OCEAN RIDGES, that emits fluids blackened by sulphide precipitates at temperatures of up to 350ºC or more. Mineral deposits from the fumarole build chimney-like structures as much as 10 metres in height and 40 centimetres wide that emit a smoky plume of metallic sulphide particles.

blanket sand *n*. a widely distributed ORTHOQUARTZITE deposited by a marine transgression on a stable shelf.

-blast or **-blastic** *suffix denoting* a feature completely formed during metamorphic RECRYSTALLISATION, for example PORPHYROBLAST.

blasto- *prefix denoting* a rock texture that has been modified during metamor-

61

phism but is still recognisable, for example a metamorphosed igneous rock in which RELICT PHENOCRYSTS remain has a *blastoporphyritic* texture.

blastoid *n*. a class of ECHINODERMS of Silurian to Late Permian occurrence, possibly evolved from CYSTOIDS. They were attached to the sea floor by a stalk of cylindrical plates, surmounted by a calyx or cup, and commonly had fivefold radial symmetry. Blastoids are characterised by paired internal respiratory structures underlying the food grooves (called *hydrospires*). Blastoids are not abundant fossils except locally, associated with a REEF FACIES, in some LIMESTONES. Some blastoids are useful as INDEX FOSSILS.

bleaching clay *n*. any clay having the capacity of removing colouring matter from oil through ADSORPTION, for example, FULLER'S EARTH.

bleach spot SEE REDUCTION SPOT.

bleb *n*. an informal term for a small, rounded inclusion of one mineral in another, e.g. a bleb of olivine poikiliti-cally enclosed in pyroxene.

Bleiberg-type lead SEE ANOMALOUS LEAD.

blende SEE SPHALERITE.

blind valley SEE KARST.

block *n*. **1.** an angular rock fragment with a diameter greater than 256 millimetres. It shows little alteration as a result of transporting agents. Compare BOULDER.
2. an angular PYROCLASTIC particle of diameter greater than 64 millimetres, ejected in a solid state. Compare VOLCANIC BOMB.
3. a FAULT BLOCK.

block caving *n*. a method of mining in which blocks of ore are caused to cave in or collapse by undercutting and blasting away the supporting pillars.

block diagram *n*. a plane figure illustrat-ing the three-dimensional geological structure of an area. A rectangular block of the Earth's crust is drawn in three-dimensional perspective to show a top surface and at least two vertical cross-sections. If an orthorhombic projection is drawn, each of the three sides of a rectangular block is fore-shortened equally. In isometric block diagrams, equal parts of the three faces of a cube are seen.

block faulting *n*. a type of normal or gravity faulting in which the crust is separated into structural units (FAULT BLOCKS) of different orientations and elevations. FAULT-BLOCK MOUNTAINS form in this way.

block field, felsenmeer, blockmeer or **stone field** *n*. an accumulation of coarse detritus on level mountain-top surfaces or gently sloping mountain sides. Block fields consist mainly of local rocks, often angular, broken by frost shattering of the bedrock, the degree of angularity depending greatly on the rock type. The time required for the formation of block fields varies with rock type and climate, and especially with the number of times the freezing point is passed. Compare BLOCK STREAM. See also FROST ACTION.

blocking temperature *n*. **1.** the tempera-ture below which the INDUCED MAGNETI-SATION of a magnetic mineral grain becomes a stable thermoremanent magnetisation, i.e. the magnetisation is 'frozen in' to the mineral. The blocking temperature is usually a few tens of degrees below the Curie temperature for that mineral (SEE NATURAL REMANENT MAGNETISATION).
2. the temperature below which radiogenic argon effectively cannot diffuse out of the crystal lattice in which it was formed (by decay of ^{40}K).

block lava SEE LAVA FLOW.

blockmeer SEE BLOCK FIELD.

block stream or **rock stream** *n*. an accumulation of rock debris, usually at the head of a ravine and following valley floors. Block streams may begin from summit block fields extending downslope, or from large TALUS accumulations. They are regarded as being of periglacial origin, but the mechanism responsible for their formation and possible movement is still open to question. Compare BLOCK FIELD.

bloom *n*. **1.** an EFFLORESCENCE.
2. the oxidised or decomposed part of a mineral vein or bed that has been exposed. See also TARNISH.
3. the sudden proliferation of organisms in bodies of fresh or marine water, e.g. ALGAL BLOOM, PLANKTON bloom.

blowhole *n*. **1.** a vertical cylindrical fissure in a sea cliff extending from the bottom of the inner end of a SEA CAVE upwards to the surface. A blowhole is formed if a joint, which is a line of weakness, extends from the tunnel end to the top of the cliff. A fountain-like effect is produced as waves and tides force water and air into it.
2. an opening through a snow bridge into a CREVASSE, the presence of which is usually indicated by a current of air.
3. a small vent on the surface of a LAVA FLOW.

blowout *n*. **1.** a general term for various hollows formed by wind action in regions of sand or light soil. It is also applied to the accumulation of a material derived from the hollow, e.g. a blowout dune.
2. the abrupt and uncontrollable escape of gas, oil or water from a well.

blue asbestos see CROCIDOLITE.

blue-green algae see CYANOBACTERIA.

blue ground *n*. an informal name given to the diamond-bearing BRECCIA filling KIMBERLITE pipes. Only the unweathered material is blue: weathered surfaces are yellow. See YELLOW GROUND, DIATREME.

Blue John *n*. a variety of FLUORITE with distinctive purple and colourless or pale yellow banding caused by HYDRO-CARBON impurities, found in Derbyshire. It has been used for decorative purposes, carved into vases and dishes.

blue mud see TERRIGENOUS DEPOSITS.

blueschist *n*. a METAMORPHIC ROCK that characteristically contains GLAUCOPHANE, a blue sodic AMPHIBOLE. Blueschists are formed as a result of high-pressure, low-temperature METAMORPHISM of rocks of BASIC igneous composition. The MINERAL ASSEMBLAGE containing glaucophane, lawsonite and QUARTZ, with or without GARNET, identifies the *glaucophane-schist facies* (or *blueschist facies* as it is sometimes called). See METAMORPHIC FACIES.

blue vitriol see CHALCANTHITE.

BM see BENCH MARK.

body chamber *n*. the final chamber of a CEPHALOPOD shell, in which the animal lives. See AMMONOID, Fig. 1, NAUTILOID, Fig. 62.

body waves see SEISMIC WAVES.

boehmite *n*. an orthorhombic mineral, AlO(OH), a dimorph of diaspore; its colours may be grey, brown or reddish. It is an important constituent of some BAUXITES.

bog *n*. a type of wetland characterised by spongy, poorly drained peaty soil. Bogs are often divided into:
(a) those of cool regions, dominated by sphagnum and heather; bogs of boreal regions, e.g. Canada, with trees such as tamarack and black spruce on them, are called *muskegs*.
(b) *fens*, which are dominated by grasses, sedges and reeds.
(c) *tropical tree bogs*, in which the peat is composed almost entirely of tree remains.

Bogs of cool regions form in depressions created by glacial ice and in small lakes in glaciated regions. Colonisation by sphagnum and subsequent poor drainage contribute to a process that may eventually fill the body of water with vegetation. At the stage where surface vegetation is still floating and not coherent, the bog is called a *quaking bog* because of the surface instability.

Peat bogs are not generally found in lowland tropical areas because the high temperatures facilitate rapid decay of organic matter. Tropical peat bogs may develop, however, in areas of very high rainfall and with groundwater of very low mineral content. The peat from such bogs is composed of the remains of seed plants rather than sphagnum. Compare MARSH, SWAMP.

bog burst *n.* the rupture of a BOG that has become swollen because of retention of water by fallen vegetation. Conditions of great rainfall rapidly increase the pressure of the swelling and the bog may 'burst', emitting black organic matter that spreads over its surroundings.

Bøggild see LABRADORESCENCE.

boghead coal *n.* a nonlayered sapropelic coal similar to CANNEL COAL in its physical properties, e.g. having a high content of volatiles but composed mainly of algal matter rather than spores. Compare TORBANITE.

bog iron ore *n.* a general term for a poor-quality iron ore found as layered or nodular deposits of hydrous iron oxides at the bottom of swamps and bogs. The deposits, consisting chiefly of goethite, FeO(OH), are formed by precipitation of iron from the ambient water and by the oxidising action of iron bacteria and algae or by atmospheric oxidation.

bog manganese see WAD.

bole *n.* a reddish WEATHERING product formed on the surface of BASIC lava flows. It is a fossil LATERITE and indicates formation under tropical or subtropical climatic conditions.

bolide *n.* a large extraterrestrial body that impacts with the Earth to form a CRATER.

bomb see VOLCANIC BOMB.

bomb sag *n.* an indentation made in layered, usually VOLCANICLASTIC, sediments as the result of the impact of a VOLCANIC BOMB.

bond *n.* the linkage between atoms in minerals and other types of chemical compound. See COVALENT BONDS, HYDROGEN BOND, IONIC BOND, VAN DER VAALS BONDS.

bone beds *n.* any sedimentary stratum, usually a thin bed of limestone, sandstone or gravel, in which fossil bones, teeth or scales are abundant. Bone beds suggest unusual features of deposition, such as a lack of sediment supply behind a transgressive system or current action preventing deposition, or even mass extinction because of salinity changes, earthquakes or outbursts (blooms) of certain organisms. They may be an economic source of phosphates; see COLLOPHANE.

bone phosphate of lime *n.* tricalcium phosphate, $Ca_3(PO_4)_2 \cdot H_2O$. See also COLLOPHANE.

boninite *n.* magnesium-rich MAFIC andesitic VOLCANIC ROCK. Boninites are glassy rocks containing microphenocrysts of one or more PYROXENES and OLIVINE. Textures characteristic of rapid growth are common, for example skeletal crystals. Boninites are normally found in an ISLAND-ARC setting, and it has been suggested that they are products of the early stages of SUBDUCTION. See BASALT, ANDESITE.

book structure *n.* in ore deposits, a

formation showing an alternation or interleaving with GANGUE.

boomer *n*. a marine seismic energy source in which capacitors are charged to high voltage and then discharged in a body of water through a transducer, a component of which consists of two metal plates. Abrupt separation of the plates produces a low-pressure region between them into which water rushes, thus generating a pressure wave.

borate *n*. a mineral type containing the borate radical $(BO_3)^{3-}$. The commonest are BORAX, COLEMANITE, KERNITE and ULEXITE.

borax *n*. a monoclinic mineral, $Na_2B_4O_5(OH)_4 \cdot 8H_2O$, mined as an ore of boron. It is found in arid regions where it forms as an evaporite in playa deposits and as a surface efflorescence. In a dry environment its clear, transparent crystals lose water and become chalky white tincalconite.

bornhardt *n*. a large INSELBERG. Bornhardts are dome-shaped monoliths, examples of which are found on every continent but Antarctica. They are best-developed in the tropics.

bornite *n*. copper-iron sulphide, Cu_5FeS_4, an ore of copper occasionally occurring with chalcopyrite and chalcocite. Bornite is a fairly common mineral, but crystals of it are rare. Freshly exposed surfaces of the mineral show a bronzelike colour that quickly tarnishes in air to a purplish iridescence, giving rise to the name *peacock ore*.

borolanite *n*. a local name for a variety of NEPHELINE SYENITE occurring near Loch Borolan in Assynt, northwest Scotland.

bort see INDUSTRIAL DIAMOND.

boss *n*. **1.** a knoblike mass of PLUTONIC IGNEOUS rock, which may be circular or elliptical in plan. Its contacts are vertical or steeply inclined.
2. a knoblike excrescence or protuberance on some organ of a plant or animal. See also TUBERCLE.

botryoidal *adj*. (of certain minerals, e.g. smithsonite) having an appearance resembling a bunch of grapes. Compare RENIFORM.

bottom load see BED LOAD.

bottomset bed see DELTA.

boudinage *n*. a structure found in greatly deformed sedimentary and metamorphic rocks in which an originally continuous rigid (COMPETENT) layer between relatively plastic (less competent) layers is stretched and thinned until rupture occurs; the rigid band, thus separated into long cylin-

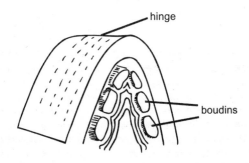

Fig. 8 **Boudinage**. Boudin structures in a deformed layer

drical pieces, resembles in cross-section a string of sausages. In the pinched parts, or points of separation, quartz or calcite is usually crystallised. See Fig. 8 on the previous page. Where the pieces have not separated completely the structure is known as *pinch-and-swell*.

Bouguer anomaly see GRAVITY ANOMALY.

boulder *n*. a separated rock mass larger than a COBBLE, with a diameter greater than 256 millimetres. It is rounded in form or shaped by ABRASION. Boulders are the largest rock fragments recognised by sedimentologists.

boulder clay see TILL.

boulder fan see INDICATOR BOULDER.

boulder pavement *n*. I. a surface of TILL consisting largely of BOULDERS and worn down by GLACIER movement.
2. an accumulation of BOULDERS, once part of a glacial MORAINE, remaining after finer materials have been removed by wind, waves and currents.
3. a DESERT PAVEMENT composed of BOULDERS.

boulder train see INDICATOR BOULDER.

Bouma sequence *n*. the 'ideal' series of structures in a TURBIDITE deposit, comprising from top to bottom:
E, homogeneous MUD
D, mud and SILT, interlaminated
C, sand and silt, a unit showing small-scale cross-laminations (see CROSS-BEDDING), ripples or convoluted laminae
B, sand, planar LAMINATIONS
A, sand, MASSIVE, or sometimes graded, unit (see GRADED BEDDING).
It is rare for all five units to be present: one or more members may be missing from the bottom or top of the sequence. Those present, however, are in the correct order.

bounce marks see TOOL MARKS.

boundary stratotype *n*. that point in a particular sequence of rock strata that is used as the standard for recognition and definition of a stratigraphical boundary. Compare STRATOTYPE.

boundary waves *n*. a mode of wave propagation along the interface between media of different properties. Where the interface or boundary is a free surface, such as an earth-to-air surface, the mode is a *surface wave* (see SEISMIC WAVES).

boundstone *n*. a LIMESTONE, the original components of which were bound together during deposition by encrusting or reef-building organisms. It is equivalent to BIOLITHITE. Boundstone is one of a number of terms used in the DUNHAM textural classification of limestones: the others are: MUDSTONE, WACKESTONE, PACKSTONE and GRAINSTONE.

bourne *n*. an intermittent stream that appears in a region underlain by CHALK when the water table rises high enough for water to flow on the surface in a normally DRY VALLEY.

Bowen's reaction series see REACTION SERIES.

box fold see FOLD.

boxwork *n*. a skeletal framework of intersecting plates composed of a hydrated iron oxide (any of several are possible) that has been deposited along fractures and in cavities from which intervening material has been dissolved by groundwater-related processes. Boxwork is common in the oxidised zone of sulphide ores (see GOSSAN) and on the ceilings of caves. In such sites of occurrence, several different boxwork structures may exist separately or become joined together in a complex.

bp *abbreviation for* before present.

brachia *n*. (*sing.* **brachium**) the arms of a CRINOID, see Fig. 21. They are flexible and may be simple or branched, with a food groove (AMBULACRUM) on the oral side. The arms are supported by calcareous plates known as *brachials*.

brachials see BRACHIA.

brachial valve *n.* the VALVE of a BRACHIOPOD shell carrying the BRACHIA, which are projections of the LOPHOPHORE, or feeding organ. The brachial valve is usually the smaller of the two. It is sometimes known as the *dorsal valve*. See Fig. 9 below.

brachidium *n.* a calcareous framework supporting the LOPHOPHORE (feeding organ) in some articulate BRACHIOPODS. It may take the form of calcareous ribbons joined in a loop or in helical spirals (*spiralia*), or of two curved plates or rods known as *crura*. See Figs. 9, 10, below and overleaf.

brachiopod *n.* any marine invertebrate of the phylum Brachiopoda. They are sessile animals that secrete an external shell consisting of two bilaterally symmetrical VALVES of unequal size (Fig. 9 below). They are usually

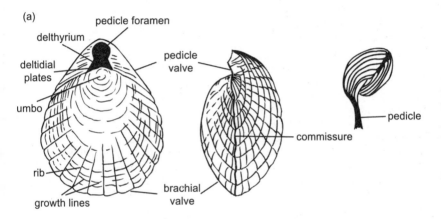

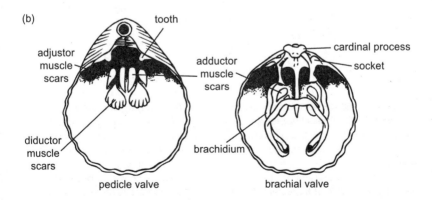

Fig. 9 **Brachiopod**. (a) external structures and mode of life of a rhynchonelliform brachiopod. (b) internal structures

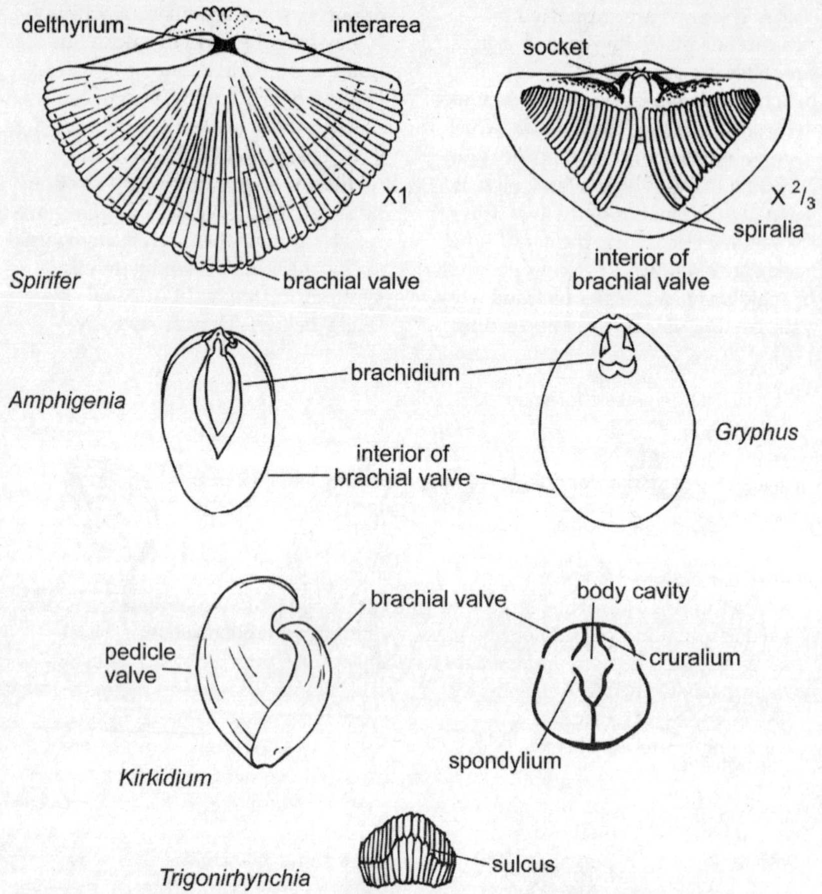

Fig. 10 **Brachiopod**. Shell shape and internal structures in some rhynchonelliform brachiopods

attached to the sea floor by a stalk, the *pedicle*. In some forms the pedicle is reduced or absent and the shell may lie free or be held in soft sediment by spines.

Brachiopods are divided into three subphyla: the *Linguliformea* (see Fig. 11 overleaf) in which the valves are held together by muscles only; the *Craniiformea*; and the *Rhynchonelliformea* (see Fig. 10 above), typical forms of which have muscles and a hinge structure consisting of two TEETH on the PEDICLE VALVE that fit into sockets on the BRACHIAL VALVE.

Some forms have calcareous structures inside the valves for the attachment of muscles or support of the lophophore, see Fig. 10 above. Brachiopods appeared in the Lower CAMBRIAN;

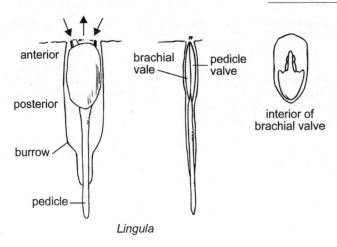

Fig. 11 **Brachiopod**. Structure and mode of life of *Lingula*, a linguliform brachiopod

they declined in the TERTIARY and are now a minor group. They have been used as zonal indices in the ORDOVICIAN, SILURIAN, CARBONIFEROUS and CRETACEOUS.

brackish *adj.* (of water) with salinity intermediate between that of normal fresh water and normal sea water.

Bragg equation *n.* an equation describing the way in which an X-ray beam will be diffracted by the atoms in a crystal lattice. The equation is $n\lambda = 2d_{hkl} \sin\theta$, where n is a whole number, λ is the wavelength, d_{hkl} is the spacing between successive planes of atoms and θ is 90° minus the ANGLE OF INCIDENCE. It is used in determining the internal structure of minerals by X-ray methods.

braid plain *n.* a broad, flat area or plain produced by the coalescence of BRAIDED STREAMS.

braided stream *n.* a stream that divides into numerous channelways that branch, separate and rejoin, thus becoming a tangle of channels, islands and sandbars resembling a complex braid. Braided streams occur under conditions of abundant BED LOAD and highly variable discharge. These conditions are provided where a river in a semi-arid region receives most of its discharge from precipitous headwaters. Braiding also occurs where glaciers are melting and the debris supply is consistently large, whereas the discharge varies as the ice melts and refreezes. Examples of this type of braiding are found wherever melting glaciers occur, e.g. in Iceland. Information concerning ancient environments is often based on recognition of braided stream deposits in older sediments.

branch fault see AUXILIARY FAULT.

Bravais lattice see CRYSTAL LATTICE.

breached anticline *n.* an ANTICLINE the centre of which has been eroded to the extent that its limbs form erosional scarps that flank the breach and face upward. See also INVERSION OF RELIEF.

breadcrust bomb *n.* a VOLCANIC BOMB with a cracked exterior resembling breadcrust. Its appearance is the result of expansion caused by vesiculation of the interior after surface solidification. See Fig. 92.

break *n.* **1.** a general term for a sudden change in a slope.

2. an abrupt change in the LITHOLOGY or faunal content of a stratigraphical sequence.

3. ARRIVAL TIME.

breaker or **breaking wave** *n.* a wave in which the velocity of the crest exceeds the rate at which the body of the wave is moving forward, thereby causing a curl to form at the top of the wave front, which plunges forward and collapses. Conditions for breakers are set as waves decrease in velocity when moving into shallow water; as a consequence of this velocity change, there is a crowding together of the waves and a steepening of their fronts. *Breaker depth* is the still-water depth taken at the point of wave break. *Breaker height* is the average height from trough to crest of breaking waves. Following the breaking of a wave, a surge of water, or *swash*, rushes on to the shore. After the swash reaches its maximum forward position, the water returns down the seaward slope as *backwash*, leaving behind a thin line or ridge of debris (sand, seaweed) called a *swash mark*.

breccia *n.* a coarse-grained (RUDACEOUS) clastic rock composed of broken angular rock fragments enclosed in a fine-grained matrix or held together by a mineral cement. Unlike CONGLOMER-ATES, in which the fragments are rounded, breccias consist of fragments that were not worn by abrasion prior to their embedment in a matrix. *Sedimentary breccias* are relatively rare but may indicate scree, terrestrial mudflows or submarine landslides. *Tectonic* or *fault*

breccias are composed of fragments produced by rock fracturing during faulting or other crustal deformation. See also VOLCANIC BRECCIA.

bridge see NATURAL BRIDGES AND ARCHES.

Brigantian *n.* the top stage of the VISEAN (CARBONIFEROUS) in Britain and western Europe.

bright spot *n.* in seismic prospecting, an unusually strong signal on a seismic PROFILE, frequently indicating a pocket of natural gas.

brine *n.* salt water, especially a highly concentrated aqueous solution of common salt (NaCl). Natural brines occur in salt lakes, underground or as sea water and have commercial importance as sources of NaCl and other salts such as chlorides and sulphates of potassium and magnesium.

brittle behaviour *n.* a response to STRESS in which a material ruptures easily with little or no PLASTIC DEFORMATION (sometimes defined as fracture occurring when total STRAIN is less than 5 per cent). Compare DUCTILE BEHAVIOUR. Temperature and confining pressure affect the way in which materials deform: rocks that display brittle behaviour at high crustal levels can deform in a ductile manner at depth.

brittle micas *n.* hydrous alumina-silicate minerals similar to ordinary MICAS except that the CLEAVAGE flakes are less elastic and the HARDNESS is greater. They contain calcium as an essential constituent, e.g. margarite, that may be regarded as the calcium analogue of PHLOGOPITE.

brittle star see OPHIUROID.

bromoform *n.* tribromomethane, $CHBr_3$, specific gravity 2.9. It is used as a HEAVY LIQUID.

bronzite *n.* a brown or green magnesium-rich orthopyroxene, intermediate in composition between enstatite and hypersthene. It often has a distinctive, submetallic bronzelike lustre. See also PYROXENE.

brookite *n.* a brown or reddish lustrous orthorhombic mineral, TiO_2. It is trimorphous with anatase and rutile, and occurs in veins and DRUSY cavities.

brown coal see LIGNITE.

brucite *n.* a hexagonal mineral, $Mg(OH)_2$, that commonly occurs as thin plates and as fibrous masses in SERPENTINE and metamorphosed DOLOMITE.

Brunhes normal epoch see POLARITY EPOCH.

Brunton compass or **Brunton pocket transit** *n.* an instrument consisting of a compass, mirror, folding-open sights and spirit-level clinometer. It is used in mining, topographic and geological surveys for levelling, reading horizontal and vertical angles, and for determining the magnetic bearing of a line. In field mapping of bedrock, it is used to determine STRIKE and DIP.

bryophyte *n.* a small, nonvascular plant without true roots but which may have differentiated stems and leaves. Mosses and liverworts are bryophytes. Compare PTERIDOPHYTE.

bryozoan or **polyzoan** *n.* any aquatic invertebrate of the phylum Bryozoa characterised mainly by colonial growth and an encrusting, branching or fanlike structure, forming a colony (*zooarium*) a few centimetres across. Bryozoans are mainly marine and are found at a range of depths from tide level to ABYSSAL. The organism's skeletal material may be calcium carbonate or chitin, but only the calcareous forms are found as fossils. Range, ORDOVICIAN to present; some are useful in strata correlation.

B shell see SEISMIC REGIONS.

B-tectonite see TECTONITE.

B-type lead see ANOMALOUS LEAD.

bubble point *n.* the point on a PHASE DIAGRAM representing a state of equilibrium between a relatively large quantity of liquid and the last vestige (or *bubble*) of vapour.

Bubnoff unit *n.* a standard unit of measurement of geological time-distance rates, e.g. geological movements and increments, defined as 1 micron/year (or 1 metre/Ma^{-1}).

buccal slits *n.* notches in the edge of the PERISTOME of a regular ECHINOID that accommodate soft tissue related to the jaws. See Fig. 27.

Buchan-type metamorphic sequence *n.* a succession of MINERAL ASSEMBLAGES observed in PELITIC rocks of the Buchan area of northeast Scotland, characterised by the appearance of ANDALUSITE, CORDIERITE, STAUROLITE and then SILLIMANITE with increasing METAMORPHIC GRADE. This sequence is believed to have developed under lower pressures than those of the BARROVIAN-TYPE METAMORPHIC SEQUENCE.

buchite *n.* a glassy HORNFELS produced by partial melting of the COUNTRY ROCKS close to the contact with an INTRUSION, or by partial or complete melting of a XENOLITH.

bulk density or **apparent density** *n.* the density of an object or material, determination of which includes the volume of its pore spaces.

bulk modulus, volume elasticity or **modulus of incompressibility** *n.* a numerical constant that describes the elastic properties of a solid or fluid by relating the change in volume when the substance is subjected to pressure on all surfaces. Thus:

bulk modulus (k) = pressure/strain

$$= \frac{P}{v_0 - v_n/v_0},$$

where v_0, the original volume of a material, is reduced by an applied pressure P to a new volume v_n; the strain may be expressed as the change in volume, $v_0 - v_n$, divided by the original volume, v_0. The bulk modulus is the reciprocal of COMPRESSIBILITY.

bullion n. **1.** a carbonate or silica concretion found in the roofs of some coal seams, often formed about a nucleus of preserved plant material. Its diameter ranges from several centimetres to a metre. Compare COAL BALL.
2. a nodule of some material (clay, shale, ironstone) that generally holds a fossil.

Burdigelian n. a STAGE of the Lower MIOCENE.

Burgess shale n. a bed of black SHALE in the Middle Cambrian of British Columbia, Canada, bearing an important fossil fauna. The fossils are preserved as carbon films on the BEDDING PLANES where bedding and cleavage coincide. All are forms of soft-bodied animals. Among the 70 genera and 130 species identified are sponges, jellyfish, annelid worms and primitive arthropods. See CARBONISATION, FOSSIL, KONSERVAT LAGERSTÄTTEN.

burrow n. a pipelike cavity in sedimentary rock, made by an animal that lived in the soft sediment. Burrows, often filled with clay or sand, may be along the BEDDING PLANE or may penetrate the rock. The porosity in a sedimentary rock that results from animal activity is called *burrow porosity*.

bushveld see SAVANNA.

butane n. either of two odourless, colourless, inflammable paraffinic hydrocarbons, formula C_4H_{10}, that occur in natural gas and crude petroleum. The straight-chain compound is denoted *n-butane*; the branched-chain form is *isobutane*.

butte n. an isolated flat-topped hill with steep slopes, especially in the western United States and Canada, representing an erosional remnant of horizontal sedimentary rocks. A butte resembles a MESA but is smaller, with a more limited summit.

b.y. *abbreviation for* billion years, $= 10^9$ years. See GA.

bysmalith n. a roughly vertical and cylindrical body of intrusive igneous rock, cross-cutting adjacent sediments. It has been suggested that bysmaliths are formed by CAULDRON SUBSIDENCE. The term is rarely used.

byssus n. fibrous threads secreted by some forms of BIVALVE to attach themselves to a firm surface.

bytownite n. a bluish to dark-grey member of the plagioclase feldspar series, occurring in basic and ultrabasic igneous rocks. The composition is An_{70}–An_{90} (see FELDSPAR, Fig. 33). The colour is often because of the presence of INCLUSIONS.

Cc

c see CARAT.

cable-tool drilling, percussion drilling *n*. a standard method of drilling and an earlier method of drilling for oil. A chisel-like bit is suspended by a cable to a lever at the surface, and the lever alternately lifts and drops the bit, causing it to chip away the rock, which is removed by a bailer. Compare ROTARY DRILLING.

cadicone see AMMONOID.

Calabrian *n*. a STAGE of the Lower PLEISTOCENE in southern Europe.

Cainozoic see CENOZOIC.

calamine *n*. **1.** in the British Isles, a former term for SMITHSONITE.
2. in North America, HEMIMORPHITE.

calamite *n*. a giant arboreal SPHENOPSID with a jointed, ribbed trunk and leaf whorls at the joints. Height to *c*. 12 metres; Carboniferous to Permian. See also FOSSIL PLANTS.

calaverite *n*. a tin-white or bronze-yellow monoclinic mineral, gold telluride, $AuTe_2$. It resembles silver more than gold but yields a button of gold when the tellurium is burned off by heat.

calc- *prefix denoting* the presence of calcium. See CALCAREOUS, CALCIFEROUS.

calc-alkali series *n*. the BASALT-ANDESITE-DACITE-RHYOLITE suite of IGNEOUS ROCKS that is typically found in the ISLAND ARCS and active CONTINENTAL MARGINS of DESTRUCTIVE PLATE BOUNDARIES. The series is generally SUBALKALINE and silica-oversaturated and contains a higher proportion of Al_2O_3 than the tholeiitic association: the basaltic member of the calc-alkali series is typically a high-Al basalt with about 17 per cent Al_2O_3 (see also THOLEIITIC BASALT). There is no trend towards iron-enrichment in intermediate members of the series.

calcarenite *n*. a term, used in the Grabau classification, for a rock composed of grains of calcium carbonate with or without a carbonate cement. A clastic limestone.

calcareous *adj*. containing calcium carbonate. When used with a rock name, it generally implies that as much as 50 per cent of the rock is calcium carbonate.

calcareous algae *n*. ALGAE that secrete a calcareous skeleton. They are very diverse, some forming irregular, encrusting masses, others erect branching growths, while some are planktonic. Calcareous algae are known from the Neoproterozoic to the present. Calcareous algae are important components of CORAL REEFS and have contributed to reef structures from the CAMBRIAN onwards. Compare CYANOBACTERIA.

calcareous deposits see PELAGIC DEPOSITS.

calcareous tufa see TUFA.

calc-flinta see CALC-SILICATE HORNFELS.

calciferous *adj*. denoting presence of calcium carbonate.

calciclastic *adj*. relating to a clastic carbonate rock.

calcification *n*. production of hard parts of organisms during life by deposition of CALCIUM CARBONATE ($CaCO_3$) in the form of CALCITE or ARAGONITE, or REPLACEMENT by calcium carbonate during fossilisation. See FOSSIL.

calcify *v.* **1.** to make hard or strong by the deposition of calcium salts.
2. to become hard or stonelike because of the deposition of calcium salts.

calcilutite *n.* a term proposed by Grabau for limestones formed of silt and clay-sized detrital particles.

calcirudite *n.* a term proposed by Grabau for limestones formed of detrital calcite particles greater than 2 millimetres diameter.

calcisiltite *n.* a LIMESTONE composed mostly of detrital, silt-sized carbonate particles that have been consolidated. See also ARENITE, LUTITE, RUDITE.

calcite *n.* a trigonal mineral, $CaCO_3$, the stable form of CALCIUM CARBONATE at all those temperatures and pressures obtaining at or near the Earth's surface; because of this, it is possible that all other forms convert to it over geological time. Calcite is a widely distributed rock-forming mineral, the chief constituent of limestone and most marble, and is common in the shells of invertebrates. Its crystals, which have perfect rhombohedral cleavage, are usually white or grey; the transparent variety is ICELAND SPAR. It is polymorphous with ARAGONITE, vaterite and some other exotic forms.

calcium bentonite see BENTONITE.

calcium carbonate *n.* the chemical compound $CaCO_3$. See CALCITE, ARAGONITE.

calcrete *n.* **1.** also called **caliche**, a LIMESTONE precipitated as surface or near-surface crusts and nodules by the evaporation of soil moisture in semi-arid climates; it may be combined with sand and gravel.
2. a calcareous DURICRUST. Compare SILCRETE.

calc-schist *n.* a metamorphosed, argillaceous LIMESTONE showing a schistose structure.

calc-silicate hornfels *n.* a rock produced by the thermal metamorphism of impure LIMESTONES containing CLAY and SILICA as well as CALCITE. The rock typically becomes very hard and minerals such as TREMOLITE, DIOPSIDE and Ca-Al GARNET are produced. Because of the extreme hardness the rock is sometimes called *calc-flinta*.

calc-silicate rock *n.* a type of METAMORPHIC ROCK in which calcium and silicon are the dominant constituents and which is derived from quartz-bearing dolomites and limestones, or from carbonate rock metasomatised by siliceous solutions from abutting granitic intrusions. Minerals characteristic of these rocks include EPIDOTE, calcic PLAGIOCLASE, DIOPSIDE, grossular-ANDRADITE GARNET and sphene.

calc-sinter see TRAVERTINE.

calc tufa see LIMESTONE.

caldera *n.* a very large, bowl-shaped volcanic depression the horizontal dimension of which is much greater than its vertical dimension. It is usually formed by a combination of the explosion and collapse of the top of a volcanic cone or group of cones. Such subsidence can be caused by the emptying of the underlying magma chamber during eruptions. In the case of explosive volcanoes with silica-rich magma this may happen suddenly following major eruptions of pumice. In basaltic volcanoes, as in Hawaii, *summit calderas* develop progressively by drainage of the magma chamber during major EFFUSIVE activity. Calderas may later fill up with water, for example Crater Lake in Oregon and Askja in Iceland. Calderas range in size from one kilometre to more than 20 in diameter. They are commonly roughly circular or elliptical in outline. The very large calderas associated with

SUPERVOLCANOES sometimes have angular outlines with indentations, their shape apparently being controlled to a degree by faults and joints in the underlying crust. These have sometimes been referred to as *volcanic-tectonic depressions*; however, the rapid extrusion of large amounts of lava seems to have been the principal reason for their collapse. See VOLCANIC CRATER, VOLCANO, SECTOR COLLAPSE.

Caledonian orogeny *n*. a number of episodes of DEFORMATION that formed a MOUNTAIN BELT, the remnants of which are now found in Spitzbergen, Scandinavia, the northern part of Britain and in Greenland, Newfoundland and the northern Appalachians of North America. The term 'Caledonian orogeny' is restricted to the various phases of deformation during the SILURIAN and DEVONIAN resulting from the collision between the North American continent (LAURENTIA), Eastern AVALONIA, BALTICA and other fragments of GONDWANA. This period of time includes the final closure of the ocean IAPETUS. In Britain, the northern boundary of the orogeny is formed by the Moine Thrust Zone, and its southern boundary is the Welsh Borderland Fault Zone. The Grampian orogeny (470–460 Ma) predated the Caledonian orogeny and involved major deformation and metamorphism of the Moine and Dalradian rocks of the Scottish Highlands and northern Ireland (see MOINE SUPERGROUP, DALRADIAN SUPERGROUP). The TACONIAN OROGENY and ACADIAN OROGENY in North America are equivalent to the Caledonian orogeny in Britain and Scandinavia.

Caledonides *n*. the orogenic belt formed by the CALEDONIAN OROGENY. It extends from southern Wales across Scotland and northeastward through Scandinavia and from Greenland down through the Appalachian Mountains on the east coast of North America.

calice *n*. the cup-shaped depression in

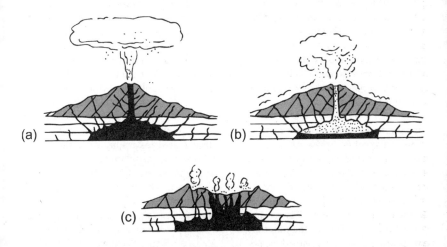

Fig. 12 **Caldera**. (a) and (b) eruptions partially empty the magma. (c) the summit of the volcano collapses into the magma chamber, forming a caldera

the upper surface of a CORALLITE, which contained the polyp (see RUGOSE CORAL, Fig. 77).

caliche *n.* **1.** a layer, chiefly of calcium carbonate at or near ground surface, in certain areas of scant rainfall. It is formed when groundwater containing dissolved calcium carbonate moves upward and evaporates, leaving a crust.
2. alluvium cemented with sodium salts in the nitrate deposits of northern Chile. See also DURICRUST, CALCRETE 1, CHILE SALTPETRE.

Callovian *n.* the top STAGE of the Middle JURASSIC.

callus *n.* a calcareous deposit over the inner lip of some GASTROPOD shells, see Fig. 38.

calving *n.* the breaking off and floating away of large masses of a glacier as it moves into the sea. See ICEBERG.

calyx or **aboral cup** *n.* a cup-shaped structure of calcareous plates that forms the lower part of the THECA of a CRINOID (Fig. 21). It may be flexible or rigid and is formed of two circlets of five plates, *basals* (lower) and *radials* (upper), sometimes with an extra ring known as *infrabasals*. Some crinoids have additional plates, the *inter-radials*, inserted between the radials, and the lowest arm plates (BRACHIALS) may form part of the calyx.

camber *n.* a superficial feature observed in areas near horizontal and relatively thin rocks that comprise a COMPETENT bed and are underlain by weaker strata. The weaker strata will sometimes show a general valleyward increase of dip, to which the competent bed adjusts by sagging. This can be seen along the edges of the outcrop.

Cambrian *n.* the oldest period of the Palaeozoic Era, having a duration of about 50 Ma and beginning about 544 Ma ago. The period is divided informally into Early Cambrian, Mid-Cambrian and Late Cambrian, the corresponding series in Britain being Comley, St Davids and Merionith. Rocks of the Cambrian system are the oldest in which fossil remains are sufficiently abundant and distinct, because of hard parts, as to provide reliable geological information. Most marine invertebrate phyla were represented in the earliest Cambrian. TRILOBITES and BRACHIOPODS are used as index fossils. GRAPTOLITES and CONODONTS are found in Middle and Upper Cambrian rocks. Marine plants were mainly green or red ALGAE. CYANOBACTERIA formed STROMATOLITES. The presence of fossil marine organisms in most of the rocks records Cambrian TRANSGRESSIONS: sea levels were rising through much of the period. Since the oldest rocks of this series occur in northern Wales, the series was named 'Cambrian' after the ancient name for Wales. Southern Scandinavia, however, serves as the Cambrian-referred section because northern Wales is poorly fossiliferous, and the boundary STRATOTYPE is located in Newfoundland. Cambrian rocks occur on all continents, with the most complete Cambrian sedimentation and faunal developments in western North America and Siberia. Palaeocontinental reconstructions show that the Cambrian units were predominantly in the Southern Hemisphere; GONDWANA and AVALONIA were polar to 60°S, BALTICA and Siberia at 30°S and LAURENTIA equatorial.

camerae *n.* (*sing.* **camera**) the gas-filled chambers of a CEPHALOPOD shell. See also AMMONOID, NAUTILOID.

Campanian *n.* a STAGE of the Upper CRETACEOUS.

camptonite *n*. an ALKALINE LAMPROPHYRE, composed mainly of PHENOCRYSTS of AMPHIBOLE (barkevikite and/or kaersutite), titanaugite, OLIVINE and titanium-rich BIOTITE, in a groundmass of LABRADORITE, amphibole and pyroxene with subordinate ALKALI FELDSPAR and FELDSPATHOIDS. Accessory apatite, magnetite, calcite and zeolites may also occur.

Canada balsam *n*. a naturally occurring resin at one time used to cement a thin slice of rock to a glass slide for use with a PETROLOGICAL MICROSCOPE. Synthetic resins are now widely used, but all these cements have a refractive index of 1.54, allowing comparison with the refractive indices of the various minerals using the BECKE TEST. See INDEX OF REFRACTION; see also RELIEF.

Canadian *n*. an obsolete name for a series in the Lower Ordovician of North America.

Canadian Shield *n*. one of the largest areas of Precambrian rock in the world, forming part of the North American CRATON. The Canadian Shield occupies about 5 000 000 square kilometres, covering a large part of northern Canada, centred on Hudson's Bay and extending down into the Lake Superior region of the United States. Complexly deformed, metamorphosed and structurally heterogeneous, it is dominated by gneisses, granitic rocks, greenstones (metamorphosed basic igneous rocks) and metamorphosed sedimentary rocks. It is rich in deposits of iron, nickel, copper, silver and gold. See SHIELD.

cancrinite see FELDSPATHOID.

cannel coal *n*. a variety of bituminous coal of fine-grained uniform texture. It is composed mainly of microspores and probably developed in lakes where floating spores were deposited. It has a dull or waxy lustre and conchoidal fracture. Because of its high content of volatiles, it ignites easily and burns with a bright, smoky flame, the feature responsible for its early name – *candle coal*. Cannel coal grades into TORBANITE.

canyon or **cañon** *n*. a deep, steep-walled gorge cut by a river or stream, generally into bedrock. Canyons are most frequently found in arid or semi-arid regions where the effect of stream action greatly outweighs WEATHERING. See also SUBMARINE CANYON.

capacity *n*. **1.** the ability of a stream or wind current to transport debris, as gauged by quantity per unit time at a given point. Stream capacity increases with an increase in slope and discharge, and a decrease in width. Compare COMPETENCE.
2. the yield of a well, reservoir or pump.
3. the ability of a particular soil to retain water.

cap carbonate *n*. LIMESTONE found above late Proterozoic glacial deposits, one of the lines of evidence used in support of the SNOWBALL EARTH hypothesis.

capillarity *n*. a manifestation of surface tension by which the portion of the surface of a liquid coming in contact with a solid is depressed or elevated, depending upon the attraction between the molecules of the liquid and the molecules of the surface. It is illustrated by the rising of water in very fine tubes or the drawing up of water between grains of rock.

capillarity conductivity *n*. the ability of an unsaturated soil or rock to transmit fluid. When water is the fluid, the conductivity, which is zero for dry material, increases with moisture content to the maximum value, which is equal to the PERMEABILITY coefficient.

capillary *n*. a mineral habit (see CRYSTAL HABIT) composed of flexible, threadlike crystals, e.g. millerite (NiS) and chalcotrichite (a fibrous variety of cuprite, Cu_2O).

capillary fringe, zone of capillarity, tension-saturated zone *n.* a zone within an AQUIFER, below the VADOSE (unsaturated) zone and immediately above the WATER TABLE, in which interstices contain water. The water is at less than atmospheric pressure, although part of the water below the water table, and is held above that level by surface tension (CAPILLARITY).

capillary water *n.* water of the CAPILLARY FRINGE.

cap rock *n.* **1.** a hard covering atop the salt covering in a SALT DOME, consisting mainly of GYPSUM and ANHYDRITE, some calcite and sulphur in varying amounts. Large sulphur deposits occur in the cap rock of certain salt domes in the Gulf Coast area of the United States. Experiments imply that cap rock sulphur was formed through the action of SULPHUR BACTERIA.
2. IMPERMEABLE strata such as SHALE or EVAPORITES that overlie the porous horizons containing oil and/or gas. See POROSITY.

capture see STREAM CAPTURE.

Caradoc *n.* a SERIES of the Upper ORDOVICIAN.

carat (**c**, **Ct**) *n.* **1.** a unit of weight for gemstones. The metric carat equals 200 milligrams.
2. also called **karat** (**K**), a value indicating the proportion of pure gold in an alloy. Twenty-four-carat gold is *pure* or *fine gold*. Ten carat gold is 10/24 pure.

carbon-14 or **radiocarbon** *n.* a radioactive isotope of carbon, ^{14}C, having a mass number of 14 and a half-life of 5730 years. It occurs in nature as the result of reaction between atmospheric nitrogen, ^{14}N, and neutrons produced by cosmic-ray collisions. Carbon-14 is used in dating substances directly or indirectly associated with the carbon cycle, provided they are not over 50 000 years old. See DATING METHODS.

carbonaceous *adj.* (of rock sediment or other material) consisting of or containing carbon, or resembling it in some aspect.

carbonaceous chondrites *n.* STONY METEORITES containing CHONDRULES in a fine-grained, volatile-rich matrix composed mainly of clay-like hydrous SILICATES, with inclusions of OLIVINE, PYROXENE, magnetite and pyrrhotite (an iron sulphide that can contain nickel). They may contain more than 3 per cent carbon in the form of organic compounds, such as HYDROCARBONS, amino acids and fatty acids. Some researchers suppose an extraterrestrial biological origin, but it has been shown that these can be formed by abiogenic processes. Carbonaceous chondrites are considered to be the least altered of all the chondritic meteorites and are thought to represent the composition of undifferentiated planetary material.

carbonado see INDUSTRIAL DIAMOND.

carbonate *n.* **1.** a mineral type containing the carbonate radical $(CO_3)^{2-}$. CALCITE, ARAGONITE and DOLOMITE represent three groups of carbonate minerals. Compare BORATE, NITRATE.
2. a sediment composed of calcium, magnesium and/or iron, e.g. LIMESTONE, DOLOMITE.

carbonate compensation depth (**CCD**) *n.* the depth below which calcium carbonate does not accumulate on the ocean floor. The rate of supply of biogenic $CaCO_3$ is just balanced by the rate of solution. The CCD can vary widely, between about three and five kilometres depth, being deeper in low latitudes.

carbonation *n.* **1.** a process of chemical weathering during which minerals containing calcium, iron, magnesium, potassium and sodium are transformed into carbonates of these metals by carbon dioxide.
2. the saturation of a fluid with CARBON DIOXIDE.

carbonatite *n.* an igneous rock, which may be extrusive or intrusive, composed essentially of CARBONATE minerals. Most carbonatites are composed of CALCITE, DOLOMITE or ANKERITE with minerals such as APATITE, PHLOGOPITE, MAGNETITE or AEGERINE and a wide range of minor accessory minerals. They are strongly enriched in a variety of trace elements, including REE, Ba, Mo, Nb, Zr, Y, Sr, Cl and F. *Intrusive carbonatites* are typically found in RING COMPLEXES associated with IJOLITES, SYENITES and FENITES, while *extrusive carbonatites* are usually associated with olivine-poor NEPHELINITES.

carbon dioxide *n.* a colourless, odourless gas, CO_2, that forms 0.03 per cent of the Earth's ATMOSPHERE. CO_2 is abundant in VOLCANIC GASES but is removed from the atmosphere by photosynthesis and by solution in sea water. A vast amount of CO_2 is contained in the oceans and is exchanged between ocean water and atmosphere in a continuous process. A proportion of the CO_2 in sea water is removed by biological precipitation of calcium carbonate (LIMESTONES), while fixation of carbon in living tissues by photosynthesis eventually contributes to the formation of FOSSIL FUELS like COAL and oil. Atmospheric CO_2 has been increasing over the past 200 years, principally as a result of the burning of fossil fuels and destruction of forest cover, leading to concern about the consequences of the GREENHOUSE EFFECT.

Carboniferous *n.* the time interval between 360 and 286 Ma. It was defined by Conybeare in 1822 for type strata in Wales, which contained great COAL deposits and the LIMESTONE underlying it. The period is divided into the Lower Carboniferous (DINANTIAN) and Upper Carboniferous (SILESIAN) subsystems (called the MISSISSIPPIAN and PENNSYLVANIAN periods in North America). The Carboniferous is best known for coal deposits but is also important for oil and gas.

Most of the present continental masses formed two great supercontinents during the Carboniferous: Europe, Scandinavia and North America formed the northern land mass LAURASIA, and GONDWANA formed the southern one (Asia was still separate). The southern part of Laurasia lay in the tropics; Gondwana was moving north.

Sedimentation in Britain during the Dinantian (360 Ma–320 Ma) was controlled by back-arc extension, producing rifting and positive, emergent features, such as the Wales-London-Brabant High, the Southern Uplands, the Scottish Highlands and marine platforms on submerged basement highs. Negative features include the fault-bounded basins in Northern England and Central Scotland. Dinantian marine sediments are of two basic types: limestone, much of it fossiliferous, and CLASTIC rocks such as SHALES and SANDSTONES. The limestones were deposited on the marine platform areas and on ramps in some basins, away from sources of TERRIGENOUS sediments. Within the basins MUDSTONES accumulated. In southwestern Britain, in the CULM Basin, black shales accumulated, together with LAVAS from submarine

VOLCANIC activity. Terrestrial deposition was dominated by fluvio-deltaic systems, sands and muds being derived from the emergent areas. There was extensive volcanic activity in the Midland Valley of Scotland, with outpourings of basaltic lava.

During the Silesian (320 Ma–286 Ma) rifting was of less importance, and a more uniform sedimentary environment developed, dominated by terrestrial deposits formed by extensive fluvio-deltaic systems. Cyclical sedimentation, also seen in the Dinantian, developed widely. Throughout the Upper Carboniferous there are repeated cycles of marine limestones, shales, sandstones and coal, with the limestones representing periodic flooding of the swamps by the sea. The cyclic patterns of sedimentation in the Carboniferous are related to eustatic changes in sea level, driven by ice-volume changes in the southern polar ice cap (MILANKOVITCH cycles). See CYCLOTHEM.

During the Carboniferous, various phases of the VARISCAN OROGENY affected a belt through central Europe, southwest England and North America, related to the closure of the ocean between Gondwana and Laurasia. In the later stages of the orogeny (Late Carboniferous, Early PERMIAN) large granite intrusions were formed under southwest England, associated with extensive mineralisation.

The flora that had appeared during the Devonian became plentiful and varied during the Lower Carboniferous. Although the later Devonian forests were populated by some true trees, the first widespread trees with woody trunks appear in the Lower Carboniferous. These include lycopod trees (lower vascular plants), such as Lepidodendron and Sigillaria, and arboreal SPHENOPSIDS, such as CALAMITES, which comprised much of the 'coal forests' of the Lower Carboniferous. Ferns were plentiful and of various kinds, and cycdadophytes (seed ferns) were even more common at this time. Other flora included CORDAITES (forerunners of modern conifers). The first seed plants appear, and GYMNOSPERMS emerge as trees for the first time. Psilophytes (primitive land plants without leaves) die out in the Carboniferous.

Marine faunas include goniatites (AMMONOIDS having characteristic sutures) and BRACHIOPODS that decline in number and forms, except for the productids (spiny brachiopods), which evolve rapidly during this period. Both groups are used as INDEX FOSSILS. Crinoids and blastoids (ECHINODERMS) expand into a great diversity, especially on the North American mid-continent shelf. FUSULINIDS (relatively large protozoans) appear for the first time during the Upper Carboniferous, being used as index fossils. Larger fauna include shark-like fish and large amphibians, and the first reptiles emerge in Late Carboniferous time. Freshwater life (including BIVALVES) abounds during this period. Amphibians and a diverse insect fauna inhabited the swamp forests. Four hundred forms of insect, mainly primitive, are known from the lower and middle Late Carboniferous. More than 20 of these, including cockroaches, exceeded lengths of 100 millimetres; one dragonfly-like insect had a wingspread of 730 millimetres. Although a warm, moist climate is indicated by the abundant Upper Carboniferous plant growth, towards the end of the period widespread glaciation became established in the Southern Hemisphere.

carbonisation *n*. the decomposition of organic matter under water or sediment, during which the hydrogen, oxygen and nitrogen in the original tissues are lost by distillation, so that only a thin film of carbon remains; this residue may retain many features of the original organism. Plants, arthropods and fish have been so preserved. Animals that lack preservable hard parts, e.g. sponges, jellyfish and annelid worms, may also be preserved in this way. Perhaps the best-known and most important carbonised remains are those of the BURGESS SHALE fauna in western Canada. See FOSSIL.

carbon monoxide *n*. a colourless, odourless, poisonous gas, CO, produced when carbon burns with insufficient air, such as during the incomplete burning of FOSSIL FUELS.

carbon ratio *n*. **1.** the ratio of the fixed carbon content in a coal to the fixed carbon plus volatile matter.
2. the ratio of ^{12}C, the most common carbon isotope, to either ^{13}C or ^{14}C. If unspecified, the term is usually $^{12}C/^{13}C$.

Carborundum *n*. the trade name for a manufactured substance (silicon carbide) used as an abrasive in, for example, the preparation of THIN SECTIONS.

carbuncle see GARNET.

cardinal process *n*. a calcareous boss projecting from the HINGE LINE in the BRACHIAL VALVE of a BRACHIOPOD. It is a point of attachment for the DIDUCTOR MUSCLES.

cardinal septum *n*. one of the six first-formed septa (walls) in a CORALLITE; see RUGOSE CORAL, Fig. 77.

Carlsbad law see TWINNING.

carnallite *n*. a white orthorhombic mineral, $KMgCl_3 \cdot 6H_2O$, occurring in marine salt deposits, apparently as an alteration product of pre-existing salts. It is a source of potassium for fertilisers.

carnelian or **cornelian** *n*. a translucent, reddish semiprecious variety of CHALCEDONY; as it becomes brownish, it grades into sard. The colour is caused by colloidally dispersed HEMATITE.

Carnian *n*. the lowest STAGE of the Upper TRIASSIC.

carnotite *n*. a highly radioactive, bright-yellow mineral ore of vanadium and uranium, $K_2(UO_2)_2(VO_4)_2 \cdot 3H_2O$. It is of secondary origin, occurring as loosely coherent masses in sandstone or as an earthy powder, particularly around petrified wood.

cascade *n*. a series of descending recumbent FOLDS associated with GRAVITY SLIDING. They are found when a bed or an inclined surface buckles as it slides downwards under the effect of gravity. See also GRAVITY COLLAPSE STRUCTURES.

casing-head gas *n*. an unprocessed natural gas produced from an oil-containing reservoir. It is so named because it is most often produced under low-pressure conditions through the casing head of an oil well.

cassiterite *n*. a brown or black tetragonal mineral, SnO_2, the major ore of tin. Significant deposits occur in high-temperature HYDROTHERMAL veins and in PLACERS as well as in granite and pegmatites.

cast *n*. a positive representation of organic remains, derived from different types of mould or the impressions of original forms. It results when any substance, artificial or mineral, fills the void created by the dissolution of the original hard parts of an original form. (Compare REPLACEMENT.) A *natural cast* is produced when a natural mould is filled with mineral substance while still in the embedding rock. Nùmerous fossils are preserved in this manner. See FOSSIL, MOULD.

cataclasis *n*. one of the processes involved in dynamic METAMORPHISM in which rock DEFORMATION is produced by the crushing or shattering of brittle rock, e.g. along FAULTS or FAULT ZONES, to the extent that the mineral composition and texture of the original rock are still recognisable. A metamorphic rock produced by such a process is a *cataclasite*, e.g. a TECTONIC BRECCIA. A structure or a material that exhibits the effects of severe mechanical stress during metamorphism is described as *cataclastic*; such features include bending, breaking and cracking of the minerals. See also MORTAR STRUCTURE.

cataclasite see CATACLASIS, CRUSH CONGLOMERATE, MICROBRECCIA.

cataclastic rock see CATACLASIS.

catastrophism *n*. an hypothesis that proffered recurrent, violent worldwide events as the reason for the sudden disappearance of some species and the abrupt rise of new ones. Cuvier, a French biologist, supported the doctrine, explaining most extinctions and landscape changes as consequences of the Biblical deluge. Although catastrophism had wide acceptance in the scientific community, it was short-lived and was supplanted by Hutton's principle of UNIFORMITARIANISM. A revived form of *new catastrophism* has emerged in recent years with the recognition of the intense global environmental effects of major impacts of extraterrestrial bodies and eruptions from SUPERVOLCANOES. See also BOLIDE.

catazone see INTENSITY ZONE.

catchment area *n*. **1.** the waterproofed area of a storage reservoir.
2. the RECHARGE area and all other areas that contribute water to an AQUIFER.
3. the DRAINAGE BASIN.

catena *n*. a soil series consisting of a group of soils that developed from similar parent material but show different profiles because of variations in topography and drainage conditions. Inasmuch as each topographic setting affects the nature of the SOIL PROFILE, a *soil catena* holds both lithological and topographical implications.

cathodoluminescence (**CL**) see Appendix 4.

cation *n*. a positively charged ION that moves towards the negative electrode (*cathode*) during electrolysis. See BASE EXCHANGE.

cat's-eye *n*. any of several gemstones displaying CHATOYANCY. *Precious*, or *oriental*, *cat's-eye* is a greenish variety of CHRYSOBERYL; *quartz cat's-eye* is a PSEUDOMORPH after a fibrous mineral such as crocidolite.

cauldron subsidence *n*. the subsidence of a more or less cylindrical block of COUNTRY ROCKS within ring fractures as a result of magmatic activity. MAGMA fills the space round and above the subsiding block to form ring dykes (see RING DYKES AND CONE SHEETS, also RING COMPLEX) and bell-jar intrusions (where the ring dyke is capped by a flat-lying sheet). These are PERMITTED INTRUSIONS. If the ring fractures reach the surface, the magma can erupt as LAVA. Large-scale subsidence (compare STOPING) may result from an earlier catastrophic eruption of VOLCANIC ROCKS that largely emptied the MAGMA CHAMBER; see CALDERA.

cave *n*. a natural cavity or system of chambers beneath the surface of the Earth, large enough for a person to enter. If a group of caves is connected, or the same underground river or stream flows through a cave group, it is referred to as a *cave system*. Caves are formed by the action of waves against cliffs, or they may be cut into glacier bottoms by meltwater streams. They may also be formed in lava. (See LAVA TUBE.) The largest caves and caverns, however, are formed mostly in limestone and dolomite. See KARST.

cave breccia *n.* angular fragments of limestone that have broken off the walls and roof of a cave, and are cemented with calcium carbonate. See also COLLAPSE BRECCIA, SOLUTION BRECCIA.

cave coral *n.* a coral-shaped cave deposit of calcite.

cave marble see CAVE ONYX.

cave onyx or **cave marble** *n.* a compact, banded cave deposit of calcite or aragonite. It resembles true onyx and can take a high polish. Compare ONYX. See also DRIPSTONE, FLOWSTONE, ONYX, MARBLE, TRAVERTINE.

cave pearls *n.* small, spherical concretions of calcite or aragonite formed by the accretion of layers around a nucleus, such as a grain of sand.

cavernous *adj.* (of volcanic rocks and limestones) characterised by caverns, cells or large pore spaces.

c-axis *n.* a vertically oriented crystallographic axis used for reference in crystal descriptions. See also A-AXIS, B-AXIS. See CRYSTAL SYSTEM, Fig. 23

cay *n.* a small insular bank of sand, rock, mud or coral, or a range of low-lying reefs or rocks in the West Indies. Compare KEY.

Cayugan *n.* a series of the Upper Silurian of North America.

CCD see CARBONATE COMPENSATION DEPTH.

celestite *n.* an orthorhombic mineral, $SrSO_4$, of the BARITE group, a source of strontium.

celsian see FELDSPAR.

cement *n.* 1. chemically precipitated mineral matter that is part of the CEMENTATION process.

2. powder manufactured from a fired limestone-clay mixture that forms concrete when mixed with water, sand and aggregate.

cementation *n.* the process by which clastic sediments are converted into sedimentary rock by precipitation of a mineral CEMENT between the sediment grains, forming an integral part of the rock. Silicon is the most common cement, but calcite and other carbonates, as well as iron oxides, also undergo the process. It is not clear how and when the cement is deposited; a part seems to originate within the formation itself, and another part seems to be imported from outside by circulating waters.

cementstone *n.* an argillaceous LIMESTONE in which the relative proportions of clay and $CaCO_3$ are such that it can be burnt for CEMENT without the addition of other materials.

Cenomanian *n.* the lowest STAGE of the Upper CRETACEOUS.

cenote *n.* a natural well or reservoir, common in Yucatan, Mexico, formed by the collapse of a limestone surface, exposing water underneath. In ancient Yucatan, precious objects were thrown into cenotes as offerings to the rain gods.

Cenozoic, Cainozoic or **Kainozoic Era** *n.* the Cenozoic ('recent life') Era covers the Earth's history during the last 65 Ma. It is subdivided into the Tertiary (65 Ma–1.8 Ma) and Quaternary (1.8 Ma to present) periods. The Tertiary comprises the PALAEOCENE, EOCENE, OLIGOCENE, MIOCENE and PLIOCENE Epochs. The Quaternary includes the PLEISTOCENE and HOLOCENE (last 10 000 years) Epochs. An alternative scheme subdivides the Tertiary into two periods, the Palaeogene (65 Ma–24 Ma) and the Neogene (24 Ma–1.8 Ma). Most British Cenozoic formations have not been deeply buried and consist chiefly of thick deposits of marine and terrestrial sedimentary rocks.

During the Early Tertiary, Britain lay about 40° to 50° north of the Equator, and there was intense igneous activity in northern Britain, the Faeroe Islands and Greenland. This took the form of the eruption of thick sequences of FLOOD BASALTS, the intrusion of DYKE swarms, and the formation of layered INTRUSIONS such as the Skaergaard in east Greenland, and central igneous complexes such as those of Skye and Mull, off the west coast of Scotland. This activity is related to the opening of the North Atlantic Ocean.

During the Tertiary the continents were moving towards their present positions, as the final disruption of GONDWANA took place. Australia and South America separated from Antarctica, and as the TETHYS Ocean closed and India moved north to collide with Asia, tectonic activity formed the Alps and Himalayas (see ALPINE-HIMALAYAN OROGENIC BELT). On the west coast of North and South America, there was a continuation of the orogenic activity that had begun in the MESOZOIC, with extensive volcanic activity and the intrusion of granitic batholiths. In the North Pacific region, there was no seaway through the Bering Strait area

Age at base (Ma)	Britain	NW Europe	Central Alps
0.01	Flandrian	Holocene	Holocene
0.11	DEVENSIAN	WEICHSELIAN	WÜRM
0.15	Ipswichian	Eemian	Riss/Würm
0.25	WOLSTONIAN	SAALIAN	RISS
0.28	Hoxnian	Holsteinian	Mindel/Riss
0.35	ANGLIAN	ELSTERIAN	MINDEL
?	Cromerian	Cromerian Complex	Günz/Mindel
0.78	Beestonian		
1.00	(hiatus)	Bavelian	
1.10	(hiatus)	MENAPIAN	GÜNZ
1.30	(hiatus)	Waalian	Donau/Günz
1.60	BEESTONIAN	EBURONIAN	DONAU
1.80	Pastonian	Tiglian	Biber/Donau
1.90	BAVENTIAN	"	"
2.10	Antian	"	"
2.20	Thurnian	"	"
2.40	Ludhamian	"	"
2.50	WALTONIAN	PRAETIGLIAN	BIBER

Fig. 13 **Cenozoic**. Divisions of the Pleistocene epoch. The major glacials (in capitals) and interglacial stages for Britain, northwestern Europe and the Alps, with approximate ages in years BP. The age of the base of the Cromerian is uncertain, and there is a gap in the stratigraphical record between the interglacial stage at the beginning of the Beestonian and the glacial stage at its base

to the Arctic until the end of the Miocene, when it opened briefly and then was closed again until the end of the Pliocene.

The Cenozoic was characterised by a general global cooling, interrupted by periods of warming. The mild climate of the Palaeocene and Eocene Epochs became cooler with the onset of a cooling trend in the Oligocene. A brief warming interval in the Middle Miocene was followed by increasing cold until the onset of continental glaciation in the Pleistocene. Antarctica, however, was already glaciated as early as Miocene time, glaciation having begun there about 20 Ma ago. During the Pleistocene Epoch, called the 'Great Ice Age', glaciers and ice sheets covered the northern landmasses at least four times. Antarctica was also covered by ice. Longer warm interglacial periods separated these massive ice advances. Deep-sea sediments record 18 cold-warm cycles during the Quaternary. See ICE AGE.

Angiosperms, which appeared in the Cretaceous, replace the gymnosperms. Compositae appear in the Palaeocene and grasses appear in the Eocene; these plants and the Leguminosae undergo rapid evolution, diversification and dispersal during the Tertiary. Orchidaceae also arise in the Tertiary. Eocene floras from Alaska and Greenland reach North America during the Miocene. Temperature decrease during the Quaternary fixes the modern zones of vegetation.

The Foraminifera evolve during the Cenozoic, reaching a maximum in the mid-Eocene; several attain a large size (e.g. NUMMULITES, 25 to 50 millimetres). Arthropods become more significant throughout the Cenozoic, and several groups of land snails evolve during the Tertiary. (Most early Tertiary genera of gastropods are now extinct, the majority of living species having originated in the Middle Tertiary.) Turtles, snakes and lizards continue into the Cenozoic. Teleost diversity increases, and most living groups of birds appear to have evolved in the early Tertiary. Mammalian evolution is the most notable biological occurrence of the Tertiary, represented by the hoofed animals, carnivores, primates and Australian marsupials. Some evolutionary branches (seals and whales) return to the sea. The Tertiary–Quaternary boundary is partially defined by *Equus*, *Elephas* and *Ursus*. The giant elk, woolly mammoth and cave bear appear in the early Holocene. Hominids appeared at the beginning of the Quaternary and *Homo sapiens* in the Holocene.

centre of symmetry see CRYSTAL SYMMETRY.

centrifugal force *n.* a body constrained to move along a curved path reacts against the constraint with a 'force' directed away from the centre of curvature of the path; this reaction is called the centrifugal force. See also CORIOLIS FORCE.

centripetal drainage pattern see DRAINAGE PATTERN.

cephalon *n.* the part of the dorsal exoskeleton of a TRILOBITE that covered the head, see Fig. 89. It is divided into a central, raised section, the GLABELLA, on either side of which are the fixed and free cheeks and a line of weakness, the *facial suture*.

The facial suture runs from the anterior margin of the cephalon past the inner side of the eye towards the back of the cephalon. It may cut the posterior margin (an *opisthoparian suture*) or the lateral margin (a *pro-*

parian suture) or run through the GENAL ANGLE (a *gonatoparian suture*, but this is rare). After death, the free cheeks may separate from the cephalon along these sutures, leaving a central portion comprising the glabella and fixed cheeks called the *cranidium*.

cephalopod *n.* a member of the Cephalopoda, a class of highly organised marine MOLLUSCS of which squid, octopus, cuttlefish and nautilus are living representatives. Cephalopods are characterised by a head with eight, 10 or more tentacles around the mouth. The univalve shell, composed mainly of ARAGONITE, may be external, internal or absent. It may be wholly chambered when external or partly chambered when internal; the chambers (CAMERAE) are gas-filled and are used for buoyancy, the gases being adjusted by means of the SIPHUNCLE. The shell takes the form of a cone and may be variously coiled or straight.

FOSSIL cephalopods range from the CAMBRIAN to the present. Extinct forms outnumber the living; the greatest cephalopod diversity was attained in the late PALAEOZOIC and MESOZOIC. The principal subclasses are the Nautiloidea, (NAUTILOIDS), Ammonoidea (AMMONOIDS) and the Coleoidea, of which the Belemnitida (BELEMNITES) are an important fossil group.

ceratite see AMMONOID.

cerioid see COLONIAL CORAL.

cerussite *n.* an orthorhombic mineral, $PbCO_3$, of the aragonite group. A common secondary mineral of lead, it is often found in the weathered zone of the lead deposits as an oxidation product of galena.

chadacrysts see OIKOCRYST.

Chadian *n.* the lowest STAGE of the VISEAN (CARBONIFEROUS) in Britain and western Europe.

chain silicate see INOSILICATE.

chalcanthite or **blue vitriol** *n.* a blue triclinic mineral, $CuSO_4 \cdot 5H_2O$, that occurs in oxidised zones of copper deposits.

chalcedony *n.* a cryptocrystalline variety of silica with a compact fibrous structure and waxy lustre. It may be translucent or semitransparent and occurs in a variety of colours. Chalcedony is often found as a deposit, lining or filling cavities in rocks. See also AGATE.

chalcocite *n.* a black or dark lead-grey mineral, Cu_2S, with metallic lustre. It is commonly massive but also occurs infrequently as pseudo-monoclinic orthorhombic crystals. Chalcocite is an important ore of copper, occurring most frequently as the result of SECONDARY ENRICHMENT.

chalcophile see AFFINITY OF ELEMENTS.

chalcopyrite *n.* a brassy or golden-yellow tetragonal mineral, $CuFeS_2$. It is the most widely occurring copper mineral and one of the most important sources of that metal. It is the principal primary copper mineral in PORPHYRY copper deposits. Although small crystals are sometimes found, it is usually massive.

chalk *n.* a soft, earthy, fine-grained white to greyish pelagic LIMESTONE of marine origin, composed almost entirely of biochemically derived low-Mg calcite that is formed mainly by shallow-water accumulations of minute plants, in particular coccoliths, and a variable component of bioclasts, including foraminifers. Chalk deposits, such as those exposed in cliffs along both sides of the English Channel, are typical rocks of the Upper CRETACEOUS age.

chalybeate *adj.* containing or impregnated with iron salts, e.g. a mineral spring.

chamosite *n.* a greenish-grey mineral

closely related to the CHLORITE group that is common in sedimentary IRONSTONES.

Champlainian *n*. a SERIES in the Middle Ordovician of North America.

chance packing see PACKING.

Chandler wobble *n*. a small continuous variation of the Earth's rigid body motion that causes it to deviate from pure spin. The wobble completes a cycle in about 428 days.

channel *n*. the bed of a stream or waterway. Channel segments are rarely straight: they may be braided or meandering or any gradation between the two.

channel bar see RIVER BAR.

channel capacity *n*. the maximum flow volume that a given channel can sustain without overflowing its banks. See BANKFULL STAGE.

channel-fill deposit *n*. a deposit in a stream channel, consisting largely of BED LOAD materials. The deposit occurs in those places where the transport capacity of the stream is not great enough to remove the material it receives.

channel-mouth bar *n*. a bar built at the place where a stream enters a body of standing water; a decrease in the stream's velocity at this point results in a build-up of sand and gravel.

channel pattern *n*. the configuration of a stream or river course, which may be braided, meandering, sinuous or straight. A river may follow a relatively straight course for some distance and a meandering course at a different location.

channel sample *n*. a rock sample usually selected across the face of a rock body or vein to provide an average value.

channel sand *n*. sand or sandy material deposited in the bed of some channel cut into the underlying rock. Such sand

may contain valuable minerals, ore or gas. See also SHOESTRING SAND.

channel storage *n*. the volume of water in a stream channel above a given reference point at a given time.

characteristic fossil *n*. a fossil species or genus that is distinctive of a STRATIGRA-PHICAL UNIT. It is either peculiar to that unit or is especially abundant in it. Compare INDEX FOSSIL.

Charnian *n*. a stratigraphical division of PRECAMBRIAN sedimentary and volcanic rocks in the English Midlands/Welsh border area.

charnockite *n*. a coarse-grained rock of approximately granitic composition and GRANOBLASTIC texture, containing FELDSPAR, ORTHOPYROXENE and QUARTZ. The quartz characteristically has a blue OPALESCENCE because of the presence of exsolved needles of RUTILE, while the feldspar may be brownish in colour. Charnockites have GRANULITE FACIES assemblages characterised by anhydrous minerals (unlike the hydrous minerals of GRANITES). The origin of charnockites, magmatic or metamorphic, is still debated: some may have been formed by crystallisation of an anhydrous magma under PLUTONIC conditions, others by high-grade regional METAMORPHISM of igneous rocks.

chart datum *n*. the level of reference to which readings on a chart, e.g. soundings, are related.

chatoyancy *n*. an optical property by which certain minerals, such as cat's-eye (CHRYSOBERYL), produce in reflected light a band of light resembling the eye of a cat. The phenomenon is the result of the reflection of light from aligned fibres or tubular channels.

chatter marks *n*. small, curved cracks found on glaciated rock surfaces and beach pebbles, usually 1 to 5 centimetres in length, but they may be submicroscopic or as long as 47 centimetres. Glacially formed examples occur chiefly on hard rocks (e.g. granite) and are produced under a glacier by the pressure of irregularly moving boulders. Chatter marks are commonly found in nested arrangements, with the fractures at right angles to the direction of glacial movement. See also GLACIAL SCOURING.

Chattian *n.* the top STAGE of the OLIGOCENE.

Chautauquan *n.* the uppermost SERIES in the Devonian of North America.

cheeks see CEPHALON.

chemical potential (μ) *n.* a function (or property) of components in a system that is a measure of the tendency of material in that system to flow or react. Chemical potential is analogous to gravitational potential in that the lowest potential is associated with the most stable state. The chemical potential of any component in a set of phases at equilibrium must be the same in all, i.e. the chemical potential of component 1 in phase A is equal to the chemical potential of component 1 in phase B (μ_1 A = μ_1 B). For example, in a hydrous MAGMA system at equilibrium, the chemical potentials of water in the complex silicate melt, of water in the associated vapour and of water in BIOTITE crystals suspended in the melt must all be equal to one another.

chemical remanent magnetism see NATURAL REMANENT MAGNETISATION.

chemical weathering see WEATHERING.

chenier *n.* a sandy or shelly BEACH RIDGE on mudflats, formed by reworking of the mudflat sediments to concentrate coarser material in ridges. They are called 'cheniers' after the belts of oak trees that grow on such ridges in the Mississippi delta. Cheniers are also found in the Thames estuary.

chernozem soil or **black soil** *n.* a soil of sub-humid climates, characterised by a deep, dark-coloured layer, rich in humus and carbonates, that grades downward to a layer of lime accumulation.

chert *n.* a dense, extremely hard microcrystalline or cryptocrystalline siliceous sedimentary rock, consisting mainly of interlocking quartz crystals, submicroscopic and sometimes containing chalcedony and opal (amorphous silica). It is typically white, black or grey and has an even to flat fracture. Chert occurs mainly as nodular or concretionary aggregations in limestone and dolomite, and less frequently as layered deposits (*banded chert*). It may be an organic deposit (*radiolarian chert*), an inorganic precipitate (the primary deposit of colloidal silica) or a siliceous replacement of pre-existing rocks. *Flint* is a dark-coloured variety of CHALCEDONY, occurring as nodules in chalk and having a conchoidal fracture.

chertification *n.* a kind of silification, especially by a fine-grained or cryptocrystalline quartz.

Chesterian *n.* the uppermost SERIES of the Mississippian of North America.

chevron see TOOL MARKS.

chevron fold see FOLD.

chiastolite *n.* a variety of ANDALUSITE, forming cigar-shaped crystals in the metamorphosed rocks of some contact AUREOLES. In cross-section the crystals show inclusions in the form of a cross, thought to be caused by the preferred adsorbtion of carbon particles against the prism faces as the crystal grew. See Fig. 14 opposite.

chickenwire anhydrite *n.* a gypsum-anhydrite sequence showing polygonal nodules. Such structure is commonly found as displaced growth in carbonate muds of modern supratidal

Fig. 14 **Chiastolite**. Dark inclusions in the form of a cross

environments; this is widely believed to be evidence of SABKHA deposition.

Chile saltpetre *n*. soda NITRE, $NaNO_3$, more correctly termed NITRATINE, as found in northern Chile.

chilidial plates see NOTOTHYRIUM.

chilidium see NOTOTHYRIUM.

china clay *n*. KAOLIN that has been processed for the manufacture of chinaware.

chiton *n*. a relatively simple marine mollusc of the class Amphineura, with a shell consisting of overlapping calcareous dorsal plates. Chitons are also called 'coat-of-mail' shells. They are not common as fossils; range, Ordovician to Recent.

chlorapatite *n*. a member of the APATITE group of minerals, $Ca_5(PO_4)_3Cl$.

chlorite *n*. a representative of a group of micaceous greenish minerals of the general formula $(Mg,Al,Fe^{2+})_{12}[(Si,Al)_8 O_{20}](OH)_8$. Chlorites are common in low-grade schists, for example GREENSCHISTS, or as alteration products of PYROXENE, amphiboles or biotite.

chloritisation *n*. the replacement by alteration into or introduction of chlorite.

chloritoid *n*. a micaceous mineral, $(Fe^{2+}Mg)_2(Al,Fe^{3+})Al_3O_2[SiO_4]_2(OH)_4$, that crystallises in the monoclinic or triclinic systems. It occurs in yellow-green to black scaly aggregates in low-grade, regionally metamorphosed sedimentary rocks (PELITES).

chokedamp see BLACKDAMP.

Chokierian *n*. a STAGE of the NAMURIAN (CARBONIFEROUS) in Britain and western Europe.

chondrichthyes *n*. the class of vertebrates including fishes whose skeletons are cartilaginous rather than bony, e.g. sharks and rays.

chondrites *n*. one of the two divisions of STONY METEORITES characterised by CHONDRULES. They are primitive meteorites, having crystallised some 4700 Ma ago, and make up over 80 per cent of meteorite falls. They are divided into three groups according to Mg and Fe ratio (2 to greater than 9). Compare ACHONDRITES. See also CARBONACEOUS CHONDRITES.

chondrule *n*. a small globular body of various materials, although mainly olivine and pyroxene, found as an inclusion in certain stony meteorites (CHONDRITES). Chondrules are usually less than 3 millimetres in diameter, may be porphyritic or glassy in texture and are clearly visible on polished surfaces.

Chordata *n*. a varied phylum of those animals that at some time in their development have a notochord (an internal supporting rod) and gill slits. The most important fossil groups are subphylum Vertebrata and subphylum Hemichordata to which the GRAPTOLITES (class Graptolithina) are believed to belong.

C horizon see SOIL PROFILE.

chromate *n*. a mineral containing the chromate ion $(CrO_4)^{2-}$. Crocoite, $PbCrO_4$, is a common chromate.

chromatography *n.* any of several techniques for separating and identifying mixtures of chemical components by means of selective adsorbtion, partitioning between different solvents, ion exchange or some other process.

chromite *n.* a mineral of the spinel group, $Fe^{2+}Cr_2O_4$, that occurs as small black octahedral crystals, usually in compact masses in layers in MAFIC and ULTRAMAFIC igneous rocks. It is the main ore for chromium. See also MAGMATIC ORE DEPOSIT.

chron see GEOCHRON, GEOCHRONOLOGICAL UNIT.

chronohorizon or **chronostratigraphic horizon** *n.* a stratigraphical surface, theoretically without thickness, that is the same age (*isochronous*) at all points. In practice it is a thin and characteristic interval that serves as a good time-correlation or time-reference zone. Examples include coal beds, bentonite beds and many BIOHORIZONS. Compare LITHOHORIZON. See also MARKER HORIZON.

chronolithological unit see STRATIGRAPHICAL UNIT.

chronostratigraphic unit see STRATIGRAPHICAL UNIT.

chronostratigraphic zone see CHRONOZONE.

chronostratigraphy or **time-stratigraphy** *n.* that area of STRATIGRAPHY dealing with the age and time relations of strata.

chronotaxis *n.* a similarity of time sequence, such as the correlation of stratigraphical or fossil sequences on the basis of age and chronostratigraphical position. Compare HOMOTAXIS.

chronozone or **chronostratigraphic zone** *n.* 1. a zonal unit including all rocks formed at any location during the time span of a particular geological feature or defined interval of rock strata.

2. the smallest division of chronostratigraphical hierarchy. See STRATIGRAPHICAL UNIT.

chrysoberyl *n.* a mineral, $BeAl_2O_4$, usually green, yellow, grey or brown. Alexandrite, yellow cat's eye and some colourless varieties are valued as gems.

chrysocolla *n.* a generally amorphous MINERALOID, in which Si_4O_{10} layers may be found in a very defective structure. It occurs in the oxidised zone of copper and sulphide deposits and is similar to the minor gem mineral dioptase, $Cu_6(Si_6O_{18})·6H_2O$.

chrysolite *n.* a yellow to yellowish-green gem variety of magnesium-rich OLIVINE.

chrysoprase *n.* a translucent pale bluish-green or bright green gem variety of CHALCEDONY.

chrysotile or **serpentine asbestos** *n.* a fibrous white, grey or green mineral of the SERPENTINE group, the most important type of ASBESTOS.

chute cut-off *n.* a cut-off made through one of the marsh areas of a point bar (see FLOODPLAIN), thus isolating part of the bar as an island between two river channels. Compare NECK CUT-OFF.

Cincinnatian *n.* a series in the Upper Ordovician of North America.

cinder see SCORIA.

cinder cone see VOLCANIC CONE.

cinnabar *n.* a mineral, HgS, the principal ore of MERCURY. It normally occurs in brilliant red microcrystalline or earthy masses. It was used extensively by the Chinese as the pigment vermilion.

CIPW normative composition *n.* a procedure for recalculating the chemical composition of a rock into a hypothetical group of water-free standard minerals. The merit of the normative composition, or NORM, is that the limited number of standard NORMATIVE MINERALS facilitates comparisons between rocks. The norm disregards the effects of geological processes and temperature/pressure conditions, focusing only on the bulk chemical composition of the magmatic rock. The initials represent the names of Cross, Iddings, Pirsson and Washington, the four men who devised the system.

circulation *n*. **1.** the upward and downward cell-like movement of water in an area of the ocean as a result of density variations produced by temperature and salinity changes.
2. the complete mixing of lake waters.
3. the process of pumping DRILLING MUD down the drill pipe and back to the surface. See ROTARY DRILLING.

circum-Pacific belt *n*. a belt, some hundreds of kilometres in width, that rings the Pacific Ocean. It contains a very high proportion of present-day crustal SUBDUCTION ZONES and is thus the locus of some of the Earth's most severe earthquakes and explosive volcanic eruptions. The western part of the belt is characterised by ISLAND ARCS and the eastern part by CORDILLERA mountain ranges.

cirque, corrie or **cwm** *n*. a semicircular basin or indentation with steep walls originating from an ordinary valley head. It is situated high on a mountain slope and is associated with the erosive activity of a mountain glacier. Walls of the cirque are cut back by removal of rock in the headwall, loosened especially by the freezing action of water that has penetrated via the BERGSCHRUND. The resulting rock material, embedded in the glacier, gouges a concave floor that may contain a small lake (*tarn*) if the glacier disappears. The expansion of neighbouring cirques produces ARÊTES, horns and COLS.

cirri *n*. (*sing*. **cirrus**) small, flexible branches on the stem or base of some CRINOIDS, which they use for temporary fixation.

citrine *n*. a yellow to yellowish-brown crystalline quartz, sometimes resembling topaz in colour.

CL see cathode luminescence, Appendix 4.

clade *n*. a branching lineage resulting from the dichotomous splitting (branching) of an earlier lineage. The branching produces two new taxa, each of which is a MONOPHYLETIC group. The point of branching is called a *node*, and a clade is any node and its descendants.

cladistics or **phylogenetic systematics** *n*. a taxonomic method of grouping organisms by determination of common descent, pioneered by W. Hennig. Common descent can be demonstrated by the shared possession of new evolutionary features (*derived characters* or *synapomorphies*) evolved in the ancestor and passed on to the descendant CLADE. Those characteristics common to all clades within an evolving group are *symplesiomorphies* and cannot be used in cladistic classification.

cladogenesis *n*. the splitting (branching) of a lineage. It is assumed that an ancestral lineage always branches dichotomously, producing two descendant CLADES.

cladogram *n*. the tree-like branching diagram produced by a cladistic analysis. A cladogram shows the closeness of descent by the arrangement of the CLADES on the diagram. The closer the relationship, the nearer the clades are to one another.

Clapeyron equation see CLAUSIUS-CLAPEYRON EQUATION.

clarain see LITHOTYPE.

clarke *n*. the abundance of an element in the crust of the Earth, named after F. W. Clarke, an American geochemist. The *clarke of concentration* is the ratio of the amount of an element in a rock to the average amount in the Earth's crust. This ratio is often used to show the relative enrichment of an element in an ore body (see ORE).

clast *n.* **1.** an individual fragment of a larger rock mass removed by the physical disintegration of the larger mass.
2. a constituent of a BIOCLASTIC or a PYROCLASTIC ROCK.

clastic *adj.* **1.** pertaining to a sediment or rock composed chiefly of fragments derived from pre-existing rocks or minerals. See also DETRITAL, SEDIMENTARY ROCK.
2. describing the texture of such a rock.
3. pertaining to the fragments (CLASTS) of which a clastic rock is composed.

clastic dyke *n.* a tabular body consisting of clastic materials that cuts across the bedding of a sedimentary formation and is derived from overlying or underlying beds, usually by liquifaction of the clastic material; in particular, a SANDSTONE DYKE.

clastic ratio or **detrital ratio** *n.* the ratio, in a stratigraphical section, of the amount of clastic material (sandstone, conglomerate) to that of nonclastic material (limestone, dolomite). Compare SAND-SHALE RATIO.

clastic rock or **fragmental rock** *n.* **1.** a sedimentary rock composed mainly of fragments broken loose from parent material and deposited by mechanical transport, e.g. sandstone, shale, conglomerate. See also EPICLASTIC.
2. PYROCLASTIC ROCK.
3. BIOCLASTIC ROCK.
4. cataclastic rock (see CATACLASIS).

Clausius-Clapeyron equation *n.* an equation that expresses the relation between the vapour pressure of a liquid and its temperature:

$$\frac{dP}{dT} = \frac{\Delta H}{T \Delta V}$$

where P is the pressure, T the temperature, ΔH the change in heat content and ΔV the change in volume.

clay *n.* **1.** a DETRITAL mineral particle of any composition, having a diameter less than 0.004 millimetres. See ARGILLACEOUS, CLAY MINERALS
2. a smooth, earthy sediment or soft rock composed chiefly of clay-sized or colloidal particles and a significant content of CLAY MINERALS. Clays may be classified by colour, composition, origin or use. See SENSITIVE CLAY, BALL CLAY.

clay ironstone *n.* a fine-grained sedimentary rock, grey or brown, consisting of clay and iron carbonate (SIDERITE); it occurs in concretions or thin beds. Clay ironstone is usually associated with carbonaceous strata, in particular overlying coal seams in COAL MEASURES of the United States and Great Britain. See IRONSTONE.

clay minerals *n.* silicates of aluminium, iron and magnesium, belonging to the PHYLLOSILICATE group. Their crystal structure consists of sheeted layer structures, with strong inter- and intrasheet bonding but weak interlayer bonding. Clay minerals can be classified into six groups; the kaolinite-serpentinite group, the illite group, the smectite group, the vermiculite group, the chlorite group and the palygorskite-sepiolite group. The *kaolinite-serpentinite group* includes minerals close to $Al_2Si_2O_5(OH)_4$, among which are ANAUXITE, DICKITE, HALLOYSITE, KAOLINITE and necrite. They form during PEDOGENESIS. The *smectite group* minerals are close to $Al_2Si_4O_{10}OH_2 \cdot nH_2O$, with sodium, magnesium and iron variously substituted. The group includes BEIDELLITE, MONTMORILLONITE and SAPON-

ITE. Smectites are swelling clays and expand in the presence of water. The *vermiculite group* result from the alteration of mica or chlorite, with a typical formula being $Mg_3(Si_3Al)O_{10}(OH)_2$. Vermiculites are swelling clays. The ILLITE *group* are a group of mica-like clay minerals. The *chlorite group* consists of magnesium, aluminium, iron silicates and includes CHLORITE, dombassite and sudoite. The *palygorskite-sepiolite group* are magnesium-rich, fibrous clays and include PALYGORSKITE (attapulgite) and SEPIOLITE. Allophane and imogolite are non-crystalline hydrated silicates with properties similar to those of clays. In the formation of clays, the type of clay produced depends, during early stages of weathering, on the mineral composition of the parent rock, but in later stages it is climate-dependent.

claypan *n.* a dense, clayey subsoil layer that is hard when dry but may be plastic when wet. Its clay content is the result of concentration by downward-percolating waters. Compare HARDPAN.

clay plug *n.* a mass of clay and silt that may fill an OXBOW LAKE, converting it into a marsh. It is perhaps more accurate to say that the marsh constitutes a clay plug, inasmuch as it is resistant to incursion by a future stream channel.

claystone *n.* sedimentary rock of indurated clay-sized silicate materials, having the texture and composition of shale but lacking its lamination and fissility. Compare MUDROCK.

cleat *n.* a JOINT in COAL. The joint surface may be coated with a mineral film, for example PYRITE. Compare END.

cleavage *n.* **1.** the property of some minerals to break along planes related to the molecular structure of the mineral and parallel to actual or possible crystal faces. Cleavage is defined by the quality of the break (excellent, good, etc) and its direction in the mineral, i.e. the name of the crystal form it parallels. Some of the more important types of cleavage are: *basal*, or parallel with the basal plane, e.g. mica; *cubic*, or parallel with the faces of a cube, e.g. galena; *octahedral*, or parallel with the faces of an octahedron, e.g. fluorite; *prismatic*, or parallel with the faces of a prism, e.g. augite; *rhombohedral*, or parallel with the faces of a rhombohedron, e.g. calcite. Compare FRACTURE.

2. the characteristic or tendency of a rock to split along parallel, closely spaced planar surfaces, e.g. SLATY CLEAVAGE. It is produced by DEFORMATION or METAMORPHISM and is independent of BEDDING. Compare SCHISTOSITY.

cleavelandite *n.* a white lamellar variety of ALBITE, found as large crystals and in pegmatite veins.

Clerici solution *n.* an aqueous solution of thallium malonate and thallium formate that is used as a HEAVY LIQUID. It is not recommended for use nowadays because it is highly toxic and can be absorbed through the skin. See also BROMOFORM, METHYLENE IODIDE.

climate *n.* weather conditions of a specified region averaged over a long time interval. Factors such as temperature, atmospheric pressure, precipitation, humidity and wind are influenced by latitude, altitude and position relative to land and sea and to the main circulation belts of the atmosphere and oceans.

climatic optimum *n.* a period of milder climate, during the early Holocene, some 8000 to 5300 years ago, during which most middle-latitude valley glaciers disappeared. According to calculations based on pollen distribution and oxygen-isotope ratios in Greenland ice cores, many regions of northwestern Europe were 2–3ºC warmer, on average, than they are today.

climbing ripples see CROSS-BEDDING.

clinker *n*. **1.** a mass of fused incombustible matter, commonly found when COAL is burned.
2. a jagged fragment of LAVA.

clinoenstatite *n*. monoclinic variety of the PYROXENE group mineral ENSTATITE, $MgSiO_3$.

clinometer *n*. an apparatus used with a geologist's compass for measuring angles of slope, in particular for determining DIP.

clinopyroxene *n*. a member of the PYROXENE group, normally containing significant calcium; it crystallises in the monoclinic system.

clinozoisite *n*. a monoclinic mineral, $Ca_2Al_3O(SiO_4)(Si_2O_7)(OH)$ of the EPIDOTE group. It forms a complete solid solution series with epidote.

clint *n*. a unit of limestone pavement bounded and defined by enlarged joints (GRIKES).

Clinton ore *n*. a PHANEROZOIC IRONSTONE of the type found in the Clinton formation (Middle SILURIAN) in the Appalachian mountain belt of the USA. These are MASSIVE, red, oolitic iron ores (with HEMATITE, CHAMOSITE and SIDERITE) associated with shales and limestones. Fossils contained within them are usually partly replaced by iron minerals. All the oolitic ironstones are cemented chiefly by calcite and dolomite. See also MINETTE.

closed basin *n*. a region that drains into a depression or lake within its environment and from which water escapes only by evaporation.

closed structure *n*. a structure, such as an ANTICLINE or SYNCLINE, shown on a map to be entirely closed by one or more contour lines.

closed system *n*. a system in which there is no transfer of matter or energy into or out of the system during a particular process.

close-grained *adj*. (of a rock) with fine constituent particles tightly packed together.

close fold see FOLD.

close packing see PACKING.

closure *n*. the vertical distance between an ANTICLINE's highest point and its lowest closed structure contour on a STRUCTURE CONTOUR map of a subsurface anticline. The information is used to estimate gas or oil reserves.

cnidarian *n*. any of the Cnidaria, a phylum of aquatic, mainly marine, multicelled invertebrates that were formerly known as *coelenterates*. They are characterised by a radially symmetrical body structure with an outer (*ectoderm*) and inner (*endoderm*) layer of cells separated by a jelly-like substance, the *mesogloea*. There is a single body cavity (the *enteron*) with one opening, which is both mouth and anus, surrounded by tentacles. In the ectoderm are stinging cells called *nematocysts*. Most cnidarians alternate between a free-swimming reproductive stage (*medusa*) and a fixed polyp stage (see ALTERNATION OF GENERATIONS), but in some groups one or other stage is dominant. Range, Precambrian to present. Cnidarians include jellyfish, sea anemones and corals. Only the CORALS (class Anthozoa) are important as FOSSILS.

coal *n*. a carbon-rich, combustible, stratified organic sedimentary rock composed of altered and/or decomposed plant remains of non-marine origin, combined with varying minor amounts of inorganic material. Coal is non-crystalline, brittle and dull to brilliant in lustre. With increasing compaction, the colour changes from light brown to black, and the specific gravity increases from 1.0 to 1.7. Coals are classified according to their constituent plant materials (COAL TYPE), degree of metamorphism as shown by the ratio of fixed carbon content to volatile matter (RANK) and amount of impurities (ash and sulphur) present (GRADE). Coal occurs in rocks of the early Proterozoic but did not become widespread until the development of woody land plants in the Devonian. Pre-Devonian coals are composed of algal remains. Sedimentary rocks of the Carboniferous age hold the greatest abundance of coal-bearing rocks. Large deposits of lignite also occur in rocks of early Tertiary age. Coal beds are commonly found with shales and fine-grained sandstones.

Coal is formed in paludal (swamp) environments in fresh or brackish water, and coal formation proceeds in stages, beginning with the accumulation of large masses of plant debris in a humid environment. Such accumulations require anaerobic depositional conditions and a slowly subsiding basin. PEAT, a spongy mass of partially decayed vegetation, is the first stage of coal formation. When covered with sediment, peat may gradually be converted to LIGNITE (brown coal). The major effect of pressure on coalification occurs during this transition, where overburden pressure reduces porosity and moisture by perhaps 50 per cent. Although pressure is not excluded as a factor affecting the transformation of lignite into BITUMINOUS COAL and bituminous into ANTHRACITE, the main factors involved are temperature and time. Time is critical because a suitable temperature must be maintained for extensive periods before the degree of metamorphism, i.e. rank, can be increased. The rate at which coal rank increases with depth depends on the geothermal gradient and on the heat conductivity of the ambient rocks. See also BANDED COAL, LITHOTYPE, SUB-BITUMINOUS META-ANTHRACITE.

coal ball *n.* a concretion of mineral (secondary carbonate) and plant material embedded in coal seams, ranging from pea to boulder size. Compare BULLION.

coalfield *n.* an area containing coal deposits.

coal gas *n.* a fuel made by the carbonisation of a high-volatile BITUMINOUS COAL. Approximate composition by volume: 50 per cent hydrogen, 30 per cent methane, 6–8 per cent carbon monoxide, 7–8 per cent carbon dioxide, nitrogen and oxygen, and 2–4 per cent olefins.

coalification *n.* the physical and chemical changes that take place within a deposit of accumulated plant remains in the formation of COAL. During coalification, the carbon percentage increases as water and volatile hydrocarbons are expelled from the deposits.

Coal Measures *n.* a stratigraphical term used in western Europe (first used in Great Britain) for the upper part of the CARBONIFEROUS. It is equivalent to the WESTPHALIAN. See also COAL MEASURES below.

coal measures *n.* a succession of sedimentary strata consisting chiefly of CLASTIC rocks with interlayered beds of COAL. See also COAL MEASURES above.

coal plant *n.* a fossil plant found in a coal bed or a plant whose altered substance contributed to the formation of coal beds, e.g. *Lepidodendron* and *Sigillaria*. See SCALE TREES.

coal seam *n.* a layer or bed of COAL.

coal type *n.* a classification of COAL according to its constituent plant materials (MACERALS).

coarse-grained *adj.* **1.** (of a crystalline rock) having individual minerals of an average diameter greater than 5 millimetres.

2. (of a sediment or sedimentary rock) having individual constituents easily visible to the naked eye. Various limits have been suggested.

coarse sand or **coarse-grained sand** *n.* sand composed of grains between 0.5 and 1 millimetre in diameter.

coast *n.* a zone of rather indeterminate width extending landwards from the seashore. In geographical terms, it may be that section of a region that is near the coast and sometimes includes the COASTAL PLAIN.

coastal plain *n.* a broad plain sloping gradually seawards. It generally represents a strip of emerged sea floor. During the emergence, the SHORELINE of a coastal plain migrates seawards so that progressively younger rocks are found at decreasing distances from the shore.

coastline *n.* the boundary between coast and shore. Compare SHORELINE.

Coast Range orogeny *n.* the major deformation, metamorphism and volcanic activity in the Coast Mountains of the British Columbia Cordillera during the Jurassic and early Cretaceous. It generally approximates to the NEVADAN OROGENY of the United States.

cobaltite *n.* a steel-grey to silver-white mineral, CoAsS, the principal ore of cobalt.

cobble *n.* a rock fragment, rounded or abraded, between 64 and 256 millimetres in diameter. It is larger than a PEBBLE and smaller than a BOULDER.

coccolith and coccolithophore *n.* cocco- liths are small regular calcareous plates (often circular) produced by minute unicellular marine planktonic photosynthetic organisms (coccolithophores) of the division Prymnesiophyta, found predominantly in warm, low-latitude waters. After death, the coccolithophore disintegrates to release the coccoliths, which are microscopic calcareous plates of many different shapes, built of calcite or aragonite. Coccoliths are found in chalk and in the deep-sea oozes of tropical and temperate oceans.

coelacanth *n.* a member of the CROSSOP- TERYGIAN fishes. They were once believed to have been extinct since the end of the Cretaceous Period, but several specimens belonging to *Latimeria*, the only surviving genus, have been found during the past 50 years. Range, Upper Devonian to Recent.

coelenterates see CNIDARIAN.

coenenchyme see TABULATA.

coesite *n.* a very dense polymorphic form of quartz, SiO_2, stable at room temperature only at pressures exceeding 20 000 bars. It is found in impact craters. See also STISHOVITE.

cogeoid or **compensated geoid** *n.* a surface lying above or below the GEOID at a distance dV/g, where dV is the gravitational potential at the point considered and g is the gravitational acceleration. The potential, dV, refers to all matter lying above sea level, plus both the mass defect of the oceans and variations of rock density as prescribed by isostasy.

cognate inclusion see AUTOLITH.

cohesion *n.* **1.** (*physics*) the attraction between molecules of the same substance. It is this attraction in liquids that allows the formation of drops and thin films. Compare ADHESION.

2. (*geology*) that component of the SHEAR STRENGTH of a rock that is independent of interparticle friction.

col *n*. **1.** a sharp-edged or saddle-shaped pass in a mountain range formed by the headward erosion of two oppositely orientated CIRQUES.
2. a saddle-formed depression in a mountain range. See also ARÊTE.

colemanite *n*. a colourless to white monoclinic mineral, $CaB_3O_4(OH)_3 \cdot H_2O$, an important source of boron. It is usually found as crystalline deposits interstratified with Tertiary lake bed sedimentary rocks.

collapse breccia *n*. angular rock fragments produced by collapse of the rock formation above an opening, e.g. the collapse of the roof of a cave. See also SOLUTION BRECCIA.

collapse caldera see CALDERA.

collapse structure *n*. any rock structure that is the result of removal of support and consequent collapse, e.g. sinkhole collapse, collapse into mine workings or gravitational sliding on the limbs of folds.

collision zone *n*. the zone or belt where two pieces of continental crust have collided after closure of an intervening ocean. See DESTRUCTIVE PLATE BOUNDARY.

colloform *adj*. more or less rounded in form, including BOTRYOIDAL, RENIFORM and MAMILLATED as specific cases.

colloid *n*. **1.** a substance with particle size less than 0.00024 millimetres, i.e. smaller than CLAY size.
2. an extremely fine-grained material either in suspension or that can be readily suspended, commonly with special properties because of its extensive surface area.

collophane or **collophanite** *n*. any one of the massive cryptocrystalline varieties of APATITE that comprise the bulk of phosphate rock and fossil bone;

it is used as a phosphate source for fertiliser. Deposits often have a COLLOFORM structure.

colluvium *n*. unconsolidated material at the bottom of a cliff or slope, generally moved by gravity alone. It lacks STRATIFICATION and is usually unsorted (see SORTED); its composition depends on its rock source, and its fragments range greatly in size. Such deposits include cliff debris and TALUS. Compare SLOPE WASH.

colonial coral *n*. a unit consisting of attached individual CORAL organisms that cannot exist as separate animals. Colonial or *compound corals* are described as *fasciculate* if the CORALLITES are not in contact and *massive* if all the corallites are in contact. A fasciculate CORALLUM, in which the corallites branch irregularly, is called *dendroid*, while if the corallites are approximately parallel it is described as *phaceloid*. A massive corallum in which the corallites are united by their walls and have a polygonal cross-section is described as *cerioid*, while one in which the corallite walls are missing is known as *astraeoid*. (See Fig. 15 overleaf.) See also RUGOSE CORAL, TABULATA, SCLERACTINIA.

colour index or **colour ratio** *n*. as used in petrology in the classification of igneous rocks, a number representing the percentage, by volume, of dark-coloured (MAFIC) mineral in a rock. Thus rocks of colour index 0 to 35 are LEUCOCRATIC, 35 to 65 MESOCRATIC, 65 to 90 MELANOCRATIC, and 90 to 100 hypermelanic or ULTRAMAFIC.

columbite *n*. an iron-black, often irridescent, orthorhombic mineral, $(Fe,Mn)Nb_2O_6$, the Nb end member of the columbite-tantalite series. It occurs in granites and related pegmatites.

columella *n*. **1.** the solid central pillar

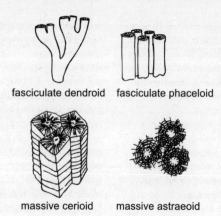

fasciculate dendroid   fasciculate phaceloid

massive cerioid   massive astraeoid

Fig. 15 **Colonial corals**. Some terms used to describe the form of the colony

forming the axis of a tightly coiled GASTROPOD shell (see Fig. 38).
2. see AXIAL STRUCTURE.

columnals *n*. the plates forming the stem of a CRINOID (Fig. 21). They may be circular or star-shaped in outline and have a central hole for an extension of the soft body.

columnar jointing or **columnar structure** *n*. long parallel columns, polygonal in cross-section, occurring most frequently in basalts but also in other extrusive and intrusive igneous rocks. The cooling of very hot lava or pyroclastic flow decreases its volume, causing cracks to form. Because the surface cools first, cracks penetrate the mass along the direction normal to the surface. The rock usually divides into columnar segments approximately perpendicular to the cooling surface. Although the columns tend to be six-sided – the ideal configuration assumed by a solid mass when it cools

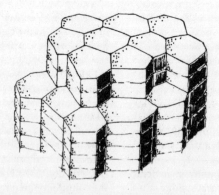

Fig. 16 **Columnar jointing**. Polygonal columns develop approximately perpendicular to the cooling surface

away from a plane surface – a degree of irregularity invariably occurs. Magnificent basaltic columns can be seen at the Giant's Causeway in Antrim, Northern Ireland, and Fingal's Cave on the island of Staffa, off the west coast of Scotland.

columnar section *n.* a graphic depiction of the sequence of rock units in a particular locality. The LITHOLOGY of a section is indicated by symbols and thicknesses are drawn to scale. See also GEOLOGICAL COLUMN.

columnar structure *n.* **1.** COLUMNAR JOINTING.
2. a columnar, near-parallel arrange-ment exhibited by aggregates of long, slender mineral crystals.
3. a primary sedimentary structure found in some calcareous shales or argillaceous limestone. It consists of columns, oval to polygonal in cross-section, and normal to the direction of bedding.

comagmatic *adj.* (of igneous rocks) regarded as having been derived from a common parent MAGMA because they share a common group of chemical and mineralogical features. See also CONSANGUINEOUS.

Comanchean *n.* a SERIES in the Lower Cretaceous of North America.

comendite *n.* a PERALKALINE RHYOLITE, low in aluminium content but enriched in titanium and sodium. The even more peralkaline variety, PANTELLERITE, is richer in iron.

comminution *n.* a process, either natural or in manufacture, by which a substance is reduced to a fine powder, or pulverised.

commissure *n.* the line along which the two VALVES meet in BIVALVES and BRACHIOPODS.

common canal *n.* a cavity along the length of a STIPE of a GRAPTOLITE colony

into which all the THECAE open and through which the individual animals are connected.

common lead *n.* any lead with a low U/Pb and/or Tb/Pb value, such that no significant radiogenic lead has been generated *in situ* from the time that the phase formed. See also ANOMALOUS LEAD.

common salt see HALITE.

community *n.* a group of organisms, living or fossil, occurring together and having a mutual relationship with each other and with the environment. Compare ASSEMBLAGE, ASSOCIATION, BIOCOENOSIS.

compactibility *n.* the property of a sedimentary material that allows a decrease in thickness or volume under load. It varies with the size, shape and hardness of the individual particles.

compaction *n.* the decrease in pore space of a sediment and consequent reduction in volume or thickness. Compaction results from the in-creasing weight of younger sediment material that is continually being deposited or pressures caused by earth movements.

compatible elements see INCOMPATIBLE ELEMENTS.

compensated geoid see COGEOID.

competence *n.* the ability of a stream or wind current to carry DETRITUS, as determined by particle size rather than amount. The diameter of the largest particle transported is the *competence value.* Compare CAPACITY.

competent *adj.* (of a sedimentary formation) strong and able to transmit compressive force much farther than a weak *incompetent* formation, i.e. the bed or stratum is able to withstand the pressure of folding without flowage or change in original thickness. Some of the factors that determine whether or not a stratum is competent are the

CRUSHING STRENGTH (resistance to crushing), the massiveness of the formation and the ability to 'heal' fractures.

component see PHASE DIAGRAM.

composite cone see VOLCANIC CONE.

composite fault scarp *n*. an ESCARPMENT the height of which is the result in part of differentiated EROSION and in part to direct movement along the FAULT. Either process, erosion or faulting, may precede the other.

composite intrusion *n*. bodies of igneous rock, such as DYKES, SILLS or LACCOLITHS, which are composed of two or more intrusions that differ in chemical and mineralogical composition. For example, an ACID phase may be injected into a more BASIC one. Compare MULTIPLE INTRUSION.

composition point see PHASE DIAGRAM.

composite profile *n*. a plot representing the surface of any relief area as viewed in the horizontal plane of the summit levels from an infinite distance. It consists of the highest points of a set of PROFILES drawn along regularly spaced parallel lines on a map.

compound alluvial fan see BAJADA.

compound corals see COLONIAL CORALS.

compound shoreline *n*. a SHORELINE along which features of both shoreline of submergence and shoreline of emergence are found.

compound vein *n*. **1.** a VEIN composed of several minerals.
2. a VEIN comprising many parallel FISSURES cut by cross-fissures.

compressibility or **modulus of compression** *n*. the capacity of a body to change in volume and density under hydrostatic pressure. It is the reciprocal of BULK MODULUS.

compression *n*. a system of external forces that tends to shorten a body or decrease its volume.

compressional wave see SEISMIC WAVES.

compressive strength see COMPRESSIVE STRESS.

compressive stress *n*. a STRESS that tends to push together on opposite sides of a plane in a direction perpendicular to that plane. The maximum compressive stress that can be applied to a material under specified conditions before rupture or fracture occurs is the compressive strength of the material. Compare TENSION.

concentric fold see PARALLEL FOLD.

conchoidal *adj*. (of a mineral or rock FRACTURE) smooth and curved, rather resembling a shell. Conchoidal fracture is characteristic of OBSIDIAN.

concordant *adj*. **1.** (of intrusive igneous bodies) having boundaries parallel with BEDDING or FOLIATION of the country rock.
2. (of strata) lying parallel with the bedding structure, i.e. structurally conformable.
3. (of radiometric ages) in agreement, although determined from more than one source or by more than one method. Compare DISCORDANT.

concordia *n*. the curved line generated on a graph by plotting the theoretical isotopic ratios of $^{206}Pb/^{238}U$ against $^{207}Pb/^{235}U$ for a range of ages through geological time. This provides a time scale with which the results of geochronological measurements can be compared. See DISCORDIA, EOCHRONOLOGY.

concretions *n*. discrete segregations or nodules of common sedimentary materials found in shales, sandstones and limestones. Concretions can form at the same time as the enclosing sediments (*syngenetic*) or during DIAGENESIS (*epigenetic*), may be rounded or irregular and may vary in size from a few millimetres up to a few metres in diameter, sometimes precipitating around some central nucleus such as leaves, seeds or shells. Many syn-genetic or early diagenetic concretions are bacterially produced. Concretions of calcite, iron oxides, pyrite and siderite are common, and are usually harder than the enclosing rock. See also CONE-IN-CONE STRUCTURE. SEPTARIAN concretions or nodules are particular masses that develop an irregular polygonal system of internal cracks. Under certain conditions of erosion, concretions can have the appearance of plants or animals and are often mistaken for fossils. See PSEUDOFOSSIL.

conductance *n*. the measure of the conductivity of a conductor, which depends on the dimensions of the conductor; it is the reciprocal of the resistance.

conduction *n*. a transfer of energy via some sort of conductor that permits the transfer of molecular activity without overall motion. It is distinct from other energy transfer forms such as radiation and heat convection. The transfer of heat, electricity and sound involves conduction. Conductivity is the conduction potential of a medium. See CONDUCTIVITY.

conductivity *n*. **1.** also called **specific conductance**, the reciprocal of resistivity. The conductivity σ of a conductor of length *l*, cross-section *A* and resistance *R* is given by

$$\sigma = l/RA$$

where σ is expressed in siemens per metre. When seats of electromotive force are absent – a condition that obtains in the Earth's oceans and atmosphere – Ohm's law is applied in the form of

$$J = \sigma E,$$

where *J* is the current density at a point of the conduction medium, σ is the conductivity of the medium and *E* is the electric field intensity at that point.

Electrical conductivity is used to measure the temperature and salinity of sea water and to determine the REDOX POTENTIAL in sediments, since it varies with the oxygen content.
2. the property of transmitting heat. The amount of heat conducted per unit of time through any cross-section of a substance depends upon the tempera-ture gradient at that section and the area of the section. Thus the quantity of heat is:

$$q = k\,\frac{dt}{ds}\,A$$

where *A* is the sectional area, dt/ds the temperature gradient and *k* is a constant – the thermal conductivity of the substance. See CONDUCTION.

cone-in-cone structure *n*. a structure composed of a series of inverted concentric cones, commonly formed of fibrous gypsum and fibrous calcite in fine-grained sediments and less frequently in coal deposits and ironstone. Some of these structures may be concretionary, but most are a response to pressure from the interior of the sediment outward. See also CONCRETIONS, PSEUDOFOSSIL.

cone of depression or **drawdown cone** *n*. a cone-shaped depression in the WATER TABLE that forms around a well or borehole from which water is being drawn, thus increasing the hydraulic gradient close to the well. See Fig. 17 overleaf. Compare DRAWDOWN.

cone sheet see RING DYKES AND CONE SHEETS.

confined aquifer see CONFINING BED.

confining bed *n*. a body of impermeable or less permeable material that may be saturated but will not allow water or other fluids to move through it under ordinary hydraulic gradients. Such a bed adjacent to one or more AQUIFERS will confine water movement to those

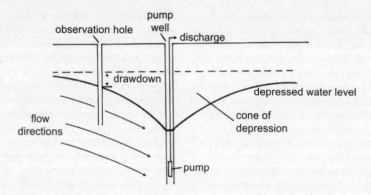

Fig. 17 **Cone of depression**. Depression of the water table when water is pumped from a well

aquifers. Compare AQUITARD, AQUIFUGE, AQUICLUDE. An aquifer bounded above and below by confining beds is a *confined aquifer*. Groundwater that is under sufficient pressure to rise above well level and the upper surface of which is an impermeable bed is called *confined groundwater*. See ARTESIAN.

confining pressure *n*. any pressure that is exerted equally on all sides, such as GEOSTATIC PRESSURE or HYDROSTATIC PRESSURE.

confluence *n*. the meeting point of two glaciers or two streams.

conformable *adj*. 1. (of rock strata or of the contact area between them) showing a continuous sequence of layers that were deposited without interruption in parallel order, one above the other. See UNCONFORMABLE. 2. the contact of an intrusive body when it is in alignment with the internal structure of the body. Compare CONCORDANT.

conformity *n*. 1. also called **conformability**, the stratigraphical continuity of adjacent sedimentary strata, i.e. they have been deposited in orderly series without evident time lapse.

2. a surface separating older strata from younger, along which neither erosion nor non-deposition is evident and with no significant hiatus. Compare UNCONFORMITY.

congelifraction, frost splitting, frost weathering or **frost wedging** *n*. rock fragmentation caused by the pressure exerted by the freezing of water contained in pores, cracks or fissures.

congeliturbation or **cryoturbation** *n*. the disturbance (stirring or churning) of soil by frost action, e.g. heaving, solifluction. It produces PATTERNED GROUND.

conglomerate or **pudding stone** *n*. a coarse-grained rudaceous clastic sedimentary rock composed of more or less rounded fragments or particles at least 2 millimetres in diameter (GRAN-ULES, PEBBLES, COBBLES, BOULDERS), set in a fine-grained matrix of sand or silt and commonly cemented by calcium carbonate, silica, iron oxide or hardened clay. Compare BRECCIA. See also FANGLOMERATE, RUDITE.

congruent melting point *n*. the temperature at a given pressure at which a solid phase becomes a liquid phase of the same composition. Most pure minerals of fixed composition, such as albite, diopside and quartz, melt congruently.

Coniacian *n.* a stratigraphical STAGE of the Upper CRETACEOUS.

conical fold *n.* a FOLD configuration that can be described by a line that is fixed at one end and rotated about that end. Compare CYLINDRICAL FOLD.

conjugate faults *n.* a set of FAULTS that develop synchronously (see Fig. 18 below). Conjugate JOINTS may form in the same manner.

connate water *n.* water trapped in the pore space of sediment at the time of its deposition. Other water that may have entered the interstices of a rock subsequent to its deposition is not connate.

conodonts *n.* small FOSSIL elements of phosphatic composition. They are tooth-shaped, occurring as simple curved cones or serially arranged compound elements that may be found in pairs. See Fig. 19 overleaf.

Fossils of animals with conodonts in the mouth parts have been found in the

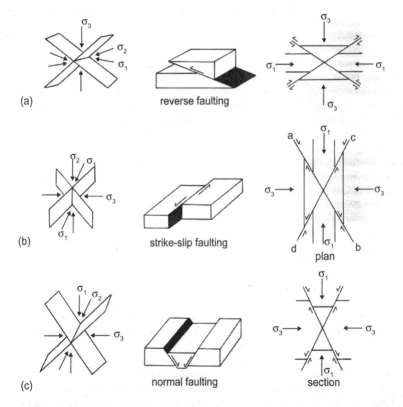

(a) reverse faulting

(b) strike-slip faulting / plan

(c) normal faulting / section

Fig. 18. **Conjugate faults**. The pattern of conjugate faults developed in response to different stress systems: σ_1 = maximum stress, σ_2 = intermediate stress, and σ_3 = compressive stress. The maximum stress bisects the acute angle between the faults

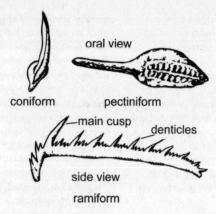

oral view

coniform   pectiniform

main cusp   denticles

side view

ramiform

Fig. 19 **Conodonts**. The basic shapes of conodont elements

Lower CARBONIFEROUS of eastern Scotland and the Lower SILURIAN of North America, and it is suggested that they have affinities with primitive chordates (see CHORDATA). Conodonts are widespread in marine rocks and are important stratigraphically. Range, Upper CAMBRIAN to TRIASSIC.

Conrad discontinuity *n*. a DISCONTINUITY in some areas of the Earth's crust where the velocities of compressional (P) seismic waves increase from about 6.1 kilometres per second to 6.4–6.7 kilometres per second. The depth of those locations where this abrupt change is noted is usually at 17 to 20 kilometres. It is believed that the discontinuity indicates a boundary between upper and lower CONTINENTAL CRUST.

consanguinous *adj*. describing the relationship between igneous rocks that are presumed to derive from the same parent magma. Such rocks are close in age and location, and usually have comparable chemical and mineralogical features. See also COMAGMATIC.

consequent *adj*. (of a geological or topographical feature) determined by pre-existing features, for example the course of a valley or stream that is determined by the initial slope of the land. See also SUBSEQUENT.

conservative margin or **conservative boundary** *n*. the margin between two lithospheric plates at which crust is neither created nor destroyed. The plates move past each other along STRIKE SLIP faults called TRANSFORM FAULTS. See also PLATE TECTONICS.

consolidation *n*. **1.** any process whereby soft or loose earth materials become firm, e.g. the cementation of sand or the COMPACTION of mud. See also OVER-CONSOLIDATED CLAY.
2. the reaction of a soil or sediment to increased surface load.

constructive plate boundary see ACCRETING PLATE BOUNDARY.

contact metamorphism see METAMOR-PHISM.

contact metasomatism see METASOMATISM.

contact zone see AUREOLE.

contemporaneous deformation or **penecontemporaneous deformation** *n*. DEFORMATION that takes place in sediments during their deposition or immediately after-wards. Small FOLDS and FAULTS may develop in soft sediments that slide down gentle slopes.

continental *adj.* formed or generated on land, as opposed to in the ocean or sea. Deposits originating in lakes, swamps or streams, or moved by land winds, are all continental.

continental accretion *n.* the theory that continents grow from a nucleus through the addition of material by sedimentation, volcanic activity and orogenesis at an active CONTINENTAL MARGIN. In PLATE TECTONIC theory accretion may be achieved when elimination of oceanic lithosphere by sugbduction leads to the collision of ISLAND ARCS or continental lithosphere with a continental margin.

continental borderland *n.* any part of the CONTINENTAL MARGIN between the SHORELINE and the CONTINENTAL SLOPE in which the topography is highly irregular and more complex than the CONTINENTAL SHELF.

continental crust *n.* the crustal rocks that underlie the continents and continental shelves; its thickness ranges from 25 to 70 kilometres. The continental crust is principally composed of rocks rich in silicon and aluminium, for which the acronym SIAL is sometimes used. Compare OCEANIC CRUST. See also CRUST.

continental deposit see TERRESTRIAL DEPOSIT.

continental displacement see CONTINENTAL DRIFT.

continental divide *n.* a drainage divide separating river systems flowing towards opposite sides of a continent.

continental drift *n.* The concept that the continents have undergone large-scale horizontal displacement during geological time. For more than 300 years the apparent fit of the bulge of eastern South America into the indentation of Africa, which can be seen on any map, has caused scientists to contemplate the movement of continents.

The first detailed and comprehensive theory of continental drift was proposed in 1912 by Alfred Wegener, a German meteorologist. Wegener postulated that throughout most of geological time there was only one continent, which he called PANGAEA; at some time during the Jurassic Period, Pangaea fragmented and the parts moved away from each other. In 1937, Alexander DuToit, a South African geologist, modified the Wegener hypothesis by supposing two primordial continents: GONDWANA in the south and LAURASIA in the north. The Wegener hypothesis, when presented, aroused much interest, debate, support and opposition for some years. It was weakened by its lack of a suitable driving mechanism for continental movement and by 1926 was no longer a subject of serious investigation.

The proponents of continental drift amassed impressive amounts of data in the matching of geological provinces of different continents. In the 1950s, interest increased as palaeomagnetic data accumulated. These data included evidence of POLAR WANDERING, *continental displacement* and *rotation*, and of GEOMAGNETIC POLARITY REVERSALS. During the early 1960s, the American geophysicist Harry Hess presented the concept of SEA-FLOOR SPREADING, in which new oceanic crust is continually being generated by igneous activity at the crests of MID-OCEANIC RIDGES. By the late 1960s, the concept of sea-floor spreading had been integrated with that of continental drift to develop the theory of PLATE TECTONICS. See Fig. 20 overleaf.

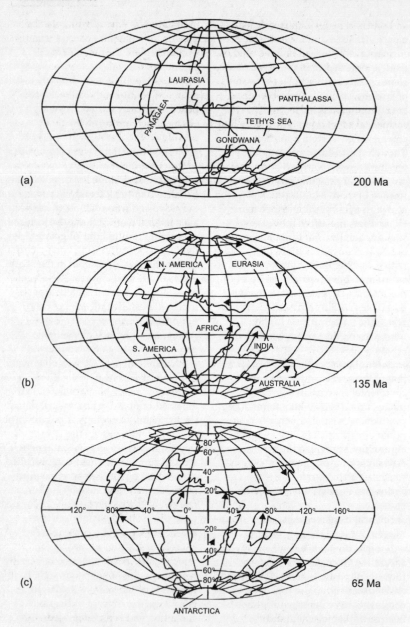

Fig. 20 **Continental drift**. The movement of the continents after the break-up of Pangaea about 200 Ma ago. The arrows show the direction of movement since drift began

continental margin *n.* the totality of the various divisions between the SHORE-LINE and ABYSSAL ocean floor. It includes the CONTINENTAL SHELF, CONTINENTAL BORDERLAND, CONTINENTAL SLOPE and CONTINENTAL RISE. Two types of continental margin can be defined. *Active continental margins* are characterised by strong SEISMIC and VOLCANIC activity. They have no continental rise but are generally bordered by an OCEAN TRENCH. An active continental margin is one form of DESTRUCTIVE PLATE BOUNDARY (see also PLATE TECTONICS). A *passive continental margin* is not affected by seismic or volcanic activity and is not a plate boundary. Most Atlantic margins are of this type.

continental platform see CONTINENTAL SHELF.

continental rise *n.* a geomorphological feature of the lower CONTINENTAL MARGIN where the vertical-to-horizontal ratio ranges from 1:50 to 1:800. It lies between the CONTINENTAL SLOPE and the ABYSSAL PLAIN or ABYSSAL HILLS, and its width ranges between 0 and 600 kilometres. Depending on its location, the rise may be divided into 'upper' and 'lower' steps or a series of steps resembling a staircase. In some regions the rise is very broad and in others nonexistent. It consists of fans of sediment derived from terrigenous silts and clays of the CONTINENTAL SHELF.

continental shelf or **continental platform** *n.* a gently sloping, shallow-water platform extending from the coast to a point (the *shelf break*) where there begins a comparatively sharp descent (CONTINENTAL SLOPE) to the ocean floor. The average width of the continental shelves of the world is 70 kilometres and their seaward slope about 0.3º. The average depth of shelf termination is about 150 metres, although this may vary greatly. Shelves may have parallel ridges and troughs or may be very flat, such as those found in the Bering Sea. Shelves off glaciated landmasses are characteristically deep basins and troughs, and hence very irregular.

continental slope *n.* the relatively steep portion of the sea floor extending from the outer edge of the CONTINENTAL SHELF to the upper limit of the CONTINENTAL RISE. Its average slope is 4°, its width is usually 20 to 100 kilometres, and its range of depth is from 100 to 200 metres to 1400 to 3200 metres. It may be terraced or smooth. Continental slopes, which are found throughout the world, may be the products of sedimentational or volcanic processes, or structures that were subjected to faulting and folding.

continuous deformation *n.* DEFORMATION by a process of FLOW rather than by FRACTURE. Compare DISCONTINUOUS DEFORMATION.

continuous reaction series see REACTION SERIES.

contourites *n.* fine sands, silts and muds deposited on the CONTINENTAL SLOPE by the action of deep-sea currents moving parallel with the slope. These deposits are usually thinly BEDDED and may be cross-laminated (see CROSS-BEDDING).

contour line or **contour** *n.* a line drawn on a map that links all points at the same height above a DATUM, usually sea level.

contraction crack see FROST CRACK.

convection *n.* **1.** the flow of water through and around heated regions of nearby PLUTONS; water circulation is initiated by thermal gradients within these regions.
2. (*oceanography*) the movement and mixing of oceanic water masses as a result of density differences.
3. (*tectonics*) a concept of the mass

movement of MANTLE material, either laterally or in CONVECTION CELLS. It is thought that the decay of radioactive isotopes within the mantle provides the heat energy required. Geophysical studies show that, given the pressures at such depths, mantle rock begins to CREEP at temperatures exceeding 1000°C. Thermal gradients cause hotter, less dense material to rise, to displace cooler, denser material and to flow parallel to the Earth's surface while cooling. When sufficiently cooled, the material descends. Model studies indicate that the rising limb of a convection cell is associated with positive gravitational anomaly areas, whereas the sinking portion of the cell coincides with a negative anomaly.

There are two theories of mantle convection. One proposes that convection cells are limited to the upper 700 kilometres of the mantle. The other holds that the cells circulate throughout the entire 2900 kilometres depth of the mantle. Evidence in support of *upper mantle convection* is stronger: no earthquakes are recorded at depths greater than 700 kilometres; secondly, the observed ratio of heat loss to heat generation at the surface of the Earth would be inconsistent with the highly efficient heat transport of the entire mantle convection. The concept of mantle convection helps explain such things as irregular distribution of heat flow within the Earth's crust and the uniformity of basalts along the MID-OCEAN RIDGES throughout the world. At present, it offers one of the most acceptable mechanisms for SEA-FLOOR SPREADING, deep-sea trenches and plate motion. See also PLATE TECTONICS, MANTLE PLUMES.

4. see MAGMATIC DIFFERENTIATION.

convection cell *n.* (*tectonics*) a pattern of mass movement of mantle material in which the outer area is downflowing and the central area is uprising because of heat differences. These cells may be of various sizes and shapes. They are not necessarily continuous and may change velocity or direction, or may stop completely. See also CONVECTION.

convergence *n.* **1.** a meeting of ocean currents or water masses that results in the sinking of the colder, denser or more saline water. Compare DIVERGENCE. **2.** *metamorphic convergence* is the close similarity between two rocks of different original characteristics following METAMORPHISM. **3.** *stratigraphical convergence* is the gradual decrease in the vertical distance between two sedimentary divisions as a result of thinning of the intervening strata. **4.** (*palaeontology*) *evolutionary convergence* is the appearance of ostensibly similar structures in organisms of different lines of descent; also called *convergent evolution*. Compare DIVERGENCE.

convergent evolution see CONVERGENCE.

convergent plate boundary see DESTRUCTIVE PLATE BOUNDARY.

converted wave *n.* a SEISMIC WAVE that has changed from a P wave to an S wave or vice versa by refraction or reflection at an interface.

convolute lamination *n.* heavily and intricately creased, crumpled or folded laminae that are contained within an undeformed layer. Parallel uncontorted layers lie above and below. The structure is thought to be the result of DEFORMATION of a partially liquefied sediment soon after deposition by means of an external shock, such as an earthquake. Convolute laminations are thought to be evidence of rapid deposition. They are common in TURBIDITES.

coordination number *n.* the number of anions that surround a cation in a stable crystalline structure. See IONIC BOND.

copper (Cu) *n.* a reddish or orange-red native metallic element. Small amounts of native copper are found in oxidised zones overlying copper sulphide deposits. Copper is a good conductor of heat and electricity, and is ductile and malleable. See CHALCOPYRITE.

coprolite *n.* a fossilised faecal pellet or casting of animal droppings (invertebrates, fish, reptiles, birds or mammals). Usually modular, tubular or pellet shaped and phosphatic in chemical composition, coprolites provide information about the kinds of organism that produced them and these organisms' habits. See FOSSILS.

coquina *n.* a carbonate rock formed almost entirely of sorted and cemented fossil debris, usually shells and shell fragments greater than 2 millimetres in diameter; *microcoquina* is similar rock composed of finer particles 2 millimetres or less in diameter. A distinction is drawn between coquina, which is a detrital rock, and *coquinoid limestone*, which is formed *in situ* and composed of coarse, shelly materials in a fine-grained matrix.

coral *n.* any of a group of bottom-dwelling, attached marine CNIDARIANS of the class Anthozoa. They may exist as solitary or colonial forms and secrete skeletons of calcium carbonate. COLONIAL CORALS may form extensive accumulations, CORAL REEFS. The principal groups with a fossil record are the Rugosa (RUGOSE CORALS), the TABULATA (tabulate corals) and the SCLERACTINIA. Range, ORDOVICIAN to present.

coralgal *adj.* consisting of CORAL fragments and other calcareous organisms, e.g. molluscs or foraminifers, bound together with algal growths and making up the preponderance of a CORAL REEF.

coralline alga *n.* a marine calcareous red alga commonly associated with CORAL REEFS. It may be a branching form or a massive encrusting type.

corallite *n.* the skeleton formed by a single CORAL polyp, which may be solitary or part of a colony. See also COLONIAL CORAL.

corallum *n.* the complete skeleton of a COLONIAL CORAL.

coral reef *n.* a wave-resistant unit constructed predominantly by HERMATYPIC CORALS and CALCAREOUS ALGAE. Coral reefs are formed where the temperature is suitable (about 18ºC minimum), water depth is shallow and salinity close to 35‰ (normal marine salinity). Reef forms include FRINGING REEFS, *patch reefs*, BARRIER REEFS and ATOLLS. See also REEF.

cordaites *n.* gymnosperms of class Pinopsida, order Cordaitales, ancestral to modern conifers, which they resembled in their soft-wood trunks and parallel-veined leaves. The leaves, however, were strap-like rather than needle-like, as in true conifers. Cordaites grew to heights of 30 metres and were prominent during the Upper Carboniferous. Their fossils are found in Carboniferous to Late Permian strata. See also FOSSIL PLANTS.

cordierite *n.* an orthorhombic silicate mineral, $(Mg,Fe)_2Al_4Si_5O_{18}$, commonly found in contact METAMORPHIC ROCKS and low-pressure regional metamorphic rocks. It displays PLEOCHROISM visible to the naked eye, appearing blue or violet when viewed parallel to the prism base and colourless when viewed vertically. For this reason it was sometimes called *dichroite*. Transparent, gem varieties are sometimes known as *water sapphires*.

cordillera *n.* a Spanish word for a mountain belt. The term *cordilleran-type mountain belt* has been used to distinguish subduction-related mountains on ocean/continent margins from intercontinental collisional mountains like the Alpine-Himalayan system, but the term *Andean-type mountain belt* is now sometimes used.

Cordilleras *n.* **1.** a mountain system in western North America, including the Sierra Nevada, Cascade Range and Rocky Mountains.
2. in western South America, the mountain system comprising the Andes and component ranges.
3. the entire chain of mountain ranges that run parallel to the Pacific coast from Cape Horn to Alaska.

core *n.* the innermost part of the Earth, which is marked by a SEISMIC DISCONTINUITY (GUTENBERG DISCONTINUITY) near a depth of 2900 kilometres, at which P-wave velocity decreases from 13.6 to 8.1 km s^{-1} and below which S waves are not transmitted. The non-transmission of S waves implies a liquid composition, an inference that has been reinforced by tidal investigations. A corresponding density discontinuity accompanies the seismic discontinuity at 2900 kilometres. It is believed that the source of the Earth's magnetic field lies in the liquid region of the core. Experimental analysis has shown that in the presence of an initial magnetic field, fluid motions within the core may produce a self-sustaining dynamo action that could induce a system of electrical currents capable of generating a magnetic field. In general terms, the processes involved in this entail convection in an electrically conducting fluid, with the result that the core acts as a dynamo that maintains and regenerates the magnetic field.

Different mechanisms have been proposed for driving the convective flow within the outer core. A thermally driven flow might be activated by the decay of radioactive isotopes. A compositionally driven flow might occur as an unmixing of dense and less dense regions. One of the more recently proposed concepts suggests that if the inner core is crystallising out of the surrounding liquid, there could be enough heat from the latent heat of crystallisation to drive the convective flow. Although the precise energy source is still equivocal, the instability basic to propelling thermal convection is caused by less dense fluid underlying denser fluid. The high density of the outer core, together with the extremely high electrical conductivity implied by the dynamo theory of the Earth's magnetic field, suggests that the core may be composed chiefly of iron. Studies of meteorites indicate that the core also contains about 10 per cent metallic nickel. In addition, shock-wave studies at core pressures suggest the presence of sulphur or elementary silicon.

The boundary of the inner core, at about 5100 kilometres, is marked by a P-wave velocity increase of about 10 per cent. As indicated by shock-wave data and calculations, the inner core is probably solid. It may be regarded as the product of the crystallisation of iron under very high pressure (3000 to 3600 kb or 300 to 360 GPa). Detailed seismic studies show the presence of a complex structure at the boundary between the outer and inner cores; the same studies also reveal that at least two seismic discontinuities may occur within the transition zone. See also CRUST, MANTLE.

core sample *n.* a continuous section or

core of rock obtained by drilling with a hollow cylinder for the purpose of obtaining geological information.

corestone *n.* a core of unweathered rock enclosed in a MATRIX of WEATHERING products that results from deep chemical weathering. Chemical weathering can penetrate more deeply along JOINT surfaces and fissures. EROSION of the weathered material may leave a BOULDER FIELD of corestones at the surface. Irregularly spaced joints may result in preferential weathering and removal of material where joints are closely spaced, leaving tower-like blocks of unweathered rock (*tors*) standing above the surrounding area.

Coriolis effect *n.* the apparent deflection of a body moving with respect to the Earth, as seen by an earthbound observer. The deflection is attributed to a hypothetical force, the CORIOLIS FORCE, but actually is caused by the Earth's rotation. It appears as a deflection to the right in the Northern Hemisphere and to the left in the Southern Hemisphere. See CENTRIFUGAL FORCE.

Coriolis force *n.* a hypothetical force introduced into the analysis of motions that are measured with respect to rotating coordinate systems; it is one of the equation terms that represents acceleration values as would be observed from the inertial frame of reference. The magnitude of the horizontal component is given by

$$2 \, \omega \sin Uv$$

where ω is the angular velocity of the Earth, U is latitude and v is the horizontal velocity of the moving body. The Coriolis force is of the same nature as 'CENTRIFUGAL FORCE', in that both are fictitious forces. The former appears to affect moving objects and the latter seems to act on stationary objects on the Earth's surface. See CORIOLIS EFFECT.

cornelian see CARNELIAN.

cornstone *n.* a concretionary LIMESTONE believed to represent a SOIL HORIZON formed under semi-arid conditions and now fossilised. See also CALCRETE.

corona *n.* a zone of MINERALISATION, often showing radial arrangement, surrounding another mineral. The term has been applied to REACTION RIMS. See also KELYPHITIC BORDER, RAPAKIVI TEXTURE.

corrasion *n.* **1.** ABRASION.
2. see ATTRITION.

correlation *n.* **1.** establishment of the lithological or chronological equivalence of STRATIGRAPHICAL UNITS.
2. in seismology, the demonstration of the identity of phases on different seismic records.

corrie see CIRQUE.

corrosion *n.* **1.** the EROSION of rocks by chemical processes, e.g. hydration, oxidation, solution, hydrolysis. The rate of corrosion of a particular area is largely determined by climate, geology, topography and vegetation. Compare CORRASION.
2. the modification (partial resorption, fusion, dissolution) of the outer parts of early-formed crystals or foreign inclusions by the solvent action of the MAGMA in which they occur.

corundum *n.* a hexagonal mineral, Al_2O_3, occurring as masses or as variously coloured, often barrel-shaped, crystals, including ruby and sapphire. It occurs as a primary constituent in igneous rocks containing feldspathoids and also results from the metamorphism of aluminium-rich rocks. The extreme hardness of corundum, 9 on the MOHS SCALE, permits it to take a long-lasting high polish. See also EMERY.

cosmic dust *n.* dust that exists between the planets, stars and galaxies. *Primordial dust* is that which remained as residue after their formation. A *secondary cosmic dust* is constantly produced in the solar system by the disruption of comets that come too close to the Sun.

cosmic erosion *n*. the wearing away or destruction of rocks on a planetary surface, a cumulative effect of impacts of hypervelocity particles from outer space.

cosmic radiation *n*. high-energy sub-atomic particles from outer space that strike the Earth's atmosphere from all directions. *Primary cosmic rays* consist of nuclei of the most abundant elements; upon entering the Earth's atmosphere, most of the primary rays collide with atomic nuclei in the atmosphere, producing *secondary cosmic rays*, which consist mainly of elementary particles, e.g. neutrons, mesons.

cosmogeny *n*. the study of the age and origin of the Universe and in particular of the Earth and solar system.

cosmology *n*. the science of the Universe, in geology relating in particular to the formation of the solar system.

costellae *n*. (*sing*. **costella**) the ribs, the outer surface of a BRACHIOPOD shell, which radiate from the umbones (see UMBO) to the COMMISSURE.

cotectic *adj*. (of conditions of temperature, pressure and composition) allowing the simultaneous crystallisation of two or more solid phases from a single liquid over a finite interval of decreasing temperature.

cotectic boundary line *n*. the line or surface on a PHASE DIAGRAM representing the corresponding phase boundary on the liquids.

cotylosaur *n*. any member of the reptile order Cotylosaurea of the subclass Anapsida. They were the earliest group of reptiles, were lizard-like in form and were probably ancestral to all other reptiles. Range, CARBONIFEROUS to Upper TRIASSIC.

coulée *n*. **1.** an anastomosing system of deep channels that formed as a result of the catastrophic release of GLACIAL meltwater from an ice-dammed lake. The best-known coulées are those cut into the Columbia plateau BASALTS in the state of Washington in the USA. **2.** (*volcanology*) a narrow flow of viscous lava, generally RHYOLITE, DACITE or ANDESITE, that forms on the steep side of a VOLCANO. It is characterised by marginal levees of solidified lava blocks that help to confine the lateral extent of the flow.

couloir *n*. **1.** a steep valley or gorge on a mountainside, especially in the Alps. **2.** a passage in a cave.

country rock or **host rock** *n*. a general term for the older rock containing an igneous intrusion or penetrated by mineral veins. Country rock may be sedimentary, metamorphic or older igneous rock. See INTRUSION.

couplet see RHYTHMITE.

Courceyan *n*. the lowest STAGE of the CARBONIFEROUS in Britain.

covalent bonds or **homopolar bonds** *n*. inter-atomic bonds in which electrons are shared between two atoms. The formation of covalent bonds depends on the number and distribution of the shared electrons in the outer shell of the atoms. Elements that do not readily gain or lose electrons to form ions will form covalent bonds, for example silicon or carbon. The Si–O bonds in the SiO_4 structural unit (or SILICA TETRAHEDRON) that characterises the SILICATE minerals are considered to be covalent. Covalent bonds are stronger than IONIC BONDS, so the strength of the bonds in silicate minerals is reflected in the relative hardness of minerals such as quartz and the FELDSPARS when compared with typical non-silicate ionic compounds like the mineral HALITE. Much of the bonding in silicates, however, takes on a character that is intermediate between ionic and covalent bonding. This is typified by the PYROXENE minerals, in which chains of silica tetrahedra are linked by metallic atoms with bonds that are partly ionic in character. Thus the bonds between the chains are weaker than the Si–O links in the chains, hence the tendency to cleave along two directions parallel to the chains. This is expressed in the two cleavages, intersecting at 87º and 93º, that are a characteristic physical property of the pyroxenes. An example of a non-silicate carbon mineral with covalent bonding is DIAMOND. See BOND.

covellite *n.* a mineral, CuS, found in the zone of sulphide enrichment; it forms as an alteration product of other copper minerals, e.g. chalcocite, chalcopyrite. It occurs most often as slaty or compact indigo-blue masses with pronounced iridescence.

cowpat bomb see VOLCANIC BOMB

cpx *abbreviation for* clinopyroxene. See PYROXENE.

crag and tail *n.* a streamlined ridge or hill resulting from glaciation. It consists of a knob (the 'crag') of resistant bedrock, with an elongate body (the 'tail') of more erodible bedrock on its lee side. Compare ROCHE MOUTONNÉE.

cranidium *n.* the part of the CEPHALON of a TRILOBITE that includes the GLABELLA and the fixed cheeks.

crater *n.* **1.** a VOLCANIC CRATER.
2. a saucer-shaped depression on the Earth's surface resulting from a meteorite impact or bomb explosion.
3. a LUNAR CRATER.

crater lake *n.* a lake formed by precipi-

tation and the accumulation of groundwater in a VOLCANIC CRATER or CALDERA.

craton *n.* a part of the CONTINENTAL CRUST that has been stable (no orogenic activity) for at least 1000 Ma. Cratons are typically formed of Lower to Middle PRECAMBRIAN igneous and metamorphic rocks with a subdued surface relief (SHIELD areas) in places overlain by largely undeformed Upper Precambrian or younger sedimentary rocks (PLATFORM areas). The oldest parts of cratons are often more than 2500 Ma old.

creep *n.* **1.** the slow, imperceptible gravity movement of soil and/or broken rock material from higher to lower levels, or the material itself that has so moved. Accompanying SHEAR STRESSES may produce permanent deformation but are too small to cause shear failure, as in LANDSLIDES.
2. slow DEFORMATION produced by long-acting stresses below the elastic limit of the material subjected to stress. Part of the deformation is elastic (non-permanent) and part is permanent.

creep recovery *n.* the gradual recovery of elastic STRAIN after STRESS is released.

crenulation cleavage *n.* a spaced CLEAVAGE in which the cleavage planes are separated by a finite distance, producing thin, flat rock units. The rock units may suffer minor internal folding (*crenulation*).

crescentic fracture or **crescentic crack** *n.* a lunate-shaped mark on a glaciated rock surface, convex in the direction from which the ice moved. It is larger than a CHATTER MARK and consists of a single mark with no removal of rock material. Compare CRESCENTIC GOUGE.

crescentic gouge or **gouge mark** *n.* a crescent-shaped groove resulting from GLACIAL PLUCKING on a bedrock surface. It is concave in the direction from which the glacier moved and comprises two fractures from which rock material has been gouged. Compare CRESCENTIC FRACTURE.

crest *n.* the highest part of an ANTICLINE fold. The line connecting the highest points on a given stratum in an infinite number of cross-sections is the *crest line*. The surface formed by all crest lines is the *crest surface* or *crestal plane*.

Cretaceous *n.* the interval of geological time that began 144 Ma ago and ended 65 Ma. It is the final period of the Mesozoic Era and precedes the Tertiary Period. Its name, proposed by Omalius d'Halloy in 1822, refers to chalk (*creta* in Latin), the characteristic rock found during the period. Rocks of the Cretaceous system are distributed throughout the world and vary in thickness and lithological character. The Cretaceous was marked by the continued break-up of GONDWANA and by the subsequent widespread basaltic VOLCANISM. Further large-scale volcanism, in the form of submarine and continental flood basalts, was produced by MANTLE PLUMES. The opening of the South Atlantic began, and an abortive attempt at opening the North Atlantic took place, with the North and South Atlantic being in communication; at the same time, India detached from Africa and began to move northeastwards, and the northward movement of Africa caused TETHYS to close. Consequently, the Alps began to form in Mid-Cretaceous times. Both North and South America were moving westwards as the Atlantic opened, and the west coast of the Americas was the site of volcanic and tectonic activity.

Sea levels rose almost continuously during the Cretaceous, with large-scale marine inundation during the later part of the period (Cenomanian) resulting in extensive covering of continental areas by CHALK and CLAY. The extensive Cretaceous shallow submergence allowed major accumulations of oil and gas. In Britain the Early Cretaceous is characterised by nonmarine facies, and the Mid- and Late Cretaceous by marine deposits, namely sandstones (including the greensands), clays and chalk.

The Cretaceous sustained a greenhouse climate, and climatic zones, as indicated by floral and faunal distribution, are generally divided between the tropical to subtropical equatorial regions and the cooler, higher-latitude regions, with a low Pole to Equator thermal gradient. Warmer ocean temperatures are indicated by oxygen-isotope determination in BELEMNITES. Except for mountain glaciers, glaciation was absent during the Cretaceous. A climatic cooling took place in the Late Cretaceous.

Ammonites were predominant during the Cretaceous (some reached a diameter of 2 metres), being used as zone fossils throughout the period but becoming extinct by the end of the period. Belemnites also disappeared at this time. Bivalves and gastropods are common and, with echinoderms, crinoids, sponges, brachiopods and bryozoans, occupied a range of marine habitats. Dinosaurs, the dominant vertebrates of the Cretaceous, were found on every continent but became extinct by the end of this period. Other vertebrates included pterosaurs and an expanding range of birds. Mammals were still insignificant. Early Cretaceous land plants (conifers, gingkoes, ferns) differed little from those of the Jurassic; angiosperms, including the grasses, became important in mid-period. Late Cretaceous flora included magnolias, figs, poplars, willows and plane trees.

Many theories, ranging from METEOR-ITE impact to a major regression of the sea, have been proposed to account for the extinction of so many species at the end of the Cretaceous. Some of these are discussed under EXTINCTION.

Cretaceous-Tertiary boundary layer *n.* a distinctive clay layer found at many continental localities and in cores of sediment taken from the ocean floor. It marks the boundary between the CRETACEOUS and TERTIARY periods. The anomalous high concentrations of siderophile elements (see AFFINITY OF ELEMENTS) such as IRIDIUM found in this layer provide evidence of a major meteorite impact that took place 65 million years ago.

crevasse *n.* **1.** see GLACIAL CREVASSE.
2. a deep opening in the earth that appears after an EARTHQUAKE.
3. a break or breach in a river bank, such as a NATURAL LEVEE.

crinanite see TESCHENITE.
crinoid *n.* an animal of the class Crinoidea, phylum Echinodermata, commonly called the 'sea lily'. The animal lives in a CALYX, composed of calcite plates, which bears a number of flexible arms (BRACHIA) used for food collecting (see Fig. 21 below). The upper surface of the calyx is covered by the TEGMEN, which contains the mouth and anus. Some crinoids are anchored to the sea bed by a long stem composed of calcite discs (*columnals*). Others lose the stem early in life and move freely over the sea floor, anchoring themselves temporarily by means of CIRRI on the base (centro-dorsal plate) of the calyx. Crinoids are sometimes classed as PELMATOZOANS.

The first crinoids appeared in the CAMBRIAN, but they were rare until the Lower ORDOVICIAN. They are important as INDEX FOSSILS in the PALAEOZOIC, where their disarticulated remains may be the principal constituent of LIMESTONES (CRINOIDAL LIMESTONES). Many forms

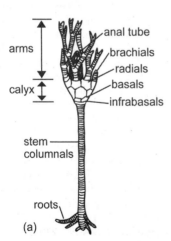
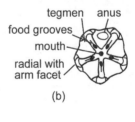
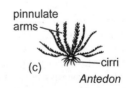

Fig. 21 **Crinoid.** (a) principal features of a crinoid. (b) oral surface. (c) a free-swimming crinoid

became extinct at the end of the PERMIAN. Range, Cambrian to present.

crinoidal limestone *n*. sedimentary rock composed almost totally of fossil crinoidal parts, especially the stem COLUMNALS.

cristobalite *n*. a high-temperature polymorph of SiO_2 found in cavities in some VOLCANIC ROCKS and in some thermally metamorphosed sandstones. It is stable above 1470ºC but can exist metastably below that temperature because reactions are sluggish. TRIDYMITE is the polymorph stable between 867ºC and 1470ºC. See also ALPHA QUARTZ, BETA QUARTZ.

critical angle *n*. the ANGLE OF INCIDENCE above which a wave is totally reflected from a boundary between two media in which it travels with different velocities. At the critical angle of incidence, the angle of refraction is 90º.

critical point *n*. the set of temperature and pressure conditions for a given substance at which liquid and gas become physically indistinguishable. The temperature and pressure at the critical point are the *critical temperature* and *critical pressure*. At temperatures above the critical temperature the substance can exist only in the gaseous state, no matter how great the pressure exerted. At this temperature, both liquid and gaseous phases merge and become identical in all properties. The critical pressure is the minimum pressure necessary to liquefy a gas at its critical temperature.

critical pressure see CRITICAL POINT.

critical temperature see CRITICAL POINT.

critical velocity *n*. the velocity at which the nature of the flow for a given fluid changes from LAMINAR FLOW to TURBULENT FLOW, or vice versa.

crocidolite or **blue asbestos** *n*. a silicate of sodium and iron, an asbestiform

variety of RIEBECKITE. The dust formed from the break-up of crocidolite fibres is highly carcinogenic.

Croixian *n*. a series of the North American Upper Cambrian, taking its name from the exposures about St Croix Falls, Minnesota.

Cromerian see CENOZOIC, Fig. 13.

cross-bedding or **cross-stratification** *n*. sedimentary layering within a BED that is inclined at an angle to the main BEDDING PLANE. The steeply dipping parts of the layers are *foresets*; if the DIP decreases towards the base of the bed, these are the *bottomset* parts. Cross-bedding forming one layer is called a *set*; two or more sets within a bed form a *coset* (see Fig. 22(a,b) opposite). The term *cross-lamination* is used for cross-stratification in which the set height is less than 6 centimetres and the individual cross-layers are less than 1 centimetre thick.

Cross-lamination/cross-bedding is mainly formed by the migration of BED FORMS, such as ripples, DUNES or SANDWAVES, as sediment is deposited on the downcurrent side but also results from the growth of DELTAS or the migration of sand bars in a river (see RIPPLE MARK). The nature of the cross-bedding reflects the shape of the bed form and thus the strength of the current that moved the sediment. Cross-bedding in which the foresets are approximately planar is called *tabular cross-bedding* and results from the migration of straight-crested bed forms. In *trough-cross-bedding* (sometimes called *festoon bedding*), the foresets are trough-shaped, resulting from the migration of bed forms with sinuous or curved crests; see Fig. 22(b). *Ripple-drift cross-lamination* (or *climbing ripples*) results from the rapid deposition of sediment; ripples climb up the backs of

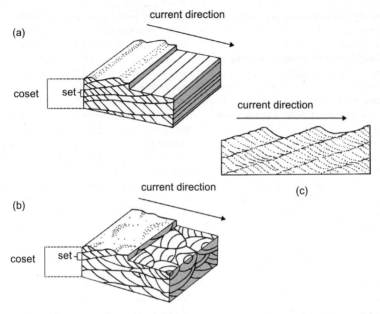

Fig. 22. **Cross-bedding** (a) tabular cross-bedding. (b) trough or festoon cross-bedding. (c) climbing ripple cross-lamination (or ripple-drift cross-lamination)

those downstream, so that the set boundaries lie at an angle and dip in the opposite direction from the cross-layers (see Fig. 22(c)). In some cosets the foresets within adjacent sets may dip in opposite directions: this is *herringbone cross-bedding*, produced where there are reversals of the current directions, for example in tidal areas. See HUMMOCKY CROSS BEDDING. Cross-bedding is useful as a WAY-UP INDICATOR: the angle between the foresets/bottomsets and the base of the bed is smaller than the angle between the foresets and the top surface, which is often an erosion surface.

cross-correlation *n*. a measure of the similarity of, or a method for the comparison of, sequences of data, e.g. a measure of the similarity of two SEISMIC WAVE forms. Compare CORRELATION.

crosscut *n*. a mining tunnel or level of access cut across the COUNTRY ROCK so as to intersect a VEIN or ore-bearing structure.

crossed polarisers or **crossed Nicols** *n*. crossed Nicols is an obsolete term that describes the arrangement of two NICOL PRISMS in a polarising microscope so that their respective planes for transmitting polarised light are at right angles to each other, in which position light emerging from one prism is disrupted by the other. The modern term, crossed polararisers, describes the same relationship using polarising filters. See PETROLOGICAL MICROSCOPE.

cross-joint see JOINT.

cross-lamination see CROSS-BEDDING.

crossopterygian *n*. a lobe-finned bony fish of the class Sarcopterygii, believed to be the predecessor of the modern lungfishes. Crossopterygians appeared in the Middle Devonian and include the COELACANTH.

cross-section or **transverse section** *n*. a plane surface cutting the longest axis of an object at right angles; the pictorial representation of the features exhibited by such a plane cut, e.g. a crosscut of an ore body or fossil or any cut that shows transected geological features. A *longitudinal cross-section* is a diagram drawn on a vertical plane and parallel to the longer axis of a given feature, e.g. a section drawn parallel to the axis of a fossil.

cross-stratification see CROSS-BEDDING.

crude oil *n*. PETROLEUM as it emerges in its natural state, before distillation or cracking.

crura see BRACHIDIUM.

cruralium see SPONDYLIUM.

crush breccia *n*. a breccia formed *in situ* as a result of movement such as faulting or folding. Compare CRUSH CONGLOMERATE, AUTOCLASTIC, FAULT BRECCIA. See also MYLONITE.

crush conglomerate, cataclasite or **tectonic conglomerate** *n*. a rock similar to CRUSH BRECCIA but the fragments of which are more rounded. It closely resembles a sedimentary conglomerate.

crushing strength *n*. the compressive load required to cause a given solid to fail by fracture. The load, divided by the cross-sectional area of the solid, is its *crushing stress*.

crust *n*. the outermost layer of the Earth, the lower boundary of which is the MOHOROVCIC DISCONTINUITY (Moho). Part of the crust is equivalent to the SIAL (the CONTINENTAL CRUST) and part to the SIMA (the OCEANIC CRUST). The entire crust represents less than 0.1 per cent of the Earth's total volume and is non-homogeneous in composition and distribu-

tion. The continental crust is distinct from the oceanic in age as well as in physical and chemical characteristics.

The most ancient parts of the continental crust are about 4000 Ma old. It is constantly modified by TECTONICS, VOLCANISM, SEDIMENTATION and EROSION, and has a thickness of 25 to 90 kilometres, with the greatest thickness beneath present-day mountain chains such as the Himalayas. The continental crust is an assemblage of igneous, sedimentary and metamorphic rocks rich in elements such as silicon and alluminium, and holds much more uranium than does the oceanic crust. GRANITE and GNEISS are dominant in the SHIELD areas.

In places the continental crust is separated into upper and lower layers by a SEISMIC DISCONTINUITY, the CONRAD DISCONTINUITY. The upper layer has a density of about 2.67×10^3 kgm^{-3} and displays 'granitic' properties, while the lower layer has 'gabbroic' properties with a density of about 3.0×10^3 kgm^{-3}. From the evidence of XENOLITHS brought to the surface in VOLCANIC pipes, however, it is thought that the lower layer is composed of high-grade METAMORPHIC ROCKS of granodioritic composition with a proportion of BASIC and ULTRA-BASIC INTRUSIONS (see GRANODIORITE).

Oceanic crust is much younger than the continental crust, the oldest region of the ocean basin being less than 200 Ma in age. The thickness of the ocean crust (5–10 kilometres) is much less than that of the continental crust; its density is in the order of 3.0 g cm^{-3}. Oceanic crust is constantly being generated along MID-OCEAN RIDGES and destroyed as it moves to a subduction zone. (See SEA-FLOOR SPREADING.) GABBROS and other rocks make up the lower layer of the oceanic crust, and BASALTS are the main material of the upper. There are three suboceanic layers as defined in terms of the velocities of P waves that travel through them. The velocity pattern in the crust is complex, however, and the idea that the velocity in these layers is constant has been seriously questioned. Instead of the absolute wave velocity being used as an indicator, therefore, the velocity gradient, i.e. the change in wave velocity with depth, is used. The oceanic crust is very uniform in structure, with sediments (increasing in thickness away from the mid-oceanic ridges) overlying two layers of igneous rocks. The upper layer, 1–2.5 kilometres thick, is composed of basaltic PILLOW LAVAS underlain by DYKES. On the basis of seismic velocities, the underlying layer, approximately 5 kilometres thick, is probably gabbro, with a thin ultrabasic, olivine-rich layer at the base. See also CORE, MANTLE.

crustacean *n.* any arthropod of the subphylum Crustacea, characterised chiefly by two pairs of antenna-like appendages in front of the mouth and three pairs behind it. Present groups are represented by shrimp, crabs, lobsters, copepods and isopods; most forms are marine. Range, Cambrian to present.

crustal flotation see AIRY HYPOTHESIS.

crustal plate see PLATE TECTONICS.

cryoconites *n.* tubular depressions that are melted into glacier ice at locations where areas of dark particles absorb the Sun's rays.

cryolite *n.* a white or colourless (transparent to translucent) mineral, Na_3AlF_6, usually massive; crystals are rare. The only important deposit, a large mass in a granite intrusion at Ivigtut, West Greenland, is now essentially exhausted. Formerly used as a source of aluminium, cryolite is now used in the manufacture of certain glass and porcelain, and as a metal-cleaning flux.

cryomorphology *n.* the subspecialisation of GEOMORPHOLOGY pertaining to the various processes of cold climates.

cryopedology *n.* the study of processes associated with intensive frost action and the occurrence of frozen ground, including techniques used to surmount or minimise problems connected with it.

cryoplanation *n.* degradation of a land surface by processes associated with intensive frost action that is augmented by the action of moving ice, running water and other agents. Compare ALTIPLANATION, EQUIPLANATION.

cryoturbation see CONGELITURBATION.

cryptic layering *n.* the gradual change in the chemical composition of those CUMULATE minerals that are members of a SOLID-SOLUTION series through a series of layers in an INTRUSION. See also LAYERING.

cryptocrystalline *adj.* (of a crystalline aggregate) so finely divided that individual crystals cannot be recognised or distinguished under an ordinary microscope. Compare MICROCRYSTALLINE.

cryptodome *n.* a dome-like uplifted area caused by the near-surface intrusion of viscous MAGMA, commonly andesitic or dacitic in composition (see ANDESITE, DACITE).

cryptoexplosion structure or **crypto-volcanic structure** *n.* a large, roughly circular impact structure, the origin of which is unknown or only supposed. Many such structures are believed to derive from crater-generating meteorites and others from uncertain volcanic activity.

cryptoperthite *n.* an extremely fine-grained, intercombined sodic and potassic feldspar in which lamellae are discernible only by means of X-rays or with the use of an electron microscope. Compare PERTHITE; see CRYPTOEXPLOSION STRUCTURE.

cryptovolcanic structure see CRYPTO-EXPLOSION STRUCTURE.

Cryptozoic *n.* obsolete term for that part of geological time equivalent to the Archaean.

cryptozoon *n.* structures from Precambrian sediments, formerly believed to be fossils. These have been largely determined to be CONCRETIONS.

crystal *n.* a homogeneous solid body in which the atoms or molecules are arranged in a regular repeating pattern that may be outwardly manifested by plane FACES. See CRYSTAL LATTICE, CRYSTAL SYMMETRY.

crystal axes or **crystallographic axes** *n.* three imaginary lines in a crystal (four in the hexagonal system) that pass through its centre; the structure and symmetry of a particular crystal are described by the relative lengths of these lines and their mutual angular orientation. The ratio of unit distances along the axes to the angle between them defines the *axial elements*. Unit distances along the crystal axes are defined by means of a reference FACE known as the PARAMETRAL PLANE. Any face can be chosen provided that it cuts all three axes (if necessary, when extended). The intercepts of the parametral plane on the crystal axes define the relative lengths of the axes, say a, b and c. The *axial ratio* is obtained by comparing the length of a crystal axis with one of the lateral axes, giving a/b: b/b: c/b, i.e. a/b: 1:c/b. Each mineral has a unique axial ratio. Crystal axes may coincide wholly or in part with axes of symmetry (see CRYSTAL SYMMETRY). See also MILLER INDICES, LAW OF RATIONAL INDICES.

crystal class or **point group** *n.* one of the 32 possible combinations of symmetry in which crystals can form; such combinations are the only possible arrangement of symmetry axes and planes in which these elements all intersect at a common point. See also CRYSTAL SYSTEM.

crystal flotation *n.* the rising of low-density crystals in magma. Compare CRYSTAL SETTLING.

crystal fractionation see MAGMATIC DIFFERENTIATION.

crystal gliding see GLIDING.

crystal habit *n.* the characteristic or general appearance of a crystal, e.g. ACICULAR, PRISMATIC, caused by the relative development of different FORMS. Some minerals are always and everywhere bounded by faces of the same crystal form, whereas in other instances a mineral at one locality may have a different appearance or habit from the same mineral at another locality.

crystal lattice *n.* the three-dimensional arrangement of atoms or ions that form a CRYSTAL. The smallest complete unit of the pattern, which can be repeated in all directions, is called the *unit cell*. The unit cell has all the symmetry elements of the crystal. Auguste Bravais, a French crystallographer, was the first to show that there can be only 14 possible basic kinds of crystal lattices, thus they are commonly called *Bravais lattices*. See CRYSTAL SYMMETRY.

crystalline *adj.* having a regular atomic or molecular structure but without developing easily discernible crystal FACES. Compare AMORPHOUS, CRYSTAL-LISED.

crystalline rock *n.* **1.** a rock consisting of minerals in a clearly crystalline state. **2.** igneous and metamorphic rocks, as distinct from sedimentary rocks.

crystallinity *n.* the degree to which a rock shows crystal development or the extent of that development. See HOLOCRYSTALLINE, HYPOCRYSTALLINE, MACROCRYSTALLINE, MICROCRYSTALLINE, CRYPTOCRYSTALLINE, CRYSTALLISED, CRYSTALLINE, AMORPHOUS.

crystallisation *n.* the gradual formation of crystals from a solution or a melt.

crystallisation interval *n.* in a cooling MAGMA, the interval of temperature between the formation of the first crystal and the disappearance of the last drop of liquid from the magma. For a given mineral, it is the temperature range over which that particular phase is in equilibrium with liquid.

crystallised *adj.* (of a mineral) occurring as well-developed CRYSTALS, i.e. showing good crystal FACES.

crystalloblast *n.* a crystal of a mineral formed completely by metamorphic processes. One type of crystalloblast, an *idioblast*, is a mineral constituent of a metamorphic rock formed by recrystallisation and bounded by its own crystal faces. A *xenoblast* has formed in a rock during metamorphism without developing its characteristic crystal faces, and a *hypidioblast* is a subhedral crystal formed during metamorphism.

crystalloblastic *adj.* (of a crystalline texture) formed by metamorphic conditions of high viscosity and regional pressure, as distinct from igneous rock textures that are the result of a sequential crystallisation of minerals under conditions of low viscosity and near-uniform pressure.

crystallographic axes see CRYSTAL AXES.

crystallography *n.* the study of crystal FORM and the internal structure to which it is related.

crystal mush *n.* MAGMA containing a high proportion of crystals.

crystal settling or **crystal sedimentation** *n.* the sinking in a magma of crystals of greater density, resulting in crystal accumulation. Compare CRYSTAL FLOTATION.

crystal symmetry *n.* the regular arrangement of the external features of crystals, which is related to the internal structure. This regularity can be described in terms of the *symmetry elements*:
(a) a crystal has a *centre of symmetry* if every face has an exactly similar face parallel to it on the opposite side of the crystal.
(b) a *plane of symmetry* divides a crystal into two equal halves that mirror each other.
(c) an *axis of symmetry* is a line about which a crystal may be rotated so that at some time during the rotation it has exactly the same appearance (i.e. it assumes a position of congruence). If this occurs every 180º the axis is called a *diad axis*. Triad, *tetrad* and *hexad axes* represent rotations of 120º, 90º and 60º respectively for the same view to occur.
(d) an *inversion axis* (*axis of rotary inversion*) occurs where rotation must be followed by inversion across a centre to bring a face to a position of congruence.

crystal system *n.* one of seven groupings or classifications of crystals according to common symmetry characteristics that permit them to be referred to the same crystallographic axis. Crystals of the *isometric* or *cubic system* are characterised by three mutually perpendicular axes of equal length. *Tetragonal system* crystals have three mutually perpendicular axes, the vertical one of which is shorter or longer than the two horizontal axes, which are of equal length. Crystals of the *orthorhombic system* are character-

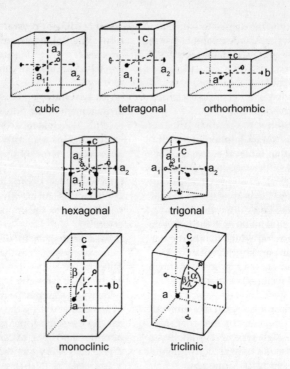

cubic   tetragonal   orthorhombic

hexagonal   trigonal

monoclinic   triclinic

Fig. 23 **Crystal system**. The seven crystal systems, showing the
relationships of the crystallographic axes

ised by three mutually perpendicular
symmetry axes, all of different relative
length. *Monoclinic system* crystals show
three unequal crystallographic axes,
two of which intersect at an oblique
angle and the third of which is perpendicular to the plane formed by the
other two. *Triclinic system* crystals lack
symmetry other than a possible centre;
they are characterised by three unequal
axes that are mutually oblique.
Crystals of the *hexagonal system* and the
trigonal system have three lateral axes of
equal length intersecting at a 60º angle
to each other, and a vertical axis of
different length perpendicular to the
other three. See Fig. 23 above.

crystal tuff see TUFF.

C shell see SEISMIC REGIONS.

CT see CARAT.

C-type granite see GRANITE.

cube *n*. (in crystallography) a six-sided
closed crystal FORM found only in the
isometric system (see CRYSTAL SYSTEM).
The faces in the cube form are all either
parallel to each other or intersect at
right angles, however the relative sizes
of the faces may vary.

cubic cleavage see CLEAVAGE.

cubic packing see PACKING.

cubic system see CRYSTAL SYSTEM.

cuesta *n*. a low, asymmetrical ridge with
a gentle slope on one side, conforming
with the dip of the underlying strata,
and a steep face on the other. Cuestas
are the result of erosion on gently

sloping sedimentary rocks. The top, which is formed on a resistant layer, remains as the cliff face is eroded back. Cuestas are found primarily on recently uplifted coastal plains and on geologically old folds or domes. See also HOGBACK.

Culm *n.* an informal and obsolete stratigraphical term used in southwest England to describe facies of the Dinantian (Carboniferous).

culmination *n.* the highest point of some geological structural feature, e.g. a dome or anticline. Several culminations may be found along a mountain range. The highest topographical point of an eroded-fold mountain may or may not be coincident with the culmination.

cummingtonite see AMPHIBOLES.

cumulate *n.* an IGNEOUS ROCK formed from an accumulation of CRYSTALS precipitated from a MAGMA. This generally takes place on the roof, walls and especially on the floor of a magma chamber where the cumulus crystals are primary precipitates and largely unmodified by subsequent crystallisation. Several types of cumulate have been defined depending on the nature of the intercumulus material:
(a) an *orthocumulate* is one in which cumulus crystals, frequently showing NORMAL ZONING, are poikilitically enclosed by new minerals nucleated within the intercumulus liquid. (See POIKILITIC, NUCLEATION).
(b) an *adcumulate* lacks intercumulus phases because of the growth of the cumulus crystals by the addition of material of the same composition, resulting in the gradual expulsion of the pore liquid. The cumulus crystals are not zoned.
(c) *mesocumulates* are intermediate in character between orthocumulates and adcumulates, with subordinate zoning

of the cumulus crystals and some intercumulus crystals present.
(d) in *heteradcumulates* the cumulus crystals are poikilitically enclosed in large, unzoned crystals of another mineral. See MAGMATIC DIFFERENTIATION.

cupriferous *adj.* copper-bearing.

cuprite *n.* a secondary copper mineral, Cu_2O, found in the oxidised zone of copper deposits associated with other copper minerals and LIMONITE. Its common occurrence as ruby-coloured isometric crystals has given it the name of 'ruby copper'.

cuprous *adj.* of or relating to copper.

Curie point *n.* the temperature above which the spontaneous magnetisation of a FERROMAGNETIC material disappears and the material behaves like a PARAMAGNETIC substance. Effectively, the material loses its magnetism above this temperature.

For FERROELECTRIC materials, the upper Curie point is the temperature above which they become depolarised. Some ferroelectric materials are depolarised below a lower Curie point, so show only a spontaneous polarisation between the lower and upper transition temperatures.

current bedding *n.* a former name for CROSS-BEDDING.

current ripple mark see RIPPLE MARK.

curvature correction *n.* an adjustment made to an observation or calculation to compensate for the Earth's curvature. In geodetic levelling, allowances are made for the joint effect of curvature and atmospheric refraction.

cusp *n.* one of a sequence of sharp, seaward-directed projections of beach material, separated at more or less regular intervals by shallow lunate troughs. The cusp-to-cusp distance depends upon the general shore contour, beach material and wave and tidal patterns; thus, a certain cusp-and-trough design is often characteristic of a particular stretch of beach.

cut-off *n*. **1.** CHUTE CUT-OFF.

2. NECK CUT-OFF.

3. a boundary in mapmaking and cross-sections, normal to the BEDDING, indicating the areal borders of a particular STRATIGRAPHICAL UNIT that is not defined by natural features.

4. an impermeable barrier placed within or beneath a dam to block or deter seepage.

cut-off grade *n*. the lowest grade of mineralised material, i.e. material of lowest assay value considered as ore in a particular deposit.

cut-off limit see ASSAY LIMIT.

cut-off spur see MEANDER.

cut-out *n*. a mass of siltstone, sandstone or shale occupying the space of an erosional channel cut into a coal seam.

cwm see CIRQUE.

cyanobacteria *n*. single-celled photosynthetic bacteria of the domain Eubacteria that live in marine or fresh water. The cells are often arranged in single or clustered filaments. They carry out oxygenic photosynthesis using photosystem II, unlike other photosynthetic bacteria. The cyanobacteria can live in very inhospitable environments, such as hot springs. They are resistant to ultraviolet radiation and can tolerate low oxygen and light levels. They are among the oldest-known organisms, and examples have been found that are approximately 3400 Ma old in structures similar to those of modern STROMATOLITES. The cyanobacteria were formerly called the *blue-green algae.*

cycle *n*. **1.** a recurring period of time, especially one in which certain events or phenomena repeat themselves.

2. a sequence of regular occurrences leading to the eventual restoration of

the original state of the process, e.g. a CYCLE OF SEDIMENTATION.

3. a series of events in which, upon completion, the last phase is unlike the initial phase.

cycle of denudation see CYCLE OF EROSION.

cycle of erosion *n*. also called **geomorphic** or **geographic cycle**, a theory of the evolution of landforms. First set forth by W. M. Davis, it assumes progressive development of land forms. The initial uplift that elevates the land is followed by dissection and denudation of the region by streams. Further erosion reduces it to an old-age surface, or a PENEPLAIN. The cycle could be interrupted at any time by further uplifts and returned to the 'youthful' stage (REJUVENATION). The theory also assumes that the history of a landscape proceeds directly from a known stage according to a predetermined framework, but it is now realised that this is tectonically unrealistic. For example, it does not take account of a dynamic equilibrium between the crustal elements of the Earth and the processes that act upon them in continuing isostatic readjustments.

cycle of sedimentation or **sedimentary cycle** *n*. a regularly repeated sequence of changes in the conditions within a sedimentary environment that is reflected in a repeated succession of beds. See also CYCLOTHEM, RHYTHMITE.

cyclic evolution *n*. the concept that in lineages of many life forms evolution took a strong and rapid initial course that was followed by a long phase of slow and moderate change. Overspecialised or inadaptive forms became extinct during a final brief episode.

cyclosilicate or **ring silicate** *n*. one of the group of SILICATES in which the SiO_4 tetrahedra form rings by linking together, each sharing 2 oxygen atoms.

Regardless of the number of such links in a ring (3, 4 or 6), all the rings have a 1:3 ratio of silicon to oxygen. BERYL, $Be_3Al_2Si_6O_{18}$, is a cyclosilicate.

cyclothem *n.* a term restricted to those CYCLES OF SEDIMENTATION found in COAL-bearing sequences in the Upper CARBONIFEROUS. Such cycles are produced by transgressive-regressive marine cyclicity, combined with compaction of previously deposited sediments and subsidence to provide accommodation space for the cycles. Each cyclothem consists of a sequence of marine LIMESTONE overlain by marine SHALE, lagoonal shale and SILTSTONE, deltaic SANDSTONE and COAL. The sequence may be formed when a DELTA builds out into a subsiding area, depositing first shales and siltstones then sandstones over the limestones of the open sea. Coal forms from the remains of the plants that become established on the top of the delta. Because most cyclothems are incomplete, a theoretical cyclothem, called the *ideal cyclothem*, is used to represent, within specified stratographic intervals and regions, the optimum succession of deposits during a full sedimentary cycle. Both experimental data and theoretical assumptions are used to construct such a cyclothem.

cylindrical fold *n.* a FOLD described geometrically as a surface generated by a line moving through space in a position parallel to itself The axis in such folds is perpendicular to the plane represented by the *girdle*, which is the projection of the plane containing all perpendiculars to the bedding. Compare CONICAL FOLD.

Cyprus-type deposit *n.* a copper deposit of the type found in the OPHIOLITE complex in Cyprus. Such deposits are believed to form within the oceanic crust at active spreading ridges. Water in sea-floor rocks, heated by magma, reacts with the sea-floor basalt and takes into solution many minerals present in the rock. As rising hot water is cooled and diluted by cold sea water, sulphides of elements such as copper, nickel, zinc and cadmium are precipitated in rock fractures. See also VOLCANIC-EXHALATIVE DEPOSITS, THERMAL SPRING.

cystoid *n.* a member of an extinct heterogeneous group of primitive ECHINODERMS belonging to the classes Diploporita and Rhombifera of the subphylum Blastozoa. They were sessile animals and were attached to the sea floor either directly or by a stalk. The THECA was ovoid and consisted of a number of calcite plates that were often arranged rather irregularly: five AMBULACRA on the upper surface radiated from the mouth to the *brachioles* (food-gathering arms). Range, Lower Cambrian to Upper Permian. See also BLASTOID.

D*d*

D see DARCY.

dacite *n*. a flow-banded, often dark-coloured IGNEOUS ROCK that is the extrusive equivalent of GRANODIORITE or TONALITE. Its principal minerals are sodic PLAGIOCLASE and QUARTZ, which occur as PHENOCRYSTS in a glassy to microcrystalline GROUNDMASS. MAFIC phenocrysts (BIOTITE and/or HORNBLENDE and/or PYROXENES) may also be present. Dacite is difficult to distinguish from RHYOLITE without microscopic examination.

Dalradian supergroup *n*. a succession of metasediments, with subordinate metavolcanic rocks and associated intrusions, that occurs within the CALEDONIAN orogenic belt of Scotland and Ireland. In Scotland they form a TERRANE lying to the northwest of the Highland Boundary Fault but not extending northwest of the Great Glen Fault. The four main stratigraphical divisions include, in upward succession, the *Grampian Group*, a sequence originally deposited as shallow marine sediments (the base of the succession is not seen but is probably more than 750 million years old); the *Appin Group*, a variety of shelf sediments with a minimum age of deposition of about 653 million years; the *Argyll Group*, the base of which is marked by the Port Askaig Tillite and the top by the 595-million-year-old Tayvallich Volcanic Formation; the *Southern Highland Group*, the topmost division, being composed largely of metamorphosed TURBIDITES but the top of the succession not clearly defined. Some authorities

regard the youngest Dalradian rocks to be of Ordovician age while others believe the entire succession to have been deposited in the Late PRECAMBRIAN. The base of the Dalradian was formerly placed at the base of the Appin Group, but most authorities now agree that there is no clear evidence of a major stratigraphical break at its junction with the underlying Grampian Group. The Dalradian was intensely deformed and regionally metamorphosed, mainly in Cambrian to Ordovician times. The intensity of metamorphism ranges from low grade in the vicinity of the Highland Boundary Fault to high grade in the northeast of Scotland, and the Dalradian is historically important as the area within which the BARROVIAN metamorphic zones were defined. See also BUCHAN-TYPE METAMORPHIC SEQUENCE, SNOWBALL EARTH.

Danian *n*. the bottom STAGE of the PALAEOCENE.

darcy *n*. a unit, symbol D, expressing the PERMEABILITY coefficient of a rock for the flow of, for example, water, gas or oil. A darcy is the permeability that permits a flow of one millilitre of fluid of one centipoise viscosity to flow in one second through a cross-sectional area of one square centimetre under a gradient of one atmosphere per second of flow-path. The customary unit is the *millidarcy* (md), equivalent to 0.001 darcy.

Darcy's law *n*. a law stating that the flow rate of a fluid through a porous material varies directly with the products of the pressure gradient causing the flow and the permeability (hydraulic conductivity) of the material. The formula that is derived from Darcy's law assumes laminar flow and negligible inertia. It is used in studies of water, gas and oil from underground structures. The formula can be stated as

flow rate = $KA(h_L/l)$

where K is the hydraulic conductivity, A is the cross-sectional area through which flow takes place, h_L is the head loss and h_L/l is the hydraulic gradient.

dark mineral n. any one of a group of rock-forming minerals generally rich in iron and/or magnesium, showing a dark colour in hand specimen and varying degrees of colour in thin section. Examples are BIOTITE, AUGITE, HORNBLENDE and OLIVINE.

dating methods n. various methods used to determine when a substance was formed. The particular method chosen depends on the nature of the material and the approximate magnitude of age.

The *radiometric dating technique* is based on the decay of radioactive isotopes that have very long HALF-LIVES. The age can be calculated from the measured proportions of parent and daughter isotopes using the AGE EQUATION: $t = (1/\lambda) \, l_n \, (1 + d/p)$, where t is the age of the sample, λ is the decay constant (which has a characteristic value for each radioactive isotope), l_n is log to the base e and d/p is the present ratio of daughter to parent isotope. The decay constant is related to the half-life T (the time taken for half the atoms to decay) thus:

$T = 0.693/\lambda.$

The reliability of the calculations depends on how accurately the half-life of the radioactive element is known,

whether any of the parent isotope and/or the daughter isotope has been removed from or added to the rock after formation, whether corrections can be made for any daughter isotope present when the rock was formed, and the accuracy of the assumption that the rock was formed during a relatively short time interval in comparison with its age.

Radiometric methods can be used to determine the age of crystallisation of an IGNEOUS ROCK from a MAGMA, the time of metamorphic recrystallisation (with the formation of a new mineral assemblage), the time of uplift and cooling of fold mountains and the age of SEDIMENTARY ROCKS in which new minerals (AUTHIGENIC minerals) formed during deposition.

The principal radioactive elements used in radiometric dating are listed in Fig. 24 overleaf. Either of the two isotopes of uranium may be used in the *uranium-lead method*, which is useful for rocks over 100 Ma old. It is less useful for younger rocks because the rate of production of ^{207}Pb is too low. Corrections may have to be made for the amount of non-radiogenic lead incorporated into the mineral at the time of formation. Uranium-lead determinations are usually made on ZIRCON and SPHENE, which are associated with many igneous and METAMORPHIC ROCKS.

The *lead-lead method*: because of the differences in half-life of the two uranium isotopes, the ratio ^{207}Pb/^{206}Pb is time-dependent and can be used for age determinations. Corrections for the proportion of radiogenic isotopes incorporated at the time of crystallisation can be made by means of an ISOCHRON DIAGRAM, using the relative proportion of non-radiogenic ^{204}Pb.

The *potassium-argon method* can be used to date rocks ranging in age from 4600 Ma to as recently as 30 000 years. This method, too, is less accurate for younger rocks because of the relatively small amount of daughter product that has formed in them. K–Ar methods are suitable for dating igneous and metamorphic minerals such as HORN-BLENDE, BIOTITE, MUSCOVITE and NEPHELINE, and for whole-rock specimens, particularly of fine-grained rocks – LAVAS and micaceous metamorphic rocks such as PHYLLITE and SLATE. Some sedimentary rocks containing GLAUCONITE can be dated by this method, since glauconite forms at the time of deposition. Problems arise in the K–Ar method because of the diffusive loss of ^{40}Ar at temperatures well below those of metamorphic recrystallisation (see BLOCKING TEMPERATURE).

Information about argon loss can be obtained using the ^{40}Ar/^{39}Ar method. Irradiation of the specimen converts a known proportion of ^{39}K to ^{39}Ar. The specimen is then heated in stages to drive off the argon, and the proportion of ^{39}Ar/^{40}Ar at each stage is measured, giving a series of dates. If there had been no loss of ^{40}Ar since the time of initial crystallisation, the dates calculated should be constant; however, if ^{40}Ar had been lost, an AGE SPECTRUM is obtained, from which it is possible to determine the time of any heating event as well as the time of initial crystallisation.

The *rubidium-strontium method* is rarely used for rocks younger than about 20 Ma unless they have a high content of ^{87}Rb. Minerals such as muscovite, biotite, all the potassium FELDSPARS and glauconite can be dated by this method. Whole-rock specimens of igneous and metamorphic rocks that are rich in micas and potassium feldspar can also be dated. Allowance can be made for the amount of ^{87}Sr incorporated at the time of crystallisation, using an ISOCHRON DIAGRAM plotting the ratios of ^{87}Sr/^{86}Sr against ^{87}Rb/^{86}Sr (^{86}Sr is non-radiogenic). Dates can be calculated from the slope of the isochron obtained either from minerals from an individual rock (a *mineral isochron*) or from several whole-rock specimens (a *whole-rock isochron*). The whole-rock isochron may give a different age from a mineral isochron from the same rock because ^{87}Sr can diffuse out of the minerals in which it formed but be retained within the rock (usually in calcium-bearing minerals).

The CARBON-14 *(radiocarbon) method* is often used to date materials and events that are relatively recent; it is particularly valuable for use on materials less

Parent isotope	Daughter isotope	Half-life
Carbon-14 ($^{14}_{6}$C)	Nitrogen-14 ($^{14}_{7}$N)	5730 years
Potassium-40 ($^{40}_{19}$K)	Argon-40 ($^{40}_{18}$Ar)	1250 Ma
Rubidium-87 ($^{87}_{37}$Rb)	Strontium-87 ($^{87}_{38}$Sr)	48 800 Ma
Samarium-147 ($^{147}_{63}$Sm)	Neodymium-143 ($^{143}_{60}$Nd)	108 000 Ma
Uranium-238 ($^{235}_{93}$U)	Lead–207 ($^{207}_{82}$Pb)	704 Ma
Uranium-235 ($^{238}_{92}$U)	Lead–206 ($^{206}_{82}$Pb)	4467 Ma

Fig. 24 **Dating methods**. Some of the isotopes used in radiometric age dating

than 50 000 years old, as this time span is too brief for the more slowly disintegrating elements to produce measurable amounts of daughter products. The radiocarbon method has been widely used in dating archaeological finds and provides the best dating for the Holocene and Pleistocene. The method is based on the assumption that the isotope ratio of carbon in the cells of living things is identical with that in air because of the balance between photosynthesis and respiration. Although several sources of error are introduced into this method, such as the inconstant supply of ^{14}C, both in time and latitude, it has proved to be enormously useful.

Fission track dating is based on the spontaneous FISSION of ^{238}U, which releases large amounts of energy. The damage this causes to the surrounding material can be revealed by etching with hydrofluoric acid. The half-life of the process is about 8×10^{15} years, and since the number of fission tracks depends only on the amount of uranium in the sample and the time elapsed since it formed, the age can be calculated. This method is useful for dating relatively recent volcanic material (up to 5 Ma), for example VOLCANIC GLASSES and minerals that contain uranium, such as APATITE, zircon, sphene and mica.

Varves, which are pairs of thin sedimentary layers deposited within a one-year period, have been used to correlate the lake deposits of many thousands of years, back into the Pleistocene and Holocene. Each varve pair consists of a fine winter layer and a coarse summer layer. The ages of deposits are determined by counting the pairs in a sample sediment core. See RHYTHMITES.

Dendrochronology, or *tree-ring dating*, is a technique of dating and interpreting past events, especially palaeoclimates, based on tree-ring analysis. A core, extending from bark to centre, is taken from a tree. After the rings are counted and their widths measured, the sequence of rings is correlated with 'deciphered' sequences from other cores. Tree-ring counts are used to calibrate dates that have been determined by carbon-14 dating. In so doing, it becomes necessary to modify the assumption of a constant ratio of radioactive to nonradioactive carbon at the time when the organic material to be dated ceased to exchange with the atmosphere.

Ice cores, from drill holes in the polar ice caps, contain annual layers resulting from snow fall. The fall-out of acid precipitation from major volcanic eruptions can sometimes be detected by measuring the acidity of the layers. Where it is possible to calibrate a core with a known recent eruption, the layers can be counted downwards and back in time to detect the signatures of major prehistoric eruptions during a period of some thousands of years.

datum *n.* (*pl.* **data**) **1.** a fixed or assumed point, line or surface used as a base or reference for the measurement of other values.
2. that part of a bed of rock on which structure contours are represented.

A *datum level* is any level surface (usually mean sea level) from which elevations are reckoned. The term *datum plane* or *reference plane* applies to some permanently established horizontal surface that serves as a reference for water depth, ground elevation and tidal data. In seismology, the datum plane is used to eliminate or minimise local topographic effects to which seismic velocity calculations are referred.

datum plane see DATUM.

daughter element *n.* an element formed from another through radioactive decay, e.g. radium, which is the daughter element of thorium.

dead arc see REMNANT ARC.

death assemblage see THANATOCOENOSIS.

debouch *v.* (*physical geography*) to emerge from a comparatively narrow valley out upon an open plain or area. A *debouchement* or *debouchure* is a mouth or outlet of a river or pass.

debris *n.* **1.** also called **rock waste**, any surface accumulation of material (rock fragments and soil) detached from rock masses by disintegration.
2. rock and soil material dumped, dropped or pushed by a glacier, or found on or within it.
3. interplanetary material, including cosmic dust, meteorites, comets and asteroids.

debris avalanche *n.* a slide of rock debris in narrow channels or tracks down a steep slope, often initiated by heavy rains. Debris avalanches are common in humid regions.

debris cone *n.* **1.** also called **alluvial cone**, a steep-sloped ALLUVIAL FAN, usually consisting of coarse material.
2. a mound of snow or ice on a glacier, topped with a debris layer thick enough to prevent ABLATION of the material underneath.

debris fall *n.* the nearly free fall of weathered mineral and rock material from a vertical or overhanging face. Debris falls are very common along the undercut banks of streams.

debris flow *n.* a moving mass of mud, soil and rock fragments of which more than half the particles are larger than sand size, the larger particles carried by a mud/water mixture. They occur

as terrestrial and marine flows. Such flows range in velocity from less than one metre per year to 160 kilometres per hour. Compare MUDFLOW.

debris line see BREAKER.

debris slide *n.* the rapid sliding or rolling of predominantly unconsolidated earth debris without backward rotation in the movement. Compare SLUMP.

declination *n.* the horizontal angle between true north and magnetic north, which varies according to geographical location. An instrument that measures magnetic declination is called a *declinometer*.

declinometer see DECLINATION.

décollement *n.* a detachment structure of strata associated with over-thrusting. It results in disharmonic folding, i.e. independent patterns of deformation in the rocks above and below the décollement. In attempting to explain the extreme disharmonic folding in the Jura Mountains, many proposed theories have assumed the presence of a décollement.

decomposition see WEATHERING.

deconvolution *n.* a technique designed to restore a wave configuration to the shape it is assumed to have taken prior to filtering. It is applied to seismic reflection and other data for the purpose of improving the visibility and clarity of seismic phenomena.

decrepitation *n.* the crackling noise caused by the heating of minerals.

decussate texture *n.* a metamorphic texture in which PRISMATIC or other inequidimensional minerals are randomly orientated. it develops under conditions where STRAIN is negligible and is frequently observed in contact AUREOLES. See also METAMORPHISM.

dedolomitisation *n.* a diagenetic process conducted by groundwater wherein dolomite, $CaMg(CO_3)_2$, is replaced by $CaCO_3$ and Mg^{++} is progressively leached out (*calcitic dedolomitisation*). The three criteria for calcitic dedolomitisation are groundwater with a high Ca:Mg ratio, permeability of the host rock, and time. See also DOLOMITISATION.

deep or **abyss** *n.* a general term for any ocean area of exceptional depth, usually implying depths exceeding 5500 metres (3000 fathoms).

deep-focus earthquake *n.* an EARTH-QUAKE the focus of which is at a depth of 300 to 700 kilometres. Compare SHALLOW-FOCUS EARTHQUAKE, INTERMEDI-ATE-FOCUS EARTHQUAKE.

deep-sea cones see SUBMARINE FANS.

deep sea drilling project (**DSDP**) *n.* a programme sponsored by the United States government with the support of the National Science Foundation to investigate the history of the Earth's ocean basins by means of drilling in deep water from the research vessel *Glomar Challenger*. It later became JOIDES and ODP (Ocean Drilling Program).

deep-seated *adj.* (of geological features and processes) originating or located at depths of 1 kilometre or more below the Earth's surface. See PLUTONIC.

deep-well disposal or **deep-well injec-tion** *n.* the disposal of liquid waste material by INJECTION into specially constructed wells that penetrate deep, porous and permeable formations; the formations hold mineralised ground-water and are vertically confined by more or less impermeable beds. Such a method is used for the disposal of many industry-related liquid wastes.

deflation *n.* the removal of material from a desert, beach or other land surface by the action of wind. In many cases, particularly in desert regions, deflation lowers the surface, thus producing *deflation basins*.

deflection of the vertical or **deviation** *n.* the angle at a given point on the Earth between the vertical, as defined by gravity, and the direction of the normal to the reference ellipsoid, where the normal is drawn through the given point.

deformation *n.* the alteration, such as faulting, folding, shearing, compres-sion and extension, of rock formations by tectonic forces. See STRAIN.

deformation ellipsoid see STRAIN ELLIP-SOID.

deformation fabric or **tectonic fabric** *n.* the spatial orientation of the compo-nents of a rock as imposed by external STRESS. During such stress conditions, minerals are translated or new miner-als develop in common orientation.

deformation twinning or **mechanical twinning** *n.* twinning produced by deformation and gliding. A portion of the crystal structure is sheared in such a way as to produce a mirror image of the original crystal. Deformation twins may be distributed throughout the crystal or can take over most or all of a crystal. See TWINNING.

degradation *n.* a gradational process that produces a general levelling of land by removal of material through erosive processes. Compare AGGRADA-TION, DENUDATION.

degrees of freedom *n.* the number of independent variables, e.g. tempera-ture, pressure and concentration, in the different phases that must be specified in order to describe a SYSTEM com-pletely. See also PHASE DIAGRAM.

delayed runoff *n.* water from precipita-tion that penetrates the ground and later discharges into streams via springs and seeps. The term is also used for water that is temporarily stored as snow or ice.

delay time *n.* in refraction shooting (see

see SEISMIC SHOOTING), the delay times are two factors, D_1 and D_2, the sum of which is equal to the intercept time; D_1 is that component associated with the shot end of the trajectory and D_2 with the detector end. Because the depths computed from INTERCEPT TIMES represent the sum of the respective depths below the shot and below the detector, most refraction results are intrinsically ambiguous. The concept of delay time is a technique for separating the depths at the two ends of the trajectory. The depths at each end of the trajectory can be determined if the intercept time can be separated into its component delay times. If the interface between the media is horizontal, each of the two delay times is half the intercept time:

$$D_1 = z \frac{\sqrt{v^2_1 - v^2_0}}{v_1 v_0} \qquad D_2 = z \frac{\sqrt{v^2_1 - v^2_0}}{v_1 v_0}$$

(where v_0 and v_1 are the respective velocities in the two media and z is the depth to the interface).

deliquescent *adj.* capable of liquefying through the absorption of water from the air.

delta *n.* a sedimentary deposit of sand, silt or clay, formed where a river enters a body of water (the sea or a lake). Sediment is deposited rapidly to form a wedge of material reaching above water level. As the sediment of a delta compacts, the delta becomes a low swampland, the *delta plain*, a nearly level surface that becomes the landward part of the delta. Lakes or marshes may be formed along the margin or within a part of the sea enclosed by the accumulation of deltaic deposits. Deltas are crossed by many distributaries that flow through to open water, depositing bars in channels and building up NATURAL LEVEES along the banks. The form and size of a delta depend upon the preformation coastline, the hydrologic regime of the river, and the local waves and tides. Most are arcuate or shaped like the Greek letter *delta* (Δ). Some, e.g. that of the Mississippi, exhibit a *digitate* or *'bird-foot'* plan that is formed by numerous distributaries flanked by low levees.

Three sets of beds are often observable in a delta. *Bottomset beds* consist of finer sediments deposited as horizontal or gently inclined layers on the floor of the area where the delta is forming. *Foreset beds* have a steeper dip and consist of coarser sediments; these represent the advancing head of the delta and the greater part of its total bulk. *Topset beds*, lying above the foreset beds, are actually an extension of the alluvial valley of which the delta is the terminal section. Distinctions between the three beds are not always clear. They are most obvious in deltas formed by small streams flowing into protected or confined areas.

delthyrium *n.* a triangular opening in the pedicle VALVE of BRACHIOPODS through which the PEDICLE (stalk) emerges. In some forms it is partly closed by two *deltidial plates*, so that the pedicle emerges through a round opening called the *pedicle foramen* (Fig. 9). In brachiopods that have no pedicle the delthyrium may be completely closed by the deltidial plates or a single *deltidium*. See also NOTOTHYRIUM.

deltidial plates see DELTHYRIUM.

dendrite *n.* a branching pattern of deposited materials, often on LIMESTONE surfaces, that resembles a tree or fern. Dendrites are often mistaken for fossil flora but are really mineral incrustations produced on or in a rock by such minerals as pyrolusite, MnO_2, or pyrite. MOSS AGATE is an example of a mineral with a dendrite inclusion. Native copper can also form dendrites. See PSEUDOFOSSIL.

dendritic *adj.* in tree-like form.

dendritic drainage pattern see DRAINAGE PATTERN.

dendrochronology *n.* the study and correlation of tree rings for the purpose of dating events in the recent past. See DATING METHODS.

dendroid 1. *n.* a member of the Dendroidea, an order of GRAPTOLITES. **2.** *adj.* having a branching habit. See GRAPTOLITE, COLONIAL CORAL.

density *n.* **1.** the mass or quantity of a given substance per unit volume of that substance, usually expressed in kilograms per cubic metre (kg m^{-1}). **2.** the quantity of any entity per unit volume or per unit area, e.g. faunal population in a region.

density currents see TURBIDITES.

density stratification *n.* the STRATIFICATION of water in a lake as a result of density differences. It is usually caused by temperature changes but may also be caused by differences in the quantity of dissolved material, e.g. where a less dense freshwater surface layer overlies denser salt water. See also THERMAL STRATIFICATION.

dentate *adj.* having teeth or toothlike projections.

dentition *n.* the arrangement of teeth and sockets along the HINGE PLATE of a BIVALVE. It is used in the classification of bivalves. Some types of dentition are shown in Fig. 7(b).

denudation *n.* the exposure of deeper rock structures by the EROSION of the land surface. Denudation and erosion are often used synonymously, but an eroded landscape is not necessarily a denuded one. See for example KARST, DEGRADATION.

deoxidation sphere see REDUCTION SPOT.

deposit 1. *n.* any sort of earth material that has been accumulated through the action of wind, water, ice or other agents. Compare SEDIMENT. **2.** *n.* mineral deposit. **3.** *v.* to lay or put down, or to precipitate.

depositional remanent magnetisation see NATURAL REMANENT MAGNETISATION (c).

depth of compensation see PRATT HYPOTHESIS.

deranged drainage pattern see DRAINAGE PATTERN.

desert *n.* a region where precipitation is less than potential evaporation and which contains no permanent streams. Although they may be hot, temperate or cold, deserts are always characterised by a scarcity of vegetation. *Hot deserts*, where summer daytime temperatures exceed 35ºC, found mainly in the southwestern United States, Africa and Saudi Arabia, consist of mountainous areas, areas of exposed rocks and sandy areas. Sand-covered areas rarely make up more than 30 per cent of the desert surface. Wind-produced formations include sand DUNES, LOESS, DESERT PAVEMENT and VENTIFACTS. Other hot-desert features include EVAPORITES, SALINAS and SALT LAKES. Hot deserts occur in areas removed or cut off from moisture. They can occur in dry-air regions and on coasts along which there are cold-water currents and onshore winds. *Cold deserts*, found in Greenland and Antarctica, have moisture-poor atmospheres because of their very low temperatures. These deserts are characterised by PATTERNED GROUND, SOLIFLUCTION deposits and BLOCK FIELDS.

desert armour see DESERT PAVEMENT.

desert crust *n.* **1.** a hard layer containing binding matter (e.g. calcium carbonate, gypsum), exposed at the surface in a desert area. See DURICRUST. **2.** DESERT PAVEMENT.

desert pavement, desert armour or **desert mosaic** *n.* a residual layer of wind-polished, closely concentrated rock fragments covering a desert surface because of the removal by wind (DEFLATION) of loose silt and sand. Desert pavements are variously known as *serir* (Libya), REG (Algeria) and *gibber* (Australia). See DESERT CRUST. See also LAG GRAVEL, BOULDER PAVEMENT.

desert polish *n.* a slick, polished surface on rocks of desert regions, produced by wind and sand abrasion. See DESERT VARNISH.

desert rose *n.* any of various flowerlike groups of crystals, e.g. *barite roses*, developed in loose sand and sagebrush areas of deserts.

desert varnish *n.* a dark, enamel-like film of oxides of iron and manganese, formed on the surface of rocks in stony desert regions. It is thought that desert varnish is a deposit formed by the evaporation of mineral solutions that are drawn to the surface of the rock. It can be difficult to distinguish between desert varnish and DESERT POLISH.

desiccation breccia or **mud breccia** *n.* a BRECCIA composed of fragmented mud-cracked polygons together with other sediments.

desiccation cracks *n.* polygonal cracks in fine sediment produced by water loss or drying, especially a MUD CRACK. See NONSORTED POLYGON.

desilication *n.* the removal of silica from a soil or rock by the chemical dissociation of silicates. In warm climates, the silica is commonly removed from soils when large quantities of rainwater percolate through the soil. The resulting soil is an *ultisol*.

Desmoinesian *n.* a series of the Upper Middle Pennsylvanian of North America.

destructive plate boundary or **conver-gent plate boundary** *n.* a boundary between two converging lithospheric plates at which one is destroyed in a SUBDUCTION ZONE. See PLATE TECTONICS.

detached core *n.* the inner part of a FOLD that has been detached or pinched from the parent structure by some process, usually severe compression and folding. The term 'core' is used even though a fold may be composed entirely of one kind of rock.

detrital *adj.* (of mineral grains) transported and deposited as sediments that were derived from pre-existing rocks either within or outside the area of deposition. Compare CLASTIC, ALLOGENIC.

detrital ratio see CLASTIC RATIO.

detrital remanent magnetism see NATURAL REMANENT MAGNETISATION.

detritus *n.* **1.** any loose rock or mineral material resulting from mechanical processes, such as disintegration or abrasion, and removed from its place of origin. Compare DEBRIS. **2.** any fine debris of organic origin, e.g. plant detritus in coal.

deuteric or **epimagmatic** *adj.* (of changes in igneous rock) produced by reactions between primary magmatic minerals and water-saturated solutions that separate from the same magma body at a late phase of its cooling profile. See also AUTOMETASOMATISM.

Devensian see CENOZOIC ERA, Fig. 13.

deviation see DEFLECTION OF THE VERTICAL.

devitrification *n.* the conversion of glass to a crystalline state. Since all VOLCANIC GLASSES are unstable under atmospheric conditions, they *devitrify* in time to a stable crystalline arrangement; APHANITIC and spherulitic (see SPHERULITE) textures are a result of devitrification. Because of their inherent metastability, most natural glasses are not older than the CENOZOIC. See also GLASS, GLASSY.

devolatilisation *n.* the reduction of

volatile components during COALIFICA-
TION, which results in a proportional
increase in carbon content. The higher
the level of devolatilisation, the higher
the RANK of coal.

Devonian *n.* a time interval of the
Palaeozoic Era during which rocks of
the Devonian SYSTEM were formed (408
to 360 Ma ago). The period was named
after the TYPE LOCALITY of Devon, in the
southwest of England. Because this
area lacked sufficient fossils for correlation, the stages of the Devonian were
established in the fossiliferous marine
deposits of the Ardennes, in Belgium.
This system shows both marine and
continental FACIES. In Europe, the
continental facies is called OLD RED
SANDSTONE. Rocks of Devonian age are
found in all continents and are sources
of iron ore, ores of tin, zinc and copper,
and of oil and evaporites.

Certain areas of Europe have become
the world standard for the many
subdivisions of the Devonian. Ammonoid cephalopods, conodonts, brachiopods, corals and, in the Lower Devonian,
graptolites are used for the stratigraphical definition of marine deposits; in
non-marine deposits, freshwater fish
and spores are used. By international
agreement, the zone of the graptolite
Monograptus uniformis is taken as the
base of the Devonian and the upper
boundary of the zone of the ammonoid
cephalopod *Wocklumeria* as the top.

Particular types of Devonian sediments characterise various areas of
continental and marine environment,
and give distinctive rock types. Marine
deposits – mudstones and limestones,
including reef limestones – are found
in the southwest of England (parts of
Devon and Cornwall), while the
continental Old Red Sandstone deposits
(conglomerates, sandstones and

mudstones) occur farther north. Britain
lay on the southern margin of the
continent of LAURASIA at this time.

Invertebrates of the Devonian period
are essentially of types established by
the time of the Ordovician: brachiopods, corals, stromatoporoids, crinoids,
trilobites and gastropods. Molluscan
groups are well represented. The origin
of the ammonoids during the Devonian
is of great significance. Conodonts
show great diversity during this
period. Giant eurypterids (arthropods)
are found in the freshwater sediments
of the Old Red Sandstone facies. The
first spiders, millipedes and insects
appear in the Devonian.

Fish dominate among Devonian
vertebrates, with freshwater and marine
varieties proliferating. The fishes
include several kinds of armoured fish,
those with jaws, including the placo-
derms and the acanthodians, and those
without jaws, including the ostracoderms. Sharks appear in the Middle
Devonian, as do the first ray-finned
fish. Lobe-fins, primitive air-breathing
fishes, are common during the period;
lungfish make up one group of lobe-
fins, and crossopterygians the other.
Crossopterygians may be the link
between fishes and lower tetrapods,
primitive amphibians. All modern
algae are represented in the Devonian,
including marine brown, red and green
seaweeds, and freshwater algae. Plants
continue their colonisation of the land,
begun in the Silurian. Bryophytes
appear in the Lower Devonian, and
terrestrial fungi appear. Vascular land
plants found in the early Devonian
become diversified by the mid- to late
parts of the period. Psilopsids (simple
rootless plants) and horsetail rushes
(herbaceous sphenopsids) diversify, as
do SCALE TREES (lycopsids) and arbores-

cent sphemopsids. Ferns and 'tree ferns' are evolved. Primitive GYMNO-SPERMS (e.g. *Callixylon*) are known in the Devonian and are plentiful late in the period.

Evidence from the Devonian supports the theory that the present continents were formerly united in some way. In the Northern Hemisphere, the linking of North America, Greenland and Europe allows the concept of a single Old Red Sandstone landmass (LAURASIA). The union of the southern continents to the continent of GONDWANA at this time is necessary in explaining the later distribution of glacial deposits. The CALEDONIAN OROGENY, which came to a climax in the Middle Devonian, was experienced in northwestern Europe. The Antler and ACADIAN orogenies of North America occurred during the Devonian. All the disturbances were accompanied by VOLCANISM. The wide distribution of evaporite basins in the Northern Hemisphere, coals in Arctic Canada and carbonate reefs suggests the northern continent (LAURASIA) was in low latitudes during the Devonian. Growth lines of Devonian corals indicate that the Devonian year was 400 days and the lunar cycle about 30 days. Palaeomagnetic evidence shows that an equator passed from California to Labrador and from Scotland to the Black Sea. See also CONTINENTAL DRIFT, PALAEOZOIC.

dextral *adj*. inclined, pertaining to or spiralled to the right, e.g. the clockwise direction of coiling of GASTROPOD shells.

dextral fault see FAULT (d).

diabase *n*. a synonym for DOLERITE (microgabbro). Pre-Tertiary dolerites or dolerites that have been saussuritised have been referred to as diabase and the term is still used occasionally in this sense.

diachronous *adj*. relating to FORMATIONS where the boundaries cut across time planes or BIOZONES. *Diachronism* occurs when a geological event migrates systematically, so that the rock produced by the event is not everywhere the same age, although the lithologic character remains the same. For example, diachronism will be caused by the lateral shifting of FACIES boundaries during a marine TRANSGRESSION or REGRESSION, or the building out of a DELTA. Most boundaries between formations are likely to be diachronous to some extent.

diadochy *n*. the replacement of one atom or ion by another in a CRYSTAL LATTICE; see SOLID SOLUTION.

diagenesis *n*. the sum of all changes, physical, chemical and biological, to which a sediment is subjected after deposition but excluding WEATHERING or METAMORPHISM. Some diagenetic changes (HALMYROLYSIS) occur at the water-sediment interface; however, most diagenetic activity occurs after burial. Unstable forms of minerals are converted to stable forms and new (AUTHIGENIC) minerals deposited, often as concretions. Examples of diagenetic activity include the modification of CLAY MINERALS to ILLITE or CHLORITE, the recrystallisation of ARAGONITE to CALCITE, and the formation of GLAUCONITE.

COMPACTION during burial affects fine-grained sediments much more than coarse-gained sediments. Grains are pushed closer together and bent or broken, and PRESSURE-SOLUTION and redeposition occur at grain contacts, forming sutured contacts.

The pore fluids expelled during compaction have an important role in the migration of ions and organic molecules within sediments. Mineral cements that are deposited in the pore spaces may be introduced by the fluid circulating through the sediment or redistributed from materials within the deposit. Reactions that occur between the fluid and the minerals with which it is in contact, and the stability of diagenetic minerals, are affected by the E_H and pH of the circulating water. See EH, PH. Common cementing minerals include QUARTZ, calcite, LIMONITE, HEMATITE and DOLOMITE. See also LITHIFICATION, DOLOMITISATION.

diamagnetic *adj.* having a magnetic permeability slightly less than unity and a small negative magnetic susceptibility, typically of the order of -10^{-5}. The permeability of bismuth, the most diamagnetic substance, is equal to 0.99998. In a magnetic field, the induced magnetisation of a diamagnetic substance is in a direction opposite that of the applied field. Compare PARAMAGNETIC, FERROMAGNETIC.

diamond *n.* a mineral of the isometric system, a crystalline form of carbon, dimorphous with graphite. Diamond is the hardest mineral known and has perfect cleavage. Excellent crystals occur in the KIMBERLITE of South Africa. Colourless and coloured varieties (yellow, brown, grey, green and black) are valued as gemstones. See also DIATREME, INDUSTRIAL DIAMOND.

diamond anvil cell see Appendix 4.

diamond bit *n.* a diamond-studded rotary-drilling bit used for coring and drilling in very hard rock. This variety of ROTARY DRILLING, called *diamond drilling*, is a well-known method of prospecting for mineral deposits. See also INDUSTRIAL DIAMOND.

diamond drilling see DIAMOND BIT.

diapir *n.* a vertical columnar plug of less dense rock or magma that is emplaced by a process, called *diapirism*, that may be caused by buoyancy (less dense

material rising through denser rocks), tectonic forces or by a combination of the two. For example, the deformation of a succession of sedimentary strata containing highly incompetent material such as rock salt often results in the upward intrusion of the salt through the overlying layers. Diapirism forms both igneous and non-igneous intrusions (e.g. salt plugs) and is accompanied by doming. It may result in piercement of the overlying strata.

diapirism see DIAPIR.

diaplectic glass *n.* a form of natural glass formed by a shock pressure in excess of about 35 Gpa by solid-state transformation without melting during major meteorite impacts. Glass formed in this way from quartz has a slightly lower refractive index than synthetic quartz glass. See IMPACTITE, MASKELYNITE.

diaspore *n.* an orthorhombic mineral, AlO(OH), associated with BOEHMITE and GIBBSITE as a constituent of some BAUXITES. It is usually found in foliated or stalactitic pale-pink, grey or greenish aggregates.

diastem *n.* a minor break in a sedimentary sequence caused by non-deposition or slight erosion. See also UNCONFORMITY.

diastrophism *n.* all movement of the Earth's crust resulting from tectonic processes. This includes the formation of continents, ocean basins, plateaus and mountain ranges. *Diastrophic processes* are usually classified as *orogenic* (mountain-building with deformation) and *epeirogenic* (regional uplift without significant deformation).

diatom *n.* any of numerous microscopic unicellular marine or freshwater algae with siliceous cell walls. They resemble delicate glass structures and are a major food source for copepods, larvae, etc. *Diatom ooze* is an ocean-bottom deposit made up of the shells of those algae. See PELAGIC DEPOSITS.

diatomaceous earth or **kieselguhr** *n.* an ultra-fine-gained siliceous earth composed mainly of DIATOM cell walls. It is used as an abrasive and in filtration.

diatomite *n.* consolidated DIATOMACEOUS EARTH. See CONSOLIDATION.

diatom ooze see DIATOM, PELAGIC DEPOSITS.

diatreme *n.* a pipe-like VOLCANIC CONDUIT formed deep within the Earth's crust by the explosive energy of magmatic gases. Diatremes are filled with fragmental debris containing both volatile-rich igneous material and XENOLITHS of the wall-rock, including large pieces from stratigraphically higher SEDIMENTARY ROCKS that appear to have subsided into the conduit.

The diamond-bearing *kimberlite pipes* of South Africa are well-known diatremes. It is thought that diamonds are formed at depths of at least 200 kilometres and transported very rapidly to the surface in a fluidised eruption (see FLUIDISATION), so that resorbtion or inversion to a lower pressure form of carbon is prevented. See also MAAR, KIMBERLITE.

dichroism see PLEOCHROISM.

dichroiscope *n.* an instrument used to determine the PLEOCHROISM of a mineral.

dichroite see CORDIERITE.

dickite *n.* a crystallised CLAY MINERAL of the kaolinite-serpentinite group, with the same composition, $Al_2Si_2O_5(OH)_4$, as KAOLINITE, differing only in details of atomic structure and certain properties.

diductor muscles *n.* the paired muscles that open the VALVES of a BRACHIOPOD shell.

differential erosion see EROSION.

differential thermal analysis (DTA) *n.* a technique used to observe change of state and to determine the temperature at which a change occurs. Two samples with an identical heat capacity are heated and the temperature differences between them are noted. The difference between them becomes especially conspicuous when one of the samples changes state. The temperature at which this occurs is called the *transition temperature*. Differential thermal analysis is a technique used in mineral analysis, particularly for CLAY MINERALS.

differential weathering *n.* WEATHERING that occurs at varying rates because of the intrinsic differences in rock composition and resistance or differences in the intensity and degree of erosion. Differential weathering helps to develop and modify many upstanding forms, such as columns, pillars and rock pedestals.

differentiation *n.* **1.** MAGMATIC DIFFERENTIATION.

2. METAMORPHIC DIFFERENTIATION.

diffraction *n.* **1.** the phenomenon exhibited by WAVEFRONTS that, upon passing the edge of an opaque body, are modulated, thus effecting a redistribution of energy within the front. In light waves this is detectable by the presence of dark and light bands at the edge of a shadow.
2. the bending of waves, especially sound and electromagnetic waves, around obstacles in their path.
3. the generation and transmission of seismic wave energy in accordance with HUYGEN'S PRINCIPLE.

diffraction pattern *n.* the interference pattern formed by the DIFFRACTION of X-rays by planes of atoms within a CRYSTAL. Each MINERAL has a characteristic diffraction pattern, enabling it to be analysed and identified using a technique called X-RAY DIFFRACTION (XRD). See also BRAGG EQUATION and Appendix 4.

digitation *n*. a subsidiary recumbent ANTICLINE attached to a larger recumbent FOLD, somewhat resembling fingers extending from a hand.

dilatancy *n*. an inelastic increase in the bulk volume of a rock under STRESS, owing to the opening and extension of small cracks. Dilatancy begins when the stress reaches about half the breaking strength of the rock. It has been demonstrated by laboratory experiments that measurable physical changes accompany dilatancy; such effects include changes in the electrical resistivity of the rock and in the velocity of elastic waves travelling through the rock. These findings have been verified in different locations. The assumption of the growth of strain-associated microcracks in crustal rocks before an earthquake, thus causing dilatancy, serves as the basic model of precursory seismic effects. See also EARTHQUAKE PREDICTION.

dilatation or **dilation** *n*. DEFORMATION by change in volume but not configuration.

dilatational wave see SEISMIC WAVES, Fig 82, P wave.

dilatation vein *n*. a mineral deposit occurring in VEIN space created by the bulging of a crack rather than in veins formed by the replacement of wall rocks.

dilation see DILATATION.

dimension-preferred orientation *n*. the preferred orientation of flattened or elongated CRYSTAL AXES, because of crystal GLIDING, magmatic flow or dynamic recrystallisation. Compare LATTICE-PREFERRED ORIENTATION.

dimension stone *n*. stone that is cut or quarried in accordance with specific dimensions.

dimorphism *n*. **1.** the crystallisation of an element or compound into two distinct CRYSTAL FORMS, e.g. diamond and graphite, pyrite and marcasite.

2. the characteristic of having two distinct forms in the same species, e.g. male and female.

Dinantian *n*. the Lower CARBONIFEROUS subsystem in Britain and western Europe.

Dinarides *n*. the southern part of the east-west OROGENIC BELT that includes the European Alps and the Himalayas.

dinoflagellates *n*. biflagellated, single-celled micro-organisms, chiefly marine. They exhibit features of both plants and animals but are thought to be plants and are placed in the division Pyrrophyta. Range, Silurian to present.

dinosaur *n*. an informal term applied to reptiles belonging to the orders Saurischia and Ornithischia of the subclass Archosauria. Dinosaurs dominated or were prominent among Mesozoic vertebrate life forms. They are divided into the two orders on the basis of pelvic structure: the Saurischia, with the normal reptile pelvis structure, and the Ornithiscia, in which the pelvis resembles that of a bird. They ranged in size from 30 centimetres to 26 metres. They were carnivorous, herbivorous, bipedal or quadrupedal. Most were terrestrial, but there were also aquatic and semi-aquatic representatives. Range, Triassic to Cretaceous. The cause of the EXTINCTION of the dinosaurs (together with many other species) at the end of the CRETACEOUS has been the subject of much debate.

diopside *n*. a monoclinic INOSILICATE of the PYROXENE group, $CaMg(SiO_3)_2$. Its colour may be pale-green, blue, whitish or brown. *Violane* is a purple manganese-bearing variety, and dark-green varieties contain chromium. Diopside occurs in contact metamorphic rocks, especially in certain dolomitic marbles and in some basic lava.

diorite *n*. a dark-coloured, coarse-grained plutonic rock composed essentially of plagioclase feldspar (oligoclase-andesine), hornblende, pyroxene and little or no quartz; it is the intrusive equivalent of ANDESITE. With an increase in the alkali feldspar content, diorite grades into MONZONITE.

dioxide *n*. an oxide containing two atoms of oxygen per molecule of compound, e.g. MnO_2.

dip *n*. **1.** the angle in degrees between a horizontal plane and an inclined earth feature such as a rock stratum, FAULT or DYKE. Measurement is made perpendicular to the strike and in the vertical plane; it thus determines the maximum angle of inclination. See also TRUE DIP, APPARENT DIP. See FAULT, Fig. 30(a).
2. the angle between a reflecting or refracting seismic wave front and the horizontal, or the angle between an interface associated with a particular seismic event and the horizontal.
3. MAGNETIC INCLINATION.

dip calculation *n*. dip is computed from the difference in reflection times of a seismic pulse that travels along two different paths after impinging on a reflecting surface, e.g. a reflecting interface. In such a technique, a series of GEOPHONES is located equidistant from each other on either side of a SHOT POINT. A shot is fired, and a reflector, at some point *R*, will transmit the pulse back to the Geophones along different routes. The difference in reflection times between the most easterly and most westerly Geophones is a measure of the amount of dip of the reflector, i.e. the reflecting surface. See also NORMAL MOVEOUT.

dip fault *n*. a FAULT that strikes essentially parallel to the DIP of the adjacent strata, i.e. its STRIKE is perpendicular to the strike of the adjacent strata. Compare STRIKE FAULT, OBLIQUE FAULT.

dipmeter *n*. a finely defined resistivity log (see WELL LOGGING). The dipmeter sonde consists of four microresistivity pads mounted at 90° angles from each other; these are applied to the walls of a borehole. The sequence of readings at each pad will deviate slightly from the others, depending on the dip of the strata. Such measurements indicate the inclination and direction of the borehole as well as the DIP and STRIKE of the strata. This information is valuable in the plotting of oilfield structures.

dip separation *n*. the distance between formerly adjacent beds on either side of a fault surface, measured directly along the dip of the fault. Compare DIP SLIP, STRIKE SEPARATION.

dip shift *n*. the relative displacement of rock units parallel to the fault dip but beyond the fault zone itself. Compare DIP SLIP.

dip shooting *n*. a system of seismic surveying in which the chief concern is determination of the DIP, or dip and depth, of a geological marker that will serve as a reflecting interface. Although one can correlate seismic reflections along all shooting profiles and map the depths to the markers, if the reflection from a particular formation is distinctive, it is recognisable over a fairly wide area. See also DIP CALCULATION.

dip slip *n*. the component of the NET SLIP that is measured parallel to the dip of the fault plane. Compare STRIKE SLIP, DIP SHIFT.

dip-slip fault *n*. a fault along which rock displacement has been predominantly parallel to the DIP. Compare STRIKE-SLIP FAULT. See also FAULT.

dip slope *n*. the slope of the land surface that more or less conforms with the inclination of the underlying rocks. See CUESTA.

directional drilling or **slant drilling** *n*. the drilling of a well during which azimuths and deviations from the vertical are controlled. One of its purposes is to establish multiple wells from a single location, e.g. an offshore platform.

directional log see WELL LOGGING.

direct runoff see RUNOFF.

dirt cone *n*. a glacial DEBRIS CONE.

disappearing stream *n*. a surface stream that disappears underground into a SINK.

discharge *n*. **1.** the amount of water flowing down a stream channel; it is measured in cubic metres per second, and data for such measurement is always taken at a specified location along a given stream.
2. sediment discharge, i.e. the amount of sediment carried by a stream, measured in kg s^{-1}. Determination of the BED LOAD discharge is difficult, however, so estimates of total discharge may be inaccurate.

disconformity see UNCONFORMITY.

discontinuity *n*. **1.** any interruption in sedimentation that may represent a time interval in which sedimentation has ceased and/or erosion has occurred.
2. (*structural geology*) abrupt changes in rock type caused by tectonic activity.
3. see SEISMIC DISCONTINUITY.

discontinuous deformation *n*. DEFORMATION by FRACTURE rather than FLOW, as distinct from CONTINUOUS DEFORMATION.

discontinuous reaction series see REACTION SERIES.

discordant *adj*. **1.** (of intrusive igneous bodies such as dykes and batholiths) with margins that cut through the bedding or foliation of the country rock. Compare CONCORDANT.
2. (of strata) structurally UNCONFORMABLE, i.e. in which parallelism of bedding or structure is absent.
3. (of radiometric ages determined by more than one method for the same specimen) disagreeing beyond experimental error. Compare CONCORDANT.
4. (of topographical features) having dissimilar elevations. Compare ACCORDANT.

discordia *n*. a linear array of points on a CONCORDIA diagram, generated by plotting the results of measurements of isotopic ratios of lead and uranium in samples of unknown age. The discordance of this line from concordia can reflect lead loss from the system as well as changes in the ratios caused by radiometric decay of the uranium isotopes. The upper intercept of the discordia line with concordia provides an estimate of the age of the sample.

disharmonic fold *n*. a FOLD in one bed the geometry of which is unrelated to, and out of harmony with, the geometry of folds in other beds.

disintegration *n*. **1.** MECHANICAL WEATHERING.
2. RADIOACTIVE DECAY.

dislocation *n*. **1.** a lattice imperfection in a CRYSTAL structure. An *edge dislocation* occurs when a plane of atoms or ions in a crystal structure is incomplete, terminating in a line; the planes on either side are distorted along the line or edge as they take up the space vacated by the incomplete plane. A *screw dislocation* is a structural defect in which slippage has taken place in a lattice plane along an axis of rotation so that part of the lattice is distorted to produce a step in the crystal structure. This type of dislocation is important in crystal growth, as the resulting ledge provides a favoured location for the addition of new ions or atoms to the crystal surface during growth. See LATTICE DEFECT.
2. DISPLACEMENT.

dislocation metamorphism see META-
MORPHISM.

dismicrite *n*. a term used in the FOLK
CLASSIFICATION of LIMESTONE for an
orthochemical limestone mainly
composed of MICRITE with patches or
lenses of sparry calcite (SPARITE). The
texture is called *fenestral* and is thought
to be caused by disturbance of the
calcareous mud by algal activity,
animal burrowing or escaping gas
bubbles. This is sometimes known as
'birdseye' limestone.

dispersed elements *n*. TRACE ELEMENTS
that occur in different minerals in a
rock. Because their relative concentrations can differ significantly, the
dispersed element composition of a
MAGMA can impose constraints on the
nature of its source rock.

dispersion *n*. **I.** variation of the refractive index of a substance with the
wavelength of light. It is because of
this property that a prism is able to
generate a spectrum.
2. the sorting of SEISMIC WAVES into trains
of waves of different frequency because
wave velocity varies with frequency. In
most cases, the dispersion of body waves
is small, but surface waves typically
show considerable dispersion because
of the variation in elastic modulus and
DENSITY with depth in the Earth. Waves
that penetrate to greater depths (those
with longer wavelengths) travel faster
(see ELASTIC CONSTANT).

disphotic zone *n*. that level in bodies of
water at which sunlight is faint and
little photosynthesis is possible.
Compare APHOTIC, EUPHOTIC ZONE.

displacement, dislocation or **slip** *n*. the
relative movement of two formerly
contiguous points on either side of a
fault as measured on the fault surface.
Total displacement is NET SLIP. See also
DIP SLIP, STRIKE SLIP.

dissection *n*. the cutting of ravines,
gullies or valleys by EROSION, especially
by streams.

disseminated deposits *n*. MINERAL
DEPOSITS formed by the infilling of pores
and small fissures in a rock by the
action of HYDROTHERMAL fluids. They are
usually associated with ACID to INTERME-
DIATE igneous INTRUSIONS. Copper,
molybdenum and tin are typically
found in these types of deposit.

dissepiments *n*. **I.** small curving plates
lying between the septa (dividing
walls) near the margin of a CORALLITE.
They may form a broad zone called a
dissepimentarium. See RUGOSE CORAL, Fig.
77.
2. transverse bars linking the STIPES of a
dendroid GRAPTOLITE.

dissociation *n*. the reversible decomposition of a complex substance into
simpler components, as produced by
variation in physical conditions, e.g.
the decomposition of water into hydrogen and oxygen after the water is
heated to a high temperature and the
recombination of the liberated elements to form water again when the
temperature is lowered. The temperature at which a compound breaks up in
such a reversible manner is the *dissociation point*. A temperature point at which
a given dissociation is presumed to
occur is called the *dissociation temperature*; it is, in fact, a range of temperatures because of variations in pressure
or composition and, simply indicated,
is the temperature at which the rate of
a given dissociation becomes significant under given conditions.

**dissociation point, dissociation tem-
perature** see DISSOCIATION.

dissolved load, dissolved solids or
solution load *n*. that fraction of the
total STREAM LOAD that is carried in
solution. Compare BED LOAD.

dissolved solids see DISSOLVED LOAD.

distal *adj*. **1.** (of an ore deposit) formed at a significant distance from the source of its constituents.
2. (of a sedimentary clastic deposit) laid down far from its source location. See also PROXIMAL.

distributary *n*. a system or set of independent channels frequently formed by rivers or streams, such as where they enter their deltas. The pattern is well defined by the Mississippi. Compare TRIBUTARY.

distributive province *n*. the environment that encompasses all rocks contributing to the formation of a contemporaneous sedimentary deposit and those agents or factors responsible for their distribution. Compare PROVENANCE.

disturbance *n*. a lesser or minor OROGENY. Compare EVENT.

divariant *adj*. (of a chemical system) possessing two DEGREES OF FREEDOM. See also PHASE DIAGRAM.

divergence *n*. **1.** (*meteorology*) a net flow of air outwards from a given volume of the atmosphere, i.e. a decrease of mass within that given volume. Mathematically, it is the expansion of the atmospheric velocity vector field. High-pressure areas are generally *divergent* near the Earth's surface. Divergence results in a decrease of cyclonic vorticity. Compare CONVERGENCE.
2. the separation of ocean currents by horizontal flow, usually the result of upwelling. Compare CONVERGENCE.
3. (*palaeontology*) adaptive radiation.

divergent plate boundary see ACCRETING PLATE BOUNDARY.

diverted stream see STREAM CAPTURE.

divide see WATERSHED.

diviner, divining rod see DOWSING.

dodecahedron *n*. a closed crystal FORM, consisting of 12 faces, belonging to the isometric system (see CRYSTAL SYSTEM).

The crystal faces of the *rhombic dodecahedron* are rhomb-shaped; this form, common in GARNET, displays full isometric symmetry. The *pentagonal dodecahedron*, or *pyritohedron*, has five-sided faces and displays a lesser degree of symmetry; it is a common form in crystals of PYRITE.

Dogger *n*. the Middle JURASSIC EPOCH. It comprises the Aalenian, Bajocian, Bathonian and Callovian Ages.

dogtooth spar *n*. a variety of CALCITE that occurs as sharply pointed crystals somewhat resembling a dog's teeth.

dolerite or **microgabbro** *n*. a MEDIUM-GRAINED intrusive IGNEOUS ROCK of basaltic composition, composed essentially of PYROXENE, PLAGIOCLASE (usually LABRADORITE) and Fe-Ti oxides and commonly showing OPHITIC TEXTURE. The name of the principal minor mineral may be added, e.g., olivine dolerite or quartz dolerite. See also DIABASE.

Dolgellian *n*. the top STAGE of the Upper CAMBRIAN.

doline *n*. a bowl, cone or well-shaped depression in LIMESTONE areas. Dolines may be formed in a number of ways. *Solution dolines* are formed by the downward solution of the limestone at a particular site, for example the intersection of JOINTS, which creates a surface depression often floored by insoluble residues. Solution of limestone underlying superficial deposits results in the gradual subsidence of these deposits, forming a subsidence doline. *Collapse dolines* are funnel-shaped or deep shafts formed by the collapse of a cave roof. They may contain a lake if the water table is sufficiently high (see CENOTE). A doline in which a river disappears underground is an *alluvial streamsink doline*. See also KARST, UVALA.

dolomite *n.* **1.** a hexagonal mineral, $CaMg(CO_3)_2$, in crystalised form, often with curved rhombohdral faces.
2. the sedimentary rock $CaMg(CO_3)_2$. The name *dolostone* has been proposed for this to distinguish it from the mineral.

The origins of dolomite are subject to speculation. Four possible routes may exist: direct precipitation from sea water, possibly assisted by bacterial action; penecontemporaneous DOLOMITISATION in the supratidal zone; dolomitisation related to evaporite deposits; or dolomitisation by ground-water. Significant dolomitisation by GROUNDWATER can be demonstrated in the zone of mixing of fresh and saline waters close to tropical shorelines. Dolomite may thus form by direct precipitation or by REPLACEMENT of existing limestones (magnesium for calcium). Such replacement can either retain or obliterate textures and features in the limestone. It is common for certain features to be preserved and others destroyed, but there does not yet seem to be a generally accepted explanation for such selectivity.

dolomitic limestone see DOLOMITISATION.
dolomitisation *n.* the process by which LIMESTONE is altered to DOLOMITE, either wholly or in part, by the REPLACEMENT of $CaCO_3$ by $CaMg(CO_3)_2$ (the mineral DOLOMITE). The magnesium is intro-duced by circulating Mg-rich solutions derived from sea water. Replacement by dolomite is common in organisms the original hard parts of which were composed of ARAGONITE (for example, GASTROPODS and CEPHALOPODS). In some rocks, however, the calcitic material of CRINOIDS, RUGOSE CORALS or BRACHIOPOD shells remains unaltered while the MATRIX is dolomitised.

The replacement of calcite by dolo-mite involves a volume reduction of some 12 per cent, which results in an increase in POROSITY.

Dolomitisation may occur soon after deposition, for example in the fine-grained aragonite muds in the tidal zones of SABKHAS. In other cases, however, dolomitisation is a late diagenetic process, occurring after LITHIFICATION. See also DEDOLOMITISATION, DIAGENESIS.

dolostone see DOLOMITE.
dome *n.* **1.** any rounded landform or rock mass resembling the dome of a building.
2. an uplifted structure, circular or elliptical in outline; doming may be gentle or steep, e.g. salt domes.
3. see VOLCANIC DOME.
4. (in *crystallography*) a crystal FORM in which the faces are parallel to one of the horizontal CRYSTAL AXES but cut the other one and the vertical axis.

Donau see CENOZOIC ERA.
dormant volcano *n.* a volcano that is not actively erupting but that has erupted within historic time and is considered likely to do so again in the future. As noted from the imprecision of the time limits, there is no clear distinction between a dormant and an active volcano.

dorsal valve see BRACHIAL VALVE.
dorsum *n.* (*pl.* **dorsa**) the inner dorsal margin of a coiled NAUTILOID or AMMONOID shell. Compare VENTER.

dot chart *n.* a graphical technique used to compare the gravity effects of assumed subsurface structures with observed gravity values when the masses are irregular. The technique generally involves the use of a *graticule*, a transparent template that is superim-posed over a cross-section of the formation the gravity effect of which is to be determined. The template is usually a fan-shaped pattern of lines forming compartments that increase in area as the distance from the vertex increases. The vertex is placed over the gravity station or the section, and the gravity effect of any body included in the section can be determined by counting the compartments it covers on the template. Each compartment represents a vertical gravity contribution at the observing station (at the vertex), and the number of dots within the compartment indicates the amount of this effect.

double refraction see BIREFRINGENCE.

doublure *n.* the margin of the CEPHALON (head carapace) that is curved under to form a narrow rim around the ventral surface. See TRILOBITE, Fig. 89.

downdip block see DOWNTHROWN.

downfaulted see DOWNTHROWN.

downthrown *adj.* (of the side of a fault) appearing to have moved downwards relative to the other side. The amount of downward vertical displacement is called the *downthrow*. Rocks on the downthrown side of a fault are referred to as the *downdip block* and are described as *downfaulted*. Compare UPTHROWN, HEAVE.

downward enrichment see SECONDARY ENRICHMENT.

downwasting *n.* **1.** see MASS WASTING. **2.** a part of the ABLATION process during which glacial thinning occurs; this may eventually lead to stagnation or lack of forward motion.

downwelling see UPWELLING.

dowser see DOWSING.

dowsing *n.* the practice of locating groundwater, mineral deposits or other concealed objects with the aid of a forked stick, or other apparatus, called a *dowsing rod* or *divining rod*. Prior to the 19th century it was thought that groundwater flowed in definite rivers, like surface water, and that a luckily dug well would reach one of these streams and produce abundantly. Because of the uncertainty of where to dig, help was sought from *dowsers*, also called *diviners* or *water witches*. Such people were supposedly gifted with supernatural powers that enabled them to find the underground streams. The belief still persists in many regions. As a dowser walks about with a dowsing rod, it is supposed to dip sharply when he or she is over an underground stream. Although there is no scientific basis for it, dowsing has had a fairly high degree of success since the WATER TABLE in unconfined AQUIFERS is everywhere productive and can be randomly accessed.

draas *n.* giant AEOLIAN DUNE structures that may be up to 400 metres in height and have wavelengths of over 650 metres. They may form longitudinally or transverse to the prevailing wind direction and sometimes develop star-shaped forms known as *rhourds*.

drag *n.* **1.** the bending or distortion of strata on either side of a FAULT, caused by the friction of the moving blocks along the surface of the fault. These distortions generally record some components of relative movement but are not reliable as some faults may reverse their movement and some may retain drag from earlier displacements. **2.** also called **drag ore**, rock and ore fragments separated from an ore body and found in and along the fault zone.

drag fold see FOLD.

drag mark *n.* a long groove or striation produced by an object as it is dragged by a current across a soft surface. The cast or impression of such a mark, such as might appear on the underside of the overlying bed, is also called a drag mark. See TOOL MARK, SOLE MARK.

drag ore see DRAG.

drainage *n*. the removal of excess METEORIC WATER by rivers and streams. A river network is related to the geological structure of a region. In the case of young DRAINAGE SYSTEMS, it is determined by the existing rocks and superficial deposits. In mature systems, the courses may have been determined by strata that were subsequently removed. See also DRAINAGE PATTERN.

drainage basin *n*. a region drained by a particular stream or river system, i.e. an area that contributes water to a particular river or channel system. The amount of water reaching the reservoir depends on the size of the basin, total precipitation, evaporation and losses caused by absorption. The term 'CATCHMENT AREA' is often used instead by hydrologists.

drainage density *n*. the relative spacing of a stream network, expressed as the ratio of the total length of all streams within a drainage basin to the area of the basin. Thus, it is a measure of the DISSECTION, or of the TOPOGRAPHIC TEXTURE, of the area. The drainage density depends on the climate, geology and physical characteristics of the basin.

drainage pattern *n*. the particular arrangement or configuration that is collectively formed by the individual stream courses in an area. Drainage patterns reflect the influence of factors such as initial slopes, inequalities in rock hardness and structural controls, and the geological and geomorphic history of the drainage basin. The most frequently encountered drainage patterns are described here.

Dendritic patterns are the most common. They are characterised by the irregular branching of tributary streams in many directions at almost any angle but usually less than 90º.

They develop on rocks of uniform resistance to erosion and are most likely to be found on nearly horizontal sedimentary rocks or in areas of massive igneous rocks. A special dendritic pattern, the *pinnate drainage pattern*, is one in which tributaries to the main stream are subparallel and join it at acute angles, thus resembling a feather. This is believed to indicate the effect of very steep slopes on which the tributaries formed.

Trellis drainage patterns show an arrangement of subparallel streams, usually aligned along the strike of the rock formations or between parallel topographical features deposited by wind or ice. They are most prominent, however, in regions of tilted or folded sedimentary rocks.

A *parallel drainage pattern* is one in which a series of streams and their tributaries flow parallel, or almost parallel, to one another and are regularly spaced. It is indicative of an area of pronounced and uniform slope.

Barbed drainage patterns are formed by tributaries that join the main stream in sharp bends oriented upstream. These patterns can occur at or near the headwater portions of systems and are usually the result of STREAM CAPTURE that has caused a reversal of the flow direction of the main stream.

In a *rectangular drainage pattern*, both the main stream and its tributaries exhibit right-angled bends. This pattern is indicative of a stream course that is determined by a prominent fault or joint system.

Radial drainage, a pattern in which streams diverge from an elevated central area develops on volcanic cones and other types of isolated conical hills.

Centripetal patterns occur where drainage channels converge into a

drumlin

central depression. They are character-istic of sinkholes and craters, and large structural basins.

Annular drainage, a roughly ringlike concentric pattern, is found around domes that are encircled by alternating belts of strong and weak rock. The annular pattern occurs when the stream erodes the weaker rocks around the dome.

A *deranged drainage pattern* is marked by irregular stream courses that flow into and out of lakes and have only a few short tributaries; it occurs in areas of more recent glaciation, where the preglacial drainage has been obliter-ated and the new drainage has not had sufficient time to develop any coordi-nated pattern.

drainage system see DRAINAGE.

dravite *n.* a brown Mg-rich variety of TOURMALINE. It is found in metamorphic rocks, particularly in metamorphosed impure LIMESTONES, and has been reported in some basic rocks.

drawdown *n.* the lowering of the water level or potentiometric surface of a well or borehole as a result of the withdrawal of water. See also CONE OF DEPRESSION.

dreikanter *n.* a VENTIFACT with three edges.

driblet cone, driblet spire see HORNITO.

drift *n.* 1. any rock material that has been carried by glaciers and deposited directly from the ice or through the agency of meltwater. See also GLACIAL DRIFT.

2. see DRIFT CURVE.

drift curve *n.* a graph of a series of gravity readings taken at a single field station and plotted against intervals during the day. The curves show a continual variation of the gravity readings with time, known as *drift*. This is caused by imperfect elasticity of the GRAVIMETER springs, which are thus

subject to a slow CREEP over long time intervals.

drift theory *n.* one of the theories of COAL origin. It states that the plant material from which coal develops has been transported from its place of growth to another locality where it accumulates and where COALIFICATION occurs. Compare IN-SITU THEORY.

drilling fluid see DRILLING MUD.

drilling mud or **drilling fluid** *n.* a sus-pension of clay (usually BENTONITE) and BARITE in water or oil used in ROTARY DRILLING to lubricate and cool the drilling bit and carry rock cuttings to the surface. The bentonite makes the mud thixotropic (see THIXOTROPY) and the barite increases its density.

dripstone *n.* CALCITE or other MINERAL DEPOSITS formed in caves by dripping water. These include STALACTITES AND STALAGMITES and similar deposits formed by flowing water. See also CAVE ONYX, FLOWSTONE, TRAVERTINE.

drive pipe *n.* a pipe driven into a bored hole to prevent caving in or to exclude water.

dropstones *n.* large rock fragments in laminated marine MUDROCKS or varved lake deposits, dropped by floating ice (see RHYTHMITES). Typically, the LAMINAE are disturbed by the impact of the CLASTS on the soft SEDIMENTS.

drowned valley *n.* a valley that is partly flooded by a sea or lake. It may result from a rise of sea level, subsidence of the land or a combination of both.

drumlin *n.* an elongated elliptical hill consisting of unconsolidated material and commonly occurring in swarms. Drumlins can be up to 60 metres high and several hundred metres long. They are formed under ice sheets or very broad valley glaciers. The long axes are approximately parallel to the direction of ice movement and are typically

147

steepest and highest at the end that faced the advancing ice, and slope gently in the direction of movement. Compare ROCHES MOUTONNÉES. Most drumlins are composed of clayey lodgement TILL but some have bedrock cores. *Drumlin swarms*, or *fields*, may contain as many as 10 000 drumlins, as in the USA (central New York State and Wisconsin), Canada and northern Ireland. It is thought that drumlins are formed by the moulding action of ice sheets on water-saturated newly deposited till; there is, however, no generally accepted explanation of their origin. See GLACIER.

druse *n*. **1.** a crust of crystals that develops along the walls of a cavity or mineral vein. The crystals are EUHEDRAL and usually of the same minerals as those of the enclosing rock. See also VUG, AMYGDALE, GEODE.
2. a cavity of this sort.
3. aggregates of subhedral crystals with distinct faces, often encrusted with crystals, e.g. quartz.
4. a hole or bubble in glacial ice filled with an air-ice crystalline mixture.

drusy *adj*. **1.** (of a crystalline aggregate) having a surface covered with a layer of small crystals, e.g. drusy quartz.
2. (of rocks) containing numerous druses (in the sense of cavity). See MIAROLITIC.

dry lake *n*. **1.** a basin formerly containing a lake.
2. a PLAYA.

dry valley *n*. a valley in which streams once flowed but which now has little or no running water. It may be the result of a fall in the WATER TABLE, of a climatic change or of STREAM CAPTURE. Dry valleys are common in areas underlain by limestone and chalk. See also WIND GAP.

DSDP see DEEP SEA DRILLING PROJECT.

DTA see DIFFERENTIAL THERMAL ANALYSIS.

dubiofossil *n*. name applied to an object that may be a fossil but for which no conclusive evidence has been presented, often used for objects from the Archaean.

ductile behaviour *n*. a response to STRESS in which materials undergo considerable PLASTIC DEFORMATION before rupture occurs (i.e. if total STRAIN before fracture is greater than 8–10 per cent). Rocks are more likely to display ductile behaviour under conditions of higher temperature and confining pressure. Compare BRITTLE BEHAVIOUR.

dumpy level *n*. a levelling instrument that has a telescope rigidly connected to the vertical spindle and can therefore rotate only in a horizontal plane.

dune *n*. a mound or ridge of unconsolidated, usually SAND-sized, sedimentary particles formed by the action of a fluid medium, which may be wind or water. Dunes are larger BEDFORMS than ripples and are formed at higher current speeds (see RIPPLE MARK).

Subaqueous dunes range from about 50 centimetres to several metres high and have wavelengths from 1 to 5 metres (compare SANDWAVES). They commonly have well-developed lee-side scours and internally show trough CROSS-BEDDING.

AEOLIAN dunes may be found in coastal regions where sand accumulates at the back of a beach and in inland sand 'seas' (or ERGS) of desert regions. Desert dunes range from less than a metre to 100 metres in height, with wavelengths of 3 to 600 metres, and occur in a variety of forms. Much larger bed forms, called DRAAS, are associated with dunes, and it is common to find ripples on dune surfaces and dunes on draas. The migration of dunes and draas results in the formation of large-scale CROSS-BEDDING structures. Since large-scale cross-bedding can also be formed by fluvial bed forms, however, this alone is not enough to distinguish an aeolian environment. Other features, such as good SORTING, high SPHERICITY and ROUNDNESS, frosted grain surfaces and a lack of MICAS and CLAY MINERALS, all help to identify the desert environment.

Longitudinal or *seif dunes* (and draas) are long (10 kilometres or more) narrow ridges with sinuous crests aligned along the effective wind direction. Some seif dunes form from the elongation of one horn of a *barchan* (see Fig. 25(c) below).

Dunes formed transverse to the dominant wind direction may be straight-crested ridges or sinuous *aklé dunes*, which have sections that are alternately convex and concave downwind.

Barchans are crescent-shaped dunes

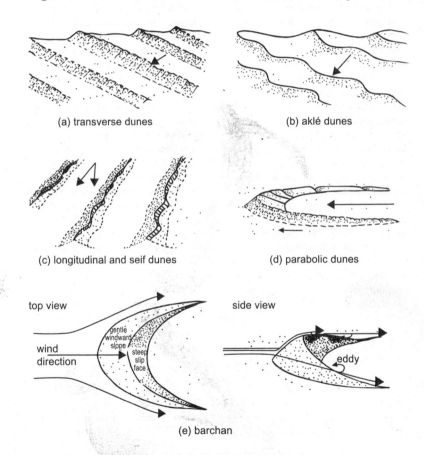

(a) transverse dunes

(b) aklé dunes

(c) longitudinal and seif dunes

(d) parabolic dunes

top view   side view

wind direction

gentle windward slope

steep slip face

eddy

(e) barchan

Fig. 25 **Dune**. The principal types of aeolian dune. Arrows show the general wind direction

that open downwind and advance in the direction of the dominant wind (see Fig. 25(e)). They are formed where the sand supply is sparse and are more widely spaced where less sand is available. With increasing amounts of sand they may merge to form aklé dunes or elongate along one horn and merge to form seif dunes.

Parabolic dunes are also crescent-shaped isolated dunes but close downwind, and many are formed round DEFLATION hollows.

dune complex *n.* a group or aggregate of mobile and fixed sand dunes the existence of which does not necessarily depend on an obstruction or topo-graphical break. A complex usually attains its maximum development on relatively flat terrain.

Dunham classification *n.* a classification of LIMESTONES proposed by Dunham (1962), based on depositional textures and mud content. Lime MUDSTONE and WACKESTONE are mud-supported and contain less than 10 per cent and more than 10 per cent grains respectively. PACKSTONE is a grain-supported lime-stone with a mud martix, and GRAINSTONE is a grain-supported limestone without mud matrix. A BOUNDSTONE is a limestone the original contents of which were bound together during deposition. Those limestones with no recognisable depositional textures are crystalline carbonates. See FOLK CLASSIFICATION.

dunite *n.* an ULTRAMAFIC rock composed essentially of olivine but which can contain such accessory minerals as pyroxene, plagioclase and chromite. It occurs as the result of accumulation of olivine crystals on the floor of layered intrusions during crystal fractionation (see MAGMATIC DIFFERENTIATION) of basic magmas and is also found as MANTLE XENOLITHS.

duplex *n.* a series of imbricate thrust wedges (see FAULT) bounded by a lower floor or sole thrust and an upper roof thrust. The individual imbricate units, which are lens-like in form and bounded on all sides by faults, are called *horses*. See Fig. 26 below. See also IMBRICATE STRUCTURE.

durain see LITHOTYPE.

duricrust *n.* a hard crust formed on the surface or within the upper horizons of a soil in a semi-arid climate, where evaporation is nearly equal to rainfall. Duricrusts are a product of deep chemical WEATHERING, when mineral deposits are precipitated within the surface layers by the evaporation of groundwater that is saturated with dissolved salts. See also CALCRETE, SILCRETE, CALICHE. Compare HARDPAN.

duripan *n.* a soil layer characterised by

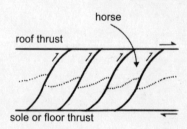

Fig. 26 **Duplex**. Cross-section showing imbricate structure between sole and roof thrusts. The dotted line shows the displacement of one horizon

cementation by silica. Duripans occur mainly in volcanic areas with arid or semi-arid climates. See also HARDPAN.

dust tuff *n.* a fine-grained PYROCLASTIC ROCK in which the average particle size is less than $1/16$ of a millimetre. See TUFF.

dyke *n.* **1.** a tabular body of intrusive igneous rock emplaced vertically or at a steeply inclined angle to the horizontal, see INTRUSION, Fig. 51. It is usually DISCORDANT with the host rock that it intrudes. It may be a COMPOSITE or MULTIPLE intrusion. Dykes may be fine-, medium- or coarse-grained, depending on their composition and the combination of their size and the length of their cooling period. They vary in width from a few centimetres to many metres, but they may be traced for several kilometres: the Cleveland dyke in the north of England probably originates in the Mull volcanic centre 200 kilometres to the northwest in the Scottish Inner Hebrides. A group of dykes is called a *dyke swarm*. They may form parallel, radial or *en échelon* arrangements. *Radial dyke swarms* are often focused on a volcanic centre, for example the 700 dykes on the island of Rum, off the west coast of Scotland. The formation of *parallel dyke swarms* appears to be related to regional crustal extension and rifting, as in the TERTIARY Thulean province of northwest Britain and Greenland. The multiple dykes

forming the sheeted dyke complex of ophiolite sequences are believed to represent the lower part of layer 2 of the OCEANIC CRUST.

2. see INTRUSION, RING DYKES AND CONE SHEETS.

3. see CLASTIC DYKE.

dyke ridge or **dyke wall** *n.* a topographical feature formed when a DYKE is more erosion-resistant than the rock on either side. A dyke ridge several kilometres long and about 15 metres high radiates from Ship Rock (a volcanic neck) in New Mexico. Even although they may resemble HOGBACKS, there is a clear distinction: dyke walls have developed from igneous intrusions whereas hogbacks are features of stratified rocks. See VOLCANIC PLUG.

dynamic metamorphism see METAMORPHISM.

dysaerobic *adj.* (of biofacies) partially or intermittently oxic benthic conditions. Compare ANAEROBIC.

dyscrystalline *adj.* (of igneous rocks) poorly crystalline.

dystrophic lake *n.* a lake in which nutrient matter and oxygen are sparse. It is rich in humic matter, which consists mainly of undecomposed plant fragments. Dystrophic conditions are often associated with acidic peat bogs. Compare EUTROPHIC LAKE, OLIGOTROPHIC LAKE.

E*e*

Earth *n*. that planet of our solar system that is fifth in size of the nine major planets and, as one of the inner planets, is third in order of distance from the Sun. The Earth's equatorial radius is 6378 kilometres, its polar radius 6357 kilometres and its equatorial circumference 40 075 kilometres. The surface area is 5.1×10^8 square kilometres, of which approximately 29 per cent is land. The mean height of land above sea level is 0.86 kilometre, with a maximum elevation of 8.9 kilometres at Mount Everest. The mean depth of the oceans is 3.8 kilometres, with the greatest depth in the Marianas Trench in the northwest Pacific. The mass of the Earth is 5.973×10^{24} kg and the mean density 5.53×10^3 kg^{-3}.

earth *n*. **1.** any of several metallic oxides that are difficult to reduce, e.g. alumina, zirconia. Compare RARE EARTHS.
2. an organic deposit that has remained generally unconsolidated, e.g. DIATOMACEOUS EARTH.
3. FULLER'S EARTH.

earth current *n*. electrical currents flowing through the ground. Some are caused by artificial sources, such as power lines, other by electrochemical reactions in the rocks. The largest of these effects are observed over sulphide or graphite ORE bodies. Ground currents caused by astronomical phenomena such as solar activity affect enormous areas and are called *telluric currents*.

earthflow *n*. a term for a gravity process and landform on a hillside that show downslope movement of unconsolidated rock and soil within defined lateral boundaries. As fluidity increases, earthflows sometimes grade into mudflows. Compare MUDFLOW.

earth hummock *n*. a low, rounded frost mound, generally 10 to 20 centimetres in height, consisting of a central core of fine material surmounted with vegetation. Earth hummocks occur in Arctic and Alpine regions, where they form in groups to produce a nonsorted PATTERNED GROUND.

earthquake *n*. a rumbling or trembling of the ground produced by the sudden breaking of rocks in response to geological forces within the Earth. According to the ELASTIC REBOUND THEORY, the elastic property of rocks permits energy to be stored during DEFORMATION by tectonic forces. When the strain exceeds the strength of a weak part of the Earth's crust, e.g. along a geological FAULT, opposite sides of the fault slip, generating elastic waves that travel through the earth. An idealised model of an earthquake source would show the rupture of the fault as originating at a point on the fault surface called the SEISMIC FOCUS or *hypocentre*. The rupture surface progresses along the fault plane until it reaches a region where the rocks are not sufficiently strained for it to continue. The length and width of the rupture surface and the location of the focus and EPICENTRE are important parameters for describing the essential characteristics of an earthquake source.

The focus mechanism of an earthquake can be described in terms of the orientation of the fault plane and the direction of slip of the blocks on either side, and this can be determined from an analysis of the initial motion of the seismic waves generated.

In recent years, the properties and origins of both earthquakes and volcanoes have been re-examined in the context of PLATE TECTONICS. According to the concept, earthquakes and volcanoes are manifestations of the motion of crustal plates. The majority of earthquakes occur in narrow zones defining the boundaries where lithospheric plates interact. Shallow earthquakes, with epicentres less than 100 kilometres below the surface, are recorded at all three types of plate boundary, but deeper earthquakes (100 to 700 kilometres) typically occur at DESTRUCTIVE PLATE BOUNDARIES, where one plate plunges beneath another at a SUBDUCTION ZONE. See also BENIOFF ZONE. Earthquakes are regarded as the most direct evidence for subduction slabs; this is deduced from the fact that the focal mechanisms of moderate and deep-focus earthquakes indicate stresses that are compatible with a bending brittle plate under gravitational force. See also EARTHQUAKE MEASUREMENT, EARTHQUAKE PREDICTION, SEISMOGRAPH, SEISMIC WAVES.

earthquake engineering or **engineering seismology** *n.* **1.** the study of the behaviour of foundations and structures, and of construction materials, when subjected to earthquake stresses. **2.** techniques that would lessen the effect of earthquakes on such structures.

earthquake intensity see EARTHQUAKE MEASUREMENT.

earthquake magnitude see EARTHQUAKE MEASUREMENT.

earthquake measurement *n.* the determination of the relative energy generated by an earthquake, which involves the measurement of both its intensity and magnitude. *Earthquake intensity*, the effects of an earthquake at a particular location, is measured by some standard of relative measurements such as damage. Some of the many factors that contribute to earthquake intensity at a certain point on the Earth's surface include the location of the focal region, the 'triggering' mechanism of the earthquake inside this region, the quantity of energy released and local geological structure.

Various intensity scales have been formulated. For many years the *Rossi-Forel scale* was the most widely used. The *Mercalli scale*, which was formulated later, was, like the Rossi-Forel scale, an arbitrary scale of earthquake intensity. Numbers from I to XII (I to X^+ is the Rossi Forel system) were assigned to conditions ranging from 'detectable only by instrument' to 'almost total destruction'.

The magnitude of an earthquake is related to the amount of STRAIN energy released, as recorded by seismographs. It is measured by the *Richter scale*, a numerical scale that describes an earthquake independently of its effect on people, objects or buildings. It has a range of values from −3 to 9. Negative and small numbers represent earthquakes of very low energy; higher values indicate earthquakes capable of great destruction. The scale of magnitude actually has no upper or lower limit but, in practice, no earthquake greater than 8.9 has been recorded. That constraint, however, is intrinsic to the Earth, not to the scale.

The Richter scale was based on the American seismologist Charles Richter's observation that the seismograms for any two earthquakes, regardless of the distance from the recording station (hence regardless of the distance from the focus), show a constant ratio between the maximum amplitude of their surface waves. Because of the enormous variation in amplitude that is possible between earthquakes, the scale range is compressed by taking the logarithm (base 10) of the ratio. Thus, an increase of 1 in the rating of an earthquake would indicate a tenfold increase in the seismic-wave amplitude. Universal application of the Richter system was realised after formulating the mathematics required to standardise seismograms from different machines in various areas of the world. See SEISMIC MOMENT.

earthquake prediction *n.* methods and techniques of forecasting the occurrence of an EARTHQUAKE, its magnitude, location and time, over either a short or a long term. Some of the most promising earthquake precursors at present include regional crustal strain, uplift and tilting of the ground surface as gauged by geodetic methods, decreases in the ratio of compressional-wave velocity to shear-wave velocity (V_p/V_s), anomalous changes in water level, and radon-gas concentration in wells. Foreshocks sometimes precede the main shock. The duration of such anomalies appears to be correlated with the magnitude of the predicted earthquake, e.g. a magnitude 5 earthquake seems to be preceded by premonitory changes for about three months, whereas an earthquake of magnitude 7 is preceded by anomalous patterns for about 10 years. The interpretation of earthquake precursors follows a basic physical model grown out of laboratory experiments. The model assumes the formation of microcracks in the rocks as strain builds up just prior to an earthquake. The opening cracks increase the volume of the strained rocks; this alters the seismic-wave velocities through the rocks and is perhaps responsible for changes in electrical resistivity of the rocks, uplift and tilting, and increased quantities of radon in well water. See DILATANCY.

The United States programme for earthquake prediction is still not sufficiently supported to realise prediction within the next decade. Lack of funds has seriously deterred the research efforts of universities and industries. Russian field experiments have yielded many, if not most, of the acknowledged anomalous precursors of earthquakes. The former USSR's strategy in prediction study was to monitor several experimental sites, as compared with the greater emphasis in the United States on monitoring specific areas in California. The Japanese are currently concentrating on surveys every few years, over ranges of more than 20 000 kilometres, and systematically observing many precursory phenomena.

The Chinese approach to prediction has certain unique features. It uses the findings of large numbers of students and peasants in remote areas. Pertinent data, based on all accepted precursory anomalies, are being obtained at 5000 points. In addition, the Chinese are giving serious consideration to pre-earthquake animal behaviour. See also EARTHQUAKE MEASUREMENT.

earthquake swarm *n.* the occurrence of a large number of lesser EARTHQUAKES in a particular region over an interval of time, perhaps several months, without the event of a major earthquake.

earthquake wave see SEISMIC WAVES.

earthquake zone *n*. an area or region in which FAULT movements occur. Such an area sometimes coincides with a location of VOLCANIC activity.

earth science *n*. a broad term for all Earth-related sciences. Although it is often used as a synonym for GEOLOGY or geological science, it includes, in its wider scope, fields such as the environmental aspects of geology, the physical geography of surface and near-surface processes, and physical OCEANOGRAPHY. *Earth system science* studies the whole Earth as a system of interacting parts, the parts including the atmosphere, the hydrosphere, the biosphere and the geosphere.

earth tide *n*. the rising and falling of the Earth's solid surface in response to the gravitational pull of the Sun and the Moon, i.e. the same forces that produce the tides of the sea. Earth tides, like ocean tides, show the greatest rise when the Moon and Sun are aligned (*spring tide*) and least rise when the Moon and Sun are perpendicular to each other (*neap tide*). However, earth tides do not show as large a displacement as ocean tides; the maximum displacement of earth tides is less than a metre.

earthy *adj*. (of a type of FRACTURE) akin to that of hard clay.

ecdysis *n*. the moulting process by which ARTHROPODS (for example, TRILOBITES) shed the rigid exoskeleton in order to form a larger one.

echinoderm *n*. any of the entirely marine invertebrates belonging to the phylum Echinodermata, having an endoskeleton composed of numerous calcite plates. Living forms have both pentamerous and bilateral symmetry. The fossil history of echinoderms dates to the Lower Cambrian, but their evolution is not clear because of an incomplete fossil record. ASTEROIDS, CRINOIDS, CYSTOIDS, ECHINOIDS and ophiuroids are important members of this phylum.

echinoid *n*. any member of the Echinoidea, a class of ECHINODERMS to which sea urchins and sand dollars belong. Echinoids characteristically have a hollow spiny hemispherical TEST composed of 20 columns of interlocking plates radiating from the APICAL DISC in the centre of the upper surface to the PERISTOME in the centre of the lower oral surface, see Fig. 27 overleaf. Five sets of columns, the AMBULACRA, carry the tube feet and alternate with the interambulacra.

Some forms adapted to an infaunal lifestyle and developed bilateral symmetry (*irregular echinoids*). In these the anus lies outside the apical system, in the posterior interambulacrum while the peristome may be central or may lie towards the anterior margin and be protected by the labrum (see PLASTRON). See Fig. 28 overleaf. The ambulacra have become petaloid in shape on the ABORAL SURFACE. Radially symmetrical *regular echinoids* were epifaunal, moving about the sea floor by means of the tube feet. Echinoids have been used as INDEX FOSSILS in the CRETACEOUS. Range, Upper ORDOVICIAN to present.

echogram *n*. a graph of the sea-floor contour made by ECHO SOUNDING. Echograms are used to establish contour maps and physiographical PROFILES.

echo sounding *n*. a method of determining depth of water or distance from an object by measuring the time lapse between the generation of an initial sound pulse and the return of the echo. With a known speed of sound in water, the depth or distance can be calculated. See ACOUSTICS, ECHOGRAM, SOUNDING.

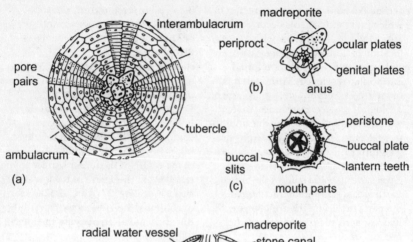

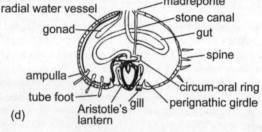

Fig. 27. **Echinoid**. The structure of a regular echinoid (*Echinus*): (a) simplified drawing of the aboral (upper) surface. (b) enlargement of the apical disc. (c) enlargement of the mouth parts on the lower (oral) surface. (d) a section cutting through an ambulacrum on the left and an interambulacrum on the right, showing the simplified internal structure

eclogite *n*. granular metamorphic rock composed essentially of GARNET (almandine-pyrope) and PYROXENE (OMPHACITE) but with a bulk chemical composition very similar to that of BASALT. The relatively high density of about 3.5×10^3 kg m^{-3} compared with about 3.0×10^3 kg m^{-3} for basalt of the same chemical composition is consist-ent with high-pressure crystallisation deep in the Earth's crust.

eclogite facies *n*. a set of metamorphic mineral assemblages in rocks of basaltic composition represented by omphacitic pyroxene and pyrope garnet. See also METAMORPHIC FACIES.

ecology *n*. the study of interrelations of living organisms with one another and with their environment, including both biological and physical factors. See also PALAEOECOLOGY.

economic geology *n*. the study and analysis of formations and materials that can be useful or profitable to people: fuels, water, minerals or metals.

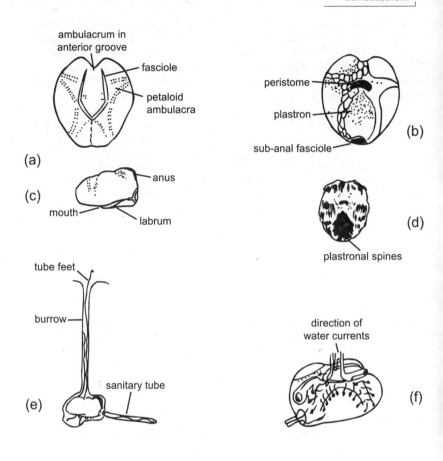

Fig. 28. **Echinoid**. External structure and lifestyle of a deep-burrowing, irregular echinoid: (a) aboral (upper) surface; (b) oral surface; (c) side view; (d) oral surface with spines; (e) life position in burrow; (f) direction of water flow over test

ecosphere *n*. those portions of the universe suitable for the existence of living organisms, in particular the BIOSPHERE.

ecotope *n*. the niche or niche space functioning as the habitat of a particular organism. See also BIOTOPE.

écoulement see GRAVITY SLIDING.

eddy *n*. a swirling or circular movement of water that develops wherever there is a flow discontinuity in a stream. Among its causes may be velocity changes, flow separation or changes in water density.

edge dislocation see DISLOCATION.

edrioasteroid *n*. a class of extinct ECHINODERMS found as fossils in Lower Cambrian to Lower Carboniferous marine rocks. Some were free-living organisms, most were attached to the substrate. They include some of the earliest echinoderms. Edrioasteroids had bodies formed of irregular, flexible polygonal plates. Their body shape varied from discoidal to cylindrical.

effective permeability *n*. the ability of earth materials, such as soils, to transmit one fluid, such as gas, in the presence of other fluids, such as water or oil. The flow parameters of each coexisting fluid form the effective permeabilities of the earth material to the fluids. See also ABSOLUTE PERMEABILITY, RELATIVE PERMEABILITY.

effective porosity *n*. the percentage of the total volume of a given rock or earth material that consists of interconnecting voids that will transmit fluids. Compare POROSITY.

efflorescence *n*. **1.** a white powdery or mealy substance produced on the surface of a rock or mineral in an arid or semi-arid region by loss of water or crystallisation upon exposure to air. **2.** the process, either through evaporation or chemical change, by which a rock or mineral becomes encrusted with crystals of such a salt.

effluent 1. *adj*. flowing out or forth. **2.** *n*. a stream flowing out of a larger stream or out of a lake. Compare INFLUENT, EFFLUENT STREAM.

effluent stream *n*. a stream the channel of which lies below the WATER TABLE and into which water from the zone of saturation flows.

effusive eruption *n*. an eruption of LAVA that occurs without explosive fragmentation. For an effusive eruption to occur, the MAGMA must have a relatively low viscosity. This is linked to silica content, so basaltic magmas, which have relatively low silica, are generally less viscous than andesites, dacites and rhyolites, and basaltic eruptions are therefore commonly effusive. The content of volatile fluids dissolved in magma also has a bearing. The boiling off of these fluids as the magma reaches the surface generates high pressures. During effusive eruptions of gas-rich magma of low viscosity, effervescence of boiling volatiles such as water and carbon dioxide leads to lava fountaining (see LAVA FOUNTAIN); in gas-rich viscous magma in which effervescence is less efficient, the presence of pressurised gas bubbles leads to explosive activity.

E_H *n*. the oxidation/reduction potential of an aqueous environment.

Eifelian *n*. the lower STAGE of the Middle DEVONIAN.

einkanter *n*. a VENTIFACT with one edge.

ejecta *n*. **1.** also called **pyroclastic material**, material flung from a volcanic vent. **2.** material, such as glass and rock fragments, thrown out of an IMPACT CRATER during its formation.

Ekman spiral see HODOGRAPH.

elastic *adj*. (of a body) capable of changing its length, volume or shape in direct response to applied stress, and of instantly recovering its original form upon removal of the stress. Compare PLASTIC.

elastica *n*. the form produced by buckling a thin elastic sheet in air. PTYGMATIC STRUCTURES approximate to elasticas. See FOLD and Fig. 34.

elastic aftereffect see CREEP RECOVERY.

elastic bitumen see ELATERITE.

elastic constant or **elastic modulus** *n*. the ratio of STRESS to the corresponding STRAIN (i.e. stress/strain = an elastic constant). An elastic constant is any one of the various coefficients, expressed in pascals, that defines the elastic properties of matter. See also BULK MODULUS, MODULUS OF RIGIDITY.

elastic deformation *n*. a DEFORMATION

that disappears upon the release of STRESS. The term is frequently used for deformation in which the relation between stress and STRAIN is a linear function.

elastic limit or **yield point** n. the maximum STRESS that a body or material can sustain and beyond which it cannot return to its original shape or dimensions. Beyond the elastic limit STRAIN is no longer proportional to the applied stress.

elasticoviscosity n. a process involving both elastic and viscous behaviour. A material is said to be *elasticoviscous* if it shows evidence of early deformation under elastic conditions and if this is followed by a continuously generated permanent strain for as long as the stress is present. The latter component is sometimes called *equivalent viscosity*, because true viscosity applies only to fluids in which the rate of flow is proportional to force, i.e. Newtonian flow.

elastic rebound theory n. a theory of earthquake genesis first propounded by H. F. Reid in 1911. In substance, it states that movement along a geological FAULT is the sudden release of a progressively increasing elastic STRAIN between the rock masses on either side of the fault. The strain is the reaction to DEFORMATION by tectonic forces. Movement returns the rocks to a condition of little or no strain. Some of the first persuasive evidence for the elastic rebound theory came from observations of slip along the San Andreas Fault in the 1906 San Francisco earthquake.

elaterite n. a dark-brown, elastic asphaltic bitumen derived from the metamorphism of petroleum.

electrical logging see WELL LOGGING.

electrical resistivity see RESISTIVITY.

electromagnetic prospecting n. a method of GEOPHYSICAL EXPLORATION in which electromagnetic waves are generated at the surface. When such waves encounter a conducting formation or ore body, currents are induced in the conductors, and these currents generate new waves that are radiated from the conductors and detected by surface instruments. Inhomogeneities in the electromagnetic field, as observed on the surface, would indicate variations in the conductivity beneath the surface and imply the presence of anomalous masses or substances.

electron diffraction pattern n. the interference pattern observed on a screen when an electron beam is directed through a substance, each substance having a unique pattern. Electron diffraction patterns supply basic crystallographic information.

electron microprobe see Appendix 4.

electron microscope n. an instrument in which a beam of electrons from a cathode is focused by magnetic fields so as to form an enlarged image of an object on a fluorescent screen. The electron beam functions in a manner similar to that in which a beam of light functions in an optical microscope. Because of the shorter wavelength of electrons, the electron microscope has a very high resolving power. See also Appendix 4.

electrum n. a natural alloy of gold and silver.

element n. a substance that can be broken into simpler parts only by radioactive decay and not by ordinary chemical means. An element consists of atoms that all have the same atomic number.

elevation n. **1.** the vertical distance from mean sea level to some point on the Earth's surface, i.e. height above sea level. In present-day surveying terminology, 'elevation' refers to heights on the Earth, whereas ALTITUDE refers to heights of spatial points above the Earth's surface.

2. any topographically elevated feature.

elevation correction *n.* in seismic surveys, an adjustment of observed reflection or refraction time values with respect to an arbitrary reference datum. Adjusted time values may be higher or lower than the true values, depending on whether the data were lower or higher than the seismic shot. See also GRAVITY ANOMALY.

Elsterian see ANGLIAN GLACIATION, CENOZOIC, Fig. 13.

elutriation *n.* **1.** the natural sorting processes that separate CLASTIC particles into coarser and finer fractions. This may refer either to the processes occurring during FLUVIAL transport or to subaerial transport, for example in PYROCLASTIC FLOWS or surges.

2. a process for separating finely divided particles into sized fractions in accordance with the rate at which they rise or sink in a slowly rising current of water or air of known and controlled velocity. Elutriation is used in the mechanical analysis of a sediment and in the removal of a substance from a compound.

eluvial *adj.* **1.** pertaining to ELUVIUM, a deposit formed by the disintegration of rock in place.

2. (of a phase of a dune cycle) marked by dune degradation rather than growth. Soil-forming processes, creep and slope wash predominate. This phase is initiated when vegetation becomes heavy enough to halt deflation. Compare AEOLIAN.

3. (of a deposit) formed by rock disintegration at the place of origin; no stream transport is involved.

eluvial placer see PLACER.

eluviation *n.* the downward movement of materials in suspension or solution that are being carried through the soil by descending soil water. Compare ILLUVIATION, LEACHING.

eluvium *n.* an accumulation of disintegrated rock found at the site where the rock originated. Compare ALLUVIUM.

elvan *n.* a Cornish mining term for a DYKE rock of granitic composition, such as a quartz porphyry (see GRANITE).

embayment *n.* **1.** the incursion of a downwarped region of stratified rocks into other rocks.

2. the penetration of a CRYSTAL by another.

3. the CORROSION of a CRYSTAL or XENOLITH by the MAGMA in which it has formed.

emerald *n.* a deep green variety of BERYL, highly valued as a gemstone. It crystallises in hexagonal prismatic forms. Its colour is attributed to the presence of chromium.

emergence *n.* an exposure of land areas that were formerly under water; the exposure may be a result of uplift of the land or decline of the WATER LEVEL. Evidence of land emergence is demonstrated by sands containing marine shells and fish remains at many levels in the stratigraphic column.

emery *n.* a finely granular, dark-grey to black impure variety of CORUNDUM mixed with magnetite or hematite, useful as a polishing and grinding material.

emplacement *n.* **1.** the INTRUSION of igneous rock into a particular place or position. See also FORCEFUL INTRUSION, STOPING, MIGMATITE.

2. the development of ore deposits in a particular location.

Emsian *n.* the top STAGE of the Lower DEVONIAN.

emulsion *n.* a colloidal dispersion of one liquid in another.

enantiomorphism, *n*. (*crystallography*) the existence of two chemically identical crystal forms that are mirror images of each other.

enantiotrophy *n*. a condition or relation between polymorphs in which one may convert to the other at a critical temperature and pressure. See INVERSION, POLYMORPHISM.

enargite *n*. a grey, black, or greyish-black mineral, Cu_3AsS_4, occurring as orthorhombic crystals with a metallic lustre. This ore of copper closely resembles STIBNITE.

enclave *n*. a piece of rock enclosed within an IGNEOUS ROCK and differing in texture and/or composition from the enclosing rock. See INCLUSION, compare XENOLITH.

end *n*. the jointing system in COAL perpendicular to CLEAT.

end member *n*. one of the two extremes of a series, e.g. types of sedimentary or igneous rocks, minerals or fossils.

end moraine see MORAINE.

endogenetic, endogenic or **endogenous** *adj*. (of processes) originating below the Earth's surface. Compare EXOGENETIC.

endomorphism or **endometamorphism** *n*. alteration of the composition of an igneous MAGMA by the complete or partial assimilation of the COUNTRY ROCK it invaded.

endogenic see ENDOGENETIC.

endogenous see ENDOGENETIC, VOLCANIC DOME.

end product *n*. any stable DAUGHTER ELEMENT of radioactive decay. The term is also used to mean the end products of weathering.

en échelon *adj*. (of geological features, such as faults or folds) in staggered or overlapping arrangement of linear features. *En échelon* faults are parallel, relatively short faults that overlap so that the movement at the end of one is taken up at the beginning of the next one.

engineering geology or **geological engineering** *n*. application of the geological sciences to civil engineering procedures so that geological factors pertinent to the location and construction of engineering projects will be recognised and provided for.

engineering seismology see EARTHQUAKE ENGINEERING.

englacial or **intraglacial** *adj*. contained, embedded or carried within a GLACIER or ice sheet, e.g. till or drift.

enrichment *n*. the action of natural agencies that increases the relative amount of one constituent mineral or element contained in a rock. This may be because of the selective removal of other constituents or the introduction of increased amounts of other constituents from external sources. The process may be:
(a) *mechanical*, e.g. the transport of light material, such as quartz, away from heavier material, such as gold.
(b) *chemical*, e.g. downward-filtering copper-containing solutions converting chalcopyrite, $CuFeS^2$ (34.5 per cent copper), to covellite, CuS (66.4 per cent copper). In this process the copper ions in solution replace the original iron. Such a process is often called SECONDARY ENRICHMENT.

enstatite *n*. a rock-forming orthorhombic mineral of the orthopyroxene group, $MgSiO_3$, an important primary constituent of basic igneous rocks. It is also present in many METEORITES, both metallic and stony. With the substitution of iron for magnesium, enstatite grades into BRONZITES and, with the addition of more iron, into HYPERSTHENE.

enterolithic *adj*. (of sedimentary structures) having small intestine-like or ropy FOLDS, one of the primary sedimentary structures described from EVAPORITE units. Such folds represent crumpling in an evaporite, caused by the swelling of ANHYDRITE during the hydration to GYPSUM. See also CHICKENWIRE ANHYDRITE.

entrainment n. 1. the process of picking up and transferring or carrying along, such as the collection and movement of sediment by water currents.
2. the trapping or incorporation of bubbles in a liquid.

entrenched meander see INCISED MEANDER.

entropy n. a property, related to the second law of thermodynamics, that can be equated with the disorder in a SYSTEM. Given a system of different particles, the more even the distribution, the greater the particle disorder and the greater the entropy. Thus, in a STRATIGRAPHICAL UNIT of different kinds of rock, as the composition approaches that of a single component, the entropy approaches zero. In terms of energy, the more evenly distributed the concentration of some form of energy in a system, the greater the entropy and the less the opportunity to perform work. For example, hot magma intruded into cooler rocks causes a flow of heat into the wall rocks, where work is performed in volumetric expansion. Without the uneven concentration of thermal energy, which creates an energy gradient, no work could be performed.

envelope n. a term often used for the outer part of a FOLD. Compare CORE.

environmental geology n. the application of geological concepts to problems created by human activities and their effects on the physical environment.

environment of deposition n. an area within which a SEDIMENT is deposited

under a particular set of physical, chemical and biological conditions. Compare FACIES. Depositional environments can be very broadly classified as ALLUVIAL (including deltaic, see DELTA), LACUSTRINE, DESERT, SHORELINE (CLASTIC and CARBONATE), shallow marine (clastic or carbonate), deep clastic seas (TURBIDITE basins), PELAGIC and GLACIAL (continental and glacio-marine).

Eocene n. the epoch of the Tertiary period between the Palaeocene and Oligocene Epochs. See CENOZOIC.

eon n. a division of geological time, the longest time unit, next in order above ERA. The Phanerozoic Eon includes the Palaeozoic, Mesozoic and Cenozoic Eras. The term is also used for an interval of 10^9 years. See also GEOCHRONOLOGICAL UNIT.

eonothem see STRATIGRAPHICAL UNIT.

Eötvös effect n. the vertical component of Coriolis acceleration, observed when taking gravity measurements while in motion. The velocity over the surface of the meter that is recording the gravity data adds vectorially to the velocity caused by the Earth's rotation; this changes the centrifugal acceleration and thus the apparent gravitational attraction. The Eötvös correction in mgal (MILLIGAL) for a meter with a velocity of K knots at an azimuth angle α and latitude ϕ is:

$$E = 7.503K \cos \phi \sin \alpha + 0.004154K^2$$

The Eötvös uncertainty, dE, in terms of direction uncertainty, $d\alpha$, and velocity uncertainty, dK, is

$$dE=(7.503K \cos \phi \cos \alpha) \, d\alpha + (7.503 \cos \phi \sin \alpha + 0.008308K) \, dK.$$

Eötvös torsion balance n. a type of instrument used in geophysics for measuring curvature and gradients in a gravitational field. This differs from the Cavendish torsion balance, which measured curvature alone; many present-day instruments measure gravity differences directly by determining the gravitational acceleration. The Eötvös balance consists of two equal weights at different heights, connected by a rigid frame. The system is suspended by a tension wire in such a way that it can rotate freely in a horizontal plane about the wire. The gravity gradient is indicated by the amount of torque causing rotation; the higher the torque of the wire, the greater the gradient. See also GEOPHYSICAL EXPLORATION.

Eötvös unit n. a unit of gravitational gradient or curvature, 10^{-6} mgal/cm.

epeiric sea n. a shallow inland sea.

epeirogeny or **epeirogenesis** n. the uplift and subsidence of large portions of the Earth's crust that have produced broader topographic features of continents and oceans, such as basins and plateaux, in contrast to OROGENY, which is responsible for mountain ranges. Epeirogeny does not involve severe deformation, only localised tilting and warping. See also DIASTROPHISM.

ephemeral stream n. a stream that flows (carries water) only during periods of precipitation and briefly thereafter. Its channel is dry most of the year, and it may be referred to as an *arroyo* or *gully*. Such streams occur in arid and semi-arid regions but may be found in any area where the channel bed is at all times above the WATER TABLE. Compare INTERMITTENT STREAM.

epibole see ACME ZONE.

epicentre n. a point on the Earth's surface directly above an earthquake focus. See SEISMIC FOCUS.

epiclastic adj. 1. (of sediments) consisting of weathered or eroded fragments of pre-existing rocks, where the fragments have been transported and deposited at the surface, i.e. any CLASTIC ROCK.
2. (of rocks) formed of such sediments, e.g. sandstone and conglomerate. Compare AUTOCLASTIC.
3. pertaining to the processes (WEATHERING, EROSION, transport and deposition) involved in the production of DETRITAL rocks. Compare PYROCLASTIC.

epicontinental adj. located on the continental shelf or on the continental interior.

epidiorite n. a metamorphosed DOLERITE or allied BASIC igneous rock. Epidiorite was used as a field term by geologists working in the Scottish Highlands but is little used now. Such rocks are more commonly named according to the textures and the major minerals (e.g. ACTINOLITE-, ALBITE-, EPIDOTE-SCHIST).

epidote n. a green monoclinic mineral, $Ca_2(Al,Fe)Al_2O(SiO_4)(Si_2O_7)(OH)$. It is common in regional and contact metamorphic rocks, especially in GREENSCHIST FACIES.

epidote-amphibolite facies see AMPHIBOLITE FACIES.

epifauna n. the animals, attached or free moving, that live on the sea floor.

epigenesis n. 1. changes, exclusive of WEATHERING and METAMORPHISM, that affect SEDIMENTARY ROCKS after they are compacted.
2. alteration in the mineral PROFILE of a rock because of external effects taking place at or near the Earth's surface.

epigenetic adj. 1. formed or originating at or near the Earth's surface. Compare HYPOGENE.
2. (of minerals) introduced into pre-existing rocks. Such deposits may occur as tabular or sheetlike lodes or in various other forms. Compare SYNGENETIC.

epimagmatic see DEUTERIC.

epitaxy n. the oriented overgrowth of

one CRYSTAL on another. Compare SYMPLECTIC INTERGROWTH.

epitheca *n.* the outer wall of a CORALLITE. See RUGOSE CORAL, Fig. 77.

epithermal deposit see HYDROTHERMAL DEPOSIT.

epizone see INTENSITY ZONE.

epoch *n.* **1.** an interval of geological time (geochronological division) longer than an AGE and shorter than a PERIOD during which the rocks of a particular series were formed. See also GEOCHRONOLOGICAL UNIT.
2. an inexact short space of geological time, such as a GLACIAL EPOCH.

equant *adj.* **1.** also called **equidimensional** (of a crystal) having the same or nearly the same dimensions in all directions. Compare TABULAR, PRISMATIC.
2. (of a sedimentary particle) having a length less than 1.5 times its width.
3. (of a rock) with most of the grains equant.

equidimensional see EQUANT.

equigranular *adj.* (of rock texture) having all MINERAL grains approximately the same size.

equiplanation *n.* development of the terrace-like surfaces by the reduction of land without the loss or gain of material and without reference to a base level, for example marine lacustrine or fluvial terraces. Compare ALTIPLANATION, CRYOPLANATION.

equipotential surface *n.* a surface along which the gravity is constant at all points and for which the gravity vector is normal to all points on the surface. The GEOID is an equipotential surface. See also GEOPOTENTIAL.

equivalent *adj.* agreeing or corresponding in geological age or position within the stratigraphical PROFILE. Strata in different regions that have been formed at the same time, or contain the same fossil types, are said to be equivalent.

equivalent viscosity see ELASTICOVISCOSITY.

era *n.* a geological (geochronological) time unit during which rocks of the corresponding ERATHEM were formed. It ranks one order of magnitude below EON. See also GEOCHRONOLOGICAL UNIT.

erathem *n.* the second largest formal chronostratigraphical unit generally recognised, formed by the grouping together of a number of systems. It consists of the rocks formed during an ERA of geological time. The Palaeozoic erathem, for example, is composed of the systems from the Cambrian to the end of the Permian. See STRATIGRAPHICAL UNIT.

erg or **koum** *n.* an undulating plain (*sand sea*) occupied by complex sand dunes produced by wind deposition, such as is found in parts of the Sahara. Compare HAMMADA, REG.

Erian *n.* a series of the Middle Devonian of North America.

erionite *n.* a fibrous ZEOLITE mineral, $(Na_2K_2Ca_2)(Al_8Si_{28}O_{72}) \cdot 28H_2O$, that occurs in volcanic ash in the Karain area, Turkey, where it is the cause of many cases of fatal malignant mesothelioma (lung cancer). See ASBESTOS.

erosion *n.* the wearing away of any part of the Earth's surface by natural agencies. These include mass wasting and the action of waves, wind, streams and glaciers. Fundamental to the process of erosion is that material must be picked up and carried away by such agents. Evidence for erosion is widespread: the retreat of marine cliffs, deposition of fluvial material and the cutting of great canyons, e.g. theGrand Canyon. *Differential erosion* is a difference in the erosion rate on individual features as well as over large areas. Rate variation is caused by the differential resistance of certain outcrops, e.g. ROCK PEDESTALS. Large-scale differential erosion can be seen in the formation of HOGBACKS. Compare DENUDATION.

erratics *n.* glacially transported stones and boulders. Erratics may be embedded in TILL or occur on the ground surface. They range in size from pebbles to huge boulders weighing thousands of tons. Their transport distances range from less than a kilometre to more than 800 kilometres. Erratics composed of distinctive rock types can be traced to their point of origin and serve as indicators of glacial flow direction. Certain erratics found in sediments of a type that would preclude glacial transport may have been deposited by icebergs.

eruption see VOLCANIC ERUPTION.

eruption column *n.* a rolling mass of atomised LAVA, partly condensed water vapour, dust and ash emitted from a VOLCANO during an eruption and rising to great heights.

escarpment or **scarp** *n.* a high, more or less continuous, cliff or long steep slope situated between a lower more gently inclined surface and a higher surface. Escarpments, which may be formed by faulting or erosion, occur as plateau margins, or one may form the steep face of a CUESTA.

E shell see SEISMIC REGION.

esker *n.* a long, narrow, sinuous or straight ridge formed of stratified glacial meltwater deposits, usually including large amounts of sand and gravel. Most eskers are probably deposited in channels beneath or within slow-moving or stagnant ice. Their general orientation runs at right angles to the ice edge. Some eskers originate on the ice and have ice cores. Typical eskers show a fairly constant relationship between their individual dimensions. Long eskers never exceed 400 to 700 metres in width and 40 to 50 metres in height; smaller eskers of 200 to 300 metres may be 40 to 50 metres wide and 10 to 20 metres high. Well-known esker areas are found in the northeastern United States, Canada, Sweden and Ireland. Compare KAME. See also DRUMLIN.

essential mineral *n.* a mineral component that is basic to, or required in, the classification of a rock. Compare ACCESSORY MINERAL.

essexite *n.* a coarse-grained alkaline IGNEOUS ROCK essentially composed of PLAGIOCLASE (LABRADORITE) and titan-augite with variable amounts of ALKALI FELDSPAR, sodium-rich AMPHIBOLE, BIOTITE and NEPHELINE, the latter often altered to ANALCIME. Common accessory minerals are OLIVINE, APATITE, TITANITE and MAGNETITE.

estuary *n.* **1.** a widened mouth of a river valley where fresh water intermixes with sea water and where tidal effects occur.
2. an elongated portion of a sea that is affected by fresh water.
Estuaries are formed where a deeply cut river mouth is drowned following a land subsidence or a rise in sea level.

etch figure *n.* a technique used for revealing CRYSTAL SYMMETRY in which solvent is applied to a crystal surface. The shapes of etch markings that appear on the crystal faces vary with the solvent and with the mineral species; for a given solvent, however, the etching effect is the same for all faces of a crystal form.

etching *n.* in GEOMORPHOLOGY, the DEGRADATION of the Earth's surface by DIFFERENTIAL WEATHERING, MASS WASTING and *sheetwash* (broad continuous sheets of running water). Etching is particularly significant in areas of diverse rocks that exhibit extreme contrast in their weathering rate. *Interstream degradation* may play the most important role in the etching concept of erosion.

ethane *n.* a colourless, odourless, flammable gas, C_2H_6, of the alkane series, present in natural gas and crude petroleum.

Eubacteria *n.* the domain of life that contains bacteria with eubacterial rRNA and the cell walls of which are lipids formed from glycerol esters. See ARCHAEA.

Eucarya *n.* the domain of life that contains organisms composed of eucaryotic cells, the cell walls of which are composed of glycerol fatty acyl diesters and the cells of which contain eucaryotic rRNA.

eucrite see GABBRO.

eugeosyncline see GEOSYNCLINE.

euhedral, automorphic or **idiomorphic** *adj.* (of mineral crystals) fully developed and showing well-formed crystal faces. Compare ANHEDRAL, SUBHEDRAL.

euphotic zone *n.* an ocean zone in which light penetration is sufficient for photosynthesis. It includes the upper 70 metres of the ocean. The zone is shallowest over the continental shelves because light is blocked by the great quantities of suspended sediment there. Compare APHOTIC.

Eurasian-Melanesian belt *n.* a belt of tectonic activity extending from the Mediterranean to Sulawesi, where it joins the CIRCUM-PACIFIC BELT.

eurypterid *n.* any of an extinct group of brackish-water ARTHROPODS, probably predaceous, closely related to horseshoe crabs. Their flattened segmented bodies were covered with chitin, and some grew to lengths of almost three metres. The cephalothorax had two pairs of eyes, four pairs of walking legs, one oarlike pair of legs and one pair of pincers. Eurypterid fossils are not used for zonation. Range, Ordovician to the end of the Permian.

eustatic *adj.* pertaining to worldwide changes in sea level, as distinct from local changes.

eutaxitic texture *n.* a streaky appearance of WELDED TUFFS in which lapilli, pumice fragments (FIAMME) and glass shards are flattened into a disc-like form more or less parallel to the plane of deposition. Eutaxitic texture is created by the simultaneous welding and compaction of hot, glassy PYROCLASTIC fragments. The most intense welding and compaction occur at the lower portion of the flow but not right at the base. See also IGNIMBRITE.

eutectic *adj.* pertaining to a system of two or more solid phases and a liquid where the components are in such proportion that the melting point of the system is the lowest possible with these components.

eutectic point *n.* the lowest temperature at which a mixture can be maintained in a liquid phase or state, the lowest melting or freezing point of an alloy.

eutrophication *n.* the process whereby an aging aquatic ecosystem, such as a lake, supports great concentrations of ALGAE and other aquatic plants because of a marked increase in concentration of phosphorus, nitrogen and other plant nutrients. ALGAL BLOOMS on the surface prevent the light penetration and oxygen absorption necessary for underwater life.

eutrophic lake *n.* a lake characterised by large quantities of dissolved plant nutrients (phosphorus and nitrogen) and a seasonal deficiency of oxygen in the bottom layers because of deposits containing significant amounts of rapidly decaying organic material. Compare OLIGOTROPHIC, DYSTROPHIC LAKE.

euxinic *adj.* (of an environment) characterised by large masses of stagnant, deoxygenated water that thus fosters reducing processes. The Black Sea is the standard example of a land-locked euxinic sea and is also called the Euxinic or Euxine Sea.

evaporite *n.* a sediment deposited from a saline solution as a result of extensive or total evaporation of the water. During the evaporation of sea water, CARBONATES will precipitate first, then SULPHATES (GYPSUM or ANHYDRITE), HALITE and then potassium and magnesium salts. Lateral and vertical zonation of the evaporite minerals is common. Some deposits have rhythmic successions of varve-like layers that may be related to winter/summer variations in the evaporation rate or to alternating precipitation and partial solution, which leaves a residue of the least soluble salts. Larger-scale cycles may relate to periodic marine incursions. Most evaporite deposits have been affected by metasomatic replacement caused by circulating brines. Although they are very soluble, evaporite minerals have been found in rocks 3500 Ma in age. The variety and abundance of minerals so produced depend on the initial composition of the bodywater. Nearly 70 evaporite minerals are known: 27 are sulphates, 27 are borates and 13 are halides. Sea water is the originating solution of most evaporite deposits. Non-marine saline lake and PLAYA evaporite waters generally have initial compositions quite different from sea water, and, as a result, non-marine evaporite deposits can contain minerals rarely formed from sea water, e.g. mirabilite, glauberite, borax. See METASOMATISM, BAR THEORY, SABKHA.

event *n.* **1.** SEISMIC EVENT.

2. any occurrence in which probable tectonic importance is implied by some geological evidence.

evolute see AMMONOID.

exfoliation *n.* the separation of successive thin, onion-like shells (*spalls*) from bare surfaces of massive rock, such as GRANITE or BASALT. It is common in regions of moderate rainfall. Geologists have attributed some small-scale exfoliation to diurnal temperature variations. Large-scale exfoliation or SHEETING can be caused by a decrease that occurs in compressional forces when previously covered rocks are exposed. The slow development at the exposed rock surface of CLAY MINERALS, which involves an increase in volume, can also produce separation. The resulting differential moisture content between the surface and inner layer causes flaking. See also SPHEROIDAL WEATHERING.

exhalant slit or **slit-band** *n.* a narrow indentation in the aperture of some GASTROPOD shells (those of the Archaeogastropoda) that accommodates the exhalant siphon far from the mouth. The calcified trace of the exhalant slit on the shell is called the *selenizone*. See Fig. 38.

exhalative mineral deposit see VOLCANIC-EXHALATIVE DEPOSIT.

exinite see MACERAL.

exogenetic, exogenic or **exogenous** *adj.* **1.** (of geological processes such as weathering and denudation) acting at or near the Earth's surface.

2. (of rocks, landforms and ores) derived through such processes. Compare ENDOGENETIC.

expansive clays *n.* (*civil engineering*) clay composed mainly of MONTMORILLONITE, which swells on absorbing water.

experimental petrology *n.* the study of igneous and metamorphic processes by reproducing under controlled conditions in the laboratory the pressures and temperatures of the different environments in which the rocks form. Both natural and synthetic SYSTEMS are investigated in this way. The results are usually portrayed on PHASE DIAGRAMS.

exploratory well, test well or **wildcat well** *n.* a well drilled in an unproved area or to an untried depth, either to seek a new pool of gas or oil or for the possibility of broadening the area of a known field.

explorer's alidade see GALE ALIDADE.

explosive index *n.* the percentage of pyroclastics found among all the solidified products of a particular VOLCANIC ERUPTION.

explosivity index *n.* an informal scale of 0–8 based on the erupted volume, plume height, duration of climactic phase and other characteristics of a VOLCANIC ERUPTION designed as a measure of the environmental effect of volcanic eruptions.

exposure *n.* an area where ROCKS can be seen free from SOIL and vegetation cover. The term is not used for detached BOULDERS. Compare OUTCROP.

exsolution *n.* the separation of a homogeneous solid solution into distinct CRYSTAL phases. For example, if an initially homogeneous high-temperature ALKALI FELDSPAR cools slowly, it exsolves into an Na-rich PLAGIOCLASE FELDSPAR phase in an alkali feldspar host. The two feldspars usually segregate within the original crystal as thin lamellae, forming the intergrowth PERTHITE.

extinction *n.* the dying out of a plant or animal species. This may have arisen from a variety of causes, for example increased competition for certain niches, variation in the physical environment, such as climatic changes or fluctuations in sea level, that affect the range of habitats available. *Background extinction* is the steady, small loss of species over time, but there are points in the geological record when the number of fossil species has decreased markedly and rapidly – sometimes called *mass* or *multiple extinctions*. Probably the best-known mass extinction occurred at the end of the CRETACEOUS, when about 85 per cent of oceanic PLANKTON species died out and, on land, many reptiles declined and disappeared, including the DINOSAURS. There are many theories about the cause of mass extinctions. Some involve cosmic phenomena, such as variations in the amount of cosmic radiation reaching the Earth's surface or periodic encounters with comets or showers of large METEORITES. Alternative hypotheses emphasise the terrestrial processes that affect the BIOSPHERE, such as variations in the salinity of the oceans or VOLCANIC ERUPTIONS that send dust high into the atmosphere, thus initiating climatic changes. It has been suggested that the end-Cretaceous event was caused by the impact of a giant extraterrestrial body that produced an extensive dust cloud, causing global cooling. Evidence supporting this hypothesis comes from a thin but very widespread layer of clay at the CRETACEOUS-TERTIARY BOUNDARY that is unusually rich in iridium, an element rare in terrestrial rocks. There are also mineral grains in the boundary layer that show the effects of SHOCK METAMORPHISM.

extinction angle *n.* the angle through which a THIN SECTION of an ANISOTROPIC mineral must be rotated to bring one of its VIBRATION DIRECTIONS into line with the vibration plane of the POLARISER of a PETROLOGICAL MICROSCOPE. In this situation, and with the ANALYSER in position, light travels directly through the mineral without being deviated and is cut out at the analyser, so the mineral appears black – it is *in extinction*. Since the vibration planes of the polariser and analyser are parallel to the cross-wires of the eyepiece, the extinction angle is measured by setting some crystallographic direction, for example a prismatic CLEAVAGE (parallel to the C-AXIS) parallel to the cross-wires and then rotating the stage until extinction occurs. This will happen four times during a complete rotation of the stage. If the extinction angle is zero, the mineral is said to have *straight extinction*. The extinction angle is useful in mineral identification. Minerals that have been strained or distorted may show *undulose* (wavy) *extinction* (a band of extinction sweeps across the crystal as the microscope stage is rotated between crossed polarisers).

extraformational *adj.* (of CONGLOMERATES and BRECCIAS) containing CLASTS that come from a source outside the area of deposition. Compare INTRAFORMATIONAL.

extraordinary ray *n.* that one of the two light rays transmitted through a UNIAXIAL mineral that vibrates in a plane containing the OPTIC AXIS. The extraordinary ray varies in refractive index and velocity according to its direction of transmission. See ORDINARY RAY, VIBRATION DIRECTIONS, INDICATRIX.

extraterrestrial geology see ASTROGEOLOGY.

extrusion *n.* **1.** an emission of LAVA on to the Earth's surface.
2. the rock formed by this. Compare INTRUSION.

extrusive volcanic rocks *n.* rocks that result from the cooling and solidification of igneous materials erupted at the surface of the Earth. Such rocks are differentiated from PLUTONIC or intrusive rocks, which have solidified from magma emplaced below the Earth's surface. Extrusive rocks tend to cool rather rapidly so they are generally fine-grained except in cases such as very thick LAVA FLOWS. Compare INTRUSIVE ROCK. See MAGMA.

F f

fabric *n.* **1.** the texture and structure of a rock body. The fabric supplies information about the geological processes that produced a particular rock. For example, the spatial orientation of the particles and crystals of which a sedimentary rock is composed implies certain conditions of deposition and consolidation. In a metamorphic rock, platy or prismatic minerals may show a PREFERRED ORIENTATION relating to the STRESS conditions under which the rock formed.
2. the physical description of a SOIL according to the spatial arrangement of its particles and voids.

fabric diagram or **petrofabric diagram** *n.* an equal-area or stereographic projection of components of a rock fabric (*fabric elements*) used in structural PETROLOGY.

face *n.* **1.** the principal or most evident side or surface of a landform, such as a cliff face.
2. one of the planar bounding surfaces of a CRYSTAL; a rational face. See also FORM.
3. the surface on which mining operations are in progress or the place that was last worked.

facet *n.* **1.** one of the small, polished plane surfaces on a cut GEMSTONE.
2. a nearly planar surface produced on a rock by the action of water or sand.
3. any plane surface resulting from faulting or EROSION and cutting across the general slope of the land.

facial suture see CEPHALON.

facies *n.* an ASSEMBLAGE or ASSOCIATION of mineral, rock or fossil features reflecting the environment and conditions of origin of the rock. It refers to the appearance and peculiarities that distinguish a rock unit from associated or adjacent units. Compare LITHOFACIES.
  The term 'facies' is used in so many contexts that the reader of current geological literature should be certain of the kind of facies that is meant. See also BIOFACIES, PETROGRAPHIC FACIES, SEDIMENTARY FACIES, METAMORPHIC FACIES.

facies change *n.* a lateral or vertical change in the characteristics of sediments or rocks. It reflects a change in the depositional environment.

facies evolution *n.* gradual lateral or vertical changes in the type of rock or fossils found in contemporaneous sedimentary deposits. Such variations reflect a change in the depositional environment.

facies fauna *n.* a group of animals characteristic of specific FACIES and thus adapted to life in the particular environment of that facies.

facies fossils *n.* assemblages of fossils, or single species or genera, that represent a particular palaeoenvironment and, therefore, a correspondingly restricted distribution.

facies map *n.* a type of map on which is shown the distribution of SEDIMENTARY FACIES that occur within a given STRATIGRAPHICAL UNIT, e.g. a lithofacies map. See also ISOPLETH.

facies tract *n.* a system of interconnected SEDIMENTARY FACIES of the same age, including the areas of EROSION from which the sediments of the facies originate.

facing direction *n.* the direction in which the beds in a FOLD become younger, defined normal to the hinge of the fold but parallel to the AXIAL PLANE (see Fig. 29 below).

failure *n.* the sudden rupture or loss of COHESION in a rock or other material stressed beyond its yield strength.

family *n.* in biological classification, a category between order and GENUS. A family may contain a single genus or a number of related genera.

Fammenian *n.* the top STAGE of the Upper DEVONIAN.

fan *n.* **1.** ALLUVIAL FAN.
 2. a fan-shaped mass of solidified LAVA.

fanglomerate *n.* a CONGLOMERATE composed of an ALLUVIAL FAN deposit that has become cemented.

fasciculate see COLONIAL CORAL.

fasciole *n.* either of two ribbon-like bands on the TEST (skeleton) of an irregular ECHINOID (see Fig. 28) that carried small spines covered by cilia. One fasciole surrounds the anterior ambulacrum, the other fasciole surrounds the anus. The movement of the cilia directed water currents towards the mouth or away from the anus.

fathogram see FATHOMETER.

fathometer *n.* a type of echo sounder. The graphic record produced by a fathometer is called a *fathogram*; it is a type of ECHOGRAM. See ECHO SOUNDING.

fault *n.* a FRACTURE in earth materials along which the opposite sides have been relatively displaced parallel to the plane of movement. The surface along which movement takes place is known as the *fault plane* (or *fault surface*, since it is not necessarily a plane). Movement may occur on a number of closely spaced faults within a FAULT ZONE rather than on a discrete surface. Fault lengths may range from a few centimetres to hundreds of kilometres, and displacement may be of comparable magnitude along the fault plane. Evidence of fault movement is often provided by SLICKENSIDES (grooves and scratches on the fault plane) or the presence of FAULT BRECCIA, FAULT GOUGE or MYLONITE. In sedimentary strata, DRAG may be manifested in fault zones, a result of frictional resistance to slippage. There may be no surface indication of faulting in an area where the soil cover is deep. Block displacement on the opposite sides of a fault plane is usually recorded in relation to sedimentary strata or to DYKES and VEINS. The apparent movement of a fault may be very different from the

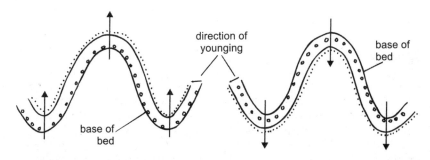

Fig. 29 **Facing direction**. The direction of facing is shown by the long arrows

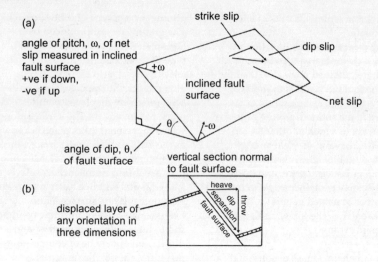

Fig. 30 **Fault**. The terms used in describing (a) total displacement or net slip and (b) the separation of a layer across a fault surface. Dip separation, throw and heave are measured in the vertical section perpendicular to the fault surface. Strike separation is the offset in the hoizontal section

actual movement if EROSION has obliterated the evidence. One basic rule in ascertaining the movement of a fault is that a fault must always be younger than the youngest rock it cuts and older than the oldest lava or undistributed sediment across it. Movement along faults may occur as anything from continuous CREEP to abrupt displacement of up to a few metres that may take place at intervals of many years apart. During quiet intervals that separate these movements, the STRESS increases until frictional forces along the fault plane are overcome. Most EARTHQUAKES result from the release of energy associated with such abrupt movement along faults.

Faults can be described according to the attitude of the fault plane and the relative DISPLACEMENT of the blocks on either side. The dip of the fault plane is

the angle between the fault plane and the horizontal. A dip of less than 45° is described as a *low-angle fault*, while a *high-angle fault* has a dip greater than 45°. Where the fault plane is not vertical, the side overlying the fault plane is called the HANGING WALL and the side underlying the fault plane is the FOOTWALL. The terminology used to describe fault displacement is shown in Fig. 30 above. The principal types of fault are as follows:

(a) a *normal fault* (or *gravity fault*) is a high-angle fault in which the hanging wall has moved downwards relative to the footwall (Fig. 31 opposite). Movement on conjugate sets of normal faults will result in the formation of graben flanked by HORSTS (see Fig. 32 overleaf, also CONJUGATE FAULTS). *Step faults* or HALF-GRABEN are formed where the downthrow of a set of normal faults is

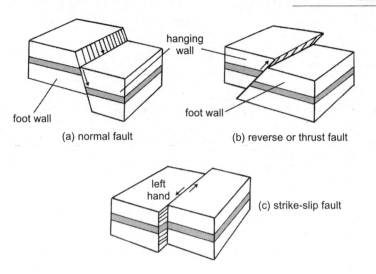

(a) normal fault

(b) reverse or thrust fault

(c) strike-slip fault

Fig. 31 **Fault**. The principal types of fault. The arrows show the direction of movement

in one direction. This is commonly accompanied by tilting of the fault blocks. (See Fig. 32(b).)

(b) a *reverse fault* is a high-angle fault in which the hanging wall has moved upward relative to the footwall (Fig. 31(b)).

(c) a *thrust* (or *thrust fault* or SLIDE) is a low-angle reverse fault. The hanging wall block above the thrust is known as the *thrust sheet*. Faults often develop on the limb of a major fold, sub-parallel to the axial plane. If one develops on the underlying limb of an ANTICLINE or ANTIFORM, it is known as a *thrust*, while a *lag* develops on the underlying limb of a SYNCLINE or SYNFORM (see Fig. 32(c)). The DISPLACEMENT on a thrust may be very large, for example the horizontal displacement of about 16 kilometres on the Moine Thrust in the northwest of Scotland or a displacement of some 40 kilometres on the Glarus Thrust in the Alps (see also NAPPE, DUPLEX). Normal

and reverse faults can be classed as DIP-SLIP FAULTS, since the dominant direction of displacement is parallel to the dip of the fault plane.

(d) in a *strike-slip fault* the principal displacement is parallel to the STRIKE of the fault plane (see Fig. 31(c)). The fault plane is usually vertical, or nearly so, and the FAULT LINE straight or only gently curved. These faults are sometimes called *wrench faults* or *transcurrent faults*. TRANSFORM FAULTS are a type of strike-slip fault occurring at plate boundaries (see PLATE TECTONICS). The sense of movement on a strike-slip fault is described as *left-lateral* (or *left-handed* or *sinistral*) if the block on the side of the fault opposite to the observer appears to have moved to the left, while if the block has apparently moved to the right the displacement is *right-lateral* (*right-handed* or *dextral*).

(e) *rotational faults* occur where the amount of displacement varies along

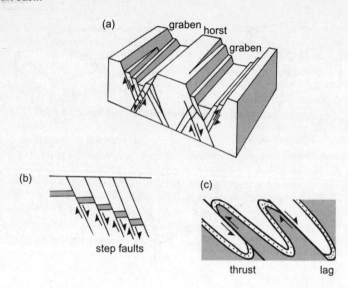

Fig. 32 **Fault**. (a) graben and horst formed by movement on conjugate sets of normal faults. (b) step faults. (c) the difference between a thrust and a lag

the fault. The rocks on either side of the fault behave like rigid blocks rotating about an axis at right angles to the fault plane. A *hinge fault* occurs where a fault dies out and the end of the fault acts as a hinge. A *scissors fault* or *pivot fault* occurs where the sense of displacement changes between opposite sides of the axis about which the blocks rotate. It is likely that most faults have some component of rotational movement. See also DIP FAULT, STRIKE FAULT, OBLIQUE FAULT, OBLIQUE-SLIP FAULT, OFFSET, LISTRIC FAULT.

fault basin *n*. a depression separated from the surrounding area by FAULTS.

fault block *n*. a unit of the Earth's crust either completely or partly bounded by FAULTS. During faulting and tectonic activity it functions as a unit. See also FAULT-BLOCK MOUNTAIN.

fault-block mountain *n*. a mountain formed by block FAULTING, i.e. isolated by faulting and categorised as struc-tural or tectonic. Good examples are found in the Basin and Range province of Nevada, in the USA, while the Ruwenzori Massif, a HORST in the East African Rift Valley, is the highest nonvolcanic mountain in Africa. The geomorphic characteristics of fault-block mountains include more or less rectilinear borders, a general lack of continuity of formation, fresh FAULT SCARPS and occasional seismic revival. The uplifted blocks may have been stripped of younger formations that covered them and are thus RELICT landforms. The normal fault boundaries of block-fault mountains are evidence of crustal extension, whereas fold mountains are products of crustal compression. See also GRABEN.

fault breccia *n*. a BRECCIA composed of ANGULAR fragments resulting from the shattering or crushing of rocks during movement along a FAULT or in a FAULT ZONE.

fault escarpment see FAULT SCARP.

fault gouge *n*. a fine-grained, claylike substance formed by the grinding of rock material as a FAULT develops; it may also be formed by decomposition caused by circulating solutions.

faulting *n*. the action or process of fracturing and DISPLACEMENT that produces a FAULT. Evidence of faulting may be demonstrated by such criteria as:

(a) discontinuity of structures;

(b) repetition or omission of strata;

(c) the presence of features characteristic of fault planes;

(d) sudden changes in sedimentary facies;

(e) physiographical features, such as springs, aligned along the base of a mountain range.

fault line or **fault trace** *n*. the intersection of a FAULT with the ground surface. It usually varies in form from straight to moderately sinuous. See also FAULT SCARP, FAULT-LINE SCARP.

fault-line scarp *n*. a cliff, basically parallel to the FAULT LINE, that has retreated because of extensive EROSION, leaving the actual fault line buried beneath sediment. The fundamental difference between a FAULT SCARP and fault-line scarp is that the former is produced by movement of the FAULT, whereas the latter is produced by differential resistance to erosion on either side of the fault line. An *obsequent fault-line scarp* faces in the opposite direction from the original fault scarp and occurs at a lower stratigraphical level. In this case, the original fault scarp has been completely eroded away. Erosion of softer materials on the uplifted block leaves the resistant rocks of the downthrown block standing topographically higher.

fault plane see FAULT.

fault scarp or **fault escarpment** *n*. a cliff formed by an upthrown FAULT BLOCK and trending along the FAULT LINE. It is the evidence of the latest movement along the FAULT. Compare FAULT-LINE SCARP.

fault set *n*. a group of parallel or near-parallel FAULTS associated with a particular deformational event. Two or more interconnecting fault sets constitute a *fault system*.

fault surface see FAULT.

fault system see FAULT SET.

fault trace see FAULT LINE.

fault zone *n*. a zone consisting of a network of many small FAULTS, or a zone of FAULT BRECCIA or MYLONITE. Fault zones may be hundreds of metres wide.

faunal break *n*. an abrupt change in a stratigraphical sequence from one faunal fossil ASSEMBLAGE to another.

faunal province *n*. a region characterised by an ASSEMBLAGE of fossils having a distinct identity at the species, genus and sometimes higher levels. The faunas are widely distributed within the province. The Malvinokaffric Province of the Devonian of Gondwana contains a distinctive group of marine faunas common to southern Africa, the Falkland (Malvinas) Islands and southern South America.

faunal succession *n*. the chronological sequence of life forms through geological time. Fossil fauna and flora succeed one another in a definite recognisable order, as EXPRESSED in the LAW OF FAUNAL SUCCESSION: like ASSEMBLAGES of fossils indicate like geological ages for the strata containing them.

faunule *n*. **1.** also called **local fauna**, a collection of FOSSIL fauna gathered from a bed extending over a very limited geographical area.

2. an ASSOCIATION of fossil fauna found

in a single stratum or a few contiguous layers in which members of a single community predominate.

fayalite *n*. a silicate mineral, Fe2SiO4, the iron-rich END MEMBER of the OLIVINE group.

feather joints or **pinnate fractures** *n*. a set of EN ECHELON extension JOINTS intersecting a FAULT plane. The acute angle between the joints and the fault plane gives the sense of DISPLACEMENT of the BLOCKS. The name reflects the resemblance of the joints and the fault, as seen in cross-section, to the barbs and shaft of a feather.

feeder *n*. **1.** the channel through which MAGMA passes from the MAGMA CHAMBER to some INTRUSION.
2. an opening in a rock through which mineral-bearing solutions can travel.
3. TRIBUTARY of a river.

feldspar *n*. the most important group of the rock-forming minerals that make up about 60 per cent of the Earth's crust. Feldspars are essential constituents of most IGNEOUS ROCKS; the kind and amount of feldspar present is used in classification. They frequently occur in METAMORPHIC ROCKS and in many SEDIMENTARY ROCKS, more commonly in the ARENACEOUS rocks than ARGILLACEOUS. They are TECTOSILICATES (i.e. three-dimensional framework silicates) in which each SiO_4 tetrahedron shares all four oxygens with adjacent tetrahedra. To allow cations such as Na^+, K^+ and Ca^{2+} to enter the lattice and maintain the electrical balance, there is substitution of some Al^{3+} in place of Si^{4+}. The composition of most feldspars can be represented on a ternary diagram with $KAlSi_3O_8$ (potassium feldspar, Or), $NaAlSi_3O_8$ (albite, Ab) and $CaAl_2Si_2O_8$ (anorthite, An) as END MEMBERS (Fig. 33 opposite). There are two principal groups of feldspars: the ALKALI FELDSPARS, with compositions between Or and Ab, and the PLAGIOCLASE feldspars, with compositions between Ab and An (Fig. 33 opposite). The alkali feldspars contain less than 5–10 per cent of the An molecule in their structure. At high temperatures there is complete SOLID SOLUTION between the K and Na end members, but at lower temperatures the SOLID SOLUTION is more limited, and in feldspars that cool slowly (for example, those in plutonic rocks), unmixing occurs, with the development of perthitic intergrowths (see PERTHITE). Three series of alkali feldspars have been defined:
(a) the high-temperature series, SANIDINE Ab_0–Ab_{63}, ANORTHOCLASE Ab_{63}–Ab_{90} and high-albite Ab_{90}–Ab_{100}, is characteristic of extrusive rocks, where rapid cooling prevents unmixing.
(b) the lower temperature series, ORTHOCLASE Or_{100}–Or_{85}, orthoclase cryptoperthite Or_{85}–Or_{20} and low-albite Or_{20}–Or_0, occurs in plutonic rocks.
(c) the lowest temperature series, MICROCLINE Or_{100}–Or_{92}, microcline-cryptoperthite Or_{92}–Or_{20} and low-albite Or_{20}–Or_0, occurs in low-grade metamorphic rocks and in some granites.

Sanidine and orthoclase are monoclinic, while anorthoclase, albite and microcline are triclinic. Sanidine and orthoclase show simple TWINNING; microcline and anorthoclase typically show multiple twinning in two directions approximately at right angles, which gives a cross-hatched or 'tartan' appearance. The twinning in anorthoclase is on a finer scale than that of microcline. Albite shows multiple twinning in one direction, which appears as a set of fine parallel lines. ADULARIA is a very low-temperature form of K-feldspar. It is monoclinic and rarely shows any twinning.

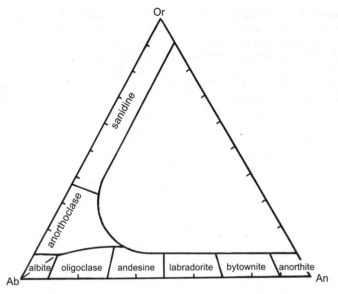

Fig.33 **Feldspar**. A composition diagram for feldspars

The plagioclase feldspars, with compositions between albite and anorthite, may contain up to 10 per cent of the K-feldspar molecule in their structure. The composition range of the different varieties of plagioclase is shown in Fig. 33 above. The composition is generally shown in terms of the proportion of the anorthite molecule so, for example, anorthite is An_{90}–An_{100}, albite is An_0–An_{10}. Sodium-rich plagioclases are found in ACID igneous rocks and the lower grade metamorphic rocks, whereas Ca-rich plagioclases are typical of BASIC and ultrabasic igneous rocks and high-grade metamorphic rocks. All the plagioclases are triclinic and show multiple twinning in the form of parallel lamellae, which are much broader in the Ca-rich forms. Compositional ZONING is common in the plagioclases of igneous rocks. As with the alkali feldspars, there is continuous solid solution at high temperatures (high-albite to high-anorthite). If rapidly cooled, these feldspars do not show unmixing. Slow cooling of lower-temperature forms results in the formation of perthitic intergrowths, of which the most important forms are PERISTERITE, BØGGILD intergrowths (in plagioclase of the composition An_{40}–An_{60}, i.e. labradorites – see LABRADORESCENCE) and Huttenlocher intergrowths (An_{70}–An_{85}). The last two are submicroscopic intergrowths.

Celsian and hyalophane are rare barium varieties of feldspar (containing more than 2 per cent BaO). Celsian contains more than 90 per cent of the $BaAl_2Si_2O_8$ molecule, while hyalophane contains less than 30 per cent. Most feldspar is colourless, white or light grey but may also be brown, yellow, red, green or black. MOONSTONE, a gem variety, is milky opalescent; other gem varieties are AVENTURINE and AMAZONITE, a deep-green microcline.

feldspathic *adj.* (of a rock or mineral aggregate) containing FELDSPAR. The percentage of feldspar present must usually be within certain limits. For example, a feldspathic SANDSTONE contains from 10 to 25 per cent feldspar and is intermediate in composition between a quartz sandstone and an ARKOSIC SANDSTONE. See SUBARKOSE.

feldspathoid *n.* a rock-forming mineral chemically and structurally similar to the FELDSPARS but containing less SiO_2. In those igneous rocks that are silica-undersaturated, i.e. that crystallise from MAGMAS containing too little silica to form feldspars, feldspathoids take the place of feldspars. Feldspathoids never occur in the same rock with quartz. The principal feldspathoids are: NEPHELINE, $(Na,K)AlSiO_4$, and KALSILITE, $KAlSiO_4$ (up to 25 per cent of the kalsilite molecule can substitute in nepheline); LEUCITE, $KAlSi_2O_6$; the *sodalite group*: SODALITE, $Na_8(AlSiO_4)_6C_2$; NOSEAN, $Na_8(AlSiO_4)_6SO_4$; HAÜYNE, $(Na,Ca)_{4-8}(Al\ SiO_4)_6(SO_4)_{1-2}$; and LAZURITE, $(Na,Ca)_8(Al,SiO_4)_6(SO_4,S,CU_2$; and cancrinite-vishnevite, $(Na,Ca,K)_{6-8}(AlSiO_4)_6(CO_3,\ SO_4,Cl)_{1-2} \cdot 1{-}5H_2O$. Cancrinite is the carbonate-rich member while vishnevite is sulphur-rich. See SILICA SATURATION.

felsenmeer see BLOCK FIELD.

felsic *adj.* **1.** an acronymic word derived from FELDSPAR and SILICA and applied to light-coloured silicate minerals such as quartz, feldspar and feldspathoids. **2.** rocks that contain an abundance of any of these minerals. Compare MAFIC.

felsite *n.* **1.** any light-coloured aphanitic igneous rock composed chiefly of quartz and feldspar. When PHENOCRYSTS (usually of quartz and feldspar) are present, the rock is called a felsite PORPHYRY. **2.** a rock showing FELSITIC texture.

felsitic or **felsitoid** *adj.* (of IGNEOUS ROCKS) having a compact, very fine-grained texture of microcrystalline or cryptoc-rystalline aggregates not distinguish-able by the unaided eye. Sometimes felsitic is restricted to light-coloured quartz-feldspar rocks and APHANITIC is used for dark-coloured rocks. Compare FELSIC.

felspar *n.* obsolete synonym for FELDSPAR.

felty or **felted** *adj.* (of texture) character-ising dense HOLOCRYSTALLINE igneous rocks or the dense holocrystalline groundmass of porphyritic igneous rocks consisting of tightly and flatly pressed microlites. These are generally FELDSPAR, interwoven in an irregular fashion. Many ANDESITES and TRACHYTES contain crowded feldspar microlites that show subparallel arrangement as a result of flow, with the microlite interstices occupied by microcrypto-crystalline material. This texture is also called PILOTAXITIC or TRACHYTIC.

femic *adj.* an acronymic term derived from iron (Fe) and magnesium (Mg) and applied to the group of standard normative dark-coloured minerals of which these elements are basic compo-nents. These minerals include the OLIVINE and PYROXENE molecules and most normative ACCESSORY MINERALS. See CIPW classification (normative composi-tion). The corresponding term for ferromagnesian minerals actually present in a rock is MAFIC. Compare SALIC.

fen see BOG.

fenestral see DISMICRITE.

fenite *n.* a syenitic rock produced by alkali METASOMATISM in the contact zone around a CARBONATITE/IJOLITE intrusion. Fenites usually contain ALKALI FELDSPAR and AEGERINE, with or without alkali AMPHIBOLE. The process is called *fenitisation*. (The name is derived from the Fen ring complex in Norway.) See SYENITE.

fenitisation see FENITE.

fenster see NAPPE.

Ferrel's law *n.* a statement of the effect of the CORIOLIS FORCE in the Earth's wind systems. It is the dynamic concept that currents of air or water in the Northern Hemisphere are deflected to the right and those in the Southern Hemisphere to the left. This deflection, the result of the Coriolis force, is nearly balanced by the pressure gradient but, because of friction, is not completely balanced. Compare GEOSTROPHIC MOTION. A *Ferrel cell* is the name given to a mid-latitude atmospheric CONVECTION CELL. A similar tropical cell is a *Hadley cell.*

ferricrete *n.* 1. an amalgam of surface sand and gravel cemented into a mass by iron oxide.
2. a ferruginous DURICRUST. Compare CALCRETE, SILCRETE.

ferriferous *adj.* 1. (of a mineral) containing iron.
2. (of a sedimentary rock) containing more iron than is usually the case. Compare FERRUGINOUS.

ferrimagnetism *n.* a weak ferromagnetism displayed by some substances in which the MAGNETIC DOMAINS are aligned antiparallel but the components are not equal. There is a resultant spontaneous magnetisation. See FERROMAGNETIC, compare ANTIFERROMAGNETISM. MAGNETITE and other iron spinels show this type of behaviour.

ferroelectric effect *n.* a spontaneous polarisation that some crystals display, (i.e. they are electrically charged) that can be reversed by the application of an electrical field. Compare PIEZOELECTRIC EFFECT. The spontaneous polarisation is affected by temperature change; see CURIE POINT.

ferromagnesian *adj.* containing iron and magnesium. In petrology, it is applied to certain dark silicate minerals, particularly AMPHIBOLE, OLIVINE and PYROXENE, and to igneous rocks containing these minerals as dominate components. Compare FEMIC, MAFIC.

ferromagnetic *adj.* having a magnetic permeability very much greater than unity. Certain elements (iron, nickel, cobalt) and alloys with other elements (titanium, aluminium) exhibit relative permeabilities up to 10°C. A ferromagnetic substance, e.g. iron, at a temperature below the CURIE POINT, can show a magnetisation in the absence of an external magnetic field. The magnetic movements of its atoms have the same orientation. Compare DIAMAGNETIC, PARAMAGNETIC.

ferruginous *adj.* 1. pertaining to iron or containing it, e.g. a sandstone cemented with iron oxide, or with a cement containing iron oxide.
2. (of a rock) red or rust-coloured from the presence of ferric oxide.

festoon bedding see CROSS-BEDDING.

fetch *n.* 1. the distance that prevailing winds travel over a body of water, thus generating a wave system. Wave height is determined by wind velocity and duration.
2. the area of generation of a given wave system. Coastlines, frontal systems or isobar curvature may be used to define the boundaries of an area of fetch.

fiamme *n.* elongate flattened PUMICE fragments, characteristic of WELDED TUFFS and IGNIMBRITES. See EUTAXITIC TEXTURE.

fibroblastic or **nematoblastic** *adj.* (of a type of fabric in metamorphic rock bodies) characterised by minerals of even grain size and fibrous habit. The fabric results from solid-state CRYSTALLISATION during METAMORPHISM.

fibrous texture *n*. an array of circular, needle-like crystals found in certain mineral deposits, such as asbestos.

fiducial time *n*. a time noted or marked on a seismic record to correspond to some arbitrary time. The marks may help in synchronising different records or may indicate some reference base.

field classification *n*. the examination or preliminary analysis of fossils or hand specimens of rocks or minerals, commonly with the aid of a hand lens.

field geology *n*. that part of GEOLOGY that is practised by direct observation in the field.

filiform *adj*. having the form of a thread.

fill *n*. 1. artificial deposits of soil, rock or various debris materials used for building embankments, filling in soggy ground, extending a shoreline into a lake or filling unused mine workings. 2. sediment deposited by any agent in such a manner that it fills or partially fills a valley.

filter *n*. a device used in a SEISMOGRAPH circuit to control the frequency characteristics of the recording system. Low frequencies are usually excluded in order to prevent GROUND ROLL and noise of other types from interfering with reflection of SEISMIC WAVES. High frequencies are attenuated in order to remove wind noise and other extraneous effects.

filter pressing *n*. a theoretically possible process of MAGMATIC DIFFERENTIATION wherein the melt in a mushlike, crystal-rich MAGMA separates from the crystals, either by draining or by being pressed out. It might occur on the floor of MAFIC intrusions, where the weight of the accumulating crystals forces some of the melt out of the underlying crystal mush.

fine gold see CARAT.

fine-grained *adj*. 1. (of a SEDIMENTARY ROCK) a general term that has been interpreted to include sedimentary rocks with a texture such that the average diameter of the particles is less than 1/16th millimetre (SILT size and smaller). 2. (of an IGNEOUS ROCK) having such a texture that particles have an average diameter of less than one millimetre. See APHANITIC.

fines *n*. 1. very small particles in a mixture of particles of various sizes. 2. finely crushed ore, mineral or coal.

fine sand *n*. SAND with particles ranging from 0.125 to 0.25 millimetre in size.

finger lake *n*. a long narrow lake occupying a deep trough, thought to have formed as a result of glacial erosion concentrated in preglacial valleys. Examples may be found in the Scottish Highlands, the English Lake District and Wales, as well, as in northern Italy and the USA (the Finger Lakes of New York State).

fiord see FJORD.

fireclay or **refractory clay** *n*. a CLAY that can withstand high temperatures without disintegrating or turning pasty. Fireclays are non-plastic clays, composed dominantly of kaolinite with illite, quartz and carbonaceous material. *Flint clays* are non-plastic fireclays that are extremely hard, microcrystalline and composed predominantly of KAOLIN. They have flint-like characteristics. Much fireclay is derived from underclays beneath COAL beds, but not all underclays are fireclays. Fireclays form where surface conditions permit most minerals, except kaolinite and illite, to be leached out. Fireclay is rich in hydrous aluminium silicate and is used widely to manufacture clay crucibles and firebrick, and as a binder in moulding sands.

firedamp *n*. a combustible and highly

explosive gas consisting chiefly of methane, formed especially in coal mines.

fire fountain *n*. a LAVA FOUNTAIN.

fire opal see OPAL.

firn *n*. a form of snow intermediate between freshly fallen snow and ice. A *firn field*, more accurately called *névé*, is a wide expanse of GLACIER surface over which snow accumulates and becomes firn. After lasting through one summer melt season, snow becomes firn by alternate melting and freezing, condensation and compaction. The point at which firn becomes glacial ice is not defined by universal standards. The transformation is effected when densities of 0.4 to 0.55 are reached, although higher densities are often recorded. The density at which the compacting material sometimes becomes impervious to water is 0.55. Although some literature uses névé and firn synonymously, many specialists do not equate them. Firn is the term used for the process and the material, whereas névé is used as a locational term for the position of the snow accumulation area.

firn limit see FIRN LINE.

firn line *n*. the lower limit of FIRN in summer; above it is the zone of net accumulation, below it the zone of net loss. The firn line is the line at which the average velocity of glacial flow is generally greatest.

first arrival *n*. the first energy to reach a SEISMOGRAPH from a source of seismic activity. First arrivals on reflection records are used to obtain information about a WEATHERED LAYER. They are often used as a basis for REFRACTION studies.

fissility *n*. the property of splitting or dividing readily along closely spaced parallel planes, e.g. BEDDING in SHALE or CLEAVAGE in SCHIST. Any SEDIMENTARY ROCKS, especially fine-grained varieties, tend to part parallel to the stratification, thus exhibiting *bedding fissility*.

fission *n*. the spontaneous or induced division of the nucleus of an atom into nuclei of lighter atoms, accompanied by the release of great amounts of energy. Compare FUSION.

fission-track dating see DATING METHODS.

fissure *n*. an extensive cleft, break or fracture in a rock formation, volcano or the Earth's surface.

fissure eruption *n*. a VOLCANIC ERUPTION from a FISSURE or series of fissures rather than from a central vent. In contrast to central vent eruptions, which occur in many tectonic settings, fissure eruptions favour tensional environments. The MID-OCEANIC RIDGES form the most extensive fissure system. The products of fissure eruptions are generally basaltic LAVAS, such as those that form the OCEANIC CRUST or vast (continental) FLOOD BASALT plateaux, although fissure eruptions of IGNIMBRITE have also occurred. See also VOLCANIC ERUPTION, SITES OF.

fissure polygon see NONSORTED POLYGON.

fissure vein *n*. a MINERAL DEPOSIT of veinlike shape, characterised by clearly defined walls as opposed to extensive country-rock REPLACEMENT.

fixed carbon *n*. a solid, finely divided black elemental carbon the presence of which distinguishes COAL from PEAT. The higher the proportion of fixed carbon, and the lower the content of VOLATILES, the higher the RANK of coal. The fixed carbon content of a coal is calculated by subtracting the moisture, volatile and ash percentages from 100. The fixed carbon content is a measure of the combustible carbon in a coal.

fjord or **fiord** *n*. a long, narrow steep-sided inlet of mountainous glaciated coasts. It results where the sea invades a deeply excavated glacial trough after the glacier has melted. The side walls are characterised by hanging valleys (see GLACIATED VALLEY) and high waterfalls. The threshold at the seaward end may be bedrock or the site of the terminal MORAINE. If BEDROCK, it marks the place where the glacier's erosive strength decreased. Fjords are found along the northwest coast of Scotland, in Norway, Greenland, Alaska, British Columbia, Patagonia, Antarctica and New Zealand.

flagstone *n.* a rock, such as micaceous sandstone or shale, that can be split along bedding planes into slabs suitable for paving (*flagging*).

flame structure *n.* a SEDIMENTARY STRUCTURE in which flame-shaped pods of mud have been compressed and forced upwards into an overlying layer. See also LOAD CASTS.

flame test *n.* a method used to detect the presence of certain elements in a substance. A sample of the given material is heated over a Bunsen burner and the colour of the flame observed. Particular elements impart a characteristic colour to the flame, e.g. sodium, yellow; barium, green; strontium, crimson.

Flandrian see CENOZOIC ERA, Fig. 13.

flank eruption see VOLCANIC ERUPTION, SITES OF.

flaser bedding *n.* discontinuous curved lenses of finer sediment (mud or silt) deposited in the troughs or draped over ripples in cross-laminated sands (see CROSS-BEDDING, RIPPLE MARKS). If the proportion of silt and mud increases, with the sand occurring in isolated ripple form-sets, the structure is known as *lenticular bedding*. Flaser bedding is characteristically produced where rippled sand is moved by the tides on intertidal mud flats.

flaser gabbro *n.* a coarse-grained cataclastic gabbro, or gabbroic augen gneiss, dredged from oceanic fracture zones, in which flakes of chlorite or mica swirl around lenses (AUGEN) of quartz and feldspar; recrystallisation and formation of new minerals is apparent.

flaser structure *n.* a fabric structure of dynamically metamorphosed rocks in which large ovoidal grains that have survived deformation are surrounded by highly sheared and crushed material (MYLONITE). The general appearance is that of a flow texture. Compare AUGEN.

flexible see TENACITY.

flexible sandstone see ITACOLUMITE.

flexural-flow fold *n.* a FOLD that is produced by simple SHEAR displacement distributed evenly throughout the beds. The DISPLACEMENT occurs parallel to the bedding surface so the thickness of the LAYERING does not change. Rocks that have a relatively uniform character, for example without well-marked bedding surfaces, are more likely to be affected. Compare FLEXURE-SLIP FOLD.

flexure see HINGE.

flexure-slip fold *n.* a FOLD in which slip occurs on discrete planes within the structure (compare FLEXURAL-FLOW FOLD), by simple shear parallel to the bedding surface. There is no change in thickness and the fold is a parallel fold. This type of folding generally affects rocks with well-marked bedding planes.

flint see CHERT.

flint clay see FIRECLAY.

flinty-crush *n.* an old term for very fine-grained cataclastic rocks with a flinty appearance. See MYLONITE.

float *n.* a term for loose rock fragments that are often found in the soil on a slope, even in the absence of clean outcrops. Insofar as the fragments cannot creep uphill, their source must be upslope from the location in which they are found. Thus floats are useful in fixing the upper contact of their source rocks. *Float ore* is a type of float composed of fragments of VEIN material, usually found downhill from the outcrop containing the vein. See also PLACER.

flocculation *n.* the process of forming aggregates or compound masses of particles, e.g. the settling of clay particles in water. In mining technology, ore particles are coagulated by reagents that promote the formation of aggregates, after which excess water is removed.

flood basalt *n.* formed from basaltic lava of low viscosity erupted in large volume generally from fissure eruptions. Accumulations of flood basalt flows submerge pre-existing topography, producing flat-topped lava plateaux. The Deccan plateau in India, covering 250 000 square kilometres, was created 65–50 Ma ago by basaltic flood eruptions. Similar lava floods created the Columbia River Plateau of the northwestern USA (17–6 Ma), the Karroo flood basalt province of South Africa (200–160 Ma) and the Thulean flood basalt province (66–52 Ma), parts of which are found on either side of the North Atlantic Ocean, in the northwest British Isles, the Faeroe Islands, Iceland, Greenland and Baffin Island. Flood basalt eruptions are associated with tensional environments and rifting, and most are related to the break-up of continents (for example, the Thulean province). Many continental flood basalts are tholeiitic basalts, but intermediate and alkali basalts also occur. Some provinces also contain picritic lavas (for example, the Thulean

province and the Karroo province). See PICRITE, KOMATIITE.

flood basin *n.* **1.** a tract of land that is inundated during the highest of floods. **2.** a flat area between a sloping plain and a natural levee where swampy vegetation often grows.

floodplain *n.* that portion of a river valley, adjacent to the river, that is built up of ALLUVIUM deposited during the present disposition of the stream flow. It is covered with water when the river overflows in flood periods. Meandering streams are typical features of floodplains. During the process of LATERAL EROSION, the form of a meandering stream is altered by reduction, trimming and cutting through until all that remains is a crescentic mark, a *floodplain meander scar*, indicating the former position of a river MEANDER on a floodplain. The beginnings of a floodplain are represented by lunate or sinuous strips of coarse ALLUVIUM along the inner bank of a stream meander. These are called *point bars*.

floodplain meander scar see FLOODPLAIN.

flow *n.* **1.** the movement or rate of movement of water, or the moving water itself.
2. any rock DEFORMATION that is not immediately removable without permanent loss of COHESION.
3. the mass movement of unconsolidated material in the semifluid or still plastic state, e.g. a MUDFLOW.
4. see LAVA FLOW.
5. see GLACIER FLOW.

flowage fold or **flow fold** *n.* a FOLD that occurs in relatively plastic rock in which the thickness of an individual bed is not constant. The rocks have flowed towards the SYNCLINE trough, and the DEFORMATION shows no apparent surfaces of a slip.

flow banding see FLOW LAYERING.

flow breccia *n.* a LAVA FLOW containing fragments of solidified lava that have become joined or cemented together by the still-fluid parts of the same flow.

flow cast *n.* a LOAD CAST, a SOLE MARK consisting of a ridge or other raised feature formed on the underside of a sand bed by sand that displaced soft plastic sediment. A flow cast may be used to determine the top and bottom of a bed. See also FLUTE MARK.

flow cleavage see SLATY CLEAVAGE.

flow differentiation see MAGMATIC DIFFERENTIATION.

flower structures *n.* a group of up-wardly diverging faults formed in a STRIKE-SLIP zone. *Negative flower structures* are found in depressed regions resulting from TRANSTENSION, and *positive flowers* are uplifted zones caused by TRANSPRESSION.

flow fold see FLOWAGE FOLD.

flow-foot breccia *n.* a BRECCIA formed by the disintegration of lava as it flows into water. See LAVA DELTA.

flowing artesian well *n.* an ARTESIAN well in which the hydrostatic pressure is sufficient to raise the water above the land surface. Compare NONFLOWING ARTESIAN WELL.

flow layering or **flow banding** *n.* a feature of an IGNEOUS ROCK expressed by layers of contrasting colour, texture or sometimes mineralogical composition, formed by MAGMA or LAVA FLOW. *Flow layers* vary in thickness from less than one millimetre up to many centimetres. Suspended crystals existing at the time of solidification may be segregated into crystal-rich and crystal-poor zones. These, as well as mineral streaks or inclusions, indicate the direction of flow before CONSOLIDATION and are called *flow lines*. Flow layers may be planar, folded into smooth open forms or intricately contorted. *Planar layers*

originate with LAMINAR FLOW within the moving magma; *folded* or *contorted layers* indicate a transition from totally laminar to nonlaminar or TURBULENT FLOW as the moving magma responds to local changes, such as obstruction. See also BANDING, RHYOLITE.

flow line see FLOW LAYERING.

flow plane *n.* the plane along which displacement occurs in igneous and metamorphic rocks. In IGNEOUS ROCKS it is a *flow layer*, in METAMORPHIC ROCKS it is usually subparallel to the FOLIATION visible in hand specimens. See FLOW LAYERING.

flowstone *n.* any MINERAL DEPOSIT formed by water flowing on the floor or walls of a cave. See DRIPSTONE, TRAVERTINE, CAVE ONYX.

flow texture see FLOW LAYERING, RHYOLITE, TRACHYTE.

flow unit *n.* one of a stack of beds or sheets of LAVA or PYROCLASTS formed during a single eruption from the same VOLCANO.

fluid inclusion *n.* a minute (less than 10^{-2} millimetre diameter) bubble of liquid enclosed in a mineral. Such INCLUSIONS may also contain a bubble of gas and one or more crystalline phases that were probably formed as a liquid cooled from high temperatures. See GEOLOGICAL THERMOMETRY.

fluidisation *n.* the EROSION, ENTRAINMENT and transport of solid particles by hot gas, so that the whole behaves like a fluid. This process is thought to be important in the formation of DIATREMES and transport of PYROCLASTIC FLOWS (see also IGNIMBRITE).

flume *n.* **1.** a deep, narrow gorge or defile containing a stream, torrent or series of cascades.
2. an apparatus used to reconstruct and study the formation of a SEDIMENTARY STRUCTURE in the laboratory.

fluorapatite *n.* **1.** a mineral of the APATITE group, $Ca_5(PO_4)_3F$. It is a common ACCESSORY MINERAL in IGNEOUS ROCKS, high-temperature HYDROTHERMAL veins and METAMORPHIC ROCKS. It also occurs as COLLOPHANE in marine deposits.
2. an APATITE mineral in which the amount of fluorine exceeds that of chlorine, hydroxyl or carbonate.

fluorescence *n.* the property of emitting light as a result of absorbing electro-magnetic radiation. Some minerals exhibit this property when subjected to light rays of a particular wavelength. Some respond to ultraviolet light, some to X-rays and others to cathode rays. Many minerals that are drab to the naked eye become brilliant with colour when placed under the appropriate wavelength of radiation. The property of fluorescence is utilised in mineral-ore prospecting and in detecting the presence of oil at a surface.

fluoride *n.* a compound of fluorine, such as methyl fluoride, CH_3F.

fluorine dating *n.* a method of determin-ing the related ages of Pleistocene or Holocene fossil bones, from the same excavation, based on the gradual combination of fluorine in groundwater with the calcium phos-phate of buried bone material. See also DATING METHODS.

fluorite *n.* a transparent to translucent halide mineral, CaF_2, of the isometric system. Fluorite is colourless and completely transparent when pure but is also found in blue, purple, yellow and green shades of colour. It occurs in medium- and high-temperature hydrothermal veins in association with lead, zinc and silver sulphides. Its most important use is in the manufacture of hydrofluoric acid. See also BLUE JOHN.

fluorspar *n.* the commercial name for FLUORITE.

flute *n.* an asymmetrical spoon- or scoop-shaped groove or depression in a sedimentary bed, elongated parallel to the current flow. Such impressions are formed by non-laminar sediment-bearing currents that can scour a soft stream bottom or sea bed. Flutes often occur in groups, arranged in an *en-échelon* pattern. They are most often observed in counterpart as FLUTE MARKS.

flute mark *n.* a structure found on the underside (the SOLE) of some sandstone or siltstone beds, especially in environ-ments where turbidites have been common. A flute mark is a discontinu-ous bulbous shape formed by the filling of a FLUTE. See also SOLE MARK.

fluting *n.* **1.** a differential EROSION in which the surface of an exposed coarse-grained rock, such as granite, becomes ridged or corrugated with flutes.
2. the process of generating a flute by the abrasive effect of water on a soft or muddy surface.
3. the formation by a GLACIER of grooves or furrows on the face of a rock mass that is blocking the advance of the glacier.

fluvial *adj.* **1.** of or pertaining to rivers.
2. living or growing in a river or stream.
3. produced by the action of a river or stream, e.g. fluvial deposits. See also FLUVIATILE.

fluvial cycle of erosion *n.* the sculpture and reduction of a landmass to base level by running water, specifically the action of rivers and streams.

fluviatile *adj.* pertaining to, belonging to or peculiar to rivers, especially the physical products of river action, e.g. a *fluviatile dam*, which is a dam formed in a stream channel by tributary-depos-ited sediment.

fluviation *n.* the sum of activities and processes in which rivers or streams are the agents.

fluvioglacial see GLACIOFLUVIAL.

flux *n.* **1.** (*physics*) the rate of flow of fluid, particles or energy. **2.** a quantity expressing the strength of a force field in a particular area. **3.** (*chemistry, metallurgy*) a substance used to refine metals by combining with impurities to form an easily removable molten mixture.

flux density *n.* the thermal, magnetic or electric flow per unit of cross-sectional area. *Radiation flux density* (electromagnetic radiation) involves a volumetric density rather than a flow across a surface.

fluxing ore *n.* an ORE smelted chiefly because it contains *fluxing agents* rather than for its metal content. Fluxing agents do not have to be added during the reduction of such ores.

fluxstone *n.* in metallurgical processes, a limestone, dolomite or other rock used to lower the FUSION temperature of the ORE or to combine with any impurities present to form a fluid SLAG.

flying magnetometer see AIRBORNE MAGNETOMETER.

flysch *n.* a sedimentary deposit typically consisting of a thick sequence of interbedded MARINE SHALES and GREYWACKE SANDSTONES that were deposited by TURBIDITES and display GRADED BEDDING. Flysch is thought to be derived from the EROSION of rapidly rising fold mountains and is itself deformed in the later stages of the OROGENY. It is therefore a syntectonic deposit (compare MOLASSE). The term was first defined in relation to Alpine rocks, but its use has been extended to other OROGENIC BELTS.

focal sphere *n.* an arbitrary sphere drawn about the SEISMIC FOCUS, or hypocentre, of an EARTHQUAKE, to which body waves recorded at the Earth's surface are referred.

focus see SEISMIC FOCUS.

foid *abbreviation for* FELDSPATHOID.

foidite *n.* a VOLCANIC ROCK in which FELDSPATHOIDS form more than 60 per cent of the light-coloured MINERALS. Normally the rock would be named according to the most abundant feldspathoid, for example nephelinite. *Foidolite* is the equivalent term for PLUTONIC rocks.

fold *n.* a bend or buckle in any pre-existing structure in a rock as a result of DEFORMATION. Folds are best displayed by structures that were formerly approximately planar, such as LAYERING or BEDDING in sedimentary and igneous rocks, or FOLIATION, SCHISTOSITY and CLEAVAGE in metamorphic rocks. Some terms defining the various components of folds are shown in Fig. 34 opposite.

An ANTIFORM is a fold with limbs that converge (or close) upwards, forming an arch; if there are older rocks in the core of the fold it is called an ANTICLINE (see Fig. 34). A SYNFORM is a fold with limbs that close downwards; if there are younger rocks in the core of the fold it is termed a SYNCLINE. A fold that closes sideways is called a *neutral fold*.

A *plunging fold* is one in which the HINGE LINE is not horizontal; the angle between the hinge line and the horizontal is the PLUNGE of the fold.

Folds may be described in terms of the attitude of the AXIAL SURFACE: in an *upright fold* the axial surface is essentially vertical; an *overturned fold* has an inclined axial surface and one limb inverted; and a *recumbent fold* has an essentially horizontal axial surface (see Fig. 34).

The angle between the fold limbs as

seen in profile is a measure of the tightness of the fold, see Fig. 34. Folds with a narrow hinge zone are described as *angular*, while if the hinge zone is broad they are described as *rounded*. *Symmetrical folds* have limbs of equal length, *asymmetrical folds* do not.

Folds with straight limbs and angular hinges may be described as *chevron folds* if the limbs are symmetrical and *kink bands* or *kink folds* if the limbs are markedly asymmetrical. *Conjugate folds* are apparently related folds with converging axial surfaces. Kink bands may form conjugate sets, while conjugate folds with rounded hinges are *box folds* (see Fig. 35 overleaf).

Parallel folds are folds in which the thickness of the layers (measured at right angles to the layer) is constant throughout the fold. *Concentric folds* are parallel folds in which the PROFILES of the fold surfaces lie on arcs of circles. *Similar folds* are folds in which the thickness of the layers is greater in the hinge zone than on the limbs. *Drag*

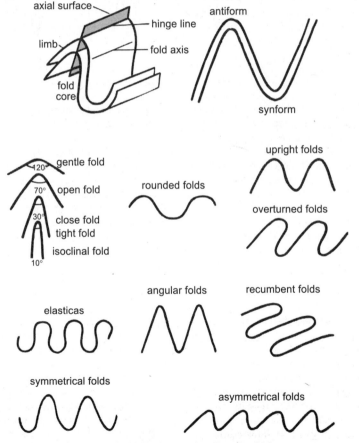

Fig. 34 **Fold**. Terms used to describe folds

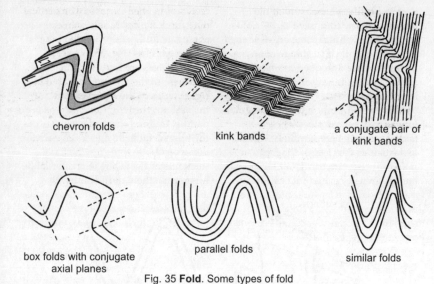

chevron folds

kink bands

a conjugate pair of kink bands

box folds with conjugate axial planes

parallel folds

similar folds

Fig. 35 **Fold**. Some types of fold

folds are minor folds found in several different types of structures. They may be found as small anticlines and synclines on the limbs of larger folds, as small folds in incompetent horizons sandwiched between competent horizons and as small crumples in fold cores.

fold belt see OROGENIC BELT.

fold mountains *n*. mountains that have been formed by large-scale and profound folding. Studies of typical fold mountains, such as the Alps, indicate FOLD involvement of the deeper crust and upper mantle as well as of the shallow upper crust.

fold system *n*. a set or group of congruent FOLDS that have been produced by the same tectonic episode.

foliation *n*. **1.** a planar arrangement of dimensionally orientated minerals in METAMORPHIC ROCKS formed by the recrystallisation and segregation of minerals growing under conditions of elevated pressure and shearing stress.

This is a characteristic texture of rocks formed by regional METAMORPHISM. See also SCHISTOSITY, GNEISS.

2. also called **banding**, a layered structure produced in the ice of a GLACIER by plastic DEFORMATION.

Folk classification *n*. a classification of limestones proposed by Folk (1962) and based upon three components: ALLOCHEMS – particles such as ooliths, skeletal grains and pellets; MICRITE – microcrystalline lime mud; *sparry calcite* – chemically precipitated cement filling pores. Limestones are named firstly by allochem and secondly by the micrite or sparry calcite content. Those limestones with over two-thirds lime mud matrix are named for the increasing allochem content – micrite and DISMICRITE with 0–1 per cent allochems, fossiliferous micrite with 1–10 per cent allochems, and biomicrite with 10–50 per cent plus allochems. Limestones with over two-thirds spar cement plus allochems are BIOSPARITES. Specific form

allochems can be used in naming limestones, for example a carbonate cemented oolite is an oosparite.

fool's gold *n.* a popular name for any of several minerals that resemble gold and have been mistaken for it, including CHALCOPYRITE; the name is normally used to describe PYRITE (FeS$_2$).

footwall *n.* the mass of rock below a fault plane, ore body or mine working, in particular the wall rock beneath a fault or inclined vein. Compare HANGING WALL.

foram *abbreviation for* foraminifer (see PROTOZOA).

foramen *n.* an opening.

Foraminifera see PROTOZOA.

Forbes bands *n.* an obsolete term describing GLACIER BANDS composed of alternating bands of milky 'bubbly' ice and dirty regelation ice. See also OGIVE.

forceful intrusion *n.* a mode of MAGMA emplacement in which space for INTRUSION is forcibly created. It is obvious in the case of many BATHOLITH intrusions that wall rocks have been pushed aside. This is shown by the deflection of structures in the wall rocks away from the regional trend towards parallelism with the walls of the PLUTON and/or updoming of the wall rocks.

fore arc see FRONTAL ARC.

foredeep *n.* an elongate crustal depression that borders an OROGENIC BELT or ISLAND ARC on the convex side. The foredeep is filled with sediment derived from the orogenic belt.

foreland *n.* an area marginal to an OROGENIC BELT towards which FOLDS have been overfolded and thrusts have moved. The side from which the surface rocks have moved is the *hinterland*.

fore reef *n.* the seaward side of a REEF. Parts of it may be slopes covered with

reef DEBRIS; other areas may be vertical walls built by deposits of marine organisms.

foreset see CROSS-BEDDING.

foreset bed see DELTA.

foreshock *n.* a small tremor that precedes a larger EARTHQUAKE by an interval ranging from seconds to weeks or even longer. Foreshocks of an earthquake all have approximately the same EPICENTRE. As a rule, major earthquakes occur without detectable warning in the nature of minor foreshocks. There is evidence, however, that some major earthquakes are preceded by foreshocks, e.g. the North Idu (Japan) earthquake of 1930. Studies have shown that the tendency for foreshocks to occur is limited to particular seismic zones.

foreshore or **beach face** *n.* that zone of a shore or beach that is regularly covered with tidal water. Compare BACKSHORE.

foresight *n.* a sight on a new survey point reckoned in a forward direction for the purpose of determining the bearing and elevation of the point. Compare BACKSIGHT.

form *n.* a set of crystal FACES each of which is similarly oriented with respect to the CRYSTAL AXES. PINACOIDS are forms consisting of two parallel faces cutting only one axis; a *basal pinacoid* cuts only the c-axis. PRISMS are forms consisting of 3, 4, 6 or more faces parallel to the c-axis. DOMES are parallel to a horizontal axis and cut the other two, and *pyramids* cut all three axes. Pyramids can enclose solid figures on their own and are thus known as *closed forms*, but pinacoids, prisms and domes must be combined with other forms to complete a crystal; they are known as *open forms*. The isometric CRYSTAL SYSTEM, with its high degree of symmetry, has a variety of closed forms, such as the CUBE,

OCTAHEDRON and DODECAHEDRON, which can occur on their own or in combinations, e.g. the dodecahedron combined with the icositetrahedron often found in garnets. See Fig. 36 below.

formal unit *n.* a STRATIGRAPHICAL UNIT that is defined and named in accordance with the rules and guidelines of an established system. Compare INFORMAL UNIT.

formation *n.* a fundamental lithostratigraphic unit (see STRATIGRA-PHICAL UNIT) used in the local classification of strata. Formations are classified by the distinctive physical and chemi-cal features of their rocks that distinguish them from other formations although formations do not have to be lithologically homogeneous. The boundaries of formations may be *sharp*, *gradational* or *interdigitating*. In Britain, a formation should be mappable on a 1:50 000 map scale. The names of formations are often taken from geographical names of places where they were originally described. These are combined with the names of the predominating rock comprising the formation, e.g. Clyde Sandstone Formation.

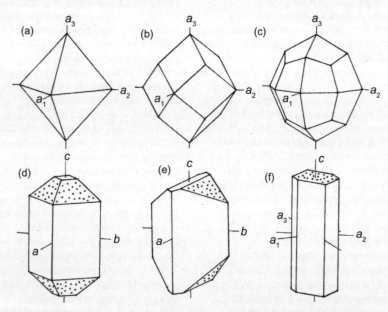

Fig. 36 **Form**. The crystals in the upper row all belong to the isometric (cubic) system. Form (a), the octahedron, commonly occurs naturally in diamond; (b), the rhombic dodecahedron, frequently occurs in garnet; and (c), the icosotetrahedron (or trapezohedron), is the usual habit of analcime. These are all closed forms but may occur in combinations with other isometric forms. The dotted areas in (d) are pyramid faces and are combined with a basal pinacoid and prism faces as in the mineral vesuvianite. In (e) the dotted areas are dome faces combined with forms common in andalusite, and in (f) the dotted area is a basal pinacoid face combined with prism faces in the typical habit of apatite

form sets *n*. isolated sand ripples in a finer-grained sediment (see RIPPLE MARK, also FLASER BEDDING).

forsterite *n*. a whitish to yellowish mineral, Mg_2SiO_4, the magnesium END MEMBER of the OLIVINE mineral group.

fosse *n*. a long narrow depression between a GLACIER and the sides of its valley. It is the result of the more rapid rate of melting that occurs here because of the added effect of heat absorbed or reflected from the valley walls.

fossil I. *adj*. ancient or extinct. A *fossil fern* connotes an ancient, extinct variety of fern; *fossil* RAIN PRINTS or *hail prints* are ancient markings. Certain geological forms that have become covered or changed are referred to as *fossil features*. For example, a *fossil cliff* is an ancient sea cliff.

2. *n*. the remains or traces of once living organisms that were buried by natural processes and, subsequently, permanently preserved. Fossils may have the following forms:

(a) *original soft parts unaltered*: this rarely occurs and requires very special conditions. Examples are the preservation of woolly mammoths in permafrost, animal remains in tar pits, animal and human remains in peat bogs and the encasement of insects in fossil resin, or AMBER.

(b) *original hard parts unaltered*: most plants and animals have some hard parts that can be fossilised. Examples are the calcitic shells of ECHINODERMS, aragonitic shells (most molluscs), phosphatic remains of BRACHIOPODS and chitinous skins of certain animals. However, aragonite recrystallises to calcite in time, with the consequent destruction of the finer structures. Aragonite shells occur in Mesozoic and younger rocks.

(c) *original hard parts altered*: the original hard structures of many organisms may undergo alterations with time. The nature of such alterations depends on the original material of the organism's environment and the conditions under which its remains were deposited.

(i) CARBONISATION or *distillation*: the reduction of organic tissue to a carbon residue.

(ii) PERMINERALISATION or *petrification*: the impregnation of porous bones and shells, after burial in sediment, by mineral-bearing solutions. This type of fossil is called a *petrifaction*.

(iii) REPLACEMENT or MINERALISATION: this occurs when the original hard parts are removed by solution and some other mineral substance is deposited in the voids created. Common replacement minerals include CALCITE, SILICA and LIMONITE. See also MOULD, CAST.

3. evidence of organisms. Although the animal itself is not preserved, a trace, or ichnofossil, provides evidence of the existence of a life form. Common evidence includes tracks and trails, borings and burrows. Other evidence includes coprolites (fossil faecal pellets) and GASTROLITHS (stomach stones). See also PSEUDOFOSSIL.

fossil assemblage see ASSEMBLAGE.

fossil community *n*. an ASSEMBLAGE whose individuals inhabited the same location in which their FOSSILS are found and in which the fossil forms are present in about the same numbers as when the organisms were alive, thus indicating no postmortem transport.

fossil fuel *n*. a general term for any HYDROCARBON that can be used for fuel, especially PETROLEUM, NATURAL GAS and COAL. The name is derived from their subsurface origin as well as their presence in fossiliferous areas.

fossil ice *n.* **1.** ice remaining from the geological past in which it was formed. For example, fossil ice is present along the coastal plains of northern Siberia, where remaining Pleistocene ice and the preserved remains of mammoths within this ice have been found.
2. comparatively old ground ice in a PERMAFROST region or underground ice in a region where the present temperatures are not low enough to have been responsible for it.

fossiliferous *adj.* (of rocks or strata) containing or bearing FOSSILS.

fossilisation see FOSSIL.

fossil man see HOMINID.

fossil plants *n.* see Fig. 37 below for a summary of the major plant groups that are found as fossils. The CYANOBACTERIA (or blue-green algae) are among the oldest known fossils. Their possession of Chlorophyll a, unlike all other photosynthetic bacteria, has allowed them to be placed among the plants. However, current practice suggests that they are more appropriately placed within domain EUBACTERIA. Fossil STROMATOLITES containing filamentous structures that resemble cyanobacteria occur in the 3500 Ma Warrawoona Group in Western Australia.

Fossil plants other than cyanobacteria are first known from marine deposits of the Proterozoic (late Precambrian) and include unicellular and multicellular forms of the red, brown and green seaweeds. Plants were not able to

Division	Description	Range
Subkingdom Algae		
Cyanophyta	Cyanobacteria	Precambrian to Recent
Rhodophyta	Red seaweeds	Precambrian to Recent
Phaeophyta	Brown seaweeds	Precambrian to Recent
Chlorophyta	Green algae	Precambrian to Recent
Subkingdom Embryophyta		
Bryophyta	**Mosses and Liverworts**	**Devonian to Recent**
Tracheophyta	**Vascular plants**	**Silurian to Recent**
Pteridophytina	Spore-bearing plants	
Psilopsida	Protracheophytes	Ordovician? to Recent
Lycopsida	Club mosses	Silurian to Recent
Sphenosida	Horsetails	Devonian to Recent
Pteropsida	Ferns	Devonian to Recent
Progymnosperms		Devonian to Carboniferous
Spermatophytina	Seed-bearing plants	
Pteridosperms	Seed ferns	Devonian to Cretaceous
Pinopsida	Conifers	Carboniferous to Recent
Ginkgoopsida	Ginkgos	Permian to Recent
Cycadopsida	Cycads	Permian to Recent
Angiospermae	Flowering plants	Upper Jurassic to Recent

Fig. 36 **Fossil plants**. The principal plant groups found as fossils

colonise the land until the late ORDOVICIAN, with the earliest evidence other than spores coming from the mid-SILURIAN. Spores and pollen grains are abundant in rocks from the Upper Silurian onwards and are useful in the correlation of rocks of continental facies (see PALYNOLOGY). Land plants evolved rapidly during the Devonian and were able to form the Earth's first rainforests during the Carboniferous. Later developments include the evolution of the ANGIOSPERMS (flowering plants) during the Upper Jurassic, followed by the evolution of the Poaceae (grasses) during the Creta-ceous.

fossil soil see PALAEOSOL.

fossil wax see OZOCERITE.

fossula *n*. (*pl*. **fossulae**) an obvious gap between septa (see SEPTUM, walls) in a RUGOSE CORAL. The fossulae are nor-mally associated with the first-formed (*primary*) septa, for example the CARDINAL SEPTUM and ALAR SEPTA (Fig. 77, page 376).

fractional crystallisation *n*. **1.** see MAGMATIC DIFFERENTIATION.
2. the separation of crystals with different stability ranges from a solution by precipitation, as effected by varying temperatures and pressures.

fractionation see MAGMATIC DIFFERENTIA-TION.

fracture *n*. **1.** the way in which a mineral breaks, other than along its planes of CLEAVAGE. Fracture can be described as CONCHOIDAL, *even* (of a flattish surface), *uneven*, HACKLY and EARTHY.
2. one of the ways in which rocks yield to deforming movements in the Earth's crust, e.g. cracks, JOINTS, FAULTS or other breaks.

fracture cleavage *n*. a CLEAVAGE formed by a series of closely spaced parallel

FRACTURES within a rock. It is not necessarily accompanied by RECRYSTALLISATION and the minerals are not aligned in a particular orientation. Fracture cleavage is currently called *spaced cleavage*.

fracture zone *n*. a zone in which faulting has occurred. It runs parallel to the FAULT line and may be oceanic or continental. A fracture zone can be the site of intense seismic and volcanic activity.

fragmental rock *n*. **1.** see CLASTIC ROCK.
2. see PYROCLASTIC ROCK.
3. see BIOCLASTIC ROCK.

fragmental texture *n*. a SEDIMENTARY ROCK texture characterised by broken particles, which distinguish it from a rock of crystalline texture. PYROCLASTIC ROCKS such as TUFF have fragmental texture.

franklinite *n*. an isometric mineral of the SPINEL group, $(Zn,Mn^{2+},Fe^{2+})(Fe^{3+},Mn^{3+})_2 O_4$, an ORE of zinc found as octahedral black crystals with a metallic lustre or in massive aggregates.

Frasch process *n*. a method of mining elemental sulphur in which super-heated water is injected into the sulphur deposits. The sulphur is melted by the water and is then pumped to the surface.

Frasnian *n*. the lower stage of the Upper DEVONIAN.

frazil ice *n*. a semi-soft aggregate of ice in the form of small crystal spikes and plates. It forms along the edges of turbulent or rapidly flowing streams but is also found in turbulent sea water.

free-air effect see GRAVITY ANOMALY.

free-air gravity correction see GRAVITY ANOMALY.

free water *n*. **1.** also called **gravitational water**, water in rock or soil, in excess of that held by capillary forces, that can

move freely in response to gravity.
2. water that is not chemically bound to a substance and can be removed without altering the structure or composition of the substance.

fresh water *n*. water containing less than 0.2 per cent dissolved salts; it is not necessarily potable.

friable *adj*. (of a rock or mineral) that can be disintegrated into individual grains by finger pressure.

fringing reef *n*. an organic REEF that grows directly against coastal bedrock and actually constitutes the SHORELINE, commonly found along tropical coasts. A fringing reef has a rough, shelflike surface that is exposed at low tide, and its seaward side drops sharply to the sea floor. Compare BARRIER REEF.

frontal arc or **fore arc** *n*. the region on the trench side of the volcanic arc at a DESTRUCTIVE PLATE MARGIN.

frost action *n*. the process of repeated freezing and thawing of water, effective in the mechanical WEATHERING and breakup of rock. Its effectiveness depends on the presence of confined spaces and the frequency of the freeze-thaw cycle. In cold regions, it is largely responsible for TALUS and rock glaciers.

frost crack, contraction crack or **ice crack** *n*. a more or less vertical crack in rock or frozen ground with an appreciable ice content. Frost cracks are produced by thermal contraction and commonly intersect to form a polygonal net.

frost-crack polygon see NONSORTED POLYGON.

frost heaving *n*. the upward expansion of ground caused by the freezing of water in the upper REGOLITH. If the surface is horizontal, the regolith after thawing will return to its original position. If the surface slopes, the regolith upon freezing will rise

perpendicular to the slope but upon thawing will have a downslope component producing CREEP. Frost heaving is responsible for the movement of rocks to the surface, for spongy, soft ground in the spring and for damage to road surfaces.

frost line *n*. **1.** the maximum depth of frozen ground in regions without PERMAFROST.
2. the greatest depth of PERMAFROST.
3. that altitude below which freezing does not occur, e.g. in mountainous tropical regions.

frost mound or **soil blister** *n*. a general term for rounded or conical forms (HUMMOCKS, knolls) in a PERMAFROST region. It contains a core of ice and indicates a localised and seasonal upwarp of the landscape resulting from FROST HEAVING and/or HYDROSTATIC PRESSURE of GROUNDWATER. See also PINGO.

frost splitting see CONGELIFRACTION.

frost stirring *n*. CONGELITURBATION in which there is no mass movement.

frost weathering see CONGELIFRACTION.

frost wedging *n*. a type of CONGELIFRACTION in which ice, acting as a wedge, prises apart jointed rock.

Froude number *n*. a dimensionless number that is used to define states of flow in liquids:

$$Fr = \frac{U}{\sqrt{gh}}.$$

U is the mean flow velocity, g is acceleration caused by gravity and h is the depth of flow. Tranquil flow regimes have Froude numbers of less than 1, while rapid flow conditions have Froude numbers greater than 1.

frozen ground or **gelisol** *n*. ground whose temperature is below the freezing point of water and usually holds some amount of water in the form of ice.

F shell see SEISMIC REGIONS.

fulgurite *n*. glassy tubes of lightning-fused rock, most common on mountain tops. Although fulgurites may form from any kind of rock, the largest have been formed from unconsolidated sand. 'Fossil' fulgurites, or *palaeofulgurites*, have been found in Permian dune-bedded sandstones in the Isle of Arran, Scotland.

fuller's earth *n*. a clay with a high adsorptive capacity, composed mostly of the smectite mineral MONTMORILLONITE. It is widely used as a bleaching agent, as a filter and for removing grease from fabrics.

fumarole *n*. a vent at the Earth's surface that emits hot gases. Fumaroles are found at the surface of LAVA FLOWS, in and around the CALDERAS and craters of active volcanoes, and in areas where hot intrusive, igneous rock bodies occur. The gas temperature within a fumarole may be as high at 1000°C. Steam and CO2 are the main gases; sulphur dioxide, hydrogen sulphide, nitrogen, carbon monoxide, argon, hydrogen and other gases may be present but vary greatly in their relative proportions even in the same fumarole. Deep MAGMAS and the heating of GROUNDWATER are the sources of fumarole gases. A SOLFATARA is a fumarole in which sulphur gases are the dominant constituents after water. MOFETTES are fumaroles rich in carbon dioxide with temperatures much below the boiling point of water.

fungi *n*. (*sing.* **fungus**) organisms belonging to the kingdom Fungi, consisting of single-celled or multicelled eucaryotes lacking chlorophyll and depending on living or dead organic material for food. Fungi form chitinous spores and have chitinous cell walls. fossil fungi date from the Ordovician onwards, although some unicellular fossils from the Precambrian may be fungi.

fusain see LITHOTYPE.

fusellae see SICULA.

fusibility *n*. the property of being convertible from a solid to a liquid state by means of heat; also, the degree to which a substance is fusible. The relative fusibility of a substance compared with a standard scale may be determined by blowpipe methods.

fusiform *adj*. spindle-shaped; elongated and tapering at both ends.

fusion *n*. **1.** the process of liquefying a solid by the application of heat. **2.** the joining or coalescence of two or more substances, objects, etc, as by partial melting. **3.** the combination of two light atomic nuclei to form a heavier nucleus. The reaction is accompanied by the release of a massive quantity of energy. Fusionable elements are among those of low atomic weight, e.g. hydrogen. Compare FISSION.

fusulinid *n*. a foraminifer of the family Fusulinidae, similar in shape to a grain of wheat. They are important guide fossils in the Pennsylvanian system. Range, Ordovician to Triassic. See PROTOZOA.

Gg

Ga *abbreviation for* Gigayear, one billion (10^9) years.

gabbro *n.* a coarse-grained intrusive IGNEOUS ROCK composed essentially of calcic plagioclase and pyroxene, commonly with small amounts of other ferromagnesian minerals, especially olivine. Magnetite, ilmenite and apatite are the most frequent ACCESSORY MINERALS. It is the plutonic equivalent of DOLERITE and BASALT. In olivine-gabbro the pyroxene is normally AUGITE, but in gabbros with affinities with tholeiite, orthopyroxene may accompany the clinopyroxene. If orthopyroxene makes up more than 50 per cent of the total pyroxene content, the rock is termed NORITE, in which case olivine would not normally be present. The plagioclase of gabbro has an anorthite content of more than 50 per cent. If the anorthite content is less than this, the rock grades into DIORITE; if the anorthite content is very high (calcic bytownite or anorthite), the term EUCRITE is used. See also TROCTOLITE, ALLIVALITE, LAYERED INTRUSION.

Gale alidade or **explorer's alidade** *n.* a lightweight compact ALIDADE with a low pillar and a reflecting prism through which the ocular may be viewed from above. In petroleum geology, it is often used in conjunction with the *Stebinger drum*.

galena *n.* an isometric mineral, PbS, that commonly occurs as cubes with a bright silvery metallic lustre. It is the main ore of lead but often contains appreciable amounts of silver as an impurity. The latter is a frequent by-product of lead smelting. Galena is a typical hydrothermal mineral in low- to medium-temperature deposits; it is often associated with other metallic sulphides and such minerals as calcite, quartz, barite and fluorite.

gallery *n.* **1.** a horizontal passageway in a cave.
2. a horizontal conduit or channel constructed to intercept water.
3. an underground excavated space produced during mining.

gamma decay *n.* a type of radioactivity in which certain unstable atomic nuclei emit excess energy by a spontaneous electromagnetic process. Unstable nuclei that undergo gamma decay are themselves the products either of alpha and beta decay or of some other nuclear process, such as neutron capture in a nuclear reactor. See also DATING METHODS, RADIOACTIVE DECAY, GAMMA RAYS.

gamma-ray logging see WELL LOGGING.

gamma rays *n.* the quanta of energy (photons) emitted by nuclei of radioactive substances. They are electromagnetic waves of the same type as X-rays but have shorter wavelengths and higher frequencies. Gamma rays have a longer range than alpha or beta particles. They may be attenuated by the use of heavy metal shields, e.g. lead.

gangue *n.* those minerals of low economic value within an ore body. Typical gangue minerals include quartz, barite and calcite. Compare GAS STREAMING.

ganister see SEAT-EARTH.

garnet *n.* a group of isometric nesosilicate minerals with the basic formula $A_3B_2(SiO_4)_3$, where A = Ca, Mg, Fe^{2+} or Mn^{2+} and B may be Al, Fe^{3+}, Mn^{3+}, V^{3+} or Cr^{2+}. Although all members of the garnet group conform to the same basic formula and crystallise in similar isometric habits (common forms are the dodecahedron and the trapezohedron, often in combination), they exhibit a great range of chemical variation. There are two main isostructural groupings, named mnemonically:

(1) the *pyralspite group*: pyrope (Mg,Al), almandine (Fe,Al) and spessartine (Mn,Al), and (2) the *ugrandite group*: uvarovite (Ca,Cr), grossular (Ca,Al) and andradite (Ca,Fe, Ti).

Other varieties include *goldmanite* (Ca,V) and *hydrogrossular* ((Ca,Al)OH) in which OH is probably substituting for SiO_4. Garnet has a vitreous lustre and is found in all colours but blue. The best-known variety is the dark-red one, PYROPE, widely used as a gemstone. Red-to-violet almandine is called *carbuncle*, and *hessonite* is a cinnamon-coloured or yellow variety of GROSSULAR. Some garnet shows angular fracture that, together with its hardness (61/2–71/2 on MOHS' SCALE), makes it a good abrasive. Garnet is common in metamorphic rocks (gneisses and schists and some types of contact metamorphism) and is stable across a wide range of temperatures and pressures. It occurs in some mantle XENOLITHS.

garnierite *n.* $(Ni,Mg)_3Si_2O_5(OH)_4$, a nickel-bearing variety of serpentine. It is an apple-green alteration product of Ni-rich PERIDOTITE, mined as Ni ore in New Caledonia, Russia and Australia.

gas see NATURAL GAS.

gas cap *n.* free gas that has accumulated above an oil reservoir. It forms whenever the available gas exceeds the amount that will dissolve in the oil under the existing conditions of temperature and pressure.

gas coal *n.* BITUMINOUS COAL containing a high percentage of volatile gas-yielding constituents, which makes it suitable for the manufacture of flammable gas. Gas coals generally have about 30–34 per cent volatiles.

gaseous transfer *n.* the removal of the more volatile chemical constituents of a MAGMA during separation of a gaseous phase, generally under conditions of low CONFINING PRESSURE.

gas field *n.* an individual GAS POOL or two or more closely related pools.

gas hydrate *n.* a combination of water and a gas formed at high pressure and/or low temperature. The gas is locked into a lattice of water ice. Methane hydrate is found in some areas of the ocean floor and under waterlogged tundra.

gas-oil ratio (GOR) *n.* the volume quantity of gas produced with the oil from a particular oil well, expressed in terms of cubic feet of gas per barrel of oil.

gas pool *n.* an economically viable subsurface accumulation of natural gas. Compare GAS FIELD.

gas sand *n.* a sand or sandstone that contains a large or significant quantity of NATURAL GAS.

gas streaming *n.* the upward flow of bubbles of gas or fluid in a MAGMA. During magmatic differentiation escaping gas bubbles may help to extract liquid from a crystal mush, resulting in closer packing of CUMULATE crystals.

gastrolith or **stomach stones** *n.* highly polished, rounded stones believed to have been a part of the digestive system of certain extinct reptiles, such as plesiosaurs. The gastroliths aided in grinding the animals' stomach contents. See FOSSIL.

gastropod *n.* any MOLLUSC of the class Gastropoda. Most are aquatic, marine or freshwater; there are also terrestrial forms. Modern varieties include winkles, whelks, limpets, snails and slugs. Gastropods secrete a single CALCAREOUS shell, closed at the apex. The shell typically is spirally coiled, either dextral or sinistral, although in some forms only the PROTOCONCH is coiled and the fully grown shell is cap-shaped. The palaeontological classification follows shell shape and symmetry, whereas the zoological classification is based on the character of the soft parts. Gastropods range from the CAMBRIAN to the present and are most abundant in the TERTIARY period where they are important as zonal indices.

gas well *n.* a well capable of producing natural gas or one that produces chiefly natural gas. In some areas the term is defined on the basis of GAS-OIL RATIO.

Gault *n.* a marine clay FACIES of the British Lower CRETACEOUS, equivalent to the APTIAN and ALBIAN stages. The Gault Clay Formation is the formal STRATI-GRAPHICAL UNIT.

geanticline *n.* an uplift of regional extent, comparable in size to a

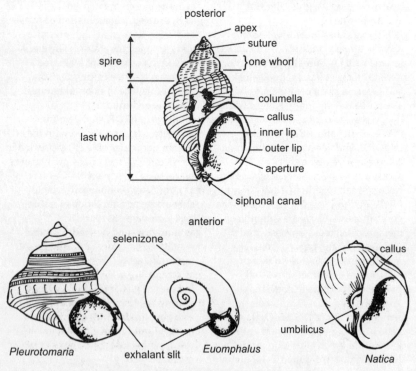

Fig. 38 **Gastropod**. General structure of gastropods

GEOSYNCLINE; it may be either inside or outside a geosyncline.

Gedinnian *n.* the lowest STAGE of the DEVONIAN.

Geiger-Mueller counter *n.* an instrument for detecting radiation. It consists of a tube (the *Geiger tube*) filled with gas at sufficiently high pressure to prevent the spontaneous flow of electrons. When a high-energy particle enters the tube and ionises the gas, a current flows. Each current pulse, which is recorded by attached electronic equipment, represents the passage of a high-energy particle through the Geiger tube.

gel *n.* a jellylike colloidal dispersion of a solid within a liquid.

gelation *n.* **1.** the process of gelling. **2.** solidification by cold; freezing.

gelifluction see SOLIFLUCTION.

gelisol *n.* FROZEN GROUND.

gem see GEMSTONE.

gemmology *n.* the study of gemstones, including their description, origin, source, means of identification and methods of evaluation.

gemstone or **gem** *n.* any precious or semiprecious stone, particularly after it has been cut and polished.

genal angle *n.* the angle between the posterior and lateral margins of the CEPHALON (head carapace) in TRILOBITES (Fig. 89). It is usually rounded, but some forms have spines (*genal spines*).

genotype *n.* **1.** the genetic composition of an organism in contrast to its physical characteristics (PHENOTYPE). **2.** that species on which the original description of a GENUS is primarily based.

genus *n.* (*pl.* **genera**) a category in plant and animal classification between FAMILY and SPECIES. A genus consists of a group of related species or a single species.

geobarometry *n.* the determination by various methods of the pressure at which geological processes occur. Minerals that are stable within particular pressure ranges are used to determine pressures exerted within regional metamorphic regimes. Changes in mineral structures can be used similarly. For example, the particular site occupied by aluminium in SILICATE structures is a function of the pressure of their formation. With an increase in pressure, aluminium shows preferred octahedral rather than tetrahedral coordination, thus making it possible to correlate the amount of aluminium in octahedral coordination with the pressure history of a rock. See also IONIC BOND.

geobotanical prospecting *n.* the study of the occurrence of plants as indicators of MINERAL or ORE presence. Those plant species tolerant of high levels of specific metal ions in the soil (*hyperaccumulators*) will indicate the presence of a particular ore. This can be detected by observation of the species on the ground or by remote sensing. Compare BIOGEOCHEMISTRY.

geocentric *adj.* **1.** related to the Earth as a centre. **2.** measured from the Earth's centre.

geochemical anomaly see GEOCHEMICAL EXPLORATION.

geochemical cycle *n.* a path followed by an individual element or group of elements in the Earth's ATMOSPHERE, HYDROSPHERE and LITHOSPHERE. During the cycle, there occur processes of both separation of elements and elemental recombination. For the lithosphere, the cycle begins with MAGMA that has been generated by partial melting in the ASTHENOSPHERE. The magma rises towards the surface of the Earth and crystallises as IGNEOUS ROCKS, either as an INTRUSION in the CRUST or at the surface if it is erupted from a VOLCANO, where much of the volatile content of the magma may enter, of atmosphere or hydrosphere as water, carbon dioxide and other gases. Decomposition of the rock at the surface by WEATHERING is followed by transport of the weathering products, which are deposited as sediment. After burial under successive layers of sediment LITHIFICATION then takes place. If caught up in mountain-building processes, METAMORPHISM may take place and eventual melting, with the generation of a new magma.

A number of different paths could be followed within this general cycle. Each element will be affected differently as cycles progress, for example, partial melting of a source rock to form a magma separates elements according to the melting characteristics of the constituent minerals. MAGMATIC DIFFERENTIATION processes produce a wide range of chemical compositions among igneous rocks and, in particular, tend to increase the concentration of Si, Al, Ca, Na and K and other lithophile elements in the more evolved magmas. Sedimentary processes are also effective in separating chemical components; for example, during the weathering of an igneous rock, magnesium released by the breakdown of ferromagnesian minerals can be removed in solution whereas iron may form insoluble iron oxides and hydroxides. Soluble cations such as $Mg2+$, $K+$, $Na+$ and $Ca2+$, which accumulate in sea water, may be removed from solution by the action of marine organisms or by direct precipitation (see EVAPORITES). Some $(SiO_4)^{4-}$ enters solution, but most silica accumulates in the form of the resistant mineral

quartz. Other elements, for example nitrogen, enter a biogeochemical cycle in which the inorganic cycling of the element is joined by an organic component as the element enters and leaves living forms.

geochemical exploration or **geochemical prospecting** *n*. the application of methods and techniques of GEOCHEMISTRY to the search for MINERAL or PETROLEUM deposits. The presence of a *geochemical anomaly* often indicates a deposit. A geochemical anomaly is a concentration of one or more elements in a rock, sediment, soil, water or vegetation that differs significantly from the normal concentration. When the anomaly is in terms of abnormal concentrations of elements or hydrocarbons in soils, it is particularly applicable to mineral or petroleum deposits.

geochemical facies *n*. an area or region characterised by particular physico-chemical conditions that affect the production and accumulation of sediments. The area is commonly distinguished by a characteristic element. In sedimentary environs, this is best illustrated in terms of oxidation potential (E_H) and acidity (pH). Certain related deposits exhibit very different mineralogical features because of different depositional environments. The sedimentary iron formations of the Lake Superior region in North America, formed during the Precambrian, are classified according to the dominant iron mineral into four principal facies: sulphide, carbonate, oxide and silicate. The respective E_H-pH diagrams of each type can indicate the depositional environment, chemical region of stability and conditions under which particular assemblages could have been formed.

geochemical prospecting see GEO-
CHEMICAL EXPLORATION.

geochemistry *n.* that branch of GEOLOGY
concerned with the quantities, distribution and circulation of chemical elements
in the soil, water, atmosphere and
interior of the Earth. Geochemistry has
as its major concern the various classes
of rocks and minerals, and their associated environments, as well as those
conditions and processes that contribute to such variety. Geochemistry
applies the principles of chemistry to
geological problems. As such, it over-
laps many fields in geology such as
MINERALOGY, PETROLOGY and HYDROLOGY.

geochron or **chron** *n.* the smallest
interval of geological time (GEOCHRONO-
LOGICAL UNIT) that corresponds to the
lithostratigraphical CHRONOZONE. See
STRATIGRAPHICAL UNIT.

geochronological unit *n.* an uninterrupted interval of time in geological
history during which a corresponding
chronozone was formed, i.e. a division
of time delineated by the rock record.
In order of decreasing magnitude,
geological time units are the EON, ERA,
period, EPOCH, AGE and *chron*.

geochronology *n.* the measurement of
time intervals of the geological past,
accomplished in terms of absolute age-
dating by a suitable radiometric
procedure, e.g. the potassium-argon
method, or relative age-dating by
means of sequences of rock strata or
varved deposit analysis. Dating past
events in terms of tree rings is called
DENDROCHRONOLOGY. Geochronology is
concerned with areas such as orogenic
events, sea-floor spreading rates and
intervals of glaciation. Methods based
on sedimentation rates and the rate of
increase in the salinity of sea water
have been shown to be unreliable. See
also DATING METHODS.

geochronometry *n.* the measurement of
geological time by the use of
geochronological methods, such as
radiometric dating. Compare
GEOCHRONOLOGY.

geocosmology *n.* the discipline that
deals with the origin and geological
evolution of the Earth, its planetary
features and the influence of the solar
system, our galaxy, and the universe on
the geological development of the
Earth.

geode *n.* a hollow globular mineral body
that can develop in limestone and
lavas. Although the shape is commonly
nearly spherical, from 2.5 to 30
centimetres in diameter, tubular or
irregularly shaped geodes have been
found. They commonly have a hollow
interior, an outer layer of CHALCEDONY
and a DRUSY lining (usually quartz
crystals) in the interior.
(a) *geodes in limestone* form by expansion from an initial cavity filled with
fluid, often a salt solution. There is
often a clay film between the geode
wall and a surrounding limestone
matrix. The chalcedony layer is from a
silica gel that isolates the salt solution.
If the salinity of the surrounding water
decreases, the internal salinity, through
osmosis, creates outward-directed
internal pressure. As a result of this,
the geode will expand until equilibrium is attained. Finally, crystallisation
and dehydration of the silica gel occur,
succeeded by cracking, shrinking and
the penetration of water-bearing
dissolved minerals, which are deposited on the chalcedony wall. When the
initial cavity is inside a FOSSIL, such as
the opening of a BIVALVE, the geode
forms within this cavity and the fossil
is burst by the expanding geode.
Geodised BRACHIOPODS are the result of
such a process.

(b) *geodes in lava* are large VESICLES produced by the expansion of gases that have come out of solution as the lava is erupted. They are filled initially by HYDROTHERMAL fluids rich in colloidal silica. As the lava cools, the colloid becomes unstable and chalcedonic silica is deposited on the walls of the vesicle, often in banded form as agate. The agate layers are coated with quartz crystals deposited from solution at lower temperatures. In some cases the geode is entirely filled with chalcedony and quartz crystals. In other instances the centre of the geode is hollow and a spectacular display of the terminations of the quartz crystals is visible. Sometimes the crystals are in the form of AMETHYST if the quartz contains trace amounts of iron. Unfilled cavities within geodes of either type are called VUGS. See AMYGDALE.

geodesic line *n.* the shortest distance between any two points on a surface.

geodesy *n.* the study of the external shape of the Earth as a whole, which provides a geometrical framework into which other knowledge of the Earth may be fitted and which, by itself, yields information about the internal construction of the Earth. Fundamental to geodesy is the interpretation of the shape of the Earth. Because no more than one quarter of the Earth's surface is land, ordinary geometrical surveys cannot be applied to the planet in its entirety. The surfaces of the seas and shape of the Earth, however, are both controlled by the force of gravity; from this connection, a model referential figure of the Earth, the GEOID, is derived. Because of the curvature of the Earth and optical scattering by the atmosphere, direct measurements between points on the surface of the Earth are limited to rays of some 50

kilometres in length. The method followed in *geodetic surveying* is to make local geometrical measurements with respect to a reference surface that can be determined on a worldwide basis. The purely geometrical measurements need only have local validity; the worldwide reference surface, the geoid, is related to the gravitational field of the Earth.

geodetic coordinates *n.* values that specify the horizontal position of a point on a reference ellipsoid with respect to a particular geodetic datum; such values are generally expressed as *longitude* and *latitude*.

geodetic surveying see GEODESY.

geodimeter *n.* an instrument designed to measure distance by the time required for light pulses to travel to and from a mirror located at a known distance from the geodimeter.

geofracture see GEOSUTURE.

geographic cycle see CYCLE OF EROSION.

geohydrology *n.* a term used interchangeably with HYDROGEOLOGY.

geoid *n.* a model of the figure of the Earth that coincides with the mean sea level over the oceans and continues in continental regions as an imaginary sea-level surface defined by spirit level. A geoid is perpendicular at all points to the direction of gravitational attraction and approaches the shape of a regular oblate spheroid. It is irregular, however, because of buried mass concentrations and elevation differences between sea floors and continents. The geoid is characterised by a constant potential function over its entire surface. This potential function considers the combined effects of the Earth's gravitational attraction and the centrifugal repulsion that results from the axial rotation of the Earth.

geological age *n.* the age of a fossil

organism or geological event or formation, as referred to the geological time scale and expressed as ABSOLUTE AGE or in terms of comparison with the immediate environment (RELATIVE AGE).

geological engineering see ENGINEERING GEOLOGY.

geological column *n*. a composite diagram showing in columnar form the total succession of STRATIGRAPHICAL UNITS of a given region, arranged with the oldest unit at the bottom and the youngest unit at the top so as to indicate the association of the units to the subdivisions of geological time. See also COLUMNAR SECTION.

geological hazard *n*. a geological phenomenon or condition, natural or artificial, that is dangerous or potentially dangerous to the environment and its inhabitants. Typical natural hazards include earthquakes, volcanic eruptions and landslides. Ground subsidence because of over-mining is an example of an artificial hazard.

geological map *n*. a map on which is shown the surface distribution of rock units, their boundaries, their type and age relationships, and the occurrence of structural features. Geological maps are also used to indicate the presence of ores, petroleum, water, coal and other valuable subterranean materials.

geological oceanography see MARINE GEOLOGY.

geological province *n*. an extensive area characterised by similar geological history and development.

geological range see STRATIGRAPHICAL RANGE.

geological record *n*. the testimony of the history of the Earth as revealed by its BEDROCK, REGOLITH and MORPHOLOGY.

geological thermometer see GEOTHERMOMETER.

geological thermometry *n*. the determi-nation or estimation by any technique of the temperatures at which geological processes take place. For current processes, such as volcanic eruptions, direct measurements can be made. For past geological occurrences, or inaccessible ones, indirect methods are used. Various rocks, minerals or mineral groups can be used, where their presence or properties can be related to the temperature under which they are formed. Silicate minerals often record the evolutionary history of rocks. For example, the amount of substitution between compositional end members and the mode of cation distribution are functions of temperature. FLUID INCLUSIONS can be used to indicate the formation temperature of the host mineral. For magmatic systems (binary, ternary, quaternary), the temperature range of solubilities and precipitation can be determined from PHASE DIAGRAMS representing such systems. Certain isotope ratios can be used as thermal indicators; the ratio of the oxygen isotopes $^{16}O/^{18}O$ in shells is related to the temperature of the ocean water in which they lived. The concentration of trace elements may also be used, since the amount of a particular trace element in a mineral is frequently a function of temperature. See also GEOBAROMETRY.

geological time *n*. the time extending from the end of the formative period of the Earth (i.e. after it was 'established' as a planet) to the onset of human history; it is the part of the Earth's history that is recorded and described in the succession of rocks.

geological time scale *n*. a scale that subdivides all geological time into named units and all rocks formed during geological time into a sequence from the oldest to the youngest. The organisation of rocks into a scale of named units – LITHOSTRATIGRAPHY – is linked to time by CHRONOSTRATIGRAPHY, which organises the sequence of rock units into chronostratigraphic units (CHRONOZONES) that correspond to intervals of time – GEOCHRONOLOGICAL UNITS.

geology *n*. the study of the Earth. The many branches of geology include the history of the Earth as a planet since its time of formation, its internal structures, the materials of which it is made, internal processes such as heat flow and the generation of magma, plate tectonics, the evolution of life forms, external processes such as weathering and erosion of the Earth's surface and the deposition of weathering products as sediment and their lithification to form rocks. See also APPLIED GEOLOGY, EARTH SCIENCE, HISTORICAL GEOLOGY, PHYSICAL GEOLOGY.

geomagnetic field *n*. the magnetic field associated with the Earth. Although the field is generally dipolar on the surface of the Earth, the dipole is distorted at locations removed from the surface because of interactions with the solar wind. As much as 6 per cent of the geomagnetic field is non-dipolar in shape. The Earth's magnetic field (both dipole and non-dipole components) is thought to be produced by electrical currents flowing within the fluid outer core, which is principally composed of iron. The non-dipole component may be caused by eddies resulting from the interaction of the CORE with the overlying MANTLE. The core is believed to be rotating more slowly than the mantle. Local MAGNETIC ANOMALIES result from variation of the magnetic properties of rocks in the Earth's crust.

geomagnetic polarity reversal *n*. the reversal in the direction of the Earth's

magnetic field, as indicated by alternating magnetic features of a sequence of rocks that are progressively younger. Such reversals are detected through measurements of the remanent magnetisation of sedimentary or volcanic rock successions. They are also responsible for the alternating strips of normal and reversed magnetic features that parallel the MID-OCEANIC RIDGES. Within a period of perhaps less than 10 000 years the field may become very weak and then increase again to equal its former intensity but with an opposite direction. The Atlantic Ocean floor consists of rocks advancing in age with increasing distance from the MID-ATLANTIC RIDGE, which is the site of new crustal generation. Rock of alternating magnetic polarity occurs as roughly symmetrical bands on each side of the ridge. Radiometric age-dating provides evidence of the timing of geomagnetic reversals. A sequence of reversals has been established back to the Cretaceous. See also POLAR WANDERING, NATURAL REMANENT MAGNETISATION, PALAEOMAGNETISM, SEA-FLOOR SPREADING.

geomagnetic pole *n*. the location on the Earth's surface where the continuation of a line along the resultant magnetic dipolar axis intersects with the Earth's surface. In 2002 the north geomagnetic pole was located in the Arctic Ocean, offshore from Canada, but it is migrating between 10 and 40 kilometres per year in a general direction north of Alaska. The magnetic declination is the angle between a line through any point to the geographical North Pole and a line from that point to the geomagnetic north pole. See also GEOMAGNETIC FIELDS, GEOMAGNETIC POLARITY REVERSAL.

geomagnetism *n*. the branch of GEOPHYSICS concerned with all aspects of the Earth's magnetic field: its origin, changes through time, remanent magnetisation of rocks and magnetic anomalies. See also GEOMAGNETIC POLARITY REVERSAL, NATURAL REMANENT MAGNETISATION, MAGNETIC ANOMALY.

geomorphic *adj*. pertaining or related to the form of the Earth or its surface features.

geomorphic cycle see CYCLE OF EROSION.

geomorphogeny *n*. that part of GEOMORPHOLOGY that is involved with the origin and development of the Earth's surface forms and features.

geomorphography see GEOMORPHOLOGY.

geomorphology *n*. the scientific discipline concerned with surface features of the Earth, including land forms and forms under the oceans, and the chemical, physical and biological factors that act on them, e.g. WEATHERING, streams, GROUNDWATER, GLACIERS, waves, gravity and wind. As parts of geomorphology, *geomorphography* is the description of geomorphic features, and *geomorphogeny* deals with the origin and development of land features. *Geomorphic maps*, usually constructed with the aid of aerial photographs, indicate land forms of particular kinds or origins within a specific region. Compare GEODESY. See CYCLE OF EROSION.

geopetal structure or **spirit-level structure** *n*. a structure formed by the partial filling with SEDIMENT of a cavity such as a VESICLE or the body cavity of a FOSSIL (see Fig. 39 below). The remaining space may be partially or wholly filled later by minerals deposited from percolating solutions. The flat top of the sediment infill is parallel to the horizontal at the time of formation and can therefore be used to indicate the way-up of a succession of folded beds. Geopetal structures are also a guide to the original attitude of the beds in which they occur, which may not have been horizontal (for example, beds deposited on a reef front). See also WAY-UP INDICATORS.

Geophone *n*. the tradename for an acoustic detector that senses ground vibrations generated by SEISMIC WAVES. Geophones are variously arranged on ground surfaces to record the vibrations produced by explosives in seismic refraction and reflection work. They are also used by the military as detection devices.

geophysical exploration or **geophysical prospecting** *n*. the use of appropriate geophysical methods to search for natural resources within the Earth's crust or to obtain information about subsurface structure for various civil engineering works. Geophysical prospecting techniques are also used in archaeological reconnaissance, where

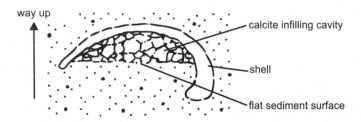

Fig. 39 **Geopetal structure**. The flat sediment surface preserved within a cavity marks the original horizon

the ground is explored for substructures such as walls and ditches. The measurement of gravitational and magnetic variations and anomalies is useful in oil and mineral prospecting. See GRAVIMETER. In seismic prospecting, SEISMIC WAVES generated by an explosive charge are used to determine underground structure. See also SEISMIC SHOOTING. Variations in the electrical properties of rocks, soils and minerals are used for detecting ore bodies. This is usually done by measuring the apparent RESISTIVITY between electrodes placed in the ground. This method is useful for locating the depths of surface soil and rocks, water-bearing strata and subsurface anomalies. Various sensors (infrared, microwave, or radiofrequency) provide relative surface information. Their application is based on the premise that all materials have particular signatures at different wavelengths. Since geophysical methods are indirect and require interpretation, they are most effective when used in conjunction with more direct methods, e.g. boring. See WELL LOGGING.

geophysical prospecting see GEOPHYSICAL EXPLORATION.

geophysics *n*. **1.** a section of EARTH SCIENCE that employs the principles and methods of physics. The scope of geophysics touches on all physical phenomena that have affected the structure, physical conditions and evolution of the Earth. Its branches include the study of variations in the Earth's gravity field, SEISMOLOGY, the influence of the Earth's magnetic field on outer-space radiation, geomagnetism, thermal properties of the Earth's interior, OCEANOGRAPHY, climatology, meteorology, VOLCANOLOGY and DEFORMATION of the Earth's CRUST and MANTLE,

including warping of the crust and shifting and displacement of continents (CONTINENTAL DRIFT). Techniques of geophysics include measurement of gravitational and magnetic fields of the Earth, elastic-wave measurement (by seismography) and the collection of data on the radiation flux in space by means of space satellites.
2. (*gravity*) *n*. the scientific discipline concerned with global gravity distribution and its significance. The objectives may be restricted geographically, as in the search for buried ore deposits, or may include large sections of the Earth's crust, as in the study of mountain belts, island arcs or rift valleys. The basic investigative method in all cases involves obtaining gravity measurements and plotting the result on a gravity map. Gravity anomalies that appear on a map can be interpreted in terms of subsurface structures. Such studies are often the first phase of mineral or petroleum exploration. See also GEOPHYSICAL EXPLORATION.
3. (*magnetic*) see GEOMAGNETICS.
4. (*solid-earth*) a scientific discipline involved with the origin, composition, structure and physical properties of the entire planet. Solid-earth geophysics obtains its data from seismic studies of earthquakes and nuclear explosions, from gravity, magnetic and heat-flow measurements, and from experimental study of the density and elasticity of rocks.

geopotential *n*. the combined differences between the potential energy of a mass at a given height and the potential energy of an identical mass at sea level. It is equivalent to the energy required to move the mass from sea level to the given height.

The difference between potentials at P_1 and P_2 is given by

$$\int_0^H g \cdot dh$$

where g is the value of gravity at height h above P_1 and H is the vertical distance between P_1 and P_2.

geoscience n. **1.** GEOLOGY.

2. EARTH SCIENCE.

geosphere n. the inorganic world, contrasted with the world of life, the BIOSPHERE. It includes the LITHOSPHERE (solid portion of the Earth), the HYDROSPHERE (bodies of water) and the ATMOSPHERE (air).

geostatic pressure or **lithostatic pressure** n. the vertical pressure exerted at a point within the Earth's crust by a column of overlying material.

geostrophic motion n. fluid flow that is parallel to the isobars (lines of equal pressure) in a rotating system, e.g. the Earth. Such a flow is the result of a balance between the pressure gradient and the CORIOLIS EFFECT. The velocity of the flow is proportional to the pressure gradient. For large-scale oceanic and atmospheric movements the geostrophic current generally represents the actual current, with about 10 per cent error. Compare FERREL'S LAW.

geosuture, geofracture or **suture** n. the boundary zone between contrasting rock masses, probably extending as deep as the mantle; the contact between continental plates that have collided.

geosyncline n. an elongate basin in which a thick sequence of SEDIMENTS is deposited. Before the concept of PLATE TECTONICS, geosynclines were thought of as subsiding troughs along a CONTINENTAL MARGIN in which the sediments that accumulated were later deformed and uplifted during mountain-building. Many different types of geosyncline were defined on the basis of the nature of the sediment fill and the relationship to the CRATON. Most of this terminology has been abandoned, and it is important to realise that the term 'geosyncline' is not now used in the original sense.

It is now recognised that the thick sedimentary sequences of geosynclines are formed in a variety of tectonic settings, ranging from the CONTINENTAL SHELF, slope and rise, and the ABYSSAL basins, to the FORE-ARC and BACK-ARC BASINS and fore-arc trenches of DESTRUCTIVE PLATE MARGINS. (See FRONTAL ARC.) Sediments from a range of environments may eventually be carried into a SUBDUCTION ZONE and deformed in an OROGENY, but there is no simple connection between sedimentation and orogenesis, as was implied in the earlier geosynclinal theory. Some geologists retain the term *miogeosyncline* for the shallow-water sediments deposited on the continental shelf and the term *eugeosyncline* for the deep-water sediments deposited at the foot of the CONTINENTAL SLOPE, but others believe these terms too should be abandoned.

geotechnology n. the application of the EARTH SCIENCES to engineering problems.

geotectonic see TECTONIC.

geotextile n. permeable woven plastic sheeting used in road and other geotechnical constructions as a protective and stabilising material.

geothermal energy n. energy that can be extracted from the internal heat of the Earth. Surface manifestations of geothermal energy are found in THERMAL SPRINGS, GEYSERS and FUMAROLES.

geothermal gradient n. the change in temperature with respect to change in depth in the Earth. The geothermal gradient, measured by means of holes bored into the Earth, ranges upward from 5°C per kilometre, with an average of about 30°C per kilometre. The gradient depends on the heat flow in the region, which, in turn, is related to the thermal conductivity of the rock.

geothermal heat flow or **heat flow** *n*. the quantity of heat energy leaving the Earth, measured in milliwatts per square metre (mW m–2). Heat flow is highest over the oceanic ridges and lowest over the older parts of the ocean basins and the geologically old regions of the continents. The greater part of *oceanic heat flow* comes from MANTLE sources, while about half of *continental heat flow* arises from heat generated by radioactive decay in the CONTINENTAL CRUST. Most of the Earth's internal heat is thought to be produced by the decay of radioactive isotopes like uranium, thorium and potassium, so the rate of heat production is declining exponentially.

geothermometer or **geological thermometer** *n*. any feature of rocks, such as a mineral that forms within determined limits of temperature under specific pressure conditions, that can be used to define a range for the temperature of formation of the ambient rock. An example would be the filling temperature of FLUID INCLUSIONS. See also GEOLOGICAL THERMOMETRY.

geothermometry *n*. **1.** determination of the temperature of chemical equilibrium of a mineral, rock or fluid.
2. the study of the Earth's heat and its effect on physical and chemical processes. Compare GEOLOGICAL THERMOMETRY.

geyser *n*. a type of hot spring that intermittently ejects sprays and jets of hot water and steam. Geysers are normally restricted to areas of RECENT volcanic activity. Groundwater in a shallow hydrothermal system in rocks of low permeability is heated at depth. The pressure of the overlying column of water raises the boiling point, allowing temperatures in excess of 150º to be generated. Convection raises the deep superheated water towards the surface until it reaches a depth at which the decrease in pressure results in flash-boiling, when the resulting rapid expansion as steam causes the eruption of the overlying column of water. The hydrothermal system is then replenished by the draining of the erupted water into the vent and/or from inflow of adjacent groundwater that is then heated until the flashpoint is reached again. The term is derived from the Icelandic spouting spring Geysir. See also THERMAL SPRING.

geyserite *n*. siliceous SINTER, especially the opaline material deposited by precipitation from the waters of a GEYSER.

gibber see DESERT PAVEMENT.

gibbsite *n*. a mineral, $Al(OH)_3$, one of the chief constituents of BAUXITE.

Gibb's phase rule see GOLDSCHMIDT'S PHASE RULE, PHASE DIAGRAM.

Gigayear see GA.

gilsonite or **uintahite** *n*. a variety of natural ASPHALT.

ginkgo *n*. a GYMNOSPERM of the division Ginkgoopsida, which contains only a single living species (*Ginkgo biloba*). The known geological range is from Early Permian to Recent, maximum abundance being reached during the Middle Jurassic. Its primitive characteristics have resulted in its being called a 'living fossil'. Ginkgoes, now native to Japan and China, are also called *maidenhair trees*. See also FOSSIL PLANTS.

Givetian *n*. the top STAGE of the Middle DEVONIAN.

glabella *n*. the raised central portion of

the CEPHALON (head carapace) of a TRILOBITE (Fig. 89). It is crossed by short grooves, the *glabellar furrows*, and ends in an arched *occipital ring* at the back of the cephalon.

glacial 1. *n.* a period during an ICE AGE in which there is a considerable increase in the total area covered by GLACIERS and ice sheets. Approximately 18 000 years ago, during the last GLACIAL, ice covered about 32 per cent of the land. See also INTERGLACIAL, STADIAL.
2. *adj.* pertaining to the products or processes of GLACIERS and ice sheets.

glacial abrasion see GLACIAL SCOURING.

glacial age see GLACIAL EPOCH.

glacial boulder *n.* a large rock fragment that has been carried by a GLACIER. Compare ERRATICS.

glacial crevasse *n.* a crack or fissure in a GLACIER resulting from STRESS caused by movement. Glacial crevasses may be 20 metres wide, 45 metres deep and several hundred metres long. *Longitudinal crevasses* develop in areas of compressive stress; *transverse crevasses* develop in areas of tensile stress; *marginal crevasses* occur when the central portion of the glacier flows faster than the outer edges. Jagged ice pinnacles (SERACS) may form where crevasses intersect at the glacier terminus. Glacial crevasses may be hidden by snow bridges and may close up where the gradient is lower. See also CIRQUE, BERGSCHRUND.

glacial drift *n.* rock material deposited by glaciers or GLACIER-associated streams and lakes. The term originated prior to glacial theory, when exotic stones and earth were thought to have been 'drifted' in by water or floating ice. Drift material ranges in size from clay to massive boulders. The greatest thicknesses are found in buried valleys. Drift is divided into two types of material: TILL, which is deposited directly by glacial ice and shows little or no sorting or stratification; and STRATIFIED DRIFT, well-sorted and layered material that has been moved and deposited by glacial meltwater.

glacial epoch or **glacial age** *n.* any portion of geological time, from the Precambrian onwards, in which the total area covered by GLACIERS greatly exceeded that of the present.

glacial erosion *n.* the reduction of the Earth's surface by processes of which GLACIER ice is the agent, including the carriage of rock fragments by glaciers, scouring and the erosive action of meltwater streams.

glacial erratic see ERRATICS.

glacial geology or **glaciogeology** *n.* **1.** the study of the features and effects resulting from GLACIAL EROSION and deposition by GLACIERS and ice sheets.
2. the features of a region that have been subjected to GLACIATION.
Compare GLACIOLOGY.

glacial lake *n.* **1.** a lake fed by meltwater or forming on GLACIER ice because of differential melting.
2. a lake confined by a morainal dam.
3. a lake held in a bedrock basin produced by GLACIAL EROSION.
4. a KETTLE lake.

glacial mill see MOULIN.

glacial plucking *n.* an EROSION process whereby fragments of BEDROCK are dislodged and transported by a GLACIER. The fragments may be *plucked* when meltwater in the bedrock JOINTS refreezes and the resulting expansion forces blocks out of their places. The fragments may also be prised loose by the plastic flow of ice around BLOCKS, which become part of the moving glacier. Glacial plucking is most pronounced on hilltops or knobs and where bedrock is well jointed. It provides much of the rock material in GLACIAL SCOURING, although erosion caused by plucking is more extensive. Plucking is the dominant process in the formation of CIRQUES, cliffs, glacial stairways and other landforms resulting from GLACIATION.

glacial recession or **glacial retreat** *n.* a decrease in the length of a GLACIER.

glacial scouring or **glacial abrasion** *n.* an EROSION process whereby a GLACIER abrades the BEDROCK over which it moves, resulting in the production of a fine rock flour. The abrasive agent is the rock material dragged by the glacier ice. The ice itself does not erode the rock, since it is softer. Fine particles polish the bedrock to a smooth finish, larger particles produce STRIATIONS or scratches, and BOULDERS gouge deep cuts into easily eroded bedrock. All these features may be present at one time. Grooves and striations parallel the ice-flow direction. Glacial scouring usually erodes a smaller amount of material than GLACIAL PLUCKING. See also CHATTER MARKS.

glacial striation *n.* fine-cut parallel or nearly parallel lines on a BEDROCK surface made by rock fragments carried in a GLACIER or cut on the transported rocks themselves. Glacial striations cover great areas in northern North America and Europe. They are usually associated with Pleistocene deposits but also occur in Permo-Carboniferous, Ordovician and Precambrian rocks.

glacial valley see GLACIATED VALLEY.

glaciated *adj.* (of a surface) once GLACIER-covered and, more particularly, modified by GLACIAL action.

glaciated valley, glacial trough or **glacial valley** *n.* a stream valley that has been glaciated, most often to a U-shaped cross-section, and which may be several hundred metres deep. It should be noted that GLACIERS do not cut the original valley but modify or change the shape of an existing one. The formerly V-shaped valley is converted to a U-shaped one and, because a glacier cannot turn as easily as a stream can (because of its greater viscosity), the valley also becomes straighter. Smaller, shallower troughs are cut by smaller tributary glaciers. When the glaciers melt, they remain as *hanging valleys* on walls of the main glacial valley. Postglacial streams may form waterfalls from hanging valleys. See also GLACIATION.

glaciation *n.* the covering and alteration of the Earth's surface by GLACIERS and ice sheets. Valley glaciers produce erosional features, such as *hanging valleys*, CIRQUES and ARÊTES, and depositional features such as MORAINES. Areas subjected to continental glaciation (*ice sheets*) are characterised by depositional features such as ESKERS and DRUMLINS as well as moraines. Lakes, ERRATICS, GLACIAL STRIATIONS, ROCHES MOUTONNÉES, KETTLES and KAMES are common in both areas. Material from seasonal meltwaters is deposited into temporary lakes that form with the recession of continental ice sheets. This may produce RHYTHMITES, layered deposits developed by cyclic sedimentation, some of which are annual and can be used in dating and palaeo-climate determination. *Mountain glaciation* usually increases topographical steepness and sharpness, whereas *continental glaciation* produces a more subdued landscape.

glacier *n.* a large mass of perennial ice showing evidence of present or past flow; it occurs where winter snowfall exceeds summer snow melt. Most of the ice of glaciers originates as fresh, light snowflakes, but it can also derive from rime ice and supercooled water vapour. The first requirement of glacier formation is the accumulation of newly fallen snow in hollows above the snowline. These accumulation areas are called *névé*. The snow crystals are first compressed into granular snow, a process that is most effective at temperatures not appreciably higher or lower than 0°C. After one winter, the granular snow reaches a density of 0.4 to 0.55 and is now called FIRN. As it ages, the firn increases in density and crystal size. When a critical density is reached, firn becomes ice; however, the time expended in this conversion is extremely variable. Periods of 200 years are not unknown.

When the weight of the amassed ice layers becomes great enough to overcome the internal resistance of the ice and the friction between ice and ground, motion is initiated and the ice becomes a glacier. The movement of glaciers consists principally of two components: *internal plastic flow* (CREEP) and basal sliding. Under the great pressure exerted by the weight of the ice, the ice crystals are rearranged into layers of atoms parallel to the glacier surface. The gliding of these layers over one another causes ice to creep. *Basal sliding* refers to the movement of the glacier over BEDROCK under the pull of gravity. Although glacier movement usually combines these two basic processes, the contribution of each varies greatly with local conditions. *Cold glaciers*, in which the temperature of the ice is below the pressure melting point, exhibit no basal sliding because they are frozen to the bedrock, whereas the movement of certain *warm glaciers* almost totally results from this process because of the lubricating effect of

meltwater beneath them. In warm glaciers the temperature of the ice is close to the pressure melting point. A glacier may be composed partly of cold and partly of warm ice. The rate of both creep and basal sliding depends on thickness, slope and temperature of the ice. Velocities also vary within sections of an individual glacier, e.g. the ice at the head and terminus moves more slowly than the ice in between. Englacial measurements show a velocity decrease from surface to bedrock.

Glaciers may be classified morpho-logically as being of three main types: (a) *valley glaciers* are confined to a path directing their movement. CIRQUE *glaciers* are a type of valley glacier. They are short and wide, and flow down a valley, at least in part, and originate with snow accumulated in cirques at valley heads.
(b) *ice sheets* and *ice caps* extend in continuous sheets that move outwards in all directions. These glaciers are called ice sheets if the permanent ice mass is greater than 10^6 square kilome-tres, and ice caps (dome-shaped) or ice fields (flat) if they are smaller.
(c) *piedmont glaciers* are formed when valley glaciers issue from their confin-ing channels and move out over level ground. ICE SHELVES are the floating parts of glaciers that have reached sea level. Smaller forms are *ice tongues*. Very large ice shelves (e.g. the Ross Shelf) may be formed of firn rather than true glacier ice.

Glaciers are also classified according to thermal and dynamic characteristics. Warm and cold glaciers are distin-guished largely by the respective presence and absence of meltwater. The *glacier budget* is the balance between the rate of addition of snow to a glacier and the loss of ice (ABLATION) by melting, calving, etc. The *equilibrium line* is the boundary between the *accumulation zone*, where there is net addition, and the *ablation zone*, where there is net loss.

glacier band *n.* a series of layers on or within a GLACIER differing from adjacent material in colour and texture. Such layers may consist of ice, snow, FIRN, rock debris, organic matter or any mixture of these. See also OGIVE.

glacier budget see GLACIER.

glacier burst or **glacier flood** *n.* a sudden release of a great quantity of meltwater from a GLACIER or subglacial lake. Water accumulates in depressions within the ice margins and at a critical stage erupts through the ice barrier, sometimes producing a catastrophic flood. Such accumulation and release may occur at almost regular intervals. In Iceland the phenomenon, often associated with volcanic or fumarolic activity, is called a *jökulhlaup*.

glacière *n.* **1.** the French generic term for all underground ice formation. It has been accepted by writers of English and long recognised in American scientific journals. **2.** a natural or artificial cavity in a temperate climate in which an ice mass remains unthawed throughout the year; an ice cave.

glacier flood see GLACIER BURST.

glacier flow see GLACIER.

glacierisation *n.* the gradual covering of a land surface by existing glaciers or ice sheets, whereas GLACIATION refers to a previous advance of an ice sheet that no longer exists.

glacier lake *n.* water confined by the damming of natural drainage by a GLACIER or ice sheet.

glacier surge see SURGING GLACIER.

glacier tectonics *n.* the DEFORMATION of glacial masses and glacial sediment by the movement of ice, indicated by fractures and crevasses that appear in the upper ice. Fractured portions may develop as ridges perpendicular or parallel to the direction of movement. Successive bedding layers commonly found in glacial ice represent the residue of seasonal snowfalls mixed with rock debris. The clear patterns into which these layers are often folded help to determine the general pattern of the glacial movement.

glaciofluvial or **fluvioglacial** *adj.* pertaining to GLACIER meltwater streams or to deposits made by such streams.

glaciogeology see GLACIAL GEOLOGY.

glaciolacustrine *adj.* derived from or pertaining to GLACIAL LAKES or the material deposited into such lakes, particularly deposits such as KAME deltas.

glaciology *n.* **1.** the study of glaciers and of ice in all its natural occurrences, including the formation and distribution of ice, its physical and chemical properties, GLACIER flow and the interrelation between ice and climate. **2.** the structural character of glacial deposits, the materials of which are derived by the movement of ice and its differential pressures.

glass *n.* **1.** a state of matter intermediate between the densely packed, well-ordered atomic arrangement of a CRYSTAL and the disordered atomic arrangement of a gas. Most glasses are supercooled liquids, i.e. they are amorphous, metastable solids with the atomic structure of a liquid. Glasses and liquids are distinguished on the basis of viscosity. **2.** an amorphous substance resulting from the rapid cooling of a MAGMA. Within the cooling period, the strong COHESION between atoms of the magma inhibits crystal formation. OBSIDIAN is an example of igneous glass. See also DEVITRIFICATION.

glassy or **vitric** *adj.* a textural term applied to IGNEOUS ROCKS or parts of rocks that completely lack crystalline structure as a result of rapid cooling. A glassy texture is most common in VOLCANIC ROCK. Glassy specimens may be composed almost entirely of massive glass, e.g. OBSIDIAN, or the glass may be vesicular, e.g. PUMICE. See DIAPLECTIC GLASS, GLASS, PSEUDO TACHYLYTE.

glauberite *n.* a light-coloured mono-clinic mineral, $Na_2Ca(SO_4)_2$, found in EVAPORITE deposits.

Glauber's salt see MIRABILITE.

glauconite *n.* a green mineral, essentially a hydrous potassium iron silicate similar in composition to BIOTITE. It is common in marine sedimentary rocks from the Cambrian to the present. Sandstones rich in glauconite (GREEN-SAND) are especially prevalent in rocks of Cambro-Ordovician and Cretaceous age.

glauconitic sand see GREENSAND.

glauconitic sandstone see GREENSAND.

glauconitisation *n.* a REPLACEMENT process common in formations containing large quantities of GLAUCONITE (a complex hydrous potassium iron silicate). Glauconite forms in marine SEDIMENTS under slightly reducing conditions, and the process is facilitated by the presence of organic matter. Foraminifer TESTS often contain glauconite formed by alteration of the mud filling. (See PROTOZOA.)

glaucony *n.* the marine sedimentary FACIES associated with the development of GLAUCONITE.

glaucophane *n.* a light blue prismatic or fibrous mineral, $Na_2Mg_3Al_2Si_8O_{22}(OH)_2$, of the AMPHIBOLE group. Glaucophane is found in low-temperature, high-pressure METAMORPHIC ROCKS, see BLUESCHISTS. It is of interest to petrologists as a means of defining the metamorphic conditions under which the surrounding rock formed.

glaucophane schist facies see BLUESCHIST.

gley soils *n.* waterlogged soils. They develop where drainage is poor or the WATER TABLE is high. A reducing environment exists in the saturated layers, which become mottled greyish-blue or brown because of the content of ferrous iron and organic matter.

glide directions see GLIDING.

gliding *n.* a DEFORMATION-related movement that takes place along lattice planes in crystalline substances. Gliding is of two types: translation gliding and twin gliding. In *translation gliding*, displacement takes place along preferred lattice planes without reorientation or rupture of the deformed parts. In *twin gliding*, the lattice of the displaced part of the crystal is symmetrically altered with respect to the lower, undisplaced part. There are a limited number of lines parallel to which movement can take place; these are the *glide directions*.

global tectonics *n.* tectonics on a global scale, such as PLATE TECTONICS.

globigerina ooze see GLOBIGERINID, PELAGIC DEPOSITS.

globulites *n.* tiny spheroidal crystallites in glassy volcanic rocks such as obsidian. See also SPHERULITE, DEVITRIFICATION.

globigerinid *n.* a foraminiferal protozoan of the suborder Globigerinina (Jurassic to Recent) with a multi-chambered calcareous shell (see PROTOZOA). The initial chamber is invisible to the unaided eye, but the subsequent larger chambers, growing in a spiral fashion, remain attached to the first shell and are easily visible to the naked eye. Globigerinids are exclusively marine, entirely planktonic and live in open water. They are the chief constituent of *globigerina ooze* (see PELAGIC DEPOSITS).

glomerocrysts *n.* clumps or clusters of PHENOCRYSTS. A VOLCANIC ROCK containing glomerocrysts is said to have a *glomeroporphyritic* or *glomerophyric* texture.

glomeroporphyritic see GLOMEROCRYSTS.

Glossopteris *n.* a late Palaeozoic genus of FOSSIL PLANTS (order Glossopteridales) found throughout the continents that formed GONDWANA in the Permian and Triassic, i.e. South America, Australia, India and Antarctica. *Glossopteris* is an enigmatic GYMNOSPERM tree that bore clusters of simple spatulate leaves. Reproductive organs were attached to individual leaves. Glossopterids occupied high latitudes and the *Glossopteris* flora formed a conspicuous element of the south polar ecosystem.

glowing cloud see NUÉE ARDENTE.

gneiss *n.* a foliated METAMORPHIC ROCK formed under conditions of high-grade regional METAMORPHISM. Gneiss is usually coarse-grained and is characterised by a layered appearance caused by the segregation of ferromagnesian from quartzofeldspathic minerals in discontinuous layers or lenticles. The ferromagnesian minerals are commonly BIOTITE and/or HORNBLENDE; PYROXENE is less common. GARNET is often present as an ACCESSORY MINERAL.

The name given to a gneiss may reflect its mineralogy (for example, biotite-gneiss), texture or the nature of the rock from which it formed. An *orthogneiss* was derived from the metamorphism of an IGNEOUS ROCK, while a *paragneiss* was formed from a SEDIMENTARY ROCK. AUGEN gneisses are characterised by very coarse-grained

eye-shaped lenticles of quartz and feldspar. See also GRANITE GNEISS, INJECTION GNEISS.

gneissic *adj.* (of rock) showing the texture typical of gneisses.

gneissoid *adj.* having a GNEISS-like texture or structure that is not related to metamorphic processes; for example, a gneissoid granite formed by viscous magmatic flow.

goethite *n.* a yellow, red or brown mineral, FeO(OH), often the commonest constituent of LIMONITE. It occurs as a deposit in bogs and is the major constituent in GOSSAN.

gold *n.* a dense, soft yellow mineral, the native metallic element Au. It occurs principally as a native metal but may also be alloyed with silver, copper and other metals. Although a rare element, gold is widely distributed in nature. It is found as nuggets, leaves, grains and in veins with quartz. See ELECTRUM.

Goldschmidt's phase rule *n.* a modification of Gibb's phase rule (see PHASE DIAGRAM) that is applied to METAMORPHIC ROCKS. It assumes that two variables are fixed externally and that, consequently, the number of phases in a system will not generally exceed the number of components, i.e. P is $\leq$ C. When used as a mineralogical phase rule, the variables are taken as temperature and pressure; the number of phases will be the minerals, and the system is the rock that is being formed.

gonatoperian suture see CEPHALON.

Gondwana or **Gondwanaland** *n.* the southern continental cluster of the Palaeozoic and early Mesozoic, taking its name from the Gondwana system of India. Rock sequences of Permian age, containing an identical fossil flora (GLOSSOPTERIS) found in the present southern continents of Antarctica, Africa, South America, Australia and

India suggest that they were components of Gondwana. Gondwana corresponds to LAURASIA in the Northern Hemisphere. Both continental masses are thought to have been derived by the splitting of a late Precambrian supercontinent and came together to form a new supercontinent, PANGAEA, during the Permian and Triassic. See also PLATE TECTONICS, CONTINENTAL DRIFT.

goniatite see AMMONOID.

goniometer *n*. an instrument for measuring the angles between adjoining faces in a crystal. The accurate measurement of interfacial angles is important in crystallography. See LAW OF CONSTANCY OF ANGLE.

GOR see GAS-OIL RATIO.

gorge *n*. a deep, narrow, steep-walled valley, sometimes carved by stream ABRASION. It may also be a narrow passage between hills or mountains.

gossan or **iron hat** *n*. the oxidised and leached near- or above-surface portion of a mineral deposit containing sulphides, particularly iron-bearing ones such as pyrite and chalcopyrite. The term 'gossan' derives from the Cornish for 'blood' and refers to the rusty colour resulting from oxidised pyrites. It basically consists of hydrated iron oxides (LIMONITE) from which sulphur and copper have been removed by downward-percolating waters. Some highly insoluble minerals, e.g. gold, may be concentrated as a surficial residuum in a gossan. See SECONDARY ENRICHMENT.

gouge *n*. **1.** see FAULT GOUGE.
2. see CRESCENTIC GOUGE.

graben *n*. a FAULT-bounded crustal unit or block, generally elongate, that has been depressed relative to the blocks on either side. The bordering faults (on the longer side of the graben) are usually of near-parallel strike and steeply dipping. A true graben, in its initial surface form, is typically a linear structural depression. The flanking highland areas are commonly HORSTS.

grade *n*. **1.** the PROFILE of a stream channel so adjusted that no gain or loss of SEDIMENT occurs.
2. also called **tenor**, the relative quantity or percentage of MINERAL content in an ORE body.
3. a particular size range of sedimentary particles, e.g. ALLING GRADE SCALE.
4. a classification of COAL based on its content of impurities, such as ash; higher grades of coal have fewer impurities. Compare RANK, COAL TYPE.
5. see METAMORPHIC GRADE.
6. a level of functional and structural complexity in organisms, representing the common achievement of a number of major taxa (see TAXON).

graded *adj*. **1.** also called **sorted**, (of a sediment or rock) containing particles of generally uniform size.
2. (*engineering*) (of a soil) composed of particles of many sizes or having a uniform distribution of particles from coarse to fine.
3. (of a land surface) on which deposition and erosion are in such balance that a general slope of equilibrium is maintained.

graded bedding *n*. layers, each with a sharply distinct base, on which the coarsest grains of a BED lie. In each layer, the grain size decreases progressively upwards, with the finest-grained material at the top of the bed. This is followed abruptly by the coarse-grained base of the next layer. The layers are commonly interbedded, with others having parallel laminations. Graded bedding is commonly formed in deposits from TURBIDITES. It is a valuable WAY-UP INDICATOR in strata that have been strongly folded or over-turned. See also INVERSE GRADING.

graded profile see PROFILE OF EQUILIBRIUM.

graded slope see GRADED STREAM.

graded stream *n*. a stream or river the transporting capacity of which is in equilibrium with the amount of material supplied to it. Thus, a balance exists between degradation and aggradation. The downstream gradient of a graded stream is called a *graded slope*. It permits the most effective transport of load.

grade scale *n*. an arbitrary division of a basically continuous range of sedimentary particle sizes into a sequence of grades for the purpose of standardisation of terms. See WENTWORTH GRADE SCALE, PHI GRADE SCALE, ALLING GRADE SCALE.

gradient *n*. **1.** the degree of inclination of a surface; steepness of slope. In quantitative terms, it is the ratio of the vertical to the horizontal.
2. HYDRAULIC GRADIENT.
3. GRADED STREAM.
4. the change in value of one variable with respect to the change in another. See GEOTHERMAL GRADIENT.

gradiometer *n*. any instrument used to measure the gradient of a physical quantity, e.g. the gradient of a magnetic field.

grahamite *n*. **1.** a black asphaltite with a high specific gravity and a high fixed carbon content.
2. a type of mesosiderite.

grain *n*. **1.** a mineral or rock particle less than a few millimetres in diameter.
2. a separate particle of ice or snow.
3. the linear disposition of topographical features in a region.
4. the plane of splitting or separation in a metamorphic rock.

grainstone *n*. a clastic, grain-supported, mud-free limestone. Compare

PACKSTONE, MUDSTONE. See DUNHAM CLASSIFICATION.

grain-supported *adj*. (of a sedimentary rock, especially limestone) in which the quantity of granular material is sufficient to form the supporting framework, either wholly or in part.

granite *n*. in the strict sense a coarse-gained alkali-rich plutonic IGNEOUS ROCK composed principally of quartz and one or two feldspars. BIOTITE and/or HORNBLENDE are common MAFIC minerals, and MUSCOVITE may also be present. Quartz forms between 20 and 60 per cent of the FELSIC components, and ALKALI FELDSPAR forms 35 to 90 per cent of the total feldspar. The composition of the plagioclase is usually in the oligoclase/andesine range, while the alkali feldspar is the low-temperature form, i.e. orthoclase, microcline or perthitic varieties of these. Common ACCESSORY MINERALS include APATITE, ZIRCON and MAGNETITE, and sometimes TOURMALINE, TITANITE, FLUORITE and MONAZITE. Textures range from GRANULAR to PORPHYRITIC, with large EUHEDRAL phenocrysts (or *megacrysts*) of alkali feldspar. Where the proportion of plagioclase exceeds 65 per cent of the feldspar content, the term GRANODIORITE is used.

Several varieties of granite have been defined. Those that contain one (perthitic) variety of feldspar are *hypersolvus granites*, whereas two-feldspar granites are called *subsolvus granites*. *S-type granites* are PERALUMINOUS granitic rocks that contain no hornblende but commonly muscovite, GARNET and CORDIERITE, and are believed to be derived from the partial melting of sedimentary source rocks. *I-type granites* have relatively high contents of Na and Ca and hornblende is common. They are often associated with porphyry Cu-Mo MINERALISATION. They are believed to be derived from partial melting of igneous source rocks. *M-type granites* are characteristically found in association with island-arc VOLCANIC activity and are believed to have evolved from mantle-derived magmas. Large volumes of granitic rocks are associated with active CONTINENTAL MARGINS, where they form enormous BATHOLITHS. Granitic rocks of alkaline affinities occur in anorogenic environments, often associated with rift valleys. These are known as *A-type granites*. See also ALKALI GRANITE, RAPAKIVI TEXTURE, TONALITE.

granite gneiss *n*. a GNEISS derived from a sedimentary or igneous rock and having the mineral composition of a GRANITE. Granite that has been metamorphosed and shows gneissic texture, however, is known as *gneissic granite*.

granite porphyry *n*. a rock of granitic composition containing PHENOCRYSTS, usually FELDSPAR and/or QUARTZ, in a medium- to fine-grained groundmass. *Porphyritic microgranite* is a better name. See also PORPHYRY, ELVAN.

granitisation *n*. an essentially metasomatic process whereby an existing rock is converted into a GRANITE, involving the introduction and removal of chemical components on a large scale. The extent to which this process occurs is uncertain. See also ICHOR.

granite wash see ARKOSE.

granoblastic *adj*. (of an isotropic fabric) consisting of a mosaic of equidimensional anhedral grains in metamorphic rocks. If inequant grains are present, they are randomly arrayed.

granodiorite *n*. a coarse-grained plutonic rock composed mainly of quartz, potassium feldspar and plagioclase (oligoclase or andesine);

biotite and hornblende are the usual mafic components. It is the approximate intrusive equivalent of rhyodacite. Granodiorite is similar to GRANITE, except that the latter contains more alkali feldspar.

granophyre *n*. a medium- to fine-grained granitic rock, sometimes PORPHYRITIC, characterised by a micrographic intergrowth of quartz and alkali feldspar (see GRAPHIC). This texture is sometimes known as *granophyric texture*.

granular disintegration *n*. a type of WEATHERING in which rock masses, especially coarse-gained rocks like GRANITE, are broken down grain by grain. The process occurs in regions of great temperature extremes.

granular texture *n*. a rock texture characterised by the aggregation of mineral GRAINS of approximately equal size. The term may be applied to any type of rock but is especially applied to an equigranular, holocrystalline IGNEOUS ROCK the particles of which range from 0.05 to 10 millimetres in diameter.

granule *n*. **1.** a rock fragment of diameter 2 to 4 millimetres, larger than a coarse SAND grain and smaller than a PEBBLE.
2. a small precipitated GRAIN.

granulite *n*. **1.** a high-grade METAMORPHIC ROCK composed of equal-sized, interlocking GRAINS. Foliaceous minerals are absent, so strong schistosity is not developed. Ferromagnesian minerals are dominantly anhydrous.
2. a coarse granular METAMORPHIC ROCK of the GRANULITE FACIES.

granulite facies *n*. the METAMORPHIC FACIES, which in rock of basic igneous composition is represented by the mineral assemblage clinopyroxene plus hypersthene plus plagioclase. It is characteristic of deep-seated regional metamorphism at temperatures exceeding 650°C.

graphic *adj*. (of igneous rock texture) characterised by an intergrowth in which a single crystal of alkali feldspar encloses many small, wedge-shaped grains of quartz, creating an overall resemblance to writing. In *micrographic texture*, the size of the grains is visible only under the microscope. Graphic granite is a PEGMATITE characterised by graphic intergrowths of alkali feldspar and quartz. See Fig. 40 below.

graphite *n*. a naturally occurring crystalline form of carbon, dimorphous with diamond. It is steel-grey to black, soft and greasy to the touch. It occurs as six-sided tabular crystals, compact masses or flakes disseminated through metamorphic rocks such as gneisses, marbles and schists. In some of these rocks it is found in veins large enough to be mined. It is probable that such graphite was derived from carbonaceous material of organic origin. Graphite is used in pencils, electroplating, batteries and electrodes; it is also used as a lubricant. Graphite is also called *plumbago* because for centuries it was confused with GALENA, a black sulphide of lead.

graptolite *n*. an extinct marine colonial organism of the class Graptolithina, usually considered to be part of the phylum Hemichordata. Graptolites are most frequently found as carbonaceous impressions in black SHALES. They secreted an organic exoskeleton consisting of one or more STIPES (branches) growing from the SICULA and carrying THECAE that contained the individual animals (see Fig. 41(a) opposite).

The graptolite *rhabdosome* (colony) may be described as pendent if the stipes grow downwards from the sicula or *scandent* if they grow upwards with the thecae facing outwards. Other attitudes of the stipes can be described as *reclined*, *horizontal* or *declined*, see Fig. 41(b). The stipes may be uniserial, biserial, triserial or quadriserial. The principal graptolite orders are the Dendroidea (Middle Cambrian to Upper Carboniferous) and the Graptoloidea (Lower Ordovician to Lower Devonian). The more primitive *dendroid* graptolites are characterised by a many branched rhabdosome in which two types of thecae, the large *autothecae* and smaller *bithecae*, bud from a chain of *stolothecae* along the stipes, see Fig. 42(b). Most dendroids

Fig. 40 **Graphic**. Graphic granite contains quartz (black) intergrown with alkali feldspar (plain). Also known as runic granite

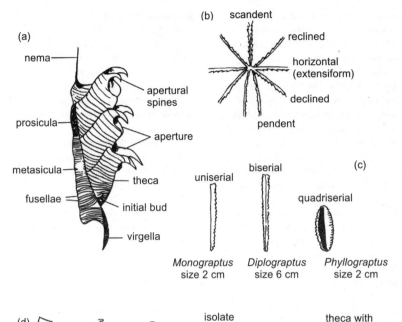

Fig. 41 **Graptolite**. Graptolite morphology: (a) features of the sicula and thecae; (b) attitude of the stipes; (c) forms of stipe; (d) variations in theca shape

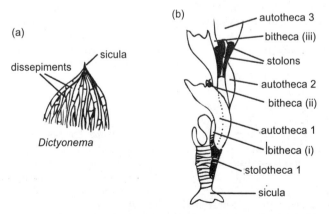

Fig. 42 **Graptolite**. Dendroid morphology: (a) part of a dendroid rhabdosome; (b) details of the structure of part of a stipe

were sessile, growing upwards from holdfasts fixed to the sea floor. Some were planktonic and hung from the nema. (See PLANKTON.)

The Graptoloidea have simpler rhabdosomes, with one or only a few stipes and one type of theca. The wide distribution and lack of any rooting structure suggests they were plank-tonic. They are important INDEX FOSSILS for ORDOVICIAN and SILURIAN rocks.

gravel *n*. **1.** an unconsolidated accumu-lation consisting of particles larger than sand (diameter greater than 2 millime-tres), i.e. granules, pebbles, cobbles, boulders or any combination of these. Gravel is the unconsolidated equiva-lent of a CONGLOMERATE.
2. the common term for DETRITAL sediments, composed mainly of pebbles and sand, found along streams or beaches.

gravimeter *n*. a device for measuring variations in the Earth's gravitational field. One type consists of a torsion balance in which the weight of a needle rotates a wire under tension. The amount of rotation of the wire needed to bring the needle to a null point gives a measure of variation in the Earth's gravitation field. Some gravimeters operate on the principle that variations in the period of oscillation of a swing-ing pendulum are a function of changes in the gravitational field. *Gravimetric surveys* on oceans often involve the use of gyrostabilised platforms. Such surveys are useful in exploring natural resources and determining local anomalies produced by mineral deposits and salt domes. See also GRAVITY ANOMALY, GEOPHYSICAL EXPLORATION.

gravitational attraction, gravitational constant see LAW OF UNIVERSAL GRAVITA-TION.

gravitational gliding see GRAVITY SLIDING.

gravitational separation *n*. **1.** the layering of oil, gas and water in a subsurface reservoir according to their respective specific gravities.
2. the postproduction separation of these materials in a gravity separator.

gravitational settling see MAGMATIC DIFFERENTIATION.

gravitational water see FREE WATER.

gravity anomaly *n*. a deviation of the gravity value from the theoretical value, after a correction has been applied to the point of the Earth's surface where the measurement was made. Gravity anomalies are the result of the Earth's non-uniform density. Of the three types of gravity anomaly defined here, the free-air and Bouguer anomalies are the most common.

As an observer moves farther from the Earth's centre, the resulting decrease in gravity is called the *free-air effect*. The *free-air gravity correction* compensates for changes in surface elevation (to include points outside the geoid). Thus, the *free-air anomaly*, Δg_0, represents the difference between the free-air gravity value (g_0) and the standard gravity value (γ), or $\Delta g_0 = (g_0 - \gamma)$, where g_0 = observed gravity + free-air correction.

A *Bouguer anomaly* remains after application of latitude and free-air corrections, and a correction (the *Bouguer correction*) that removes the effect of the mass of the Earth's material between the observation point and the geoid. The Bouguer anomaly is thus expressed as $\Delta g_0" = (g_0" - \gamma)$, where $g_0"$ is the Bouguer gravity value (observed gravity plus Bouguer correction), and γ is the standard gravity value.

Isostatic anomalies remain after correction for gravity reductions because of terrain. Such corrections are of two principal types. One assumes compensation to be affected by horizontal density variations, the other that compensation is influenced by variations in thickness of a constant-density layer. Gravity anomalies occur over mineral deposits and salt domes, and over caves where earth materials are absent or considerably less dense than the adjacent earth materials. Negative gravity anomalies occur, for example, over ocean trenches and regions that were depressed under an ice sheet during the last glacial advance, and positive anomalies over the island arc at a DESTRUCTIVE PLATE BOUNDARY.

gravity collapse structures *n*. FOLDS that develop in sedimentary layers because of gravity sliding on the flanks of an uplifted region. See Fig. 43 below. See also CASCADE, SLUMP.

gravity compaction *n*. the compaction of a sediment as a result of overburden pressure.

gravity fault see FAULT.

gravity geophysics see GEOPHYSICS.

gravity map see GEOPHYSICS 2.

gravity meter see GRAVIMETER.

gravity sliding or **écoulement** *n*. a downslope movement of a portion of the Earth's crust, usually on the flanks of an uplifted region. This mechanism, activated by gravitational forces, seems to be responsible for many shallow faults. It is also thought to be closely associated with many major mountain systems, such as the Alpine belt.

gravity tectonics *n*. tectonics in which the primary activating mechanism is believed to be downslope sliding because of gravitational effect.

great circle *n*. that circle defined by the intersection of a plane with the centre of a sphere. The shortest distance between any two points on the sphere is along the arc of the great circle joining them.

greenhouse effect *n*. the heating of the Earth's surface caused by the presence, in the lower atmosphere, of carbon dioxide, water and other greenhouse gases, such as methane, that absorb long-wavelength terrestrial radiation, thus halting or hindering its escape from the Earth's surface. See CARBON DIOXIDE.

green mud see TERRIGENOUS DEPOSITS.

greensand, glauconitic sand or **glauconitic sandstone** *n*. a greenish sand or sandstone, specifically a marine sediment composed of quartz and dark green grains of GLAUCONITE.

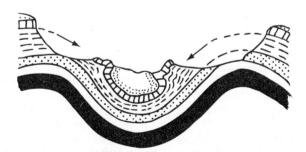

Fig. 43 **Gravity collapse structures**. Folds resulting from the down-dip sliding of sedimentary layers

greenschist *n.* a green schistose metamorphic rock the colour of which results from the presence of chlorite, epidote or actinolite. Compare GREENSTONE.

greenschist facies *n.* the METAMORPHIC FACIES in which basic rocks are represented by albite + epidote + chlorite + actinolite. The mineral assemblage is characteristic of low-grade regional metamorphism corresponding to temperatures in the range of 300 to 500°C. The greenschist facies assemblage of minerals partially or totally replaces primary magmatic minerals, e.g. ophitic pyroxene, calcic plagioclase. Relict igneous fabrics are often preserved.

greenstone *n.* any green, aphanitic, weakly metamorphosed igneous rock the colour of which is caused by chlorite, actinolite or epidote. Compare GREENSCHIST.

greenstone belt *n.* an area in Precambrian SHIELDS, especially in the Archean, occupied largely by volcanic rocks of compositions varying from basalt to rhyolite and including ultrabasic rocks such as KOMATIITE. Such areas have commonly been subjected to GREENSCHIST FACIES metamorphism, but to a large extent the rocks retain many of their original textural features and structures. Associated metasediments include metagreywacke, metapelite and iron-bearing formations. See BANDED IRON FORMATION.

greisen *n.* an aggregate of quartz and muscovite (or lepidolite) with accessory amounts of topaz, fluorite, tourmaline, rutile, cassiterite and wolframite, formed by the pneumatolytic alteration of granitic rock *in situ* or deposited in associated veins. See PNEUMATOLYSIS.

Grenville orogeny *n.* the name applied to a series of major plutonic metamorphic and deformational events during the Late Precambrian (c. 1300–1000 Ma). The OROGENY is named from the region along the southeastern border of the Canadian Shield but includes all those areas deformed during the assembly of the supercontinent RODINIA.

grey mud see TERRIGENOUS DEPOSITS.

greywacke *n.* a term that has been redefined many times since its origin and is now usually applied to a well-compacted and cemented dark-grey or greenish-grey coarse-grained SANDSTONE characterised by ANGULAR particles of QUARTZ, FELDSPAR and rock fragments embedded in a clayey matrix that forms more than 15 per cent of the rock. Greywacke commonly exhibits the GRADED BEDDING characteristic of certain marine sedimentary facies, so is believed to have been deposited by submarine turbidity currents. See TURBIDITES. Compare WACKE.

grike *n.* a vertical JOINT in LIMESTONE that has been widened by solution. Further solution may lead to the formation of a SWALLOW-HOLE. See also KARST.

groove mark *n.* a long, narrow and straight trough produced by an object being rolled or dragged along a soft sediment surface. The term is also applied to the counterpart of this groove on the underside of an overlying sandstone bed. See TOOL MARK. Compare DRAG MARK.

grossular *n.* the calcium-aluminium member of the GARNET series, $Ca_3Al_2(SiO_4)_3$. It is found mainly in crystalline limestone as a product of contact or regional metamorphism. Its colour range is from white through golden to cinnamon.

grossularite *n.* a redundant name for GROSSULAR.

groundmass or **matrix** *n.* **1.** the fine-grained material that holds the PHENOCRYSTS in a porphyritic IGNEOUS ROCK.

2. the MATRIX of sedimentary rock.

ground moraine see MORAINE.

'ground roll' *n*. a disturbance that often interferes with the clarity of reflections on seismic records. Rayleigh waves are thought to be the principal component of ground roll. See SEISMIC WAVES.

ground surge *n*. a low-density flow of PYROCLASTIC MATERIAL immediately preceding the main, denser part of a PYROCLASTIC FLOW and commonly forming the basal deposit of such a flow.

groundwater *n*. **1.** also called **phreatic water**, subsurface water in the zone of saturation, including water below the WATER TABLE and water occupying cavities, pores and openings in underlying rocks.

2. also called **subterranean water** or **underground water**, all subsurface water, as distinct from surface water.

groundwater barrier *n*. an obstruction, either natural or artificial, to the lateral movement of GROUNDWATER. There is a distinct difference in GROUNDWATER level on opposite sides of the barrier. A DYKE is an example of a natural obstruction.

groundwater basin *n*. a complete three-dimensional hydrogeological system comprising all the AQUIFERS in an area that has definable boundaries and definable recharge and discharge areas. A groundwater basin may or may not have the shape of a basin.

groundwater divide *n*. an elevation along the WATER TABLE from which GROUNDWATER flows away in opposite directions.

groundwater flow *n*. the natural or artificially produced movement of water in the zone of saturation (see WATER TABLE).

groundwater hydrology see HYDROGEOLOGY.

groundwater level see WATER TABLE.

groundwater reservoir see AQUIFER.

groundwater surface see WATER TABLE.

group *n*. the formal STRATIGRAPHICAL UNIT next in order above FORMATION; it includes two or more geographically associated formations with notable features in common. A differentiated assemblage of formations is a *subgroup*, and an assemblage of related groups or of formations and groups is called a *supergroup*.

group velocity see PHASE VELOCITY AND GROUP VELOCITY.

growth fabric *n*. the orientation of FABRIC elements of a rock that is characteristic of the manner in which the rock was formed and not dependent on the effect of stress and deformation.

grunerite *n*. a monoclinic mineral, $Fe_7Si_8O_{22}(OH)_2$, of the AMPHIBOLE group.

G shell see SEISMIC REGIONS.

Guadalupian *n*. the lower series of the Upper Permian of North America.

guard see BELEMNITE.

guide fossil see INDEX FOSSIL.

Gulfian *n*. a series of the Upper Cretaceous of North America.

gully see EMPHEMERAL STREAM.

gully erosion *n*. the EROSION of soil by running water that forms clearly defined, narrow channels that generally carry water only during or after heavy precipitation. Compare SHEET EROSION.

gumbotil *n*. a grey to dark-coloured, leached, deoxidised clay representing the B horizon of fully mature soils (see SOIL PROFILE). Gumbotil profiles develop under conditions that exist on flat and poorly drained till plains.

Gunz see CENOZOIC.

Gutenberg discontinuity *n*. the seismic velocity discontinuity at 2900 kilome-

tres, marking the MANTLE-core boundary within the Earth. At this boundary
the velocity of P waves (primary
waves) is reduced and S waves
(secondary waves) disappear. It is
believed to reflect the change from a
solid to a liquid phase (solid mantle to
molten outer core) as well as a change
in composition. A molten outer core
would block S waves entirely, since
they cannot travel through liquids. See
also SHADOW ZONE.

gutter casts see SCOUR MARK.

guyot *n*. a SEAMOUNT of more or less
circular form and with a flat top.
Guyots are thought to be VOLCANIC
CONES the tops of which have been
truncated by surface-wave action. They
occur in deep ocean, usually at depths
exceeding 200 metres, and are generally composed of alkali basalt, which
differs from oceanic tholeiite in having
a higher content of K, Na and Ti.
Guyots are found chiefly in the Pacific
basin, where they occur in groups, as
do the present volcanic island chains.
See BASALT, ISLAND ARC.

gymnosperm *n*. a member of that group
of the Spermatophytina termed
gymnosperms. The term covers those
usually tree-like plants whose seeds are
commonly held in cones or other
modified shoots and whose ovules are
not totally encased by tissue. Examples
are cycad, GINKGO, fir, pine and spruce.
Range, Devonian to Recent. Compare
ANGIOSPERM.

gypsite *n*. a name sometimes applied to
an earthy variety of GYPSUM found only
in arid regions as an powdery deposit
over an outcrop of gypsum.

gypsum *n*. a common, widely distributed mineral, $CaSO_4 \cdot 2H_2O$, frequently
associated with halite and anhydrite in
evaporites. Curved, twisted crystal
growths of gypsum found on cave
walls are called *gypsum flowers*. The
massive, fine-grained variety of
gypsum is ALABASTER. Gypsum is used
in making plaster of Paris and as a
retarder in Portland cement.

gyre *n*. a circular motion of water found
in each of the major ocean basins. It is
centred on a subtropical high-pressure
region, and its motion is generated by
the convective flow of warm surface
water, the Earth's rotation and the
effects of prevailing winds. See also
GEOSTROPHIC MOTION.

H*h*

habit see CRYSTAL HABIT.

hackly *adj.* (of fractures) rough or jagged.

hadal *adj.* pertaining to the deepest oceanic environment, in particular to ocean trenches where depths exceed 6000 to 7000 metres. Sediments in this region accumulate very slowly, and organisms are relatively scarce. See CONTINENTAL SHELF. See also ABYSSAL.

hade *n.* the angle a FAULT PLANE makes with the vertical. The term is obsolescent and it is usual to give the attitude of the fault plane in terms of the DIP, measured from the horizontal.

Hadean *n.* a stratigraphical name proposed for the pre-ARCHAEAN, corresponding to the PRECAMBRIAN PRISCOAN EON. The Hadean Eon lies between 4530 Ma (the formation of the Earth) and 3900 Ma.

Hadrynian *n.* the top division of the Canadian PROTEROZOIC.

haematite see HEMATITE.

hairstone or **needlestone** *n.* a clear, crystalline quartz containing numerous fibrous or needlelike inclusions of other minerals such as RUTILE or ACTINOLITE. See also VENUS HAIR, SAGENITE.

half-graben *n.* downfaulted blocks resulting from the formation of single sets of normal FAULTS during crustal extension. This may be compared with the GRABEN resulting from the formation of CONJUGATE sets of normal limits. See Fig. 32(a). A half-graben is bounded by a fault on one side only.

half-life *n.* the time required for the decomposition of one half of any given quantity of a radioactive element. It differs for various elements and isotopes but is always the same for a particular ISOTOPE. It cannot be accelerated or decelerated by heat, cold, magnetic or electrical fields. A nucleus is regarded as stable if its half-life is greater than the estimated age of the Earth (about 5.5×10^9 years). See also ALPHA DECAY, BETA DECAY, RADIOACTIVITY.

halide *n.* a compound in which the negative element is a halogen, e.g. HCl, HF or, as a mineral example, HALITE, NaCl.

halite, common salt or **rock salt** *n.* an isometric mineral, NaCl, that occurs most commonly in bedded deposits. It is the most abundant of the evaporites, following gypsum and anhydrite in the sequence of precipitation of salts from sea water. See also HOPPER CRYSTALS.

halloysite *n.* a CLAY MINERAL related to KAOLINITE. It has two forms, one with kaolinite composition, $Al_4(Si_4O_{10})(OH)_8$, and a hydrated form, $Al_4(Si_4O_{10})(OH)_8 \cdot 4H_2O$. The second type dehydrates to the first with the loss of interlayered molecules of water.

halmyrolysis *n.* DIAGENESIS that takes place at the water-sediment interface. Early stage of diagenesis.

halo *n.* **1.** a circular or lunate distribution about the source location of a mineral, ore or petrographic feature. **2.** a ring-shaped discoloration of a mineral that can be observed when the mineral is viewed in THIN SECTION. Most halos of this type come from radiation damage caused by the radioactive decay of mineral inclusions, e.g. the pleochroic halos found in biotite around inclusions of uranium-bearing zircon.

hammada or **hamada** *n*. a bare rock surface in a desert region, all loose rock particles having been removed by wind. Compare ERG, REG.

hand level *n*. a small, lightweight levelling instrument in which the spirit level and mirror are so positioned that the bubble may be viewed while sighting through the telescope.

hanging valley see GLACIATED VALLEY, GLACIATION.

hanging wall *n*. the overlying side of a FAULT, ORE body or mine working, in particular the wall rock above a fault or inclined VEIN. Compare FOOTWALL.

hard coal see ANTHRACITE.

hardground *n*. a sea floor lithified *in situ*. Early DIAGENESIS at the sea floor forms a hard cemented surface that is preserved as a minor HIATUS in the stratigraphical record. Hardgrounds may support marine organisms fixed or cemented to the hard sea bed.

hardness *n*. the resistance of a mineral to scratching, one of the properties by which minerals may be described. See MOHS SCALE.

hardness scale see MOHS SCALE.

hardpan *n*. a hard subsurface layer resulting from the deposition of silica, clay, oxides and hydroxides or carbonate salts in the layer of deposition. In humid areas, hardpans consist mostly of clay. CALICHE hardpans are extensive in the western United States, and their structure shows regular variation with age. Some hardpans are formed by accumulations of organic matter. Compare CLAYPAN, DURICRUST, IRON PAN.

hard water *n*. water that contains large quantities of dissolved salts, e.g. calcium and magnesium ions, that prevent the ready lathering of soap and are responsible for the formation of boiler scale.

Harker diagram see VARIATION DIAGRAM.

harmonic folding *n*. folding in which the strata remain parallel or concentric with each other and in which there are no abrupt changes in the form of the folds at depth. Compare DISHARMONIC FOLD.

Hartmann's law *n*. given two sets of intersecting SHEAR planes, the acute angle between them is bisected by the axis of greatest principal stress and the obtuse angle by the axis of least principal stress.

harzburgite *n*. a PERIDOTITE composed mainly of forsteritic olivine and magnesium-rich orthopyroxene.

Hauterivian *n*. a STAGE of the Lower CRETACEOUS.

haüyne or **haüynite** *n*. a rare isometric FELDSPATHOID mineral similar to sodalite but with calcium substituting for some of the sodium.

Hawaiian-type eruption see VOLCANIC ERUPTION, TYPES OF.

hawaiite *n*. a VOLCANIC ROCK slightly richer in alkalis and silica than ALKALI BASALT. The PLAGIOCLASE is in the andesine range, and there is a higher content of ALKALI FELDSPAR. It is defined chemically as the sodic variety of TRACHYBASALT. Hawaiite is a member of the igneous rock series: alkali basalt–hawaiite–MUGEARITE–BENMOREITE–sodic TRACHYTE.

HDR see HOT DRY ROCK.

headwall *n*. a precipitous slope at the head of a valley. The term is especially applicable to the rock cliff at the back of a CIRQUE.

headward erosion or **headwater erosion** *n*. the lengthening of a valley or gully, or of a stream, by EROSION at the valley head.

headwater erosion see HEADWARD EROSION.

heat capacity *n*. the amount of heat required to increase the temperature of a body or system at constant pressure and volume by one degree. It is usually expressed in joules per kelvin. Heat capacity differs for different materials.

heat flow see GEOTHERMAL HEAT FLOW.

heave *n*. **1.** an upward expansion of a surface that may be caused by such factors as clay swelling or FROST ACTION. **2.** CREEP in mines. **3.** the horizontal component of DIP SEPARATION measured in a vertical section at right angles to the FAULT surface.

heavy liquid *n*. a liquid with a density in the range 1.4 to 4.0 used to separate minerals and ores into heavy and light fractions by flotation. Examples are BROMOFORM, METHYLENE IODIDE, CLERICI SOLUTION.

heavy minerals *n*. **1.** dense accessory detrital minerals (e.g. tourmaline and zircon) from a sedimentary rock. These are separated from minerals with lower specific gravity, such as quartz, by means of high-density HEAVY LIQUIDS, e.g. bromoform. **2.** samples of crushed igneous or metamorphic rock that sink in a heavy liquid. Light minerals are those that float.

heavy oil *n*. crude oil that has a low API GRAVITY. Compare LIGHT OIL.

heavy spar see BARITE.

hectorite *n*. a CLAY MINERAL of the MONTMORILLONITE group, containing lithium and magnesium.

hedenbergite *n*. a monoclinic mineral, $Ca,FeSi_2O_6$, the iron-rich END MEMBER of the diopside-hedenbergite series of PYROXENE.

Heinrich cycle see PALAEOCLIMATOLOGY.

helicitic texture *n*. curved, folded or irregular lines of inclusions within a POIKILOBLAST. The inclusions are a RELICT of an earlier metamorphic texture and have not been disturbed or rotated during growth of the poikiloblast (see Fig. 44 below). Compare SNOWBALL TEXTURE.

helictite *n*. a contorted, twiglike or branching cave deposit, usually of calcite. Unlike STALACTITES, helictites are often not vertical; many grow horizontally. Some helictites are formed where water seeps slowly through the bedrock and emerges through nearly microscopic fissures by capillary action.

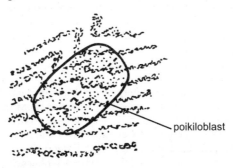
poikiloblast

Fig. 44 **Helicitic texture.** The lines of inclusions are concordant with the structures in the matrix of the rock

Helikian *n.* a stratigraphical division of the Canadian PROTEROZOIC from 1800 Ma to 1000 Ma.

helluhraun *n.* Icelandic term for PAHOEHOE lava flows.

hematite or **haematite** *n.* a mineral oxide of iron, Fe_2O_3, the principal ore mineral of iron. It occurs as individual crystals in the hexagonal system, arranged in flowerlike formations (*iron roses*), but more commonly as massive granular formations. Massive formations that are soft and earthy are known as red ochre. Hematite occurs as an oxidation product in the hydrothermal alteration of some lavas: it is rare in plutonic rocks but common in pegmatites and hydrothermal veins. Much of hematite is formed under sedimentary conditions through the DIAGENESIS of LIMONITE. It remains stable in a low-temperature metamorphic environment, where it often replaces magnetite.

hematisation *n.* the REPLACEMENT of hard parts of an organism by HEMATITE (iron oxide, Fe_2O_3). Compare PYRITISATION. Hematisation generally results from iron deposition around a FOSSIL. The shells of bryozoans, trilobites, brachiopods and crinoid fragments may be preserved in this manner.

hemera *n.* an obsolete geological time unit, corresponding to the ACME ZONE, within which the greatest abundance of a taxonomic entity is found in a defined section. It is that period of time during which a certain type of organism is at its evolutionary apex.

hemimorphic *adj.* (of a crystal) having a polar symmetry, so that the two ends have different forms.

hemimorphite *n.* a hydrated zinc silicate, $Zn_4(Si_2O_7)(OH)_2 \cdot H_2O$, an ore of zinc. Single crystals are rare; it is more often found as fibrous radiating crusts or mammillary masses. It forms in the oxidation zone of zinc sulphide and lead sulphide deposits.

Hercynian see VARISCAN.

hermatypic corals *n.* CORALS with symbiotic ALGAE (zooxanthellae), which are therefore restricted to shallow water of normal marine salinity. Hermatypic corals grow best in water less than 30 metres deep with a limit of efficient growth at 70 metres depth. Few or none grow below 100 metres depth. Reef-building corals are hermatypic. Compare AHERMATYPIC CORALS; also see CORAL REEF.

herring-bone cross-bedding see CROSS-BEDDING.

hessonite see GARNET

heteradcumulate see CUMULATE.

heterochrony *n.* the study of the evolutionary changes in the timing of development or appearance of characteristics during the life history of an individual, leading to differences in size and shape from the immediate ancestor. In *peramorphosis* an extra developmental stage is added, and in *paedomorphosis* developmental stages are lost.

heterochthonous *adj.* 1. (of a rock or sediment) formed in a place other than that in which it now resides. See also ALLOCHTHON. 2. (of fossils) transported from their original site of deposition and re-embedded elsewhere.

heteromorph see AMMONOID and Fig. 2.

heteromorphism *n.* the CRYSTALLISATION of a MAGMA to form rocks with contrasting mineralogy as a result of cooling under different pressure and temperature regimes.

heteropygous see PYGIDIUM.

Hettangian *n.* the bottom STAGE of the Lower JURASSIC.

Hexacorallia *n.* alternative name for the SCLERACTINIA.

hexagonal system see CRYSTAL SYSTEM.

hiatus *n.* a gap or interruption in the stratigraphical record, as represented by the absence of rocks that would normally be part of a sequence but were never deposited or were eroded away prior to the deposition of beds immediately overlying the break. The strata missing at a physical break represent the length in time of such a period of non-deposition. Compare LACUNA.

high-angle fault *n.* a fault with a dip exceeding 45°. Compare LOW-ANGLE FAULT.

high-grade metamorphism *n.* METAMORPHISM that proceeds under conditions of high temperature and pressure. Compare LOW-GRADE METAMORPHISM, METAMORPHIC GRADE.

high quartz see BETA QUARTZ.

Hilt's law *n.* the generalisation that, at any location in a coalfield, coal RANK increases with depth in a vertical succession.

Himalayan orogenic belt see ALPINE-HIMALAYAN OROGENIC BELT.

hinge, hinge line or **flexure** *n.* the locus of points (usually a line) of maximum bending or curvature along a folded surface.

hinge fault see FAULT.

hinge line *n.* **1.** a boundary between a stable or stationary region and one experiencing upward or downward movement.
2. see HINGE.
3. the line along which the valves interlock at the posterior margin of a shell. See also TEETH.

hinge plate *n.* the broad edge of the dorsal margin of each VALVE on a BIVALVE (Fig. 7). It contains the LIGAMENT and the DENTITION, and is the region where the valves are in contact and interlock.

hinterland see FORELAND.

histogram *n.* a diagram, usually a vertical bar graph, in which the frequency of occurrence of some event is plotted against some other measurement, such as the particle-size distribution in sediments.

historical geology *n.* a major branch of GEOLOGY concerned with the history of the Earth and its life forms, from its origins to the present. As such, it involves studies in PALAEONTOLOGY, STRATIGRAPHY, GEOCHRONOLOGY, PALAEOCLIMATOLOGY and plate tectonic movements. Compare PHYSICAL GEOLOGY.

hodograph *n.* a graphical representation of distance travelled versus time. A *tidal hodograph* is a plot of the mean tidal current cycle. In seismology, travel time is plotted as a function of distance from an earthquake epicentre. In oceanography, a special hodograph, the *Ekman spiral*, is used to indicate the progressive departure of an ocean current from its initial direction.

hogback *n.* a linear, sharp ridge on steeply dipping sedimentary rocks, the opposite slope of which may be symmetrical in shape. Hogbacks are the resistant remnants of eroded and folded layers of rocks. Hogbacks are similar to CUESTAS but have steeper rock dips and sides that are more nearly symmetrical.

holaspid see PROTASPIS.

Holkerian *n.* a stage of the VISÉAN (CARBONIFEROUS) in Britain and western Europe.

holoblast *n.* a mineral crystal that originates during METAMORPHISM and is completely formed during that process.

Holocene *n.* the most recent EPOCH of geological time, the upper division of the Quaternary Period. See CENOZOIC.

holochroal eyes *n.* compound TRILOBITE eyes in which the edges of the lenses are all in contact and covered by a single corneal membrane. They are the oldest type of trilobite eye and are found in forms from early CAMBRIAN to the PERMIAN. Compare SCHIZOCHROAL EYES.

holocrystalline *adj.* **1.** (of the texture of an igneous rock) composed completely of crystals (i.e. having no GLASS). **2.** (of a rock) having such texture.

holohyaline *adj.* (of an igneous rock) composed entirely of GLASS.

holomictic lake *n.* a lake the waters of which undergo complete mixing during periods of circulation OVERTURN.

holosymmetric *adj.* (of a CRYSTAL class) having the maximum number of symmetry elements. See CRYSTAL SYMMETRY.

holothurian *n.* a member of the ECHINODERM class Holothurioidea, which includes fossil and living forms, e.g. sea cucumbers. A holothurian has a cylindrical body with an anus at or near one end and a mouth at or near the other. Feeding tentacles surround the mouth. Holothurians are soft-bodied, except for small mineralised sclerites, and have little geological importance. Range, Ordovician to Recent.

holotype see TYPE SPECIMEN.

homeomorph *n.* **1.** a CRYSTAL that resembles another in its crystal form and habit but differs in chemical composition. Each crystal is said to be a homeomorph of the other. See PSEUDOMORPH. **2.** homeomorphs: organisms that resemble one another in external appearance although each is derived from a different ancestor.

homeomorphism *n.* a resemblance among CRYSTALS in habit and form despite differences in chemical composition.

homeomorphy *n.* a biological phenomenon in which species of different ancestry display a general external similarity but in which there is dissimilarity in detail. See also CONVERGENCE.

hominid or **fossil human** *n.* any member of the FAMILY Hominidae. Early hominids are australopithecines, the earliest of which, *Australopithecus afarensis*, appeared in East Africa approximately 4 Ma ago. By 2 Ma ago the earliest members of the GENUS *Homo* had appeared, *Homo habilis* being the first representative of the genus. Early members of the genus *Homo* probably co-existed for a time with australopithecines. *Homo erectus* was a more advanced form that first appeared about 1.8 Ma ago and spread widely into Europe, probably giving rise to *Homo sapiens neanderthalensis* and *Homo sapiens sapiens* (modern man) between 300 000 and 200 000 years ago. These two apparently co-existed for some time, but Neanderthal man, who had survived in Europe under glacial conditions 75 000 years ago, disappeared about 30 000 years ago.

homoclinal ridge *n.* an erosional feature that develops in areas of moderately dipping strata; it is characterised by a significant difference in the length of its front and back slopes as well as in their steepness. Because a homoclinal ridge is less sharply defined than a HOGBACK and smaller in area than a CUESTA, it is used as a term to describe those forms intermediate between these two.

homoclinal shifting see UNICLINAL SHIFTING.

homocline *n.* a structural state in which rock strata exhibit uniform dip in one direction, e.g. one limb of a fold. Compare MONOCLINE.

homogeneous stress see STRESS.

homology *n*. the similarity, but not complete identity, between structures in different organisms as a result of evolutionary differentiation of the same or corresponding structures derived from a common ancestor.

homoplasy *n*. the similarity or correspondence of organs or parts of different organisms that developed as a result of parallelism or convergence rather than from a common ancestor. Compare HOMOLOGY.

homopolar bonds see COVALENT BONDS.

homopycnal inflow *n*. flowing water that has the same density as that of the body it enters, resulting in easy mixing of the two systems. Compare HYPERPYCNAL FLOW, HYPOPYCNAL FLOW.

homotaxis *n*. **1.** a similarity between stratigraphical sequences in separate locations. Two rock STRATIGRAPHICAL UNITS in different areas that have a similar order of arrangement but are not necessarily contemporaneous are *homotaxial*.
2. a similarity between separately located FOSSIL sequences. Two formations that contain fossils occupying corresponding positions in different vertical sequences are *homotaxial*.

honeycomb coral *n*. a compound coral whose CORALLITES are so arranged as to resemble the structure of a honeycomb. See COLONIAL CORAL.

hook *n*. a SPIT or long, narrow cape that curves sharply landward at its outer terminus.

Hook's law *n*. the statement that the STRAIN in an elastic material is linearly proportional to the applied STRESS.

hopper crystal *n*. a type of skeletal crystallisation often shown by rock salt; the crystal has grown faster along its edges than at the centre, resulting in stepped depressions on cube faces.

horizon *n*. **1.** an interface indicating a particular position in a stratigraphical sequence. Theoretically, it is a surface with no thickness; in practice, however, it is usually a distinctive BED. It is not a synonym of ZONE.
2. SOIL HORIZON.
3. (*surveying*) a line or plane used as a reference, generally relating it to a horizontal surface.

horizontal displacement see STRIKE SLIP.

horizontal normal separation see OFFSET.

horn see ARÊTE.

hornblende *n*. a monoclinic silicate mineral, the most common member of the AMPHIBOLE group, dark green or black in colour, sometimes acicular or fibrous in parallel or sheaf-like aggregates. It is an important constituent of metamorphic rocks. It is also found in intermediate and basic igneous rocks and some acid plutonic rocks.

hornblendite *n*. a plutonic rock composed chiefly of HORNBLENDE.

horn coral *n*. a solitary species of RUGOSE CORAL.

hornfels *n*. a hard, fine- to medium-grained METAMORPHIC ROCK composed of a mosaic of mineral grains of about the same size and random orientation. Porphyroblasts or relict phenocrysts may be present in the typically granoblastic fabric; relict bedding may also be found. Hornfels is formed by thermal metamorphism, solid-state crystal growth within a contact AUREOLE surrounding an igneous intrusion.

hornfels facies *n*. an imprecisely defined term used to denote the physical conditions involved in, or the series of mineral assemblages produced by, thermal metamorphism at comparatively shallow depths in the Earth's crust.

hornito or **driblet cone** *n* a small hollow mound of SPATTER or driblet on the top of a LAVA FLOW (generally PAHOEHOE), formed by the accumulation of clots of lava spraying or oozing through an opening in the roof of an underlying LAVA TUBE. A *driblet spire* is a type of hornito resembling a thin column or spire. Compare SPATTER CONE. See LAVA.

horse see DUPLEX.

horst *n*. an elongate BLOCK that has been raised relative to its surroundings. Its long sides are bounded by steeply inclined faults that generally dip away from each other. Vertical movements induced by lateral tension or stretching are believed to be the cause in the formation of most horsts. Compare GRABEN. See FAULT.

hortonolite *n*. a generally redundant term for iron-rich OLIVINE with a composition in the range Fo50 to Fo10.

host *n*. **1.** a rock or mineral that is older than rocks or minerals introduced into it.

2. a large crystal enclosing smaller crystals of a different species.

host rock see COUNTRY ROCK.

hot dry rock (HDR) *n*. a potential source of heat energy within the Earth's crust, represented by rocks at depths of less than 10 kilometres and temperatures above 150°C. Their heat energy is derived from magmas and conduction from the deeper interior.

hot spot *n*. an anomalously hot region within the Earth's MANTLE, 100 to 200 kilometres across and persistent for millions of years. Hot spots are thought to be the manifestations of a rising plume of superheated material from deep within the mantle, possibly even near the CORE/mantle boundary (see MANTLE PLUME). The surface expression of such hot spots is a region of voluminous VOLCANIC activity, such as Hawaii, the Canaries or Iceland. The motion of lithospheric plates over hot spots results in the formation of a chain of VOLCANOES increasing in age away from the present site of activity, e.g. the Hawaiian-Emperor chain of volcanic islands, extending some 6000 kilometres across the Pacific northwest of Hawaii.

hot spring see THERMAL SPRING.

Hoxnian see CENOZOIC, Fig. 13.

Hudsonian orogeny *n*. a period of metamorphism, plutonism and deformation in the Canadian Shield

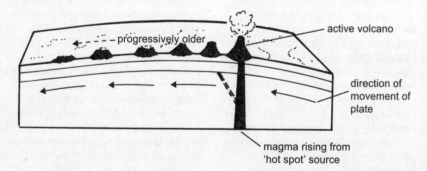

Fig. 45 **Hot spot.** A procession of volcanoes formed as a plate moves over a 'fixed' hot spot in the mantle

during the early Proterozoic (Precambrian).

humic acid *n.* black acidic organic matter occurring in soils, low-rank coals and decayed plant substances.

hummock *n.* **1.** a knoll or mound above the general level of a surface, or a knob of turf or earth in alpine regions.
2. a mound of broken floating ice pushed upwards by pressure.

hummocky cross-bedding *n.* a form of CROSS-BEDDING produced by storm conditions in mobile shallow-water sandy sediments. Convex upwards hummocks of length *c.* 1–2 metres are produced. The hummocks are internally cross-stratified with convex and concave upwards cross-strata. Mud drapes may form on top of the hummocks.

hummocky moraine see KAME-AND-KETTLE MORAINE.

hums *n.* steep-sided residual LIMESTONE hills in KARST landscapes.

humus *n.* a dark organic material in soils, produced by the decomposition of animal or vegetable matter and essential to the fertility of the earth.

Huygen's principle *n.* the statement that any particle excited by wave energy becomes a new point source of secondary waves. These interfere, and the resultant wave is identical with the original.

hyaline *adj.* GLASSY.

hyalite *n.* a colourless variety of OPAL the appearance of which varies from glass-like transparency to translucent or whitish.

hyaloclastite *n.* an aggregate of GLASSY fragments produced by rapid chilling and shattering when LAVA is erupted or flows into water or over water-saturated sediments, or under ice. See also PEPERITE.

hyalocrystalline *adj.* (of the texture of porphyritic rocks) in which the volume occupied by phenocrysts is equal or nearly equal to the volume of a glassy groundmass. Compare INTERSERTAL.

hyalopilitic *adj.* (of the texture of igneous rock) consisting of acicular microlites in a glassy groundmass. When the amount of groundmass glass is decreased or is replaced by crypto-crystalline material, the texture is described as INTERSERTAL. PILOTAXITIC texture is a felted mass of lath-shaped microlites without glass.

hybrid *n.* **1.** a rock mass the chemical composition of which is the result of the mixing of MAGMAS of different composition. See also ASSIMILATION.
2. an individual whose parents are of different species.

hydrargillite *n.* the crystallised form of GIBBSITE. This term has also been applied to aluminite, TURQUOISE and wavellite.

hydrate 1. *n.* a chemical compound produced by *hydration* (the chemical combination of water with another substance) or one in which water is a constituent of the chemical composition. See GAS HYDRATE.
2. *v.* to effect the incorporation of water into the chemical composition of a compound.

hydration shattering *n.* a process of rock disintegration caused by the wedging or prising pressure of films of water on silicate mineral surfaces. Water drawn between the grains of rocks by a type of osmosis exerts a differential pressure of sufficient strength to dislodge and separate the grains. Such a process occurs without the aid or intercession of freezing or thawing.

hydraulic action *n.* the dislodging and removal of weakly resistant material by the action of flowing water alone.

hydraulic conductivity *n.* the volume of water that will flow through a unit cross-sectional area of AQUIFER in a unit time under a unit HYDRAULIC GRADIENT and at a specific temperature.

hydraulic fracturing or **hydrofracturing** *n.* **1.** the formation and opening of FRACTURES by the action of fluids under pressure. For example, if the escape of pore fluids from a sedimentary layer is prevented by overlying relatively IMPERMEABLE strata, the pore fluid pressure increases with depth of burial until the strength of the overlying rocks is exceeded and fractures develop.
2. a process of injecting sand-charged liquids under such high pressure as to fracture a rock and inject the newly made crevices with coarse sand to keep them open. It is a method of improving the flow of oil to a well.

hydraulic gradient *n.* the ratio between the difference of elevation and the horizontal distance between two points. The rate and direction of water movement in an AQUIFER are determined by the permeability and the HYDRAULIC GRADIENT. Compare PRESSURE GRADIENT. See DARCY'S LAW.

hydraulic head *n.* **1.** the height of the exposed surface of a body of water above a specified subsurface point.
2. the difference in water level at a point upstream from a given point downstream.

hydraulic mining *n.* a method of mining in which high-pressure jets of water are used to remove earth materials containing valuable minerals.

hydrocarbon *n.* any of a class of organic compounds consisting of carbon and hydrogen only.

hydrofracturing see HYDRAULIC FRACTURING.

hydrogen bond *n.* in a bond between hydrogen and a much larger atom, such as oxygen or fluorine, the electrons tend to drift towards the larger atom to produce a small residual positive charge on the hydrogen and a corresponding small negative charge on the larger atom. Thus, in the water molecule (H–O–H) the hydrogen is attracted by oxygen atoms in adjacent molecules. A practical effect of this is to lower the boiling point of water so that it exists as a liquid at normal temperatures on the Earth's surface. Hydrogen bonding is present in many layered silicates, such as clays and micas, that contain OH (hydroxyl) groups.

hydrogen-ion concentration see PH.

hydrogen sulphide *n.* a toxic gas, H_2S, with the odour of rotten eggs; it is a natural product of the decomposition of organic matter and is one of the gases emitted during volcanic eruptions.

hydrogeochemistry *n.* the chemistry of surface and groundwaters, especially as it refers to the relationship between the chemical features and quality of the water and the geology of an area. Compare BIOGEOCHEMISTRY, LITHOGEOCHEMISTRY.

hydrogeology or **geohydrology** *n.* the study of subsurface water (GROUNDWATER). It includes the geology of water-bearing rocks, the chemistry, physics and movement of groundwater, and the laws governing groundwater movement. See DARCY'S LAW.

hydrograph *n.* a graph showing time-related variation in flow. The variation may be shown in terms of yearly, monthly, daily or instantaneous change.

hydrogrossular see GARNET.

hydrolith *n.* a rock that is chemically precipitated from water, e.g. rock salt or gypsum.

hydrological budget *n*. an accounting of the inflow into, outflow from and storage in a hydrological unit such as an AQUIFER, soil zone or lake. The groundwater budget deals specifically with the inflow-outflow budget of aquifers.

hydrological cycle *n*. the constant interchange of water between the oceans, atmosphere, land and subsurface areas of the Earth via precipitation, transpiration and evaporation, surface flow and sub-surface flow.

hydrology *n*. the science concerned with the occurrence, distribution, movement and properties of all waters of the Earth and its atmosphere. It includes surface water and rainfall, groundwater and the water cycle.

hydrolysates *n*. sediments typified by elements that are easily hydrolysed and which concentrate in fine-grained alteration forms of primary rocks. Thus, they are abundant in clays, shales and bauxites. The elements found in hydrolysates are aluminium and associated potassium, silicon and sodium. Compare OXIDATES, RESISTATES, EVAPORITES, REDUSATES.

hydrolysis *n*. (*geology*) a term generally implying a reaction between silicate minerals and water, either alone or in aqueous solution.

hydrometer *n*. a device for determining the SPECIFIC GRAVITY of a liquid.

hydrophone *n*. a pressure-sensitive SEISMIC DETECTOR that responds to sound transmitted through water.

hydrosphere *n*. the waters of the Earth, as distinguished from the ATMOSPHERE, BIOSPHERE and LITHOSPHERE.

hydrostatic level *n*. the height to which water will rise in a well under a full-pressure head. It marks the POTENTIO-METRIC SURFACE.

hydrostatic pressure *n*. the pressure exerted by a homogeneous liquid at rest. This pressure exists at every point within the liquid and is proportional to the depth below the surface. At any given point in the liquid, the magnitude of the pressure force exerted on the surface is the same, regardless of the surface orientation, and is everywhere perpendicular to the surfaces of the container. Compare CONFINING PRESSURE, GEOSTATIC PRESSURE.

hydrothermal *adj*. of or pertaining to heated water, to its actions or to products related to its actions.

hydrothermal alteration *n*. the alteration of rocks by the interaction of hydrothermal water with preformed solid phases.

hydrothermal deposit *n*. a MINERAL DEPOSIT formed from hydrothermal solutions at a range of temperatures and pressures. The minerals, which derive from fluids of diverse origin, are deposited in faults, fractures or other openings.

Epithermal deposits are formed within about one kilometre of the Earth's surface in the range of 50°C to 200°C. These deposits are typically found in volcanic rocks; the chief metals are gold, silver and mercury.

Mesothermal deposits are formed at considerable depth (1200–4500 metres) in the temperature range of 200°C to 300°C. Among the principal ore minerals in these deposits are pyrite, chalcopyrite and galena.

Hypothermal deposits are formed at great depth (3000–15 000 metres) in the temperature range of 300°C to 500°C. Ore minerals of greatest importance in these deposits are cassiterite, wolframite and molybdenite.

More modern classification schemes are based upon the type of rocks with which the mineral deposit is associated and the environment in which it formed rather than the depth/temperature scheme given above. See also MINERAL DEPOSIT.

hydrothermal stage *n.* that stage in the cooling of a MAGMA during which the residual fluid contains abundant amounts of water and other volatiles. The limits of the hydrothermal stage are variously defined in terms of phase assemblage, temperature and pressure.

hydroxide *n.* an oxide composed of a metallic element and the OH ion.

hydrozincite *n.* a minor ore of zinc, $Zn_5(CO_3)_2(OH)_6$, hydrozincite is found in the oxidised zones of zinc deposits as an alteration product of SPHALERITE.

hydrozoan *n.* any CNIDARIAN of the class Hydrozoa, including the hydra and the Portuguese man-of-war. There are marine and freshwater types, which may be solitary or colonial. Range, Lower Cambrian to Recent.

hygrometer *n.* an instrument for measuring the water vapour content of the atmosphere.

hygroscopic *adj.* having the property of absorbing moisture from the air.

hygroscopic water *n.* moisture confined in the soil and in equilibrium with that in the regional atmosphere to which the soil is exposed.

hypabyssal rocks *n.* IGNEOUS ROCKS forming minor INTRUSIONS, such as DYKES and SILLS, at relatively shallow depths in the crust. Compare PLUTONIC, VOLCANIC. Hypabyssal rocks may vary from medium-grained to fine-grained, since cooling rates depend on the size of the intrusion and temperature of the country rocks.

hyperaccumulators see GEOBOTANICAL PROSPECTING.

hyperfusible *adj.* (of a substance) capable

of reducing the melting ranges in end-stage magmatic fluids.

hyperpycnal inflow *n.* an inflow of a suspension of sediment in water that is denser than the body it enters, resulting in the development of a turbidity current. Compare HYPOPYCNAL INFLOW, HOMOPYCNAL INFLOW.

hypersthene *n.* a rock-forming mineral of the orthopyroxene group, (Mg,Fe)SiO_3. Enstatite is the magnesium END MEMBER. Hypersthene is an essential constituent of many IGNEOUS ROCKS. See PYROXENE.

hypervelocity impact *n.* the impact of an object on a surface at a velocity such that the magnitude of the SHOCK WAVES created on contact greatly exceeds the static compressive strength of the target substance. In such an impact, the kinetic energy of the object is transferred to the target material in the form of profound shock waves that produce a crater much greater in diameter than the impacting object. The required minimum velocities needed for this effect vary for different materials. See METEORITE, SHATTER CONE.

hypidioblast see CRYSTALLOBLAST.

hypidiomorphic see SUBHEDRAL.

hypocentre see EARTHQUAKE, SEISMIC FOCUS.

hypocrystalline or **merocrystalline** *adj.* (of the texture of an igneous rock) having crystalline components in a GLASSY groundmass, the crystal-to-glass ratio being between 7:1 and 5:3. Compare HYPOHYALINE.

hypogene *adj.* **1.** formed by ascending fluids within the Earth as ore or mineral deposits. Compare SUPERGENE. **2.** formed beneath the Earth's surface, e.g. granite.

hypohyaline *adj.* (of the texture of an igneous rock) having crystalline components in a GLASSY groundmass in which the crystal-to-glass ratio is between 3:5 and 1:7.

hyponomic sinus *n*. an indentation in the ventral margin of the aperture of a NAUTILOID that accommodated the funnel.

hypopycnal flow *n*. an inflow of water the density of which is less than that of the body of water it enters. Compare HYPERPYCNAL INFLOW, HOMOPYCNAL INFLOW.

hypostome *n*. a shield-shaped plate attached to the DOUBLURE on the ventral side of the CEPHALON (head carapace) of a TRILOBITE, in front of the mouth region (Fig. 89).

hypothermal deposit see HYDROTHERMAL DEPOSIT.

hypsographic *adj*. pertaining to the measurement and mapping of the Earth's topography relative to sea level. A *hypsographic curve* is used to describe and calculate the distribution of prominences of land surface and sea floor. The elevations are usually referred to as *sea-level datum*.

hysteresis *n*. **1.** a delay in the return of a body to its original shape after elastic deformation.
2. the property that a substance is said to show when its magnetisation is not reversible along the original magnetisation curve.

Ii

Iapetus *n*. an ocean that is believed to have existed in the early PALAEOZOIC between Laurentia (the North American continent), BALTICA, Siberia and the Avalonian (northern) margin of GONDWANA. Closure of this ocean took place between 460 Ma and 420 Ma, starting in the ORDOVICIAN, with the main closure event being SILURIAN (c. 425 Ma). Deformation following the closure of the ocean forms part of the CALEDONIAN OROGENY.

ice age *n*. a period of time, lasting 20 to 100 Ma, during which ice sheets occur somewhere on land. Between ice ages there are periods of some 150 Ma during which there are no ice sheets and few mountain glaciers. During an ice age, periods of GLACIAL maxima, when the ice cover expands to temperate latitudes, alternate with warmer INTERGLACIAL periods when they retreat to high latitudes.

The Antarctic ice sheet and mountain GLACIERS have existed in high latitudes during the last 15 Ma. Since the beginning of the PLEISTOCENE, 1.8 Ma ago, however, at least four glacial periods have occurred during which ice spread to the middle latitudes. The most recent GLACIATION in the Northern Hemisphere was at a maximum about 18 000 years ago, and the period from 10 000 years onwards (the HOLOCENE) is the present interglacial. See also MILANKOVICH HYPOTHESIS, CENOZOIC, LITTLE ICE AGE.

iceberg *n*. a large floating mass of ice detached from a GLACIER or ICE SHELF and carried seawards. When the terminus of a glacier moves into the sea, its lesser density subjects it to an upward force that increases as the glacier reaches farther into the sea and finally causes the forward end of the glacier to break off and form icebergs, a process called CALVING. Calving is less frequent in the extreme Northern and Southern Hemispheres. The degree of SUBMERGENCE of an iceberg depends greatly on its shape as well as on the density of the water and on the air content of the iceberg. The ratio of above surface height to submerged height for a rectangular berg is 1:7 and for a rounded berg 1:4; however, the ratio of the total ice mass above water to the submerged mass is 1:9 for all icebergs.

ice cap see GLACIER.

ice contact deposit *n*. stratified DRIFT that is deposited in contact with melting GLACIER ice, e.g. an ESKER, a KAME, a KAME TERRACE.

ice core see DATING METHODS.

ice crack see FROST CRACK.

icefield see GLACIER.

Iceland spar *n*. a chemically pure and transparent crystallised variety of CALCITE named for its occurrence in Iceland where it is found in VUGS and cavities in VOLCANIC ROCKS. It cleaves readily and exhibits clear double REFRACTION. It was formerly used for the manufacture of NICOL PRISMS for petrological microscopes.

ice rafting *n*. transport of rock fragments and other materials by floating ice.

ice sheet see GLACIATION, GLACIER.

ice shelf *n.* a sheet of very thick ice, most of which is afloat but which is attached to land along one side. Ice shelves are fed by snow accumulations and the seaward projection of land glaciers.

ice tongue see GLACIER.

ice wedge *n.* wedge-shaped, foliated ground ice formed in PERMAFROST. It occurs as a vertical or inclined sheet, dyke or vein, and originates by the growth of hoar frost or by freezing of water in a narrow crevice.

ice-wedge polygon see NONSORTED POLYGON.

ichnofossil see TRACE FOSSIL.

ichor or **residual liquid** *n.* a hypothetical fluid that has been suggested as responsible for GRANITISATION. The term initially bore the connotation of derivation from MAGMA.

ichthyosaur *n.* an extinct marine reptile of the order Ichthyosauridae. Ichthyosaurs resemble modern dolphins and porpoises, showing CONVERGENCE, both groups being specialised for life as aquatic predators. Range, Middle Triassic to Upper Cretaceous.

icositetrahedron *n.* a closed crystal form consisting of 24 faces and belonging to the ISOMETRIC system. This form, also known as the *trapezohedron*, is commonly found in crystals of ANALCIME.

ideal cyclothem see CYCLOTHEM.

idioblast see CRYSTALLOBLAST.

idiomorphic see EUHEDRAL.

idiotopic *adj.* (of the fabric of a crystalline sedimentary rock) in which most of the constituent crystals are EUHEDRAL.

idocrase see VESUVIANITE.

igneous breccia *n.* a BRECCIA composed of fragments of igneous rock or a breccia produced by igneous processes, e.g. VOLCANIC BRECCIA.

igneous rock *n.* a rock that has solidified from molten rock material (MAGMA) that was generated deep within the Earth. Igneous rocks are one of the three main groups of rocks that comprise the Earth's crust (see also SEDIMENTARY ROCK, METAMORPHIC ROCK). Magma that solidifies below the surface forms INTRUSIONS, either PLUTONIC or HYPABYSSAL. Extrusive or VOLCANIC ROCKS form from magma that solidifies on the surface. Crystal size is largely related to the length of time the rock takes to solidify. Igneous rocks are characterised by an interlocking CRYSTALLINE texture, except in the case of very rapidly cooled GLASSY rocks and the PYROCLASTIC ROCKS.

Common igneous textures include: HYALINE, APHANITIC, HYPOCRYSTALLINE, HOLOCRYSTALLINE, PORPHYRITIC, VESICULAR, AMYGDALOIDAL, OPHITIC, pegmatitic and granular (see GRANULAR TEXTURE, PEGMATITE). Igneous rocks may be described and classified on the basis of grain size, mineralogy and chemical composition, using MODES, COLOUR INDEX, SILICA CONCENTRATION and SILICA SATURATION (see NORMS).

The International Union of Geological Sciences (IUGS) has recommended classification schemes for a wide range of igneous rocks. The principal classification for plutonic rocks and volcanic rocks for which mineral modes can be determined is shown in Fig. 46 overleaf. It is based on the modal proportions of:

Q = QUARTZ and other polymorphs of SiO_2;

A = ALKALI FELDSPARS, including ORTHOCLASE, MICROCLINE, PERTHITE, ANORTHOCLASE, SANIDINE and ALBITE (An_0 to An_5);

P = PLAGIOCLASE (An_5 to An_{100}) and SCAPOLITE;

F = FELDSPATHOIDS (foids);

M = MAFIC and related minerals:

AMPHIBOLE, OLIVINE, PYROXENE, MICA opaque minerals, ACCESSORY MINERALS such as APATITE, ZIRCON, SPHENE, etc, EPIDOTE, GARNET, MELILITE and primary carbonate etc.
The sum of Q + A + P + F + M = 100%.

The QAPFM classification is not used for pyroclastic rocks, ULTRAMAFIC rocks (colour index or M > 90 per cent), CARBONATITES, LAMPROPHYRES, KIMBERLITES, CHARNOCKITES, or rocks containing more than 10 per cent modal melilite. Volcanic rocks that are too glassy or

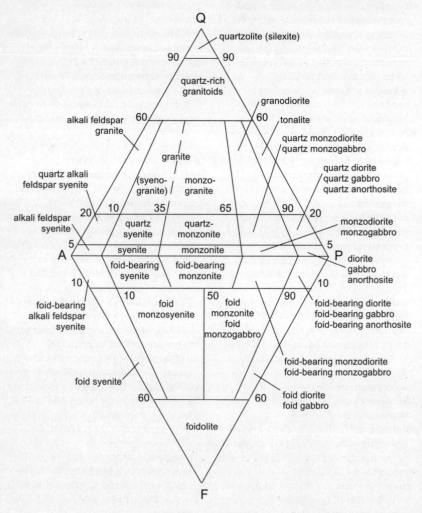

Fig. 46 **Igneous rock**. The classification of plutonic rocks based on modal mineral proportions in the QAPF diagram, with M < 90

fine-grained for modes to be determined can be classified using a total alkali/silica diagram (TAS) as shown in Fig. 48 overleaf.

igneous rock series *n.* a group of IGNEOUS ROCKS showing a continuous sequential variation in chemical composition and mineralogy from one end of the series to the other. Members of a series are thought to be related by having been derived from the same parent MAGMA.

ignimbrite *n.* a pumiceous PYROCLASTIC FLOW deposit, or rock, poorly SORTED,

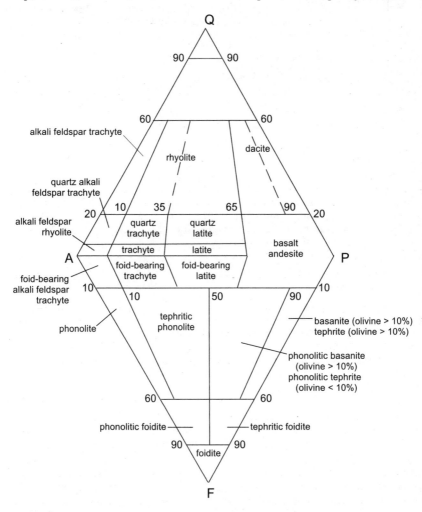

Fig. 47 **Igneous rock**. The classification of volcanic rocks based on modal mineral proportions in the QAPF diagram, with M < 90

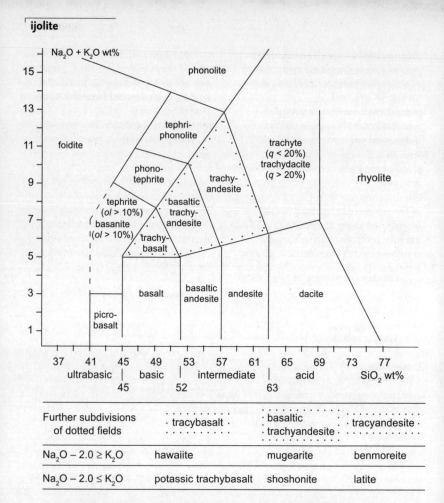

Fig. 48 **Igneous rock**. The classification of volcanic rocks based on the ratio of total alkalis to silica (weight %), or TAS

with fragments ranging from ash-size to blocks of PUMICE more than one metre in diameter and a variable proportion of lithic clasts, the latter derived from pre-existing rocks torn from the vent walls. It may or may not be welded: welded ignimbrites have non-welded zones. Welded ignimbrites characteristically have EUTAXITIC TEXTURES. Ignimbrite eruptions can be extremely voluminous: it has been estimated that some have volumes of over 1000 cubic kilometres. The largest ignimbrite deposits are associated with formation of CALDERAS or volcano-tectonic depressions. The composition may be andesitic, dacitic or rhyolitic, and some are compositionally zoned, apparently having been erupted from a zoned magma chamber. See also WELDED TUFF.

ijolite *n.* a coarse-grained, feldspar-free

alkaline IGNEOUS ROCK containing 50–70 per cent NEPHELINE along with CLINO-PYROXENE. Melanite GARNET may be present. Ijolite is the PLUTONIC equivalent of NEPHELINITE. It is frequently associated with CARBONATITE and intrusive RING COMPLEXES.

Illinoian *n.* the classical third GLACIAL stage of the Pleistocene Epoch in North America.

illite *n.* a group of MICA-like CLAY MINERALS of three-sheeted structure, widely distributed in argillaceous shales. They are intermediate in structure and composition between muscovite and montmorillonite. In illite, Mg and Fe partly replace the octahedral Al of muscovite. The sheeted layers in illite are tied together by K+ ions. Illites are formed by the hydrothermal alteration of feldspars or micas.

illuvial horizon see ILLUVIATION.

illuviation *n.* the accumulation in the subsoil of soluble or suspended material transported from an upper SOIL HORIZON by the process of ELUVIA-TION. Iron and aluminium minerals, including clays, accumulate here. A soil horizon to which material has been so added is called an *illuvial horizon*.

ilmenite *n.* a black or dark brown hexagonal mineral, $FeTiO_3$; it occurs as rhombohedral crystals but usually in massive, compact or granular aggregates. Ilmenite is a common accessory in plutonic rocks as a high-temperature segregation product. It also occurs in pegmatites and in metamorphic rocks such as gneiss and chlorite schist. It is the major source of titanium.

imbibition *n.* the tendency of granular rock or any porous substance to absorb a fluid completely under the force of capillary attraction, without the exertion of any pressure on the fluid.

imbricate structure or **imbrication** *n.* **1.**

a sedimentary structure in which GRAVEL, PEBBLES or GRAINS are stacked with their flat surfaces dipping upstream. It is most obvious in gravels and conglomerates but is also present in sands. Imbricate structure must be distinguished from the PREFERRED ORIENTATION observed in cross-bedded gravels in which tabular CLASTS lie parallel to the BED face and thus dip downstream (see CROSS-BEDDING).
2. a TECTONIC structure in which a series of lesser thrust FAULTS that overlap and are nearly parallel are all oriented in the same direction, which is towards the source of STRESS. See DUPLEX.

immature *adj.* **1.** (of a valley or drainage basin) still well above BASE LEVEL.
2. (of a coarse detrital rock) texturally immature; having a wide range of grain sizes.
3. mineralogically immature; having an abundance of feldspar and lithic fragments. Compare MATURE.

immature soil see AZONAL SOIL.

immiscible *adj.* (of two or more liquids) incapable of mixing and which, when brought into contact, form more than one phase, e.g. oil and water.

impact crater *n.* a depression formed by the impact of a meteor, asteroid or comet; such craters occur widely on the surface of planets and their satellites in the solar system. On the Earth they tend to be destroyed by surface erosion, but there is increasing evidence that such impacts were very common during the early history of the Earth. See also ASTROBLEME, SUEVITE.

impactite *n.* a collective term for rocks affected by an impact between planetary bodies. It includes brecciated and glassy rocks created by the sudden, intense pressure and heat generated by the impact of large METEORITES and comets. See also DIAPLECTIC GLASS,

LECHATELIERITE, MASKELYNITE, RINGWOODITE, SUEVITE, TEKTITE.

impermeable *adj.* (of a rock, soil or sediment) incapable of transmitting fluid under pressure. An impermeable bed is also impervious. Compare PERMEABLE, PERVIOUS.

impregnated *adj.* (of a mineral deposit) containing minerals that are EPIGENETIC and diffused through the host rock. Compare INTERSTITIAL.

incised meander *n.* a MEANDER that has become deepened by renewed down-cutting as a result of environmental change. It is usually bordered on one or both banks by vertical walls. Two types of incised meander are generally recognised: the *entrenched meander*, which shows little or no difference between the slopes of the two valley sides of the meander curve, and the *ingrown meander*, which exhibits obvious asymmetry in cross-profile, with undercut banks on the outside curve and prominent spurs on the inside.

inclination *n.* **1.** departure from the horizontal or vertical; also, the slope. **2.** (*structural geology*) see DIP. **3.** the angle of a well bore taken from the vertical at a given depth. **4.** MAGNETIC INCLINATION.

inclinometer *n.* **1.** an instrument used for measuring the angle an object makes with the vertical, for instance inclination in a well bore. Compare CLINOMETER. **2.** an instrument that measures MAGNETIC INCLINATION.

included gas *n.* gas that occurs in isolated intervening spaces in either the zone of saturation or the zone of aeration (see WATER TABLE). The included gas may also be bubbles of air, or some other gas, that are enveloped in water in either zone. These impede the water flow unless the gas dissolves in the water.

inclusion *n.* **1.** an enclave of rock within an IGNEOUS ROCK that differs in fabric and/or composition from the igneous rock. Some inclusions are blocks of wall rock that have been incorporated into the magma body by STOPING. Some are aggregates of early-formed crystals precipitated from the magma itself. Other inclusions might represent clots of heat-resistant residuum left over from partial melting processes deep within the crust where the magma originated. See also XENOLITH, AUTOLITH. **2.** FLUID INCLUSION.

incompatible elements *n.* elements that have an affinity for the molten phase when the Earth's MANTLE is melted. They include many trace elements, such as the RARE EARTH ELEMENTS thorium, uranium, lead, caesium, tantalum and niobium. Elements that show preference for the mineral phase are said to be *compatible*. The degree of compatibility of a particular element is influenced by its physical and chemical environment during melting or crystallisation.

incompetent see COMPETENT.

incompressibility modulus see BULK MODULUS.

incongruent melting *n.* melting that involves DISSOCIATION or reaction with the liquid so that one crystalline phase is converted to another plus a liquid of a different composition. Most pure minerals with a fixed composition (albite, diopside, quartz) melt congruently to yield a liquid of precisely the same composition. However, some minerals melt incongruously; for example, enstatite melts to yield a slightly more siliceous liquid plus forsterite crystals.

incongruent solution *n.* dissolution that

yields dissolved material in proportions different from those in the original.

index fossil *n.* a FOSSIL that identifies and dates those strata in which it is found. This especially applies to any fossil TAXON that is morphologically distinct, common in occurrence, geographically widespread (perhaps even worldwide) and confined to a narrow STRATIGRAPHICAL RANGE. Planktonic or nektonic fossils are preferred as index fossils. Graptolites and ammonites are two of the best index fossils. Microfossils are used increasingly as index fossils. Compare CHARACTERISTIC FOSSIL.

index horizon or **index plane** *n.* a structural surface selected as a reference in studying the geological structure of an area.

index mineral *n.* a MINERAL that occurs throughout rocks of similar bulk composition and is exposed over a distinct area of the metamorphic terrane. Such a mineral develops under a specific set of temperature and pressure conditions, thus identifying a particular METAMORPHIC ZONE. See also ISOGRAD.

index of refraction *n.* (*crystal optics*) a value expressing the ratio of the velocity of light in a vacuum or in air to the velocity of light within the crystal.

index plane see INDEX HORIZON.

index zone *n.* a body of strata, discernible by its lithological or palaeontological character, that can be followed laterally and which identifies a reference location in a stratigraphical section.

indicator *n.* **1.** some feature that suggests the presence of an ore or mineral deposit, e.g. a MAGNETIC ANOMALY.

2. a plant or animal found in a specific environment that can therefore be used to identify such an environment.

3. an INDICATOR BOULDER.

indicator boulder *n.* a glacial ERRATIC of known origin used for determining the source area and transport distance for any given TILL complex. Its critical feature is a distinctive appearance, unique mineral assemblage or characteristic fossil pattern. Indicator boulders are sometimes arranged in a *boulder train*, a line or series of rocks derived from the same BEDROCK source and extending in the direction of movement of the glacier. A *boulder fan* is a fan-shaped area containing distinctive erratics derived from an outcrop at the apex of the fan. The angle at which the margins diverge is a gauge of the maximum change in the direction of glacial motion.

indicatrix *n.* an imaginary surface that represents the variation in refractive indices of the light rays passing through a mineral. Light transmitted through an ISOTROPIC mineral travels with the same speed in all directions, so the isotropic indicatrix is a sphere. ANISOTROPIC minerals transmit light with different speeds in different directions.

The UNIAXIAL indicatrix is an ellipsoid in which the length of the principal axes is proportional to the refractive indices of the waves vibrating parallel to those axes (see Fig. 49 overleaf). Light vibrating at right angles to the OPTIC AXIS has a constant RI (n_0), which is the radius of the circular section. The RI for light vibrating in the plane of the optic axis varies between n_e and n_0, depending on the direction of transmission. If $n_e > n_0$, the mineral is optically positive, and if $n_e < n_0$ the mineral is optically negative.

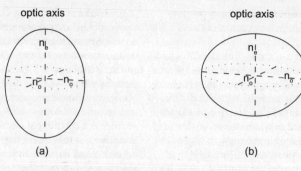

Fig. 49. **Indicatrix**. Representations of (a) the positive and (b) the negative uniaxial indicatrix

The BIAXIAL indicatrix is a triaxial ellipsoid with three major semi-axes representing the values of the principal refractive indices, n_a, n_b and n_c, with $n_a < n_b < n_c$. If n_a is the acute BISECTRIX between the two optic axes, then the crystal is optically negative; if n_c is the acute bisectrix, the crystal is optically positive. The third semi-axis, n_b, lies at right angles to the optic AXIAL PLANE and is known as the *optic normal*.

indices (*crystallographic*) see MILLER INDICES.

induced magnetisation *n.* magnetisation induced in a substance by an applied magnetic field that disappears when that field is removed. Compare NATURAL REMANENT MAGNETISATION.

induction log *n.* a continuous record of the conductivity of strata penetrated by a borehole, versus depth. See WELL LOGGING.

induration *n.* **1.** hardening of rock or rock material by heat, pressure or the introduction of some pore-filling material (cementation). See also LITHIFICATION.
2. the hardening of a SOIL HORIZON by chemical action to form a HARDPAN.

industrial diamond *n.* a DIAMOND, synthetic or natural, that is used in work such as wire-drawing or drilling, and as a general abrasive for lapping and polishing. Any diamond that is too badly flawed to have value as a gem may have industrial use. Three natural varieties of such diamonds exist: ballas, bort and carbonado. *Ballas* is composed of spherical masses of minute, concentrically arranged diamond crystals with poor CLEAVAGE. *Bort* is a grey to black massive diamond, the colour of which is the result of inclusions and impurities. The name is also used for highly flawed or badly coloured diamonds unsuited for gem purposes. *Carbonado* is a black opaque diamond with a slightly porous structure and no cleavage.

inert gases *n.* the gases helium, neon, argon, krypton, xenon and radon, a group of related elements that, because of their particularly stable atomic nucleii, are unusually chemically unreactive. They form a small proportion of the gases released from MAGMA during VOLCANIC ERUPTIONS, and, because of their inertness, they tend to accumulate in the atmosphere apart from a proportion of the lighter ones lost to space. Argon is of particular geological interest as its stable isotope, ^{40}Ar, is the decay product of radioactive potassium, ^{40}K, used in radiometric dating (see DATING METHODS) by the potassium/argon method. Radon is radioactive and constitutes a hazard in areas of active volcanism; it also poses an environmental hazard in some areas of granite intrusions where it is a decay product of radium. In both cases, being a dense gas, it accumulates in the cellars of buildings and can cause lung cancer if breathed.

inertinite group see MACERAL.

infauna *n.* those animals living buried in the sediment of the sea or lake bottom. *Infaunal animals* may bury themselves by digging burrows.

infiltration *n.* the movement of a fluid into a solid substance through pores or cracks, in particular the movement of water into soil or porous rock. Compare PERCOLATION. The *infiltration capacity* is the maximum rate at which soil under given conditions can absorb falling rain or melting snow.

infiltration capacity see INFILTRATION.

influent 1. *adj.* flowing in.
2. *n.* a stream that flows into a pond or lake, or a stream that flows into a larger stream (a TRIBUTARY). Compare EFFLUENT, INFLUENT STREAM.

influent stream *n.* a stream that contributes water to the zone of saturation, i.e. it loses water to the subsurface system; its channel lies above the WATER TABLE.

informal unit *n.* a body of rock or other STRATIGRAPHICAL UNIT that is referred to in casually descriptive but non-definitive terms using ordinary nouns, e.g. 'sandy formation', 'muddy formation'. Compare FORMAL UNIT.

infraglacial see SUBGLACIAL.

infrared *n.* the invisible part of the electromagnetic spectrum that comprises radiations of wavelengths from 0.8 to 1000 microns. See ULTRAVIOLET.

infrastructure *n.* the deeper structural

zone of an OROGENIC BELT, characterised by complex deformation and the formation of migmatites under conditions of high temperature and pressure. Compare SUPERSTRUCTURE.

ingrown meander see INCISED MEANDER.

initial dip *n.* the angle of DIP of BEDDING surfaces at the time of deposition. For example, beds deposited on a reef front may have initial dips of up to 30°. See also GEOPETAL STRUCTURE.

injection *n.* **1.** INTRUSION.
2. the forced insertion of sedimentary material into a crack, fissure or crevice in an existing rock.

injection gneiss *n.* a layered GNEISS formed by the injection of MAGMA, usually of granitic or granodioritic composition, along the SCHISTOSITY planes or other parallel layers of the host rock. This produces a rock of a veined or striped appearance. It is a type of MIGMATITE.

inlier *n.* a circular, elliptical or irregular area of older rocks surrounded by younger rocks. Inliers are found in a normal stratigraphical sequence wherever EROSION has broken through the younger strata.

inosilicate *n.* any SILICATE with a structure consisting of parallel chains of tetrahedral silicate groups linked by sharing oxygen. The PYROXENES are composed of parallel chains linked to each other mainly by iron and magnesium atoms with or without calcium; their silicon to oxygen ratio is 1:3 (SiO_3). AMPHIBOLES have a double silicate chain structure in which every second silicate tetrahedron shares an oxygen atom with a silicate tetrahedron of the other chain to give a silicon to oxygen ratio of 4:11 (Si_4O_{11}). See SILICATES.

inselberg *n.* isolated residual uplands standing above the general level of the surrounding plains in tropical regions;

they may be ridges, domes or hills. Like MONADNOCKS, they are remnants of the EROSION cycle, but even though there may occasionally be great morphological similarity between the two, they are not the tropical equivalents of monadnocks, which are features of temperate zones. Typical inselbergs rise more abruptly from the plains than do typical monadnocks. The cause of this sharp transition appears to be anomalous WEATHERING patterns related to the rock structure of a particular remnant and to the topography of the area. The sugarloafs of Brazil are renowned inselbergs. See also BORNHARDT.

insequent see INSEQUENT STREAM.

insequent stream or **insequent** *n*. a stream the course of which cannot be ascribed to control or adjustment by any apparent surface slope, weakness in the Earth's surface or rock type. Because the stream course is random, its basin develops a dendritic DRAINAGE PATTERN.

in situ *adj*. (of a rock, soil or fossil) in the situation or position in which it was originally deposited or formed.

in-situ theory *n*. the theory that COAL originates at the location where the plants that gave rise to it grew, died and decayed. Compare DRIFT THEORY.

insolation *n*. the geological effect of solar rays on the Earth's surficial materials, especially the effect of temperature changes on the MECHANICAL WEATHERING of rocks.

intake see RECHARGE.

intensity zone *n*. one of three zones in the Earth's crust defined on the basis of the interaction between an igneous INTRUSION and the COUNTRY ROCKS into which it is emplaced. This depends on the ductility contrast between magma and country rock.

In the upper, *epizone*, the country rocks display BRITTLE BEHAVIOUR and intrusive contacts are usually DISCORDANT. The country rocks in the deepest zone, the *catazone*, display DUCTILE BEHAVIOUR and deform plastically. Intrusive contacts are usually concordant. The country rocks in the *mesozone* have characteristics intermediate between those of the epizone and those of the catazone and intrusive contacts may be partly discordant and partly concordant. The location and size of the zones is not fixed, as the way in which the country rocks deform is affected by temperature, and the temperature gradient within the crust is not everywhere the same.

interambulacra see AMBULACRA.

interareas *n*. flat or curved regions on either side of the DELTHYRIUM (pedicle aperture) between the UMBO and the HINGE LINE of some BRACHIOPODS.

interbedded or **interstratified** *adj*. (of strata) being positioned between or alternated with other layers of dissimilar character, e.g. a lava flow interbedded with contemporaneous sediments. Compare INTERCALATED.

intercalated *adj*. (of layered material) existing or introduced between layers of a different type. It applies especially to layers of one kind of material that alternate with thicker strata of another material, e.g. beds of shell intercalated in sandstone.

intercept time *n*. the time-distance relationship between two media separated by a horizontal DISCONTINUITY at a certain depth in a seismic refraction survey. In the equation

$$T_1 = 2z \frac{\sqrt{V_1^2 - V_0^2}}{V_1 V_0}$$

T_1 is the intercept time, V_0 and V_1 are the respective velocities of the two media and z is the depth of the horizontal discontinuity.

interface *n*. **I.** the boundary separating the top of the uppermost layer of a sediment from the water in which the sedimentation process is taking place. **2.** the contact between fluids in a reservoir. **3.** a SEISMIC DISCONTINUITY.

interfacial angle *n*. the angle between adjoining FACES of a CRYSTAL, quoted in terms of the angle between the normals to the crystal faces.

interference colours see POLARISATION COLOURS.

interference figure *n*. an optical pattern produced when a THIN SECTION of a MINERAL is examined using convergent light and CROSSED POLARS. Mineral sections that have particular orientations are chosen. For UNIAXIAL minerals this would be one cut at right

angles to the OPTIC AXIS, for BIAXIAL minerals, either a section perpendicular to the acute BISECTRIX or perpendicular to one of the optic axes.

The uniaxial interference figure consists of a dark cross superimposed on concentric coloured bands (*isochromatic curves*) that are related to the BIREFRINGENCE of the mineral. See Fig. 50(a) below. The arms of the cross are known as ISOGYRES. Biaxial interference figures are more complex and the isogyres form curves or crosses as the microscope stage is rotated (acute bisectrix section, see Fig. 50(b),(d)). The vertices of the curves mark the point of emergence of the optic axes, but both will be in the field of view only if the angle between the optic axes ($2V$) is less than 45°.

For minerals with a large $2V$, an optic axis section is useful. The interference

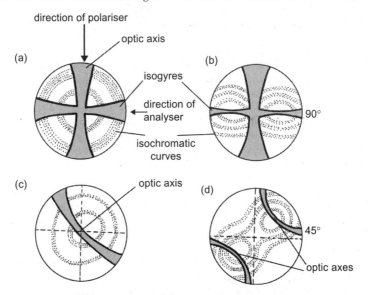

Fig. 50 **Interference figure**. (a) a centred uniaxial interference figure; (b) a biaxial interference figure in the 90° position; (c) a biaxial optic axis figure; and (d) a biaxial figure in the 45° position

figure shows only one isogyre, with or without colour bands. The curvature of the isogyre gives an estimate of the size of the 2V. Interference figures are also used to determine whether a mineral is optically positive or negative from the orientation of the fast or slow vibration directions.

interference ripple mark SEE RIPPLE MARKS.

interflow *n*. that part of the water infiltrated into the soil or unsaturated zone above an unconfined AQUIFER that makes its way into a channel or other surface RUNOFF system.

interfluve *n*. the comparatively undissected upland between adjacent streams flowing in the same direction.

interglacial I. *n*. the warmer interval between GLACIAL periods during an ICE AGE.
2. *adj*. pertaining to the warm interval between GLACIALS.

intergranular *adj*. (of the fabric or texture of an IGNEOUS ROCK) characterised by an interlocking network of randomly oriented MINERAL grains. Such texture results from the formation of abundant nuclei of all crystalline phases during undercooling of the melt, combined with adequate crystal growth rate to consume all the melt. This texture differs from INTERSERTAL texture by the absence of interstitial glass. Compare OPHITIC.

intergranular movement *n*. a movement that involves displacements between individual mineral grains. If rocks that are composed of both mineral grains and crystals are subjected to stress, the individual crystals and grains may move independently, and individual grains may maintain their shape and size. Intergranular movement is a factor in GLACIER flow; it takes place within a glacier (only near the surface) when grains of ice rotate relative to each other and slide over each other. Compare INTRAGRANULAR MOVEMENT.

intergrowth *n*. the interlocking of grains of two different minerals because of their simultaneous CRYSTALLISATION.

interior basin *n*. **I.** a depression from which no stream flows outwards; an undrained basin. Compare CLOSED BASIN.
2. an INTRACRATONIC BASIN.

intermediate *adj*. (of an IGNEOUS ROCK) between BASIC and SILICIC in composition; its silica content is generally between 52 and 60 per cent. See SILICA CONCENTRATION.

intermediate-focus earthquake *n*. an EARTHQUAKE the focus of which is located between depths of about 60 and 300 kilometres. Compare SHALLOW-FOCUS EARTHQUAKE, DEEP-FOCUS EARTHQUAKE.

intermittent stream *n*. **I.** a watercourse that flows only at certain times of the year, such as when it receives water from springs.
2. a stream that may lose its water by evaporation or absorption if its channel flows on highly porous rocks. Compare EPHEMERAL STREAM.

intermontane *adj*. lying or located between mountains.

internal waves *n*. subsurface waves that occur between layers of different density or within layers where there are vertical density gradients. They can be present in any stratified fluid and can be generated by atmospheric disturbances, shear flow or tidal forces, and flow over an irregular bottom. The wave amplitude varies with depth and is influenced by the density distribution of the fluid.

interpretive map *n*. (*environmental geology*) a map prepared for public use that classifies land according to its suitability for a particular function on the basis of its geological characteristics.

interrupted water table *n*. a WATER TABLE that inclines sharply over a GROUNDWATER barrier; there is a pronounced difference in elevation above and below it.

intersertal *adj*. (of a texture of igneous rock) in which angular interstices within an interlocking network of grains are occupied by glass. This texture forms essentially in the manner of INTERGRANULAR texture, but more rapid undercooling prevents complete crystallisation of the melt.

interstadial *n*. a warmer substage of a GLACIAL stage during which there is a temporary recession of the ice. Compare STADIAL.

interstitial *adj*. pertaining to the small spaces between the grains or crystals of a rock.

interstratified see INTERBEDDED.

interstream degradation see ETCHING.

interval velocity *n*. the distance across a specified stratigraphical thickness divided by the time it takes for a SEISMIC WAVE to traverse the distance. WELL SHOOTING is the most accurate procedure for velocity measurement. The interval velocity is the distance between successive detector locations in the well divided by the difference in ARRIVAL TIMES of the wave at two designated depths after a correction has been made for the fact that the actual wave is slanting rather than vertical. The average velocity is the total vertical distance divided by the total time.

interval zone *n*. the stratigraphical interval between two biostratigraphical zones; see BIOZONE.

intraclast *n*. a component of a LIMESTONE consisting of pieces of PENECONTEMPORANEOUS, partially lithified carbonate mud from within the basin of deposition. (See also ALLOCHEM.) The term 'intraclast' has come to be used for any sedimentary fragment of intraformational origin, for example MUDROCK fragments in SANDSTONE.

intracratonic *adj*. located within a stable continental region.

intracratonic basin or **interior basin** *n*. a depression situated within a CRATON.

intraformational *adj*. formed within the boundaries of a geological FORMATION, more or less contemporaneously with the sediments that contain it. For example, an intraformational CONGLOMERATE or BRECCIA is formed by the EROSION and redeposition of a sediment within the basin of deposition.

intraglacial see ENGLACIAL.

intragranular movement *n*. a movement in which displacement occurs within the individual CRYSTALS of a body by GLIDING and DISLOCATION. Ice crystals that are oriented in a particular fashion are deformed by slippage along layers without rupturing the CRYSTAL LATTICE. Compare INTERGRANULAR MOVEMENT.

intramicrite *n*. an allochemical LIMESTONE with more than 10 per cent INTRACLASTS in a MICRITE matrix (see ALLOCHEMS).

intrasparite *n*. an allochemical LIMESTONE composed of more than 10 per cent INTRACLASTS in a SPARITE matrix (see ALLOCHEMS).

intratelluric *adj*. 1. (of a PHENOCRYST) having grown slowly in a MAGMA beneath the surface, prior to a second stage of more rapid CRYSTALLISATION. 2. being located, or having formed or originated, deep within the earth.

intrazonal soil *n*. a soil that more greatly reflects the effects of local conditions than of broad climatic patterns.

intrusion *n*. **1.** the emplacement of MAGMA into pre-existing rock. Intrusion can take place either by DEFORMATION of the involved rock or along some structural channels such as BEDDING PLANES, CLEAVAGES or JOINTS. Major intrusions are those that occupy a large volume in relation to the surface area of their contact with the COUNTRY ROCK, e.g. BATHOLITHS, STOCKS, LACCOLITHS and LOPOLITHS. Minor intrusions present a relatively large surface area at their contacts, which tends to lead to more rapid heat loss to the country rock, hence finer-grained intrusive rock, e.g. SILLS, DYKES, sheets. See FORCEFUL INTRUSION, PERMITTED INTRUSION, PLUTON.

2. IGNEOUS ROCK formed in this manner. The term should not be applied to igneous-looking rock bodies that are the product of metasomatic REPLACEMENT.

3. the injection of sedimentary material under abnormal pressure, e.g. emplacement of a diapiric salt plug.

intrusive rock *n*. IGNEOUS ROCK formed of MAGMA that consolidates beneath the Earth's surface. The texture of the intrusive rock depends on the depth of emplacement and the dimensions of the INTRUSION. Rocks at greater depth cool more slowly, allowing the growth of large crystals, which results in a coarse texture characterised by clearly visible minerals, e.g. granite. Cooled at shallow depths and/or if the dimensions are limited, the rock tends to be more finely crystalline, and even APHANITIC.

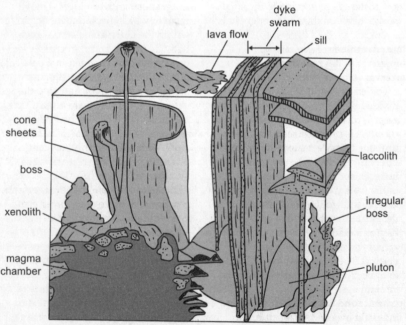

Fig. 51. **Intrusion.** The form of some igneous intrusions and related volcanic features

intumescence *n.* a swelling up as seen in ZEOLITE minerals when heated.

invariant equilibrium *n.* a phase assemblage for which there are zero DEGREES OF FREEDOM: none of the variables – temperature, pressure or composition – can be changed without causing the loss of one or more phases. See also PHASE DIAGRAM.

invasion *n.* **1.** an igneous INTRUSION.
2. TRANSGRESSION of the sea over a land area.

Inverian event *n.* an episode of DEFORMATION and METAMORPHISM affecting the BASEMENT rocks of northwest Scotland in the later stages of the Scourian, approximately 2500 Ma ago. See also LEWISIAN GNEISS.

inverse grading *n.* a variation of GRADED BEDDING in which the size of the particles increases upwards through the BED, in contrast to normal grading where particle size decreases upwards. Normal grading is much more common.

inverse zoning or **reversed zoning** *n.* in PLAGIOCLASE, the change by which the outer parts of crystals become more calcic. Compare NORMAL ZONING, ZONING.

inversion *n.* **1.** (*minerals*) the change from one polymorph to another. See POLYMORPHISM; see also ENANTIOTROPY, MONOTROPY.
2. (*structural*) the reversal of a long-continued direction of vertical movement of a part of the CRUST.

inversion axis see CRYSTAL SYMMETRY.

inversion of relief *n.* a landform in which geological structure and topography do not correlate. Uplands are composed of rocks with a synclinal structure and lowland areas are underlain by anticlinal structures.

involute see AMMONOID.

ion *n.* an atom that carries an electronic charge as a result of the gain or loss of electrons. As electrons, by definition, carry a negative charge, an atom that loses one or more electrons becomes positively charged and is termed a CATION. An atom that gains one or more electrons becomes negatively charged and is termed an ANION. Metals readily lose electrons, so they commonly occur in solution or in crystalline structures as cations whereas many non-metals, e.g. fluorine and chlorine, readily gain electrons to become anions,

ionic bond *n.* the link between atoms that have either gained or lost one or more electrons. It is the electrostatic attraction between IONS of opposite charge. In an ionically bonded mineral such as HALITE (NaCl), each ion is surrounded by oppositely charged ions and the structure is electronically neutral, see Fig. 52 overleaf. The relative size of ions is important in the structure of the CRYSTAL LATTICE along with the electrostatic charge. The negatively charged anions are generally larger than the positively charged cations, so most of the space in a crystal lattice is commonly taken up by anions, with cations fitting in the spaces between. Chlorine ions (Cl^-) have a negative charge of one and sodium ions (Na^+) a positive charge of one, so in halite an equal number of sodium and chlorine ions are required to produce an electronically neutral structure. In the halite crystal lattice, every sodium ion is surrounded by six chlorines so the sodium is said to be in *six-fold coordination* with chlorine. In fluorite (CaF_2), however, the calcium ions (Ca^{2+}) have a positive charge of two and the fluorine ions (F^-) a negative charge of one, so there are twice as many fluorine ions than calcium ions in the fluorite crystal lattice. This is reflected in the more complicated geometry of the fluorite lattice in which calcium (Ca^{2+}) is in six-fold coordination with fluorine. In the SILICATE minerals, which are based on the negatively charged SILICATE TETRAHE-DRON (Si^4), the way in which the tetrahedra are bonded and their geometric arrangement determine the number of cations required to achieve an electronically balanced crystal lattice. In the simpler silicate lattices, such as that of olivine, iron and magnesium, ions are in six-fold coordination with the oxygens of the silicate tetrahedra, see Fig. 84. In the more complex lattices of many other silicates, examples of eight-fold coordination are common. However, many of the bonds in silicates are not purely ionic but are intermediate in character between ionic and covalent bonds; a fuller explanation of the relationships between bonds, coordination, electrostatic charge and the geometry of crystal lattices is beyond the scope of this dictionary (see Appendix 5).

ionic radius *n.* the effective radius of an ION in a crystalline structure. This varies according to the electronic charge on the ion.

ion microprobe see Appendix 4.

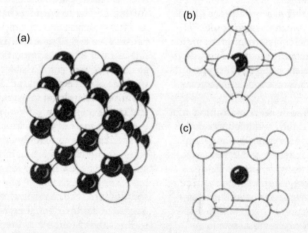

Fig. 52. **Ionic bond**. (a) the molecular structure of halite (NaCl). The smaller spheres represent positively charged sodium cations that are bonded ionically to negatively charged, relatively large chlorine anions in an isometric crystal lattice expressed in the cubic form normally found in crystals of common salt; (b) part of the crystal lattice of halite expanded to make clearer the position of the sodium ions in 6-fold coordination with chlorine. The chlorine anions are arranged around the sodium cation at the corners of an octahedron; (c) the structural relationship of calcium ions (Ca^{2+}) to the fluorine anions (F^-) in fluorite. Because the calcium ions have an electronic charge of +2 and the charge on the fluorine atoms is −1 there have to be two fluorine ions per calcium ion. This leads to a stable crystal lattice in which the fluorine ions are arranged at the corners of a cube with a single calcium ion in the centre, in 8-fold coordination

Ipswichian see CENOZOIC, Fig. 13.

iridescence *n.* the interference of light either at the surface of or within a mineral, thereby producing a series of colours as the angle of the incident light changes. See SCHILLER.

iridium (Ir) *n.* one of the PLATINUM GROUP METALS. The anomalous high concentrations of iridium and other SIDEROPHILE elements in the CRETACEOUS-TERTIARY BOUNDARY LAYER provide evidence of a major meteorite impact. See AFFINITY OF ELEMENTS.

iron *n.* a heavy, chemically active metallic element, Fe. Native iron is rare in terrestrial rocks but occurs alloyed with nickel in crystallised forms in IRON METEORITES. Iron occurs in a wide range of ores in combination with other elements. The chief ores of iron consist mainly of the oxides of this element: HEMATITE (Fe_2O_3); GOETHITE (α-FeO (OH)); and MAGNETITE (Fe_3O_4).

iron bacteria *n.* EUBACTERIA involved in the biological iron cycle that cause the deposition of iron compounds. Different groups of bacteria oxidise and reduce iron compounds in nature. Iron oxidation (Fe^{2+} to Fe^{3+}) is carried out in aerobic and neutral or acidic conditions by *Gallionella ferruginea*, *Leptospirillum ferrooxidans* and *Thiobacillus ferrooxidans*. Additionally, marine magnetotactic bacteria such as *Aquaspirillum magnetotacticum* transform iron compounds into magnetite (Fe_3O_4). Iron reduction (Fe^{3+} to Fe^{2+}) takes place under anaerobic conditions and is carried out by eubacteria such as *Geobacter metallireducens* working on ferric iron. A typical and important process carried out by iron bacteria is the biodesulphurisation of coal in oxidising conditions, such as in coal mines, coal waste dumps and abandoned mines; PYRITE (FeS_2) in coal is oxidised, giving a source of Fe^{2+} ions in acid conditions. *Thiobacillus ferrooxidans* and others then oxidise ferrous iron to ferric iron, producing a series of products that are acidic and polluting. Accumulations of iron developed through the agency of such bacteria are BACTERIOGENIC. Compare SULPHUR BACTERIA.

iron formation *n.* an iron-rich chemical SEDIMENTARY ROCK, thin-bedded or laminated and commonly containing layers of CHERT. Iron may be present as an oxide, silicate, carbonate or sulphide. Most iron formation is of Precambrian age. Compare IRONSTONE. Many terms are essentially synonymous: TACONITE, banded hematite quartzite, BANDED IRON FORMATION.

iron hat see GOSSAN.

iron meteorites, irons or **siderites** *n.* one of the three major groups of METEORITES. They consist mainly of iron, with up to 20 per cent nickel. Because they are easily distinguishable from terrestrial rocks, they are found more frequently than other meteorite types. Irons are grouped according to their structure, which can be revealed by the WIDMANSTÄTTEN PATTERN; there are three groups, grading into one another: ataxites, hexahedrites and octahedrites.

iron pan *n.* a type of HARDPAN in which there is a considerable amount of iron oxide.

irons see IRON METEORITES.

ironstone *n.* a typically non-cherty, iron-rich Phanerozoic SEDIMENTARY ROCK, often oolitic. Ferriferous minerals found in ironstones include goethite, hematite, pyrite, siderite, limonite and chamosite. See MINETTE, CLINTON ORE, BLACKBAND IRONSTONE and CLAY IRONSTONE. Compare IRON FORMATION.

island arc *n.* a curved chain of volcanic islands, many of which are located

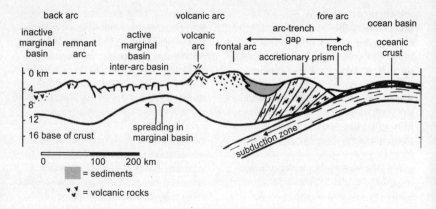

back arc   volcanic arc   fore arc   ocean basin

Fig. 53. **Island arc.** The features of an intra-oceanic island arc subduction zone

among the circum-Pacific margins, e.g. the Aleutian Islands and the islands of Japan. Island arcs are associated with deep-sea trenches (see OCEAN TRENCH); they occur on the landward side of the trench and are generally curved, with their convex side facing oceanwards. See Fig. 53 above.

iso- *prefix denoting* equality.

isoanomaly *n.* a line joining parts of equal geophysical anomaly, e.g. MAGNETIC ANOMALY, GRAVITY ANOMALY.

isobar *n.* a line connecting points of equal pressure. An *isobaric surface* is one in which every point has the same pressure.

isobaric surface see ISOBAR.

isobath *n.* 1. a line on a chart or map that connects points of equal water depth.
2. an imaginary line on a land surface along which all points are at the same vertical distance above the upper or lower surface of the WATER TABLE or an AQUIFER.

isochemical metamorphism *n.* metamorphic processes that occur without any change in the bulk composition of

the parent rock (although H_2O and CO_2 are particularly mobile and usually lost). Compare ALLOCHEMICAL METAMORPHISM.

isochemical series *n.* an assemblage of rocks displaying the same bulk chemical composition throughout sequential, mineralogical or textural changes.

isochore map see ISOPACH MAP

isochromatic curves see INTERFERENCE FIGURE.

isochron *n.* (*seismology*) a line on a chart or map passing through points at which the difference between the ARRIVAL TIMES of SEISMIC WAVES emanating from two reflecting surfaces is equal.

isochron diagram *n.* a diagram in which the relative proportions of radioactive parent and daughter isotopes are plotted against the proportion of a stable isotope (which will not have changed in abundance since the formation of the rock). The increase in the ratio of radiogenic daughter isotope to stable isotope depends on the initial ratio of radioactive parent to stable isotope and on the age of the specimen. The isotopic ratios of minerals from the same rock should lie on a straight line (a *mineral isochron*), the slope of which is proportional to the age.

The isotopic ratios for whole-rock specimens can also be plotted on isochron diagrams, and it may be found that the mineral isochron for a particular rock gives a different age from the whole-rock isochron. In such a case, the whole-rock isochron may record the initial time of crystallisation, while the mineral isochron records a later heating event. Isochron diagrams are commonly used in the $^{87}Rb/^{87}Sr$ method of radiometric dating (see DATING METHODS).

isoclinal fold or **isocline** see FOLD.

isoclinic line *n.* a line on a chart connecting points of equal MAGNETIC INCLINATION.

isodynamic line see ISOGAM.

isogal *n.* a contour line enclosing points of equal gravity value.

isogam or **isodynamic line** *n.* a line joining points of equal field intensity.

isogeotherm *n.* a line or surface within the Earth along which or over which temperatures are the same. Compare ISOTHERM.

isogonic line *n.* a line drawn through all points on the Earth's surface with the same MAGNETIC DECLINATION. Not to be confused with MAGNETIC MERIDIAN. Compare AGONIC LINE.

isograd *n.* a line on a map joining METAMORPHIC ZONES of equal grade, i.e. the points at which METAMORPHISM proceeded at similar values of pressure and temperature, as observed from rocks belonging to the same METAMORPHIC FACIES. An isograd is delineated by the first appearance of an INDEX MINERAL, or mineral assemblage, after which the next higher zone to it may be

named. For example, the first appearance of sillimanite defines the lower boundary of the sillimanite zone.

isogyre *n.* (*crystal optics*) a dark part of the INTERFERENCE FIGURE resulting from EXTINCTION; it indicates the emergence of those light components having the same direction of vibration as the vibration planes of the POLARISER and analyser.

isohyetal line *n.* a line on a map connecting points that indicate locations receiving equal amounts of precipitation.

isolith *n.* an imaginary line connecting location points of similar LITHOLOGY and excluding rocks of dissimilar features such as texture, colour or composition.

isometric system see CRYSTAL SYSTEM.

isomorphism *n.* (*crystallography*) the characteristic whereby two or more minerals in a SOLID SOLUTION series crystallise in the same form.

isomyarian see BIVALVE.

isopach map *n.* a map that shows the thickness of a unit, such as a BED or SILL, throughout a geographical region by means of *isopachs*, which are contour lines representing equal thickness. An *isochore map* is based on the vertical distance between the top and bottom of particular strata. Equal drilled thicknesses of a specified subsurface unit are connected by *isochores*. The second type of map defines more accurately the present condition of the strata rather than the thickness, which is mostly a function of the primary condition.

isopleth *n.* a line on a chart or map that connects points of the same value, e.g. depth, elevation, population.

isopycnic *adj.* of constant or equal density; this may refer to density in space (quantity or amount) or density in time (frequency). It is used in meteorology and oceanography.

isopygous see PYDIGIUM.

isoseismal line *n*. a locus of points of equal EARTHQUAKE intensity. If there were complete symmetry about the vertical through the earthquake focus, the isoseismal lines would be a family of circles with the earthquake epicentre as their centre. But because of unsymmetrical factors affecting the intensity, the curves are often far from circular in shape. Indeed, an isoseismal curve of given intensity sometimes consists of more than one closed curve.

isostasy *n*. the condition of equilibrium whereby the Earth's crust is buoyantly supported by the plastic material of the mantle. See AIRY HYPOTHESIS, PRATT HYPOTHESIS.

isostatic anomaly see GRAVITY ANOMALY.

isotherm *n*. a line on a map or chart connecting points of equal temperature. Compare ISOGEOTHERM.

isotope *n*. a particular atom of an element that has the same number of electrons and protons as the other atoms of the element but a different number of neutrons, i.e. the atomic numbers are the same but the atomic weights differ. Since the chemical properties of an element are determined by the atomic number, isotopes of an element have essentially the same chemical properties as the other atoms, but there are small differences in physical properties such as specific gravity. See also DATING METHODS, SMOW.

isotropic *adj*. (of a material) of which the properties are independent of direction. For example, an isotropic crystal will transmit light at the same speed regardless of its direction within the crystal. Compare ANISTOTROPIC.

isotropy *n*. the condition or state of having properties that are uniform in all directions.

Isua supracrustals *n*. a succession of deformed and metamorphosed SEDIMENTARY ROCKS, with basic and ultrabasic lavas and intrusions, in the ARCHAEAN of southwest Greenland. They are among the oldest terrestrial rocks in the world, with an age of about 3800 Ma.

itabirite *n*. a laminated IRON FORMATION in which original JASPER or CHERT bands have been recrystallised during METAMORPHISM into clearly discernible grains of quartz and the iron is present in thin laminae of hematite or magnetite.

itacolumnite *n*. this so-called 'flexible sandstone' is in fact a quartz-rich METAMORPHIC ROCK or micaceous SANDSTONE in which the subordinate minerals have been leached out by WEATHERING to leave loosely interlocking quartz grains.

IUGS classification *n*. an internationally adopted classification of IGNEOUS ROCKS proposed in 1989 by the International Union of Geological Sciences.

Jj

jade *n.* a hard, durable gemstone that may be either the pyroxene mineral JADEITE or the AMPHIBOLE mineral NEPHRITE. Both forms are found in white and various shades of green, brown, yellow and lavender. Jade of either mineral group is highly valued as a gemstone, especially among the Chinese who for centuries have associated it with forces basic to life and fortune.

jadeite *n.* a monoclinic mineral of the clinopyroxene group, $NaAlSi_2O_6$, with very limited substitution of Fe^{3+} for Al. It is found in compact waxy masses as metasomatic concentrations in serpentinised ultramafic rocks. Jadeite is one of the minerals referred to as JADE. Burma is its principal source. See also NEPHRITE.

jasper *n.* a compact, microcrystalline variety of quartz. Its colour is commonly red but may also be yellow, dark green or greyish-blue. Red varieties usually contain HEMATITE.

jaspilite *n.* a rock consisting mainly of red JASPER and iron oxides in alternating bands.

Jeans-Jeffrey theory *n.* a dualistic theory of the origin of the planetary system. The first such theory (1745) proposed that the impact of a massive 'comet' tore material from the Sun, which later dispersed and condensed into planets. More than a century later, the idea was modified into a close encounter of the Sun with a second star. The tidal effect of such an encounter would draw a gaseous filament from the Sun, which

would then break up into planetary fragments. Compare NEBULAR HYPOTHESIS.

jet *n.* a dense black LIGNITE found in some mudrocks. It formed from driftwood submerged in sea-floor mud. It takes a high polish and was once popular in jewellery. Because it is a variety of coal, jet will burn.

JOIDES *n. acronym for* Joint Oceanographic Investigation for Deep Earth Sampling, a programme to obtain cores of sediments from the deep-ocean bottom. It is now ODP (Ocean Drilling Program).

joint *n.* a surface FRACTURE (vertical or horizontal) in a rock without DISPLACEMENT. A *joint set* consists of a group of more or less parallel joints. A *joint system* consists of two or more joint sets with a characteristic pattern.
In folded rocks, the joints that lie at right angles to the fold hinge are known as *cross-joints*, while *longitudinal joints* are parallel to the fold hinge. If the fold hinge is horizontal they may be called *dip joints* and *strike joints* respectively. *Oblique joints* are oriented at an angle to these two sets. See Fig. 54 overleaf.

joint set see JOINT.

joint system see JOINT.

jökulhlaup see GLACIER BURST.

jökull *n.* (*glaciology*) an Icelandic term for an ice cap, a continuous sheet of ice smaller than a continental GLACIER.

Joplin-type lead see ANOMALOUS LEAD.

J-type lead see ANOMALOUS LEAD.

Jurassic *n.* the period of geological time extending from 206 to 144 Ma, occupy-

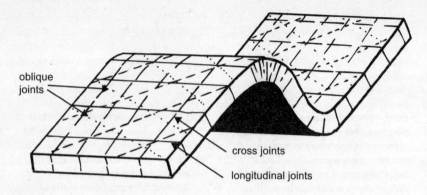

Fig. 54 **Joints**. Cross, longitudinal and oblique joints in a folded layer

ing the middle portion of the Mesozoic Era between the Triassic and Cretaceous periods. Its name derives from limestone rocks called the 'Calcaires de Jura', recognised in the Swiss-French Jura Mountains. It is divided into three series – Lower, Middle and Upper, formerly called Lias, Dogger and Malm. These are subdivided into 11 stages – in the Lower Jurassic these are, from the base, Hettangian, Sinemurian, Pliensbachian and Toarcian. In the Middle Jurassic, Aalenian, Bajocian, Bathonian and Callovian. In the Upper Jurassic, Oxfordian, Kimmeridgian and Tithonian. Jurassic strata occur in all continents and reach great thicknesses in some regions. The early Jurassic marine sediments found around the continental edges of the North Atlantic indicate the beginning of the break-up of PANGAEA, as marine conditions were established. Later evidence of ocean opening can be seen in the extensional FAULT systems and associated fault-bounded basins developed in the Lower and Middle Jurassic of Britain. Rifting in the area of the present North Sea was accompanied by volcanic activity. Through the Jurassic a general

rise in global sea level took place, related to the growth of ocean ridges, with a local fall in sea level during the Middle Jurassic in northwest Europe.

In southern Britain, Jurassic sedimentary rocks are mainly marine and include sandstones, mudrocks and limestones (including oolitic limestones, developed on extensive carbonate platforms during the Middle Jurassic). In northern Britain, the Lower and Upper Jurassic sedimentary rocks are dominantly marine, with Middle Jurassic rocks being dominantly fluviodeltaic or lagoonal and dominated by extensive sandstone sequences that form oil reservoirs, notably in the North Sea.

Among the economically important Jurassic deposits are oolitic and sideritic iron ores of marine deposits in central and western Europe, and fuller's earth in southern England. In the North Sea major accumulations of oil and gas are found in Jurassic sequences. The total absence of Jurassic sediments in places such as the east coast of North and South America suggests that the South Atlantic Ocean had not yet begun to open. The Jurassic sequence of Madagascar, off the southeastern coast of Africa, is thought to result from deposits in an ocean strait that was beginning to divide the supercontinent, GONDWANA.

Among the invertebrates, dinoflagellate microfossils are common in the Jurassic. Siliceous sponges reach their greatest abundance, and coelenterates, ECHINOIDS and BRACHIOPODS are plentiful. Jurassic brachiopods are mainly the rhynchonellids and terebratulids. BIVALVES show marked development, and many new gastropods appear at this time. Cephalopods are represented by BELEMNITES and the numerous and diverse descendants of two genera of AMMONOIDS that had survived the Triassic ammonite EXTINCTION. Ammonites serve as zone fossils for the Jurassic. Small, herring-like fish are the first TELEOSTS, with fully ossified vertebrae. Reptiles, especially the dinosaurs, are the dominant vertebrates, dinosaurs reaching their maximum size during the Jurassic.

Reptiles occupy sea and land, and develop flying forms (pterosaurs) that parallel the emergence of the first birds (Archaeopteryx) in the middle of the period. Mammals are sparsely represented by small rodent-like forms. The Jurassic flora includes the spore-bearing true ferns, seed ferns, club mosses (lycopsids), cycads, conifers and GINKGOES. The first flora showing flower characteristics similar to ANGIOSPERMS appear in the Jurassic.

The first effects of the Alpine orogeny were being felt in southern Europe. At this time, Britain lay between about 30° and 40° north. The climate was moist and humid, on the evidence of plant growth. More uniform ocean temperatures are indicated by oxygen-isotope fractionation from belemnite shells.

juvenile gases *n.* gases from the interior of the Earth that are new and have not previously been at the surface.

juvenile water *n.* water from magmatic sources within the Earth that has not previously existed as atmospheric or surface water. See CONNATE WATER.

Kk

K see CARAT.

Ka *n*. kiloyear, one thousand (10^3) years.

kainite *n*. a monoclinic mineral, $KMg(SO_4) \cdot Cl \cdot 3H_2O$, a natural salt occurring in granular masses. It occurs in EVAPORITE deposits and is used as a source of potassium and magnesium compounds.

Kainozoic see CENOZOIC.

kalsilite *n*. a hexagonal feldspathoid mineral, $KAlSiO_4$, isostructural with NEPHELINE.

kamacite *n*. an iron-nickel alloy containing between 4 and 7 per cent nickel, found in IRON METEORITES associated with TAENITE. See WIDMANSTÄTTEN PATTERNS.

kame *n*. a mound composed chiefly of stratified sand and gravel, formed at or near the snout of an ice mass or deposited at the margin of a melting GLACIER. Kames, like ESKERS, occur in areas where large quantities of coarse material are available as a result of the slow melting of stagnant ice. Meltwater must be present in amounts great enough to redistribute the debris and deposit it at the margins of the decaying ice mass. See also DRUMLIN.

kame and kettle moraine *n*. an irregular assemblage of knolls, mounds or ridges between depressions, or KETTLES, that may contain swamps or ponds. Such an undulating landscape is a type of end MORAINE that may result from slight oscillations of an ice front as it recedes. A section or area of kame and kettle moraine topography that may have developed either along a live ice front or around masses of stagnant ice is called *hummocky moraine*.

kame terrace *n*. a terrace-like body of stratified sand and gravel left by a GLACIER against an adjacent valley wall after the ice has melted. See KAME.

Kansan *n*. the classical second glacial stage of the Pleistocene Epoch in North America, after the Aftonian.

kaolin *n*. china clay, a general name applied to high-grade clay used in ceramics, as a filler in the manufacture of paper and in the rubber industry. See CLAY MINERALS.

kaolinite *n*. a phyllosilicate CLAY MINERAL, $Al_2Si_2O_5(OH)_4$, formed by the alteration of ALKALI FELDSPARS and other aluminium-bearing minerals, a major constituent of china clay; its largest deposits are beds of clay formed in lakes. DICKITE and nacrite are chemically similar.

K-Ar age method see DATING METHODS.

karat see CARAT.

karst *n*. a type of topography characterised by DOLINES, caves and caverns, DRY VALLEYS and underground drainage. It is named after Karst, the type area on the coast of Slovenia and Croatia. Development of karst topography requires the subsurface dissolution of some soluble rock, usually limestone (dolomite may suffice). The rock should be dense, highly jointed and preferably thinly bedded, since permeability along joints and bedding planes is more favourable than mass permeability. With the exception of the Yucatan, karst regions occur in areas of moderate to abundant rainfall. *Karst plains,* flat areas showing karst features, develop in regions of nearly horizontal limestone strata. A valley in karst that ends abruptly at the point where its stream disappears underground (the SWALLOW HOLE, SINK or streamsink) is called a *blind valley.* During periods of storm waters, a blind valley may become a temporary lake. In the type karst area, red soils (TERRA ROSSA) have accumulated on the floor of the dolines as a residue of the solution of the limestone. See also UVALA, HUMS, POLJE.

karst plain see KARST.

katatectic layer *n.* a layer of solution residue, usually composed of gypsum and/or anhydrite, in SALT-DOME cap rock.

Kazanian *n.* a STAGE of the Upper PERMIAN.

keel *n.* a ridge on the outer margin of the shell in some AMMONOIDS (see Fig. 1).

Keewatin *n.* a division of the Archaeozoic (Archaean) rocks of the Canadian SHIELD.

kelyphytic border or **rim** *n.* a REACTION RIM or zone of cryptocrystalline material or parallel or radial growths of fibrous minerals between minerals in plutonic rocks. This name is commonly used to describe a MICROCRYSTAL-LINE overgrowth of fibrous pyroxene or amphibole on olivine or garnet. See also CORONA.

Kenoran orogeny *n.* a name for the PROTEROZOIC orogeny that deformed the ARCHAEAN of the Canadian SHIELD, equivalent to the ALGOMAN OROGENY.

kernite *n.* a borate mineral, $Na_2B_4O_6$ $(OH)_2 \cdot 3H_2O$, that occurs in great quantities beneath the Mojave Desert, California, in a bedded series of Tertiary clays.

kerogen *n.* the solid bituminous substance in OIL SOURCE rocks that yields oil when the shales are subjected to destructive distillation.

kersantite *n.* a variety of calc-alkaline LAMPROPHYRE containing PHENOCRYSTS of biotite-phlogopite, augite or olivine and the same minerals together with PLAGIOCLASE (oligoclase or andesine) and sometimes ALKALI FELDSPAR in the groundmass.

kettle or **kettle-hole** *n.* a depression in a glacial OUTWASH drift, produced by the melting of a separated glacial ice mass that became buried either wholly or partly. Such stranded ice masses are thought to be the result of outwash accumulation on top of the uneven GLACIER terminus. Kettles range from 5 to 13 metres in diameter and up to 43 metres in depth. Most kettles are of circular to elliptical shape, since melting ice blocks tend towards roundedness. When filled with water they are called *kettle lakes.* The kettles may occur in groups or singly; when large numbers are found together, the terrain, which appears as basins and mounds, is called KAME AND KETTLE MORAINE or *kettle moraine.*

kettle moraine see KETTLE.

Keuper *n.* a red MARL FACIES of the European Upper TRIASSIC.

Keweenawan *n.* a provincial series of the Precambrian in Michigan and Wisconsin, USA.

key *n.* a term used for a member of the chain of small, low-lying islands that form the Florida Keys, off the south coast of the state of Florida, USA. They are composed of coral, oolitic limestone and carcareous sand. See CAY.

K-feldspar see POTASSIUM FELDSPAR.

kidney ore *n.* a variety of HEMATITE occurring in BOTRYOIDAL to RENIFORM shapes with a radiating structure.

kieselguhr see DIATOMACEOUS EARTH.

kimberlite *n.* a volatile-rich UNDER-SATURATED ultrabasic IGNEOUS ROCK of variable composition containing a variety of minerals (both PHENOCRYSTS and XENOCRYSTS) in a finer-grained groundmass that may contain micro-phenocrysts. Serpentinised OLIVINE is the principal constituent, with variable amounts of PHLOGOPITE, ILMENITE, SPINEL, GARNET, ORTHOPYROXENE and CLINOPYRO-XENE and carbonate (CALCITE). Kimber-lites commonly contain a variety of XENOLITHS, both crustal rocks and mantle-derived ultramafic types such as garnet lherzolite. Accessory minerals commonly include ZIRCON and, much less frequently, DIAMOND, but although they are their most important primary source not all kimberlites contain diamond. Kimberlites occur in VOL-CANIC pipes, where usually they have a fragmental texture (see DIATREMES), and in DYKES and SILLS. They are found mostly in continental CRATONS, and their location appears to be controlled by deep-seated FRACTURE systems. See also BLUE GROUND, YELLOW GROUND.

kimberlite pipes see DIATREME.

Kimmeridgian *n.* a STAGE of the Upper JURASSIC.

Kinderhookian *n.* the lowermost series of the MISSISSIPPIAN of North America.

Kinderscoutian *n.* a stratigraphical stage of the Namurian series (CARBONIFEROUS) in Britain and western Europe.

kink band, kink field see FOLD.

klint *n.* an exhumed fossil BIOHERM or coral REEF that appears as a knob or ridge because the enclosing material has been eroded away.

klippe see NAPPE.

knick point, nick point or **rejuvenation head** *n.* a break of slope in the longitu-dinal PROFILE of a river or stream caused by REJUVENATION (see also PROFILE OF EQUILIBRIUM). The knick point may be marked by a waterfall or rapids that begins to migrate upstream, and river terraces are formed that represent the former level of the valley floor or floodplain as the river begins to erode its bed more energetically.

knock and lochan topography *n.* a rocky, lowland topography of basins, grooves and elongated, smoothed ridges produced by areal scouring beneath an ice sheet. Compare KAME AND KETTLE MORAINE.

knoll see FROST MOUND.

komatiite *n.* a type of ULTRAMAFIC lava of tholeiitic affinity ranging in composi-tion from MgO-rich komatiite basalts to peridotitic komatiites containing between 18 and 32 per cent MgO, with TiO_2 less than 1 per cent and Na_2O + K_2O less than 1 per cent. Komatiites in the strict sense are restricted in their occurrence to Archaean GREENSTONE BELTS and are distinguished from PICRITE basalts by the presence of SPINIFEX TEXTURE, an array of crossed sheafs of elongated, skeletal branching olivine crystals. There is evidence that komatiite lava flows could melt the rock surfaces beneath them, indicating that they were erupted at significantly higher temperatures than modern-day lavas; this would be consistent with a higher GEOTHERMAL GRADIENT in the MANTLE during the Archean than in more recent times.

konservat lagerstätten ('conservation lagerstätten') *n.* fossiliferous deposits in which soft-part preservation is common. Preservation may result from rapid burial (*obrution*) or low oxygen concentrations. See LAGERSTÄTTEN.

koum see ERG.

KREEP *n. acronym for* a basaltic lunar rock, first observed in Apollo 12 breccias and fines. It is distinguished by an inordinately high potassium content (K), RARE EARTH ELEMENTS (REE), phosphorus (P) and other trace elements.

Kungurian *n.* the top STAGE of the Lower PERMIAN.

kupfernickel see NICKELINE.

kutnahorite see ANKERITE.

kyanite *n.* a NESOSILICATE mineral of the triclinic system, Al_2SiO_5, often found in groups of light-blue crystals; less frequent colours are white, grey and green with spots of colour. Kyanite is trimorphous with SILLIMANITE and ANDALUSITE. It occurs almost exclusively in rocks rich in aluminium and metamorphosed under high-pressure conditions: gneissic mica schists, amphibolites and eclogites. It is important as a means of identifying the grade and type of metamorphism of the host rock.

Ll

labile *adj.* **1.** unstable, apt to change. **2.** (of rocks and minerals) easily decomposed.

labradorescence *n.* the iridescence displayed by plagioclase FELDSPAR that is characterised by Bøggild intergrowths. This is a submicroscopic exsolution structure occurring in PLAGIOCLASE of the composition An_{40} to An_{60} in the ANDESINE-LABRADORITE range. The iridescence is caused by the interference of light reflected from the exsolution lamellae. Compare SCHILLER.

labradorite *n.* a PLAGIOCLASE feldspar with the composition An_{50}–An_{70} (see FELDSPAR, Fig. 33). Labradorite commonly occurs in basic IGNEOUS ROCKS and high-grade METAMORPHIC ROCKS.

labrum see PLASTRON.

laccolith or **laccolite** *n.* a dome-like CONCORDANT body of intrusive IGNEOUS ROCK. It arches the overlying rocks and sediment and has a floor that is more or less flat. Although they resemble inflated SILLS, laccoliths are massive rather than tabular and may be simple, composite or multiple. Multiple laccoliths in the form of a series of stacked intrusions are sometimes called 'cedar tree laccoliths'. Compare PHACOLITH. See INTRUSION, LOPOLITH.

lacuna *n.* (*pl.* **lacunae**) a gap or break in the stratigraphical record. The gap may be produced by destruction of the stratigraphical record through erosion or by non-deposition of any rock during a particular time interval. Lacuna and hiatus are synonymous when referring to rock that has been removed or not deposited. Compare HIATUS.

lacustrine *adj.* **1.** related to, inhabiting or produced by lakes. **2.** (of a region) characterised by lakes.

lacustrine plain see LAKE PLAIN.

Ladinian *n.* the top STAGE of the Middle TRIASSIC.

lag see FAULT.

lag deposit see LAG GRAVEL.

lagerstätten *n.* fossiliferous deposits that are exceptionally preserved in that they represent the majority of the life forms that lived in a particular area or an accumulation of large amounts of fossil remains. *Concentration lagerstätten* include deposits such as bone beds and traps (the La Brea tar pits); *conservation* (konservat) *lagerstätten* are those deposits where all life forms, including soft-bodied forms, are preserved. See KONSERVAT LAGERSTÄTTEN.

lag gravel *n.* **1.** a residuum of coarse rock particles left on a surface, especially desert, after finer material has been removed by deflation. See also DESERT PAVEMENT. **2.** also called **lag deposit**, coarse material that is dragged along a stream bottom, thus lagging behind finer material.

lahar *n.* a potentially destructive mud and debris flow of hot VOLCANICLASTIC material on the flanks of a VOLCANO, formed when water from any source combines with the hot volcanic debris, and slides downwards under gravity. The water may have been ejected from a crater lake or come from rapidly melted ice or snow or very heavy rainfall. The descent of NUÉES ARDENTES into adjacent streams can cause lahars. The clasts in lahars are characteristically angular to sub-rounded and poorly sorted, and the resulting EPICLASTIC deposit can be difficult to distinguish from that of a cold debris flow. In Indonesia, all mudflows tend to be called lahars. See also jökulhlaup under GLACIER BURST.

lake plain or **lacustrine plain** *n*. the nearly level surface marking the floor of a former lake. Because lakes are ephemeral landforms, their sites eventually are filled with inwash or their outlets lowered by erosion.

lamella *n*. (*pl*. **lamellae**) a thin plate, scale or layer.

lamellibranch see BIVALVE.

lamina *n*. (*pl*. **laminae**) the thinnest discernible layer in a sedimentary rock that differs from other layers in composition, particle size or colour, usually 0.005 to 1.00 millimetres thick.

laminar flow *n*. adjacent layers of fluid that move smoothly without mixing. Laminar flow occurs in conditions of low velocities, small channel sizes and high viscosities. Laminar flow is the normal state of flow of GROUNDWATER. Compare TURBULENT FLOW. See also REYNOLDS NUMBER.

lamination *n*. the formation of laminae or the state of being laminated, specifically the finest stratification, such as is found in shale.

lamprophyre *n*. a group of melanocratic to mesocratic porphyritic IGNEOUS ROCKS that typically occur as DYKES and small INTRUSIONS. The PHENOCRYSTS are commonly EUHEDRAL ferromagnesian minerals, BIOTITE-PHLOGOPITE and/or

AMPHIBOLE, with CLINOPYROXENE and OLIVINE; FELDSPARS and/or FELDSPATHOIDS occur only in the groundmass. Lamprophyres usually show signs of HYDROTHERMAL alteration; CALCITE, ZEOLITES and CHLORITE may be present in the groundmass. Chemically, lamprophyres vary from silica-oversaturated to silica-undersaturated varieties. They generally have a relatively high content of K_2O or ($K_2O + Na_2O$) and of P_2O_5, Ba, H_2O and CO_2 for their SiO_2 content. Lamprophyres are classified principally according to their felsic minerals, and may be divided into:
(1) the *calc-alkaline lamprophyres*, containing both PLAGIOCLASE and ALKALI FELDSPARS but no feldspathoids (minette, vogesite, kersantite and spessartite);
(2) the *alkali lamprophyres*, containing feldspars and/or feldspathoids (camptonite, sannaite and monchiquite);
(3) the *melelititic lamprophyres*, alnöite and polzenite. Alnöite is an ULTRAMAFIC lamprophyre containing phenocrysts of biotite, olivine and AUGITE in a groundmass that contains melilite, augite, biotite, GARNET and calcite, while polzenite in addition contains nepheline but lacks augite.

Groups 2 and 3 are generally associated with ALKALINE igneous complexes, including CARBONATITE/IJOLITE complexes, while group 1 is generally associated with GRANITES, SYENITES or MONZONITES, although rocks of this group may occasionally occur within CARBONATITE complexes.

lamp shell see BRACHIOPOD.

land bridge *n*. a land area that forms a connecting route between continents or landmasses. Before the acceptance of the hypothesis of CONTINENTAL DRIFT, the migration of animals and plants between continents was explained by proposing the former existence of land bridges.

landform *n.* any of the various features that make up the surface character of the Earth. It includes mountains, plains, valleys, rivers and canyons.

landmass *n.* a part of the CONTINENTAL CRUST above sea level.

Landsat *n.* unmanned NASA satellites that orbit Earth and transmit, to receiving stations on Earth spectral images in the 0.4 to 1.0 μm range. They are usually used in the determination of vegetation cover.

landslide *n.* a general term for the downward movement, under gravity, of masses of soil and rock material as a result of a variety of processes. Types of landslides include rockfalls, mudflows and slumps. See MASS WASTING.

langbeinite *n.* an isometric EVAPORITE mineral, $K_2Mg_2(SO_4)_3$. It is used as a source of potassium compounds in the fertiliser industry.

Langhian *n.* the top STAGE of the Lower MIOCENE/base of the Middle Miocene.

lapilli *n.* pyroclasts with diameters in the general range of 2 to 64 millimetres. These volcanic ejecta usually consist of old lavas and scoriae thrown out in a completely solid state. Compare SCORIA. See also PYROCLASTIC MATERIAL, PYROCLASTIC ROCK, ACCRETIONARY LAPILLI.

lapilli tuff *n.* pyroclastic deposit composed of volcanic ejecta measuring 2 to 64 millimetres in diameter. See also PYROCLASTIC MATERIAL, PYROCLASTIC ROCKS.

lapis lazuli *n.* a crystalline rock consisting of a mixture of LAZURITE with small amounts of calcite, pyroxene and other silicates. It has a rich blue or blue-violet colour and is used as a semi-precious stone and ornament.

lappets *n.* projections on either side of the aperture of the shell in some AMMONOIDS (see Fig. 1).

Laramide orogeny *n.* a series of Late Cretaceous and Early Tertiary mountain-building events that affected western North America, resulting in the folding and uplift of the Rocky Mountains. The Laramide OROGENY was part of a continued series of orogenic movements affecting the western margin of North America that began in the Mesozoic as the American plate drifted westwards over the Farallon plate. The Laramide phase extended into the Oligocene.

larvikite *n.* a coarse-grained SYENITE with large feldspar crystals that display a distinctive blue iridescence (sometimes called SCHILLER). Titanaugite is the main MAFIC mineral, and coarse-grained APATITE is a characteristic ACCESSORY MINERAL.

laser Raman spectrometry see Appendix 4.

lateral accretion *n.* materials deposited at the sides of channels, such as those where BEDLOAD materials are being moved towards the inner sides of MEANDERS. Compare VERTICAL ACCRETION.

lateral erosion *n.* the action of a meandering stream as its course snakes from side to side, undercutting the banks; the process results in LATERAL PLANATION.

lateral moraine see MORAINE.

lateral planation *n.* the reduction of the land adjacent to a river or FLOODPLAIN to a plain or nearly even surface by the processes of lateral corrasion or LATERAL EROSION by a stream.

laterite *n.* a soil residue composed of secondary oxides of iron, aluminium, or both, together with CLAY MINERALS and some SILICA. In regions of extreme WEATHERING intensity, e.g. the Amazon River Basin, even kaolinite is unstable and silica is leached from the clay mineral, leaving an amorphous residue.

latite *n.* a porphyritic extrusive rock with PHENOCRYSTS of potassium feldspar and plagioclase in approximately equal amounts, little or no quartz and aphanitic to glassy groundmass; CLINOPYROXENE is the most abundant ferromagnesian mineral, orthopyroxene, olivine and hornblende less so. Biotite is commonly present. It is the extrusive equivalent of MONZONITE. When the potassium feldspar content increases, latite grades into TRACHYTE. Latite is the potassic variety of TRACHYANDESITE.

lattice defect *n.* any irregularity in a CRYSTAL LATTICE. This may arise from the omission of an ION, the displacement of an ion to an abnormal site or the insertion of a 'foreign' ion into the lattice. Other defects can arise as irregularities in the stacking of the lattice, such as *edge* DISLOCATIONS, where a plane of atoms or ions terminates in a line within the lattice, or *screw dislocations*, in which part of a lattice plane is rotated. The techniques of creating lattice imperfections are basic to semiconductor technology. See also TWINNING.

lattice-preferred orientation *n.* the preferred orientation of crystallographic axes or planes. In metamorphic rocks, it is produced by crystal GLIDING and depends on the mineral structure and temperature, pressure and stress during DEFORMATION. In IGNEOUS ROCKS, it is primarily a function of the original form of the crystals during flow or settling. Compare DIMENSION-PREFERRED ORIENTATION.

Laurasia *n.* the northern continental unit of the later Palaeozoic and Mesozoic. Laurasia was assembled from the collisions of minor units, such as AVALONIA, and major units, such as BALTICA, with LAURENTIA. During the Carboniferous, Laurasia collided with the southern continental unit GONDWANA to produce the supercontinent PANGAEA, a single landmass containing all continental units. The present continents of the Northern Hemisphere have been derived from Laurasia by continental drifting. The name is derived from Laurentia, a palaeogeographical term for the Canadian SHIELD and its surroundings, and Eurasia. See also PLATE TECTONICS.

Laurentia *n.* the early Palaeozoic continental unit that comprised North America, Greenland and parts of northwest Scotland. See LAURASIA.

lava *n.* **1.** molten rock that issues from openings at the Earth's surface or on the ocean floor. Such openings may be located in craters or along flanks of volcanoes or in fissures. (See VOLCANIC ERUPTION, SITES OF, MAGMA) **2.** the rock formed from the solidified material.

Measurements of eruptive lava temperatures range from 1050°C to 1190°C for tholeiitic BASALT to 725°C to 850°C for DACITE. There is a wide range of lava compositions, from acidic to ultrabasic (see SILICA CONCENTRATION) and PERALUMINOUS to PERALKALINE. The VISCOSITY of lava generally decreases with decreasing SiO_2 content, increasing temperature and content of dissolved volatiles, especially water; pressure may be a factor in the case of deep submarine eruptions. BASIC lavas are less viscous than acidic lavas at the same temperatures. Viscosity is lower in lavas with higher contents of dissolved volatiles; exsolution of volatiles increases the viscosity of a lava flow and can lead to the situation where a basaltic flow consists of PAHOEHOE near the vent but changes to A'A' as it loses volatiles as it travels farther away. All these factors influence the type of volcanic eruption. See LAVA FLOW, VOLCANIC ERUPTION.

lava blister *n*. a small, hollow, relatively steep-sided swelling produced on the surface of a LAVA FLOW by gas bubbles trapped in the crust of the lava flow.

lava cascade *n*. a fall of fluid, incandescent LAVA, formed when a lava stream passes over an abrupt change in elevation.

lava cave see LAVA TUBE.

lava delta *n*. the delta-shaped feature often formed when LAVA flows into a lake or the sea. The lava in contact with the water is chilled but the continuing flow of lava from behind pushes the solidified mass outwards, often generating convex-outwards pressure ridges on the surface of the delta. Lava deltas are formed from the more viscous types of lava such as A'A' or BLOCK LAVA.

lava flow *n*. a stream of lava issuing from a vent or fissure during an EFFUSIVE eruption; the term also applies to the solidified material so produced. The velocity and dimensions of a flow depend essentially on viscosity. Thus, a basaltic lava tends to form fast-moving thin long flows, whereas the more viscous siliceous lava forms slow-moving thick short flows. Lava flows cool more rapidly than intrusive rocks, resulting in rock of a fine-grained to glassy texture. Subaerial basaltic lava flows may form extremely voluminous sheets (FLOOD BASALTS) from FISSURE ERUPTIONS or SHIELD VOLCANOES built principally of PAHOEHOE and A'A' flows. Pahoehoe flows, characterised by a smooth, glassy surface often displaying ropy structure, develop from very fluid basaltic lava. They can flow at speeds of 10 to 20 kilometres per hour, or even faster on steep slopes; however, they

tend to crust over rapidly, and the lava flows beneath the crust through LAVA TUBES. As pahoehoe courses downslope it may change to a'a', which has a chunky broken extremely rough surface formed by a mantle of clinker-like or spiny fragments. The transformation is attributed to an increase in viscosity as a result of cooling, gas loss and progressive crystallisation. Where the lava is more viscous to begin with, a'a' is formed at the vent and it moves relatively slowly downslope. Lava levees (ridges) may form at the side margins of a'a' flows, especially on steep slopes. Subaqueous basaltic lava flows form PILLOW LAVAS.

Block lava is less vesicular than a'a', lacks its spininess, and the top surface and flow front are mainly composed of large randomly oriented angular blocks. It is more viscous and, although it may be basaltic, it is typically more siliceous. Flows associated with stratovolcanoes, especially ANDESITES and DACITES, tend to form block lavas, and these frequently develop levees (or COULÉES). Extremely viscous andesite or dacite lavas may form domes or spines within the crater (for example, the dacitic spine that grew on Mount Pelée, in Martinique, before the 1902 eruption). RHYOLITE lavas are also very viscous and commonly form domes, *mesa lavas* (circular in plan but rather flat bodies) or coulées. The lava surface is rough and blocky, with ridges, called *ogives*, that are concave upflow. See also VOLCANIC ERUPTION.

lava fountain *n*. a plume of incandescent LAVA sprayed into the air by gas rapidly escaping from magma as it reaches the surface. Fountains, which are characteristic of Hawaiian-type eruptions, usually range from about 10 to 100 metres in height, but a Hawaiian one in

1959 reached 400 metres. See VOLCANIC ERUPTIONS, TYPES OF.

law of universal gravitation

1959 reached 400 metres. See VOLCANIC ERUPTIONS, TYPES OF.

lava lake *n*. a lake of molten LAVA in a crater or other depression of a VOLCANO.

lava plug see VOLCANIC PLUG.

lava shield see SHIELD VOLCANO.

lava tube, lava cave or **lava tunnel** *n*. a tunnel-like space beneath the surface of a solidified LAVA FLOW, formed when still-molten LAVA in the interior drains away after the formation of a surficial crust. Lava stalactites and stalagmites may form in these tunnels. Lava tubes are best developed in basaltic PAHOEHOE flows. They are rare in siliceous flows, rare or short in A'A' flows and absent from block lava.

lava tunnel see LAVA TUBE.

law of constancy of angle *n*. the angles between adjacent CRYSTAL FACES of a MINERAL have constant values characteristic of that mineral.

law of cross-cutting relationships *n*. an IGNEOUS ROCK, or any other geological feature, is younger than any rock it transects.

law of faunal assemblages *n*. similar assemblages of fossil fauna or flora indicate similar geological ages for the rocks in which they are embedded. See FAUNAL SUCCESSION.

law of original continuity *n*. a water-laid STRATUM, when formed, must continue laterally in all directions until it thins out as a result of non-deposition or until it abuts the edge of the original basin of deposition.

law of original horizontality *n*. SEDIMENTARY ROCKS, especially water-laid strata, are deposited parallel to the surface on which they are deposited and thus horizontally. Such rocks are assumed to have been horizontal at the start of their geological history, although they may have been folded or tilted subsequently.

law of rational indices *n*. the indices of any CRYSTAL FACE are always rational numbers and are determined by dividing the intercepts of any face into the intercepts of the PARAMETRAL PLANE and clearing fractions. See MILLER INDICES.

law of reflection *n*. the angle between an incident ray and the normal to a reflecting surface is the same as the angle between the reflected ray and that same normal, i.e. the ANGLE OF INCIDENCE equals the angle of reflection.

law of refraction or **Snell's law** *n*. when a wave traverses a boundary between two isotropic substances, in which it travels with different velocities, the wave normal alters its direction in such a way that the sine of the ANGLE OF INCIDENCE between the wave normal and the boundary normal, divided by the wave velocity in the first medium, equals the angle of refraction divided by the wave velocity in the second medium, i.e.

$$\frac{\sin i}{v_1} = \frac{\sin r}{v_2} \quad \text{or} \quad \frac{\sin i}{\sin r} = \frac{v_1}{v_2}$$

where i is the angle of incidence, r is the angle of refraction, and v_1 and v_2 are the wave speeds in the two media; the ratio v_1/v_2 is known as the *refractive index*.

law of superposition *n*. in any sequence of sedimentary strata that is not strongly folded or tilted, the youngest STRATUM is at the top and the oldest at the bottom.

law of universal gravitation *n*. every body of the universe exerts a force of attraction upon every other body of the universe, expressed as an equation,

$$F = Gm_1m_2/d^2$$

where m_1 and m_2 are the masses of the two bodies and d is the distance between their centres. G, the gravitational constant, equals 6.672×10^{-11} Nm^2 kg^{-2}.

layered intrusion *n*. a major igneous INTRUSION within which the MAGMA has crystallised as a series of layers with contrasting mineralogy and chemical composition. See CUMULATE.

layering *n*. a series of tabular or sheet-like structures within a rock, caused by differences in mineral composition, texture and/or grain size.

Laxfordian *n*. an orogenic event the main phase of which took place around 1900 Ma to 1750 Ma. The event deformed the LEWISIAN COMPLEX. See LEWISIAN GNEISS.

lazulite *n*. a bright blue monoclinic mineral, $(Mg,Fe)Al_2(PO_4)_2(OH)_2$. See also LAZURITE.

lazurite *n*. an ultramarine to violet-blue mineral of the SODALITE group, $AlSiO_4)_6(SO_4,S,Cl)_2$. It is the main constituent of LAPIS LAZULI; lazurite is an obsolete synonym for AZURITE. See also FELDSPATHOID.

leaching *n*. **1.** dissolution of soluble substances from a rock by the natural action of percolating waters.
2. the removal of mineral salts from an upper to a lower SOIL HORIZON by the action of percolating water.

lead *n*. a soft, heavy metallic element, Pb. Lead is found mostly combined with sulphur in GALENA; it is found extremely rarely in native form. See also ANOMALOUS LEAD.

lead-lead method see DATING METHODS.

lead-uranium ratio *n*. the ratio of lead-206 to uranium-238 and/or of lead-207 to uranium-235 that forms from the radioactive decay of uranium. See DATING METHODS.

lechatelierite *n*. naturally fused SILICA formed during METEORITE impacts at pressures in excess of about 50 GPa. See IMPACTITE.

lectotype see TYPE SPECIMEN.

lee *n*. **1.** the side of a prominence, such as a hill or dune, that is sheltered from the wind.
2. also called **lee-side** or **lee-seite**, the side of a hill in a glaciated region facing the direction towards which a GLACIER is moving. It is protected from ABRASION and is thus more jagged in appearance than the opposite side, the STOSS SIDE.

lee-seite see LEE.

lee-side see LEE.

Lenian *n*. the top STAGE of the Lower CAMBRIAN.

lens *n*. an ORE or ROCK body that is thick in the middle and thin at the edges.

lenticular bedding see FLASER BEDDING.

lentil *n*. **1.** a lens-shaped rock body.
2. also called **tongue**, in North America, a specially shaped member of a FORMATION. A lentil is impersistent, i.e. it thins out in all directions.

Leonardian *n*. the upper series of the Lower Permian in North America.

lepidoblastic *n*. a textural term for a schistose or foliated rock caused by parallel alignment during the re-crystallisation of platy minerals such as micas, chlorite, talc and graphite.

Lepidodendron see SCALE TREES.

lepidolite *n*. a mineral of the MICA group, $K(Li,Al)_{2-3}(AlSi_3O_{10})(O,OH,F)_2$. It is usually pink to lilac but may also be pale yellow to greyish white; it occurs in pegmatites with other lithium-bearing minerals.

leucite *n*. a tetragonal mineral of the FELDSPATHOID group, $KAlSi_2O_6$, abundant in K-rich basic lavas. It characteristically occurs as white trapezohedral crystals in a fine-grained matrix.

leucocratic *adj*. (of IGNEOUS ROCKS) light-coloured, containing between 0 and 35 per cent of DARK MINERALS, i.e. with a COLOUR INDEX between 0 and 35. Compare MESOCRATIC, MELANOCRATIC, ULTRAMAFIC.

levee *n*. **1.** a man-made embankment along a watercourse to protect land from flooding.

2. a NATURAL LEVEE.

level *v*. **1.** to determine the relative altitude of different points on the Earth's surface, generally by sighting through a levelling instrument.

2. to find a horizontal line by means of a level.

Lewisian complex *n*. the Late Archaean high-grade granulite-gneiss TERRANE of northwest Scotland.

Lewisian gneiss *n*. highly deformed and metamorphosed PRECAMBRIAN rocks of northwest Scotland, comprising the *Scourian complex* (approximately 2900–2500 Ma) and the *Laxfordian complex* (approximately 1900–1500 Ma). The Scourian and Laxfordian episodes are separated by the INTRUSION of a swarm of basic and ultrabasic DYKES – the *Scourie dykes* – in two episodes at about 2400 and 2000 Ma. The *Loch Maree group* occurs as a tract of metamorphosed sediments and tholeiitic basalt lavas in the Gairloch and Loch Maree area, formed at about 2000 Ma and then caught up in the Laxfordian episode.

lherzolite *n*. an ULTRAMAFIC plutonic IGNEOUS ROCK composed mainly of OLIVINE with ORTHOPYROXENE and CLINOPYROXENE. Some varieties contain minor amounts of chrome SPINEL or pyrope GARNET. Lherzolite is a two-pyroxene PERIDOTITE. Spinel lherzolite and garnet lherzolite are found as XENOLITHS in alkali basaltic rocks and KIMBERLITES respectively and are thought to represent upper MANTLE compositions.

Lias *n*. the Early JURASSIC Epoch. Lithostratigraphically the Lias is a GROUP.

Liesegang rings *n*. nested bands or rings produced by rhythmic precipitation within a fluid-saturated rock.

life assemblage see BIOCOENOSIS.

ligament *n*. springy material that joins the VALVES along the hinge in BIVALVES and opens the shell when the muscles relax (see Fig. 7). The ligament is described as *amphidetic* if it lies on either side of the umbones (see UMBO) and *opisthodetic* if it is posterior of the umbones. An *external* ligament lies above the HINGE PLATE and an *internal* ligament lies between the hinge plates, usually in a pit.

light mineral *n*. **1.** a ROCK-forming MINERAL with a specific gravity less than 2.80; this includes calcite, quartz, feldspars and feldspathoids. See also HEAVY MINERALS.

Note: the use of this term to designate light-coloured (LEUCOCRATIC) rock-forming minerals should be discouraged to avoid ambiguity.

light oil *n*. crude oil with a high API GRAVITY or Baumé gravity. Compare HEAVY OIL.

lignite or **brown coal** *n*. brown to black COAL that has been formed from PEAT under moderate pressure. Its texture is like woody peat and it crumbles on exposure to the atmosphere. It is intermediate in COALIFICATION between peat and SUB-BITUMINOUS COAL and dates chiefly from the TERTIARY or the end of the MESOZOIC. Although deposits of lignite are plentiful, with extensive beds in Germany, Canada, the USA and Brazil, its low calorific value (about 7–11 MJ kg–1) causes it to be inferior to BITUMINOUS COAL.

limb *n*. that area of a FOLD between adjacent fold hinges; it may be planar or slightly curved.

limburgite *n*. a porphyritic IGNEOUS ROCK allied to BASANITE with phenocrysts of olivine, titanaugite and opaque iron oxides in an alkali-rich nepheline-free glassy groundmass.

limestone *n.* a rock composed primarily of calcareous sediments consisting of calcium carbonate ($CaCO_3$), mainly as CALCITE. Organic limestones consist of shell remnants or of calcite deposits precipitated by certain ALGAE, e.g. coral limestone, crinoidal limestone, chalk. CLASTIC limestones are composed of broken fragments of pre-existing limestones, of shells or of calcite crystals. They can be classified on the basis of grain size, see ARENITE, LUTITE, RUDITE. Chemically precipitated limestones form in warm shallow seas. These are oolitic and pisolitic limestones and the (dolomitic) limestones of evaporite sequences. *Calc tufa* is another form of chemically precipitated limestone.

The principal classification of limestone that is now used was suggested by Folk and is based on the relative content of ALLOCHEMS (discrete particles of carbonates such as OOLITES, FOSSILS), microcrystalline calcite (MICRITE) and sparry calcite cement (SPARITE). The allochem variety is used as a prefix to the rock name, with either micrite or sparite as a suffix, for example OOMICRITE, OOSPARITE. Limestones are classified as *orthochemical limestones* if micrite is the main constituent, *allochemical limestones* with variable proportions of allochems and micrite or sparite, and *autochthonous reef rocks* (BIOLITHITE). A limestone classification suggested by Dunham is based on depositional textures and mud content. See FOLK CLASSIFICATION, DUNHAM CLASSIFICATION.

Limestones differ greatly in colour and texture, depending on the size of the shells or crystals they contain. The mineral dolomite, $CaMg(CO_3)_2$, is formed by the REPLACEMENT of limestone. During DOLOMITISATION, as the process is called, magnesium (Mg) is substituted for up to 50 per cent (molecular) of the calcium (Ca) in the original calcite ($CaCO_3$). The alteration process of limestone into dolomite proceeds, at least in some cases, in stages: *high-calcium limestone* – CaCO3 $\geq$ 95 per cent; *dolomitic limestone* – 10 to 50 per cent dolomite; 50 to 90 per cent calcite; *magnesian limestone* – $\geq$ 10 per cent dolomite; $\geq$ 90 per cent calcite; *dolomite* – $\geq$ 90 per cent dolomite; $\geq$ 10 per cent calcite.

limnic *adj.* **1.** pertaining to a body of fresh water. Compare LACUSTRINE.
2. (of COAL deposits) having formed inland in PEAT bogs, as distinguished from PARALIC deposits.

limnology *n.* the scientific study of fresh waters, in particular ponds and lakes. It deals with the physical, chemical, meteorological and biological conditions that pertain to such bodies of water.

limonite *n.* mainly a field term for amorphous, hydrated iron oxide, $FeO \cdot OH \cdot nH_2O$, dark brown to black; a minor ore of iron, it occurs in earthy masses of various forms. Limonite is of SUPERGENE origin and is one of the chief constituents of GOSSAN. It often contains small amounts of HEMATITE, CLAY MINERALS and manganese oxides, and is used as the pigment yellow ochre. See also BOG IRON ORE.

lineament *n.* a linear topographical feature that is thought to reflect crustal structure, e.g. FAULT lines, straight stream courses, aligned VOLCANOES. Compare LINEATION. See HOT SPOT.

lineation *n.* a general term for any ROCK feature showing a linear structure, e.g. parallel crystal arrangement, SLICKENSIDES, BOUDINAGE and FOLD axes. Compare LINEAMENT.

liptinite see MACERAL.

liquefaction *n.* the temporary transformation of a SEDIMENT or SOIL into a fluid mass following a sudden loss of SHEAR STRENGTH, resulting from an increase in pore-fluid pressure. This can be caused by a shock, such as an earth tremor, or a sudden increase in loading (resulting from rapid deposition) that changes the PACKING of a waterlogged sediment, or by an influx of water from some outside source. A variety of SEDIMENTARY STRUCTURES result from liquefaction, for example LOAD CASTS, SAND VOLCANOES, SANDSTONE DYKES. See also THIXOTROPY.

liquid immiscibility *n.* a process of MAGMATIC DIFFERENTIATION in which the MAGMA separates into two or more immiscible phases.

liquid limit *n.* the boundary between semi-liquid and plastic states of a sediment, e.g. a soil determined by the moisture content (percentage by weight of oven-dried soil) at which the soil will begin to flow if slightly disturbed. It is one of the ATTERBERG LIMITS. Compare PLASTIC LIMIT.

liquidus see LIQUIDUS-SOLIDUS.

liquidus-solidus *n.* in a temperature-composition diagram representing an equilibrium assemblage of crystals and melt, the temperature at which CRYSTALLISATION begins; i.e. the temperature above which the system is entirely liquid is the *liquidus*; the temperature below which the system is entirely crystalline is the *solidus*. See also PHASE DIAGRAM.

listric fault *n.* a curved normal FAULT in which the fault surface is concave upwards. Listric faults are commonly formed when the crust is in extension. See Fig. 55 below.

lithic *adj.* 1. pertaining to or made of stone.
2. (of a SEDIMENTARY ROCK or PYROCLASTIC deposit) containing large amounts of fragments derived from previously formed rocks.

lithic arenite *n.* a SANDSTONE containing less than 10 per cent mud matrix, over 25 per cent rock particles and less than 10 per cent feldspar.

lithic tuff see TUFF.

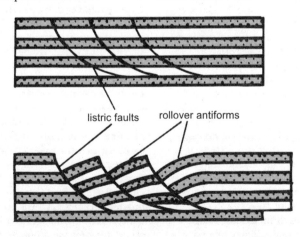

listric faults   rollover antiforms

Fig. 55 **Listric fault**. Movement on faults may result in development of hanging wall 'rollover' antiforms

lithification *n.* **1.** the conversion of unconsolidated sediments into a solid rock. It may involve a combination of such processes as COMPACTION, CEMENTATION, CRYSTALLISATION and desiccation. Laboratory experiments have shown that the presence of ductile fragments (e.g. schist) and absence of non-ductile fragments (e.g. quartz) can produce lithification by compaction alone. The composition of the average sandstone (many fragments are non-ductile) indicates that lithification must be achieved largely by the introduction of chemical precipitates as cements. Compare CONSOLIDATION, INDURATION.
2. a type of COAL-bed termination caused by a lateral increase in impurities wherein the bed changes gradually into bituminous SHALE or other rock.

lithofacies *n.* **1.** a mappable subdivision of a specified STRATIGRAPHICAL UNIT differentiated from adjacent subdivisions by LITHOLOGY.
2. the rock record of any sedimentary environment, including both physical and organic peculiarities. Compare LITHOTYPE.

lithogeochemistry *n.* a branch of GEOCHEMISTRY dealing with the chemical properties of the LITHOSPHERE (rocks, soils, sediments). Compare BIOGEOCHEMISTRY, HYDROGEOCHEMISTRY.

lithographic limestone *n.* a dense, homogeneous, very fine-textured limestone; a MICRITE. It was formerly used in lithography for engraving.

lithographic texture *n.* a texture of certain calcareous rocks characterised by uniform clay-sized particles and a very smooth appearance resembling that of a lithographic stone.

lithohorizon *n.* a surface showing lithostratigraphic change or a particular lithostratigraphic character; it is highly useful for correlation. It is usually the boundary of a lithostratigraphic unit but may be a thin marker layer within a lithostratigraphic unit. Compare BIOHORIZON, CHRONOHORIZON.

lithology *n.* the description of the characteristics of rocks, as seen in hand specimens and outcrops, on the basis of colour, grain size, texture and composition.

lithophile see AFFINITY OF ELEMENTS.

lithophysae *n.* radiating aggregates of fibrous crystals, a few centimetres in diameter, that have formed around expanding vesicles in a melt while it is still capable of flowing. They occur in OBSIDIAN of rhyolitic composition and may be superimposed on flow structures. Fragmented lithophysae are an indication of continued flowage. See SPHERULITE, ORBICULE.

lithosome *n.* a body of SEDIMENT deposited under uniform physicochemical conditions over a given interval of time; it is the lithostratigraphic equivalent of a BIOSOME.

lithosphere *n.* the solid outer shell of the Earth, including the CRUST and uppermost rigid layer of the MANTLE. It is described as a strong or rigid zone, above the ASTHENOSPHERE, or weak zone. In PLATE TECTONICS, a lithospheric plate is a segment of the lithosphere that moves over the plastic asthenosphere below. The Earth's outer shell comprises eight large lithospheric plates and about two dozen smaller ones.

lithospheric plate see LITHOSPHERE.

lithostatic pressure see GEOSTATIC PRESSURE.

lithostratigraphic unit see STRATIGRAPHICAL UNIT.

lithostratigraphy *n.* STRATIGRAPHY concerned with the organisation of strata into units based on lithological character and the correlation of these units.

lithotype *n.* a description of banded coal based on physical characteristics other than botanical origin. There are four lithotypes of coal as defined by its megascopic textural appearance. *Vitrain* occurs in thin horizontal bands, has a brilliant vitreous lustre and breaks with CONCHOIDAL fracture; *clarain* has a silky lustre and shows platy fracture; *durain* is close-textured, showing a granular or matte surface when broken; *fusain* occurs as patches or wedges and consists of powdery, somewhat fibrous strands. The differences in appearance of coal lithotypes result from differences in composition of the original plant material from which the coal was formed and in the degree of alteration, but the relationships are not well understood. See also MACERALS.

lit-par-lit *n.* a layered structure that occurs in rock, the laminae of which are penetrated by thin parallel sheets and lobes of igneous material. Compare INJECTION GNEISS. See also MIGMATITE.

Little Ice Age *n.* a period from 1550 to 1850 characterised by extended cold seasons, heavy precipitation and the expansion of glaciers. Of the many theories proposed for the occurrence of this interval, one of the more attractive seems to be that of its coincidence with a period of unexplained absence of sunspot activity. A collateral factor of the Earth might have been a sharp increase in the ALBEDO of vast areas that had been deforested just prior to the Little Ice Age; this could have worsened the conditions of the cold. Ice accretions during this period, however, were not great enough to alter world climate. See also ICE AGE.

littoral *n.* 1. the designation of the shore area between tide marks.
2. the BENTHIC zone, extending from the high-water point on the beach to the edge of the continental shelf. See also NERITIC ZONE.

littoral cone *n.* a cone of PYROCLASTIC MATERIAL formed on the surface of basaltic LAVA FLOWS as the result of the explosive expansion of steam trapped under the LAVA when it flows into the sea or a lake. See also PSEUDOCRATERS.

littoral drift see LONGSHORE DRIFT.

Llandeilo *n.* the top STAGE of the Llanvirn Series of the Middle Ordovician.

Llandovery *n.* the bottom stratigraphical series of the SILURIAN.

llano see SAVANNA.

Llanvirn *n.* the top stratigraphical series of the Middle ORDOVICIAN.

load (of a river) see STREAM LOAD.

load casts *n.* rounded lobes of sandy material that protrude into an underlying finer layer, most commonly occurring in interbedded SANDSTONES and MUDROCKS. They form because the denser sand layer sinks into the underlying mud, perhaps as a result of vibration that causes the mud to liquefy (see LIQUEFACTION). The upward-pointing wedges of mud between the load casts form FLAME STRUCTURES. Sometimes globular load casts may become detached from the overlying sandy layer and form isolated load balls, or *pseudonodules*. The mudrock commonly has a disturbed or slurried texture. Since load casts form on the base of a bed they are a type of SOLE MARK. Compare BALL AND PILLOW STRUCTURE. See Fig. 56 on the next page.

local fauna see FAUNULE.

lode *n.* a mineral deposit consisting of an entire zone of dissemination.

lodestone *n.* common name for MAGNETITE when it is used as a natural magnet.

lodgement till see TILL.

loess *n.* a fine-grained, loosely coherent blanket deposit, generally buff col-

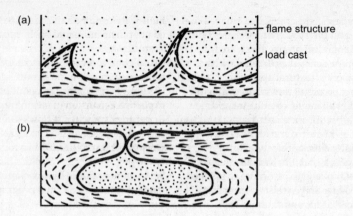

Fig. 56 **Load casts**. (a) load casts and flame structures; (b) globular load cast becoming detached from the overlying sandstone layer to form a load ball, or pseudonodule

oured and without stratification. The average grain size lies in the silt range. Because it contains vertical tubules (remains of rotted-out grass roots), loess stands in nearly vertical walls despite its weak cohesion. It is thought that loess is a windblown deposit of Pleistocene age. Most loess lies downwind from areas that were glaciated during the Pleistocene Period. The greatest loess deposits are found in China but extensive deposits occur over much of Europe, North and South America and New Zealand.

loess doll, loess kindchen or **loess männchen** *n.* a nodule or concretion of calcium carbonate found in LOESS; it bears a resemblance to a doll, a child's head or a potato.

logging *n.* any of various methods for obtaining a continuous record (a log) as a function of the depth of observations made on the rocks and fluids encountered in a well bore. See WELL LOGGING.

longitudinal cross-section see CROSS-SECTION.

longitudinal dune see DUNE.

longitudinal fault *n.* a FAULT the structure of which parallels that of the general structural trend of the region.

longitudinal joints see JOINT.

longitudinal scours see SCOUR MARK.

longitudinal stream *n.* a stream that flows in the direction of the STRIKE of the underlying rocks.

longitudinal valley *n.* the valley of a SUBSEQUENT stream that has evolved parallel to the general STRIKE of the underlying strata.

longitudinal wave see SEISMIC WAVES.

Longmyndian *n.* the Longmyndian Supergroup is a major sequence of sedimentary rocks outcropping in the Welsh borderland. It includes a high proportion of FLUVIATILE rocks as well as deposits of apparent marine origin and subordinate TUFF. It forms an important part of the late PRECAMBRIAN of England and Wales, and is younger than the URICONIAN.

longshore bar *n.* a low sand embankment formed primarily by wave action.

It lies generally parallel with the SHORE-LINE, at some distance from it, and is submerged during high tide.

longshore drift or **littoral drift** *n.* the transport of sediment along a coast by the action of waves that approach the coast at an angle, creating a dominant movement of water currents in one direction along the shore (*longshore* or *littoral currents*).

lophophore *n.* the BRACHIOPOD feeding organ.

lopolith *n.* a large CONCORDANT, typically layered, igneous INTRUSION that is planoconvex or lenticular in shape. They occur characteristically in Precambrian SHIELD areas of the continents and commonly range in composition from ULTRAMAFIC through BASIC to FELSIC and often display well-developed igneous layering, for example the Bushveldt layered intrusion of South Africa, which is exposed over an area of 65 000 square kilometres. Although the original concept was of a basin-shaped intrusion, it has long been recognised that the lower contacts generally dip irregularly but quite steeply downwards towards the centre and that many of them can be described as funnel-shaped. Accordingly, the term 'lopolith' is less widely accepted than formerly, this being reflected in names such as the Bushveldt Complex or the Stillwater Complex. These intrusions are of great economic importance as they are important sources of gold, platinum group metals and ores of chromium, nickel, vanadium, copper and tin.

lost river *n.* **1.** a stream that disappears underground in a KARST region.
2. a dried-up stream in an arid region.

Love wave see SEISMIC WAVES.

low-angle fault *n.* a FAULT dipping less than 45º. Compare HIGH-ANGLE FAULT.

low-grade metamorphism *n.* METAMORPHISM that takes place under conditions of low to moderate temperature and pressure. Compare HIGH-GRADE METAMORPHISM, METAMORPHIC GRADE.

low quartz see ALPHA QUARTZ.

low-velocity layer correction see WEATHERED LAYER.

low-velocity zone *n.* **1.** the WEATHERED LAYER.
2. the zone in the upper MANTLE of the Earth in which SEISMIC WAVE velocities are about 6 per cent slower than in the outermost part of the mantle. Its depth is defined as being between 60 and 250 kilometres. The reduction in seismic wave velocities is thought to indicate partial melting in this zone of between 1 and 10 per cent.

L-tectonite see TECTONITE.

Ludlow *n.* a series of the SILURIAN, above Wenlock, below Přidoli.

luminescence *n.* the emission of light by a substance after it is subjected to an external source of electromagnetic radiation. It is usually observed in minerals containing small amounts of impurity ions, called activators. See also PHOSPHORESCENCE, FLUORESCENCE.

lunar basalt *n.* an IGNEOUS ROCK found as great flows in the maria (see MARE) of the lunar surface. Lunar basalts exhibit the same types of fabric as terrestrial basalts but they differ compositionally. The major minerals in lunar basalts are titanium-rich clinopyroxene and very calcic plagioclase. All mare basalts contain iron-titanium oxides, usually ilmenite.

lunar crater *n.* a roughly circular depression in the lunar surface, relatively shallow and ranging in diameter up to hundreds of kilometres; lunar craters appear to have resulted mainly from METEORITE impacts although some are possibly of VOLCANIC origin.

lunar playa *n.* a small area on the Moon's surface in the EJECTA blankets surrounding some of the lunar craters. It may be a small lava flow.

lunar regolith or **lunar soil** *n.* a thin greyish layer on the Moon's surface, composed of loosely compacted fragmental material ranging in size from microscopic particles to blocks greater than a metre in diameter. It is thought to be the result of repeated meteoritic and secondary fragment impact extended over a long period of time.

lunar soil see LUNAR REGOLITH.

lustre *n.* the appearance of a MINERAL surface in reflected light. Terms used to describe the quality of lustre include ADAMANTINE, VITREOUS, silky, metallic, pearly, resinous and dull. The intensity of the reflection can be described as SPLENDENT, shining, glistening or dull. Fresh unweathered surfaces should be used.

lutaceous *adj.* **1.** (of a SEDIMENTARY ROCK) formed from mud. Compare ARGILLACEOUS. See LUTITE.
2. (of the texture of such a rock) pertaining to a LUTITE.

Lutetian *n.* a STAGE of the Middle EOCENE.

lutite *n.* a general name for any fine-grained SEDIMENTARY ROCK composed of material that was once mud, such as MUDSTONE or SHALE. More specifically, the proportion of SILT contained should be between one-third and two-thirds of the total. The terms lutite, claystone, mudstone, siltstone and shale often overlap in usage. Compare PELITE. See also ARENITE, RUDITE. If the sedimentary rock consists of detrital calcite particles more than 50 per cent of which are silt or clay size, it is a CALCILUTITE. Compare CALCISILTITE.

luxullianite *n.* a rock formed by the pneumatolytic alteration of GRANITE, with the introduction of boron. The process is known as *tourmalinisation*. Luxullianite is characterised by radiating clusters or ACICULAR black TOURMALINE crystals (SCHORL), QUARTZ and corroded reddened FELDSPAR. *Roche rock*, a distinctive black and white rock composed of randomly oriented tourmaline in a mosaic of quartz, is formed by the same process. See PNEUMATOLYSIS, see also GREISEN.

L wave see SEISMIC WAVES.

Mm

Ma *abbreviation for* 10^6 years (one million years).

maar *n.* a shallow VOLCANIC CRATER formed as the result of the explosive interaction of MAGMA with GROUNDWATER or surface water. They range in diameter from a few hundred metres to about three kilometres. Most maars are the result of single explosive eruptions, have low-rimmed cones and the crater has a depth to width diameter ratio of about 1:5 when formed. With time the crater is filled with EPICLASTIC deposits. Craters that intersect the WATER TABLE may form lakes; many examples of these occur in the Eifel area of Germany. Some maars may be the surface manifestation of DIATREMES, but the underlying volcanic pipe is exposed only by deep erosion.

Maastrichtian *n.* the top STAGE of the Upper CRETACEOUS.

maceral *n.* a petrological compositional unit of COAL, identified microscopically by thin or polished sections. Macerals are related to coal as minerals are related to rock. There are three groups of coal macerals, classified on their appearance and biological affinities. The macerals collinite and tellinite belong to the *vitrinite group* and consist of woody tissue such as stems, bark, roots and twigs. Vitrinite, formed from bark, is the main constituent of the LITHOTYPE vitrain. Alginite, sporinite, resinite and cutinite make up the *liptinite* (or *exinite*) *maceral group*. They consist of material from ALGAE, resins, leaf cuticles and the outer cases of spores and pollens. Liptinite is an important constituent in the lithotypes durain and clarain. The *inertinite group* consists of the macerals macrinite, micrinite, semifusinite, fusinite, sclerotinite and inertodetrinite. They consist mainly of organic materials similar to those in vitrinite but have been altered under oxidising conditions. Fusinite is the principal constituent of the lithotype fusain. Compare PHYTERAL.

macro- *prefix meaning* large or great.

macrocrystalline *adj.* (of a rock texture) in which the CRYSTALS are clearly visible to the unaided eye or with the use of an ordinary hand lens. Compare MICROCRYSTALLINE.

macropygous see PYGIDIUM.

maculose *adj.* (of METAMORPHIC ROCKS) spotted or knotted. Maculose texture is commonest in thermally metamorphosed rocks and results from the local growth of new minerals. See also PORPHYROBLASTS, METAMORPHISM.

madreporite *n.* the largest of the genital plates in the APICAL DISC of an ECHINOID. It has many small holes that lead into the WATER-VASCULAR SYSTEM. See Fig. 27 (page 158).

Maentwrogian *n.* the bottom STAGE of the Merioneth SERIES (Upper CAMBRIAN).

mafic *n.* (*petrology*) *an acronym*, derived from *ma*gnesium + *f*erric + *ic*, and equivalent to subsilicic or BASIC. It pertains to magnesium and iron-rich minerals such as OLIVINE, PYROXENE, AMPHIBOLE, BIOTITE, etc, and the IGNEOUS ROCKS with a high modal content of these minerals. It is the complement of FELSIC. In general, it is synonymous with DARK MINERALS. Compare FEMIC, SALIC. See also MESOCRATIC, MELANOCRATIC.

magma *n*. high-temperature molten rock, usually of silicate composition, containing volatile components in solution under pressure, especially water, carbon dioxide and hydrogen sulphide, with lesser amounts of many gases such as nitrogen and the inert gases. Magma is generated deep within the Earth's CRUST, or upper MANTLE, as a result of partial melting and is the source of IGNEOUS ROCKS. Since the volatile components of a magma are usually lost during or after CONSOLIDATION, the igneous rock formed cannot be said to entirely represent the original magma. During the process of MAGMATIC DIFFERENTIATION, liquids of a wide range of compositions may be derived from an initially homogeneous magma. Magma that consolidates below the surface forms INTRUSIVE or PLUTONIC ROCKS. That which emerges above the surface and solidifies forms EXTRUSIVE or VOLCANIC ROCK. See also MAGMA CHAMBER.

magma chamber *n*. an underground reservoir filled with MAGMA from sources within the Earth's MANTLE or as the result of crustal melting. They commonly occur a few kilometres beneath major volcanoes and provide the immediate source of lava and related volcanic products erupted at the surface.

magmatic differentiation *n*. the processes that cause an initially homogeneous body of MAGMA to evolve into magmas of different chemical composition, thus developing series of chemically related IGNEOUS ROCK from a common magma. The most important

processes can be brought together under the heading *crystal fractionation*; this depends on the complex way in which crystals separate from silicate melts. In the case of a basic magma, with a basaltic bulk composition, high-temperature phases such as OLIVINE crystallise first. As the magma cools, lower-temperature phases including PYROXENES and FELDSPARS are precipitated. In addition, the composition of the individual phases varies as the temperature drops and early-formed phases like olivine may be partly or completely resorbed depending on the SILICA SATURATION of the magma (see REACTION SERIES). MAGMA CHAMBERS are generally thought to cool largely by losing heat through the roof. This will set up thermal gradients within the chamber, resulting in movement of the magma in CONVECTION currents. Once crystals are formed, they will be carried around the magma chamber and may be extracted either by settling out on the chamber floor or sticking to its walls or roof. As a result, the composition of the residual liquid is changing constantly as a magma chamber cools so that if magma is drawn off during the process it will have a different composition, generally enriched in silica, than the original magma. Magmas drawn off at successive stages in the cooling process will be successively enriched in silica to give a series of related lava types, such as BASALT, HAWAIITE and TRACHYTE. The process is further complicated if the magma chamber is periodically replenished with fresh basaltic magma from the mantle. The separation of crystals from liquid is aided by mechanisms such as *gravitational settling* of denser phases, FILTER PRESSING, where interstitial liquid is squeezed out of the mush of crystals accumulating on the magma chamber floor because of the weight of the overlying mush and the tendency for the magma to become less dense as it differentiates (early-formed phases are in general denser than the liquid from which they are formed). The path that differentiation takes is also influenced by the depth at which it is taking place, as pressure has an influence on the order in which phases crystallise. For example, under high-pressure conditions augite can crystallise in advance of olivine. Evidence of *flow differentiation* has been observed in some PORPHYRITIC dykes. It was observed that PHENOCRYSTS tend to be concentrated in the centre of the INTRUSION where the magma flows fastest but may have a tendency to sink when the flow slows down so that the lower part of the intrusion is enriched in phenocrysts.

Differentiation resulting from LIQUID IMMISCIBILITY involves the separation of two compositionally distinct liquid phases from an initially homogeneous melt. Immiscibility between sulphide liquids and silicate liquids appears to be an important stage in the formation of many magmatic nickel-sulphide ore deposits. There is also evidence for the formation of immiscible silicate liquids at a late stage in the crystallisation of a magma, as can be seen in some lunar mare BASALTS and in basalts from the Deccan province of India. However, liquid immiscibilty is not considered to be an important process in the differentiation of most silicate magmas. See ASSIMILATION, CUMULATE, HYBRID, LAYERED INTRUSION, ZONE MELTING.

magmatic ore deposit *n.* MINERAL DEPOSITS that formed as a result of MAGMATIC DIFFERENTIATION processes. CHROMITE, ILMENITE and MAGNETITE commonly occur as layers associated with BASIC and ultrabasic IGNEOUS ROCKS in differentiated INTRUSIONS such as the Bushveld Complex in South Africa. Nickel sulphide deposits (which also contain copper, iron and platinum-group elements) may also be found in layered intrusions and associated with ultrabasic lava flows (KOMATIITES).

magmatic stoping see STOPING.

magmatic system *n.* a conceptual body of MAGMA, composed of one or several components, that is used in model systems for the determination of crystal-melt equilibria.

magmatic water see JUVENILE WATER.

magmatism *n.* **1.** the development and movement of MAGMA and its solidification into IGNEOUS ROCK.
2. the theory that much of the Earth's GRANITE has been derived through CRYSTALLISATION from MAGMA rather than through GRANITISATION, which is a metamorphic process.

magnafacies *n.* a major continuous belt of deposits that show similar lithological and palaeontological characteristics. It extends through several chronostratigraphical units and represents a distinctive depositional environment that prevailed with a shifting of geographical placement over time.

magnesian limestone see LIMESTONE.

magnesiowüstite *n.* a high-pressure MINERAL, $(Mg,Fe)O$, found as INCLUSIONS in DIAMONDS. It is considered to occur in the Earth's lower MANTLE.

magnesite *n.* a carbonate mineral of the hexagonal system, $MgCO_3$, usually found as compact white, yellow or grey porcelain-like masses. It is formed by the alteration of ULTRAMAFIC rocks through the action of waters containing carbonic acid. As a diagenetic mineral, it replaces calcite and dolomite.

magnetic anomaly *n*. a departure from the expected value of the Earth's magnetic field over a point of the Earth's surface after adjustments have been made for local inhomogeneities. A magnetic anomaly generally indicates the presence of some object with either remanent magnetisation or high MAGNETIC SUSCEPTIBILITY (the measure of the response of the material to a magnetic field) beneath the anomalous area. A change in rock type or the presence of a magnetic ore body may be indicated by such anomalies. Iron ore deposits are often detected by magnetic prospecting methods.

The continental pattern of anomalous magnetic values is, in general, irregular. The ocean is chiefly dominated by linear magnetic anomalies related to the MID-OCEAN RIDGE system. Basaltic submarine lava flows forming the OCEANIC CRUST, which typically show a high ratio of remanence to induced magnetisation, appear to be responsible for most oceanic anomalies the features of which are symmetrically arranged parallel with the axis on both sides. According to the theory of SEA-FLOOR SPREADING, these parallel anomalies are underlain by zones of normally and reversely magnetised basaltic rocks that are created at the mid-oceanic ridge axis and acquire a thermoremanent magnetisation as they cool. The magnetic polarity depends upon the polarity of the geomagnetic field at the time of cooling. Observed magnetic profiles are calibrated by matching with computer-simulated anomaly profiles. In the assignment of ages to magnetic anomalies, it is assumed that the distinctive magnetic anomaly associated with the *Gauss normal epoch* (3.40 Ma–2.49 Ma) can be matched in the simulated PROFILES. The spreading rate since then can thus be determined by measuring the distance of this anomaly feature from ridge crests. See also GEOMAGNETIC POLARITY REVERSAL, NATURAL REMANENT MAGNETISATION, GEO-PHYSICAL EXPLORATION, PALAEOMAGNETISM.

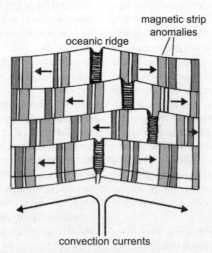

Fig. 57. **Magnetic anomalies.** Alternating strips of normal and reversely magnetised oceanic crust on either side of the mid-oceanic ridge

magnetic declination *n.* the acute angle in degrees between the directions of magnetic and geographical north.

magnetic dip see MAGNETIC INCLINATION.

magnetic domain *n.* a sub-region within a CRYSTAL within which the electron spins of the atoms are aligned in one direction, giving a net magnetic effect.

magnetic equator or **aclinic line** *n.* that imaginary line on the Earth's surface where the magnetic needle remains horizontal, i.e. where the magnetic inclination is zero. Compare AGONIC LINE.

magnetic field *n.* the region of magnetic influence that surrounds the poles of a magnet or a moving charge.

magnetic field intensity or **magnetic field strength** *n.* the force exerted by the MAGNETIC FIELD on a magnetic substance.

magnetic field reversal see GEOMAGNETIC POLARITY REVERSAL.

magnetic field strength see MAGNETIC FIELD INTENSITY.

magnetic inclination, inclination or **magnetic dip** *n.* the angle at which MAGNETIC FIELD lines dip. It is the angle that a compass needle free to rotate in a vertical plane makes with the horizontal. Inclination is positive if the north pole of the needle is below the horizontal and negative if it lies above the horizontal.

magnetic meridian see MAGNETIC NORTH.

magnetic north or **magnetic meridian** *n.* the uncorrected direction indicated by the northerly directed end of a magnetic compass needle; it is the northerly direction of the magnetic meridian at any specified point. Compare TRUE NORTH.

magnetic permeability *n.* the ratio of the magnetic flux density (B) to the magnetising field intensity (H), $\mu = B/H$ (absolute permeability). Units are henry per metre. μ_0 is the magnetic constant, i.e. the permeability of free space. Relative permeability (μ_r) = μ/μ_0.

magnetic prospecting see MAGNETIC ANOMALY.

magnetic pyrites see PYRRHOTITE.

magnetic storm *n.* a disturbance of the Earth's MAGNETIC FIELD, thought to be caused by ionised particles ejected by solar flares. Magnetic prospecting is usually suspended during such periods, particularly if the disturbance is intense.

magnetic susceptibility *n.* the ratio of the strength to which a substance has been magnetised (induced magnetisation) to the strength of the MAGNETIC FIELD causing the magnetisation.

magnetite *n.* an isometric, strongly magnetic mineral, $Fe^{2+}Fe^{3+}_2O_4$, in which there can be some substitution of Mg and Mn^{2+} for Fe^{2+} and Al, Cr, Mn^{3+} and Ti^{4+} for Fe^{3+}. Although this important ore of iron can be found in the form of octahedral steel-black crystals with a metallic lustre, it occurs more commonly as compact and granular masses. It is a common accessory mineral in igneous rocks and can be concentrated by magmatic segregation forming large ore bodies, often with a high titanium content, and is plentiful in contact metasomatic conditions.

magnetometer *n.* any of many types of instrument that measure the magnitude and changes of the Earth's magnetic field.

magnetopause *n.* the boundary between the MAGNETOSPHERE and the interplanetary medium.

magnetosphere *n.* the outer region of the Earth's magnetic field in which magnetic forces predominate over gravity in the motion of charged particles. It is an asymmetric configuration, compressed on the sunward side of the Earth and 'pluming' far outwards on the opposite side.

magnetostriction *n*. the elastic strain that accompanies magnetisation.

maidenhair tree see GINKGO.

malachite *n*. a banded emerald-green monoclinic mineral, $Cu_2CO_3(OH)_2$, a minor ORE of copper. It commonly occurs as a green film on other copper minerals and as botryoidal or reniform masses. Malachite is typically found in the oxidation zones of copper deposits, where it is produced by a sulphide-carbonate GANGUE reaction; it is valued as a gemstone and as a decorative mineral.

malleable see TENACITY.

Malm *n*. the Late JURASSIC epoch.

mamelon see TUBERCLE.

mamillated *adj*. (of minerals such as LIMONITE or MALACHITE) having large mutually interfering spheroidal surfaces.

mammoth *n*. a form of Pleistocene elephant of the genus *Mammuthus*. Mammoths were adapted to the cold climates of the Pleistocene, some having a covering of woolly hair.

manganese nodule *n*. an irregularly shaped mass, rich in manganese and iron, that occurs on the ocean floor in areas of slow sedimentation and also where RED CLAY deposits are located. Nodules commonly found at the surface of pelagic sediments range from 0.5 to 25 centimetres but generally average about 3 centimetres in diameter. Those from different parts of the ocean tend to have unique physical characteristics; all types, however, consist of many minerals and crystallites and lack an overall crystalline structure. Nodule minerals are derived from the interaction of hot upwelling metal-bearing solutions at a plate boundary with organic sediments of the ocean itself. Precipitation from the mineral solutions about some centre of accretion (e.g. a sand grain) is the apparent mechanism of nodule formation. The composition of manganese nodules shows definite regional variation. Mn/Fe ratios less than 1 are characteristic near the continents. In areas farthest from land, nodules show high nickel and copper content. In other regions they are relatively rich in cobalt. In areas of concentration, the economic potential of manganese nodules can be significant.

manganite *n*. a hydrated manganese oxide mineral, MnO(OH), of the monoclinic system. It is opaque except in minute splinters and has a sub-metallic lustre. Manganite occurs in low-temperature hydrothermal veins associated with CALCITE and BARITE.

mantle *n*. the zone within the Earth that extends below the CRUST to the CORE. The upper boundary is marked by the MOHOROVICIC DISCONTINUITY and the boundary between mantle and core by the GUTENBERG DISCONTINUITY. The upper mantle extends to a depth of 400 kilometres and is characterised by regional dissimilarities in seismic velocity PROFILE. A transition zone, extending downwards from about 400 to 1000 kilometres, is characterised in general by a rapid increase of velocity with depth. The lower mantle, between 1000 and 2900 kilometres, shows rather gradual seismic velocity increases with depth. Massive convection currents in the mantle are energised by the heat of decay of radioactive isotopes. Although it appears that this convective circulation is the agent for plate movement, much of it is not related to the movement or boundaries of the plates. (See CONVECTION, CONVECTION CELL.)

A seismic P-wave velocity of 8.2 (± 0.2) kilometres per second in the uppermost mantle, together with certain broad mineralogical and petrological constraints, indicate some combination of OLIVINE, PYROXENE, GARNET and, in restricted regions, AMPHIBOLE. PERIDOTITE (olivine-pyroxene) and ECLOGITE (pyroxene + garnet) are the two principal rock types bearing these minerals. Geophysical evidence on the density of the upper mantle favours a density close to that of peridotitic composition. A study of KIMBERLITE pipes in Africa and Siberia has revealed that the pipes carry inclusions of crustal rocks that are known to have been entrained during rapid upward transport of the magma. These pipes also contain INCLUSIONS of rocks not occurring in the vicinity, especially peridotites and eclogites, which are thought to have been brought up from depths of 100 to 200 kilometres, within the upper mantle.

In many oceanic regions, the upper 70 kilometres of the upper mantle show an S-wave velocity of about 4.6 kilometres per second, which decreases suddenly to 4.2 kilometres per second at a depth of about 100 kilometres. The zone of decreased velocity extends to about 150 to 200 kilometres; a corresponding low-velocity zone for P waves is also believed to occur. It is thought that the observed magnitude of the velocity change suggests a small degree of partial melting in the region. It was demonstrated that the density in the transition zone (400–1000 kilometres) increased much more rapidly with depth than would be expected in a uniform layer. It is thought that abnormal density changes in this zone were caused by a sequence of major phase changes. Studies of the density

distribution in the lower mantle and of shock-wave data on the densities of silicates at comparable pressures indicate the density of the lower mantle to be about 5 per cent higher than that of the composition that was assumed. A possible explanation for this is that it is composed mainly of PEROVSKITE and MAGNESIOWÜSTITE, a mineral assemblage that is intrinsically denser than the overlying assemblage is. See also CRUST, CORE, SEISMIC WAVES.

mantle plume n. a persistent column of MAGMA rising upwards from the Earth's MANTLE to the CRUST. Its surface expression is in long-lived VOLCANIC activity not directly related to SUBDUCTION or SEA-FLOOR SPREADING. See also HOT SPOTS.

marble n. a fine- to coarse-grained METAMORPHIC ROCK consisting mainly of recrystallised CALCITE and/or DOLOMITE. It is a metamorphosed LIMESTONE. Colourful streaks in marble are the result of impurities such as quartz or dolomite in the original limestone, which result in the formation of minerals such as forsterite (or serpentine). Purer varieties, without inclusions, are valued by sculptors; veined varieties are used as ornamental stone.

marcasite n. a pale-coloured, bronze-yellow orthorhombic sulphide mineral, FeS_2, that darkens in colour on exposure to air. Dimorphous with PYRITE, it occurs as massive aggregates in low-temperature hydrothermal veins and as concretions in sedimentary environments. Jewellery and trinkets made of marcasite were once popular.

mare n. (pl. **maria**) one of several large level low-lying areas on the surface of the Moon. The maria show fewer large craters than the uplands; they are composed mainly of flood basalts, probably erupted very rapidly and in flows of great extent. Although similar

287

in some respects to terrestrial basalts the lunar ones have very high titanium content, are relatively depleted in the volatile elements and highly depleted in europium. See LUNAR BASALT.

marginal basin see BACK-ARC BASIN.

marine abrasion *n.* **1.** the EROSION of a BEDROCK surface by the movement of sand that is agitated by waves.
2. the EROSION of SUBMARINE CANYONS by the gravity-related downslope movement of sediments.

marine band *n.* a horizon containing marine FOSSILS within a succession of non-marine strata, representing a brief marine TRANSGRESSION. Marine bands have been used to correlate strata in CARBONIFEROUS successions in Britain and mainland Europe. See also MARKER HORIZON.

marine geology or **geological oceanography** *n.* the study of the ocean floor and its many features. It includes submarine relief topography, petrology, the geochemistry of ocean-floor rocks and sediments, and the effect of waves and sea water on the ocean bottom.

marine terrace *n.* a relatively narrow coastal strip that is formed of deposited material sloping seawards. The slope just suffices to keep the material in slow seaward transit.

marine transgression see TRANSGRESSION.

marker horizon *n.* a thin layer within a succession of BEDS that has distinctive characteristics such as a particular FOSSIL ASSEMBLAGE or an unusual LITHOLOGY, which means it can be used for stratigraphical correlation. *Marker beds* are normally assumed to be isochronous, i.e. everywhere of the same age, but this may not always be the case.

As well as the more usual type of marker horizon, for example a MARINE BAND or a BENTONITE bed representing a

layer of VOLCANIC ash, beds that produce a characteristic set of seismic reflections or with distinctive electrical or magnetic characteristics can also be used as marker horizons.

marl *n.* a friable mixture of subequal amounts of micrite and clay minerals; a calcareous clay. Compare MARLSTONE.

marlstone *n.* an indurated rock of about the same composition as MARL, i.e. an earthy limestone; it is less fissile than SHALE.

Marsdenian *n.* a STAGE of the Namurian (CARBONIFEROUS) in Britain and western Europe.

marsh *n.* a type of wetland characterised by poorly drained mineral soils and by plant life dominated by grasses, sedges, reeds and rushes. The latter feature distinguishes a marsh from a SWAMP, where trees are the dominant plant life. Marshes are common at the mouths of rivers, especially where large deltas have formed. The delta soil favours the growth of fibrous-rooted grasses, which bind the muds, thus hindering water flow and promoting the spread of both DELTA and marsh. Marshes may also be found in areas where depressions have been left by retreating glacial ice. *Salt marshes* are flat intertidal areas that are periodically flooded by sea water. Salt-marsh grasses will not grow in permanently flooded flats. A *tidal marsh* is a marshy TIDAL FLAT bordering a sheltered coast that is regularly inundated during high tides. It is formed of mud and the roots of salt-tolerant plants. Compare BOG.

marsh gas *n.* methane (CH_4) produced as a decomposition product of the decay of vegetable substances in stagnant water.

mascon *n.* a large, high-density mass concentration beneath the lunar surface.

maskelynite *n.* a natural GLASS produced by SHOCK METAMORPHISM of plagioclase FELDSPAR. It was first described in the Shergotty ACHONDRITE meteorite.

massif *n.* a large elevated feature, usually in an OROGENIC BELT, differing topographically and structurally from the lower adjacent TERRANE. It usually consists of more than a single isolated summit and exhibits no extensive level areas. The rocks of which a massif is formed are more rigid than those of its surroundings and are usually older; they are generally thought to be rather complex structurally. Rocks forming a massif may be the mark of previous deformations and not necessarily related to its development as a massif

massive *adj.* **1.** (of homogeneous rock bodies) having great bulk. BATHOLITHS are examples of massive PLUTONS.
2. (of rocks) having homogeneous structure or texture.

mass spectrometry see Appendix 4.

mass transport *n.* the carrying of material in a moving medium, such as air, water or ice.

mass wasting *n.* a general term for the bulk, downslope transfer of masses of rock debris under the direct influence of gravity. No other medium is an agent. *Downwasting* is the thinning of a GLACIER by ABLATION.

matrix *n.* the groundmass of an IGNEOUS ROCK or the finer-grained material enclosing larger grains in a SEDIMENTARY ROCK. The term also applies to the background material in which a FOSSIL is embedded.

mature *adj.* **1.** pertaining to a particular point in the cycle of EROSION or to a stage in the development of a stream or SHORELINE.
2. (of a clastic sediment) that has evolved from its parent rock and is now characterised by stable minerals

that are well-ROUNDED and SORTED. See TEXTURAL MATURITY.

md *abbreviation for* millidarcy (see DARCY).

M-discontinuity see MOHOROVICIC DISCONTINUITY.

meander 1. *n.* a curve or loop in a stream channel.
2. *v.* to flow in a winding manner.

Many streams meander over part of their course and are relatively straight for other distances. Although some streams follow meandering courses cut in bedrock (see INCISED MEANDER), most are on FLOODPLAINS. Meandering streams have fairly light bed loads and banks that are much steeper than those of BRAIDED STREAMS. Most are close to BASE LEVEL, and only by becoming more sinuous can they lower their gradients. Others not near base level become sinuous because of difficulty in carrying fine sediments when there is no bedload. As water rounds any bend, its inertia carries it towards the concave bank, causing a slight uptilting of the water surface; this tilt tends to make the stream rush towards the opposite bank as it flows downstream, and the curvature of the bank tends to reverse. Thus the outside bank of each curve is subject to EROSION at the same time as the deficiency of water on the inner bank results in deposition. The combination of these two actions compels meander bends to migrate. Where a meander is cut off in the normal process of erosion, the remnant bank is known as a *cut-off spur* (NECK CUT-OFF). See also OXBOW LAKE.

meander cut-off see MEANDER, NECK CUT-OFF.

mechanical twinning see DEFORMATION TWINNING.

mechanical weathering or **physical weathering** *n.* weathering processes that are physical in nature and result in the fragmentation of rock without a chemical change, e.g. FROST ACTION, thermal expansion. Compare chemical weathering (see WEATHERING).

medial moraine see MORAINE.

median rift *n*. the submarine rift valley on the crest of a MID-OCEANIC RIDGE.

Medinan see ALEXANDRIAN.

medium-grained *adj*. **1.** (of an IGNEOUS ROCK) consisting of crystals that have an average diameter in the range of 1 to 5 millimetres.
2. (of a SEDIMENTARY ROCK) consisting of individual particles that average $1/16$th to 2 millimetres in diameter.

megabreccia *n*. a very coarse BRECCIA in which the larger fragments are over a kilometre long. The term 'megabreccia' is a descriptive term without genetic implication. Some megabreccias may be formed by landslides. See also MÉLANGE, OLISTHOSTROME.

meimechite or **meymechite** *n*. a variety of serpentinite with abundant (usually altered) olivine phenocrysts considered by Russian petrologists to be the equivalent of KIMBERLITE. See also PICRITE.

mélange *n*. a chaotic mixture of blocks and fragments of rock of a variety of compositions, both local and exotic, in a MATRIX that is usually composed of fine-grained sedimentary material but may be igneous or metamorphic. The rock fragments range in size from a few centimetres up to slabs several kilometres long. If the whole formation shows evidence of intense deformation, with folding and shearing of the matrix and/or blocks, it is known as a *tectonic mélange*. These are thought to be formed by the disruption of rock sequences in large-scale imbricate faulting or gravitational gliding (see IMBRICATE STRUCTURE).

A mélange formed mainly by

sedimentary processes is known as an OLISTHOSTROME. It can be difficult to distinguish between the two.

melanocratic *adj*. (of IGNEOUS ROCKS) dark-coloured, containing 60 to 100 per cent of DARK MINERALS, i.e. rocks with a COLOUR INDEX between 65 and 90. Compare LEUCOCRATIC, MESOCRATIC.

melilite *n*. a tetragonal mineral of the åkermanite–gehlenite series with the general formula $(Ca,Na)_2(Mg,Al)(Si,Al)_2O_7$. Melilite occurs in silica-UNDERSATURATED basic LAVA FLOWS.

melt *n*. (*petrology*) a fused liquid ROCK.

meltwater *n*. water resulting from melting ice and snow. Its geological effects include the depositing of drift, to form ESKERS and KAMES, and the formation of potholes. Glacial melt-water streams are often greyish-white because of a suspension of ROCK FLOUR resulting from glacial abrasion.

member *n*. a rock STRATIGRAPHICAL UNIT comprising a named lithological entity within a formation. It may be a FORMAL UNIT, INFORMAL UNIT or unnamed. Compare LENTIL, TONGUE.

Menevian *n*. the top STAGE of the St David's SERIES (Middle CAMBRIAN).

Meramecian *n*. the lower part of the Upper Mississippian of North America. It corresponds to the middle part of the VISÉAN series of Europe.

meraspid see PROTASPIS.

Mercalli scale see EARTHQUAKE MEASURE-MENT.

mercury *n*. a heavy, silvery-white liquid, the native metallic element Hg. Natural mercury is found as tiny globules in cinnabar or deposited in certain hot springs.

merocrystalline see HYPOCRYSTALLINE.

mesa *n*. an isolated flat-topped hill bounded on at least one side by a steep cliff and having an extensive summit area. Mesas are erosional remnants that persist because of a protective cover of more resistant sedimentary rocks, especially sandstones, or of lava flows or gravels. They range from 30 to 600 metres in height, and from a few hundred metres to several kilometres in length. Compare BUTTE. See HOGBACK.

meso- *prefix meaning* middle.

mesocratic *adj.* (of IGNEOUS ROCKS) containing 30 to 60 per cent DARK MINERALS, i.e. with a COLOUR INDEX lying between LEUCOCRATIC and MELANOCRATIC.

mesocumulate see CUMULATE.

mesosiderites *n.* a miscellaneous group of STONY IRON METEORITES; the metal content may exceed 40 per cent but does not form a continuous network.

mesothermal deposit see HYDROTHERMAL DEPOSIT.

Mesozoic Era *n.* the Mesozoic ('middle life') Era was a span of geological time between 245 and 65 Ma. It includes the TRIASSIC, JURASSIC and CRETACEOUS periods.

mesozone see INTENSITY ZONE.

Messinian *n.* the top STAGE of the MIOCENE.

meta- *prefix denoting* a metamorphosed igneous or sedimentary rock in which the original texture is still recognisable, e.g. metasediment, metabasalt.

meta-anthracite *n.* the highest rank of ANTHRACITE coal; it has a fixed carbon content of 98 per cent or greater.

metabentonite see BENTONITE.

metacryst see PORPHYROBLAST.

metallic bond *n.* the attraction between metal CATIONS (positive IONS) and the cloud of free electrons that characterises the structure of metallic crystals. The metal cations are closely packed and the electron cloud permeates the structure. The mobility of the electrons explains the high thermal and electrical CONDUCTIVITY of metals.

metallogenic *adj.* mineralising, especially ore-depositing. *Metallogenic epochs* are units of geological time during which particular mineral deposits were formed. A *metallogenic province* is a region in which there occurs a particular series of minerals of certain types of mineralisation, e.g. the gold-quartz veins of California.

metallogenic epoch see METALLOGENIC.

metallogenic province see METALLOGENIC.

metamict *adj.* (of a MINERAL) wherein original crystal structure has been disrupted by varying degrees by radiation from radioactive elements (e.g. uranium or thorium) contained in that mineral while the original external crystal form is retained. Consequently, in the extreme case, a metamict mineral has no CLEAVAGE and does not diffract X-rays. In many of these minerals, the crystallinity is restored by prolonged heating at a temperature below the decomposition point. Examples occur in thorite, zircon and many other minerals.

metamorphic differentiation *n.* a process in which noticeably heterogeneous rock is created with layers or pods of contrasting mineralogical composition from an initially homogeneous body during METAMORPHISM.

metamorphic facies *n.* a classification of METAMORPHIC ROCKS depending on the supposed temperature and pressure conditions of their formation, implemented according to the characteristic minerals developed in rocks of a given chemical composition under given metamorphic conditions. Thus each facies should be represented by a collection of MINERAL ASSEMBLAGES reflecting the different chemical compositions of the parent rocks. It is possible to define many sub-facies, but most petrologists use a small number of major facies that represent ranges of temperature and pressure. The facies

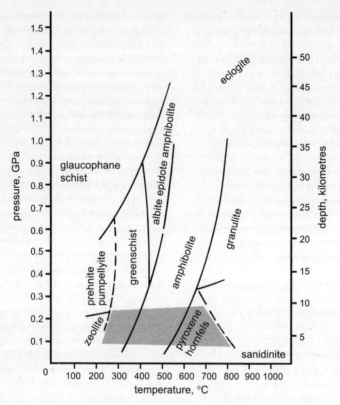

Fig. 58. **Metamorphic facie**s. The approximate pressure and
temperature ranges corresponding to the different metamorphic facies

were defined originally on the basis of
the typical mineral assemblages
observed in rocks of BASIC igneous
composition. See Fig. 58 overleaf.

metamorphic grade *n*. rocks that have
been metamorphosed under the same
conditions of temperature and pressure
may be said to have the same metamorphic grade. See also ISOGRAD,
METAMORPHIC ZONE.

metamorphic rock *n*. any of a class of
ROCKS that are the result of partial or
complete recrystallisation in the solid
state of pre-existing rocks under
conditions of temperature and pressure

that are significantly different from
those obtaining at the surface of the
Earth. The mineralogy and structure of
a particular metamorphic rock reflect
the type of METAMORPHISM that produced
the rock. Texture is a function of
change in the interrelationship between
grains grown in the solid state as
opposed to liquid crystallisation. The
shape of new crystals may be regular
or irregular, with crystal faces that
depend on the mineral type and the
presence or absence of stress. Prismatic
or platy minerals may develop a
PREFERRED ORIENTATION in response to

directed stress. A RELICT texture is one that is inherited from the pre-existing rock.

Metamorphic rocks formed by contact metamorphism are associated with an igneous INTRUSION and are limited in extent; pressure gradients are absent. The type of metamorphic rock formed is constrained by the composition of the original rock. Sandstones composed dominantly of quartz grains are recrystallised to form quartzite; limestones are converted to marble. Country rocks of pelitic composition close to the contact with the intrusion may have recrystallised completely to a tough, fine-grained rock of flinty appearance known as HORNFELS. These typically have an even-grained, GRANULAR TEXTURE. Farther from the intrusion the rocks may have a MACULOSE (spotted) texture where new minerals have grown.

The most abundant metamorphic rocks are those produced by *regional metamorphism*. Typical rocks of this type are schists and gneisses. Examination of a sequence of these metamorphic rocks shows that in addition to mineralogical changes there is a general pattern in the textures or structures of the rocks. For example, the grain size of the rocks tends to increase with increasing METAMORPHIC GRADE. In pelitic rocks, a sequence of increasing metamorphic grade may be traced from SHALE through SLATE, then PHYLLITE, SCHIST, GNEISS and eventually GRANULITE and MIGMATITE. In other sequences, quartz sandstones become quartzites, limestones become marbles, and basic igneous rocks become greenschist, amphibolites and, ultimately, eclogites. The appearance of INDEX MINERALS (biotite, garnet, staurolite, kyanite or sillimanite) indicates progressive changes in metamorphic conditions producing a series of METAMORPHIC ZONES.

metamorphic zone *n.* an area of METAMORPHIC ROCKS delimited by the appearance of certain minerals (INDEX MINERALS) in rocks of a particular composition. In any region of metamorphic rocks it may be possible to distinguish a regular arrangement of zones defining variations in the metamorphic conditions in that area. One example would be the zones showing a progressive decrease in temperature away from an INTRUSION within a contact metamorphic AUREOLE. Barrow's zones are another example (see BARROVIAN-TYPE METAMORPHIC SEQUENCE). See also METAMORPHISM.

metamorphism *n.* the processes that produce structural and mineralogical changes in any type of rock in response to physical and chemical conditions differing from those under which the rocks originally formed. Changes brought about by near-surface processes such as DIAGENESIS and WEATHERING are excluded.

Metamorphism takes place in the solid state and involves the growth of new minerals (*neomineralisation*), the recrystallisation of minerals already present, and the formation of new textures and structures within the rock. Metamorphic changes that are brought about by a rise in temperature and pressure are referred to as *prograde metamorphism*. Compare RETROGRADE METAMORPHISM. Most metamorphic processes occur with little change in the bulk composition of the rock (see ISOCHEMICAL METAMORPHISM). For metamorphism involving the migration of chemical components, see METASOMATISM.

The nature of the metamorphic processes and the type of metamorphic rock produced are controlled by interactions between temperature, pressure and shearing stress and the parent rock. Three principal types of metamorphism may be defined:

(a) *dynamic metamorphism* (also known as *dislocation metamorphism* or *cataclastic metamorphism*) is controlled chiefly by shearing stress with little long-lasting temperature change. It occurs in zones of intense deformation, such as fault zones. Mechanical DEFORMATION produces rocks ranging from BRECCIAS to MYLONITES (see also CATACLASIS, SHOCK METAMORPHISM);

(b) *contact metamorphism* (or *thermal metamorphism*) occurs primarily as a result of temperature increases under conditions of minor differential stress in a restricted area (the AUREOLE) surrounding a body of igneous rock. Because the extent of contact metamorphism is limited, the pressure is near constant. The resulting rocks are usually fine-grained and equigranular, typically HORNFELS;

(c) *regional metamorphism* results from the action of increased temperature and pressure over a wide area and is characteristic of orogenic belts. It may be subdivided into different pressure-temperature conditions according to observed sequences of mineral assemblages (see METAMORPHIC FACIES);

(d) SHOCK METAMORPHISM, where changes in rocks and minerals are brought about extremely rapidly as the result of the ultra-high pressures associated with meteorite impacts, for example the formation of diaplectic glass where the transformation from solid to glass takes place without going through a melting stage.

See also HYDROTHERMAL ALTERATION, METAMORPHIC ROCK, AUTOMETAMORPHISM, POLYMETAMORPHISM.

metaquartzite *n.* a QUARTZITE formed by metamorphic recrystallisation, as differentiated from an ORTHOQUARTZITE, the crystallinity of which is of diagenetic origin.

metasomatism *n.* a metamorphic process whereby existing minerals are transformed totally or partially into new minerals by the REPLACEMENT of their chemical constituents. This occurs by the introduction of capillary solutions, originating either outside or within the rock, that are reactive with the existing minerals. In *contact metasomatism*, there is a mass change in the composition of rocks that are in contact with an invading MAGMA. Certain fluid constituents from the magma combine with the country rock constituents to form a new series of minerals. *Pyrometasomatism* is the formation of contact-metamorphic mineral deposits at high temperatures; the process involves replacement of the enclosing rock by the addition or subtraction of materials. See also GRANITISATION.

metastable *adj.* (of a mineral) existing under conditions outside its stability range.

meteoric water *n.* water that occurs in or is derived from the atmosphere.

meteorite *n.* any extraterrestrial solid mass that reaches the Earth's surface. Meteorites range in diameter from one millimetre to many kilometres. Particles less than one millimetre in size, referred to as *micrometeorites*, along with dust make up the bulk of extraterrestrial material that survives its passage through the atmosphere. Three classes of meteorites may be distinguished:

(a) IRON METEORITES (siderites), composed entirely of metal (iron-nickel), which are the ones most frequently found;

(b) STONY IRON METEORITES (siderolites) consisting of both metal (iron and nickel) and silicate;

(c) *stones* or STONY METEORITES (AEROLITES), composed mainly of silicate minerals, which are the meteorites most frequently observed to fall. They are further divided into CHONDRITES (containing CHONDRULES) and ACHONDRITES (without chondrules). More sophisticated systems of classification are used in modern scientific studies.

TEKTITES should not be regarded as a type of meteorite. They appear to originate as droplets of melted material from the surface of the Earth material sprayed out as the result of the impact of very large meteorites although it may be that some extraterrestrial tektites have been generated by impacts on the planet Mars.

methane *n*. an odourless, colourless inflammable gas, CH_4, the main constituent of MARSH GAS and the FIREDAMP of coal mines; it is obtained commercially from NATURAL GAS.

methylene iodide *n*. a liquid compound, CH_2I_2; specific gravity 3.32. It is used as a HEAVY LIQUID in flotation processes of ore separation.

miarolitic *adj*. 1. (of plutonic rock bodies, e.g. granite) containing irregular cavities into which EUHEDRAL crystals of the rock-forming minerals project. These crystals are generally larger than those in the main body of the rock. See DRUSY; compare AMYGDALOIDAL.

2. (*petrology*) (of small angular cavities in plutonic rocks) containing projecting crystals of the rock-forming minerals.

mica *n*. a group of monoclinic phyllosilicate minerals characterised by their platy habit, perfect basal CLEAVAGE and the elastic properties of the cleavage flakes (compare BRITTLE MICAS).

The basic structure of micas consists of pairs of sheets of SiO_4 tetrahedra with the general formula $(Si_4O_{10})_n$ that sandwich between them either a layer of cations (Fe^{2+}, Fe^{3+}, Mg,Al) plus $(OH)^-$ (known as a BRUCITE sheet) or a GIBBSITE sheet consisting of Al cations and OH anions. Each pair of silicate sheets is separated from the next by a weakly bonded layer of univalent cations, K^+ or Na^+ that forms the layer along which the mineral splits readily. The general mica formula is $X_2Y_{4-6}Z_8O_{20}$ $(OH,F)_4$ where X = K or Na, Y = Mg, F^{2+}, F^{3+} or Al, and Z = Si or Al. The principal mica minerals are MUSCOVITE, BIOTITE, PHLOGOPITE and LEPIDOLITE. Micas occur in a wide range of igneous and metamorphic rocks and some sedimentary rocks. See also GLAUCONITE.

micaceous *adj*. 1. consisting of or pertaining to MICA.

2. resembling a MICA, either in lustre or in having the property of being easily split into thin sheets.

mica plate see ACCESSORY PLATES.

micrite *n*. 1. the dull, semi-opaque to opaque microcrystalline matrix of limestones, composed of chemically precipitated carbonate sediment with crystals less than 5 microns in diameter. Micrite is much finer-textured than SPARITE.

2. a limestone composed mostly of micrite matrix and with less than 1 per cent allochems, e.g. lithographic limestone.

micro- *prefix meaning* small. When joined with a rock name, it usually signifies fine-grained.

microbialite *n*. an organosedimentary deposit produced by a benthic microbial community. Microbialites are differentiated from STROMATOLITES by their lack of lamination.

microbreccia *n*. intensely fractured

rocks consisting of angular fragments in a finer-grained matrix. They have no fluxion structure, unlike MYLONITES. Fine-grained (less than 0.2 millimetre average grain size) microbreccias in which the angular fragments form less than 30 per cent of the rock are called *cataclasites*.

microcline *n.* a mineral of the alkali feldspar group, $KAlSi_3O_8$. It can be red, yellowish, blue, green or white, and usually shows cross-hatched TWINNING under the microscope. Microcline is found as compact aggregates in granitic pegmatite and metamorphic rocks formed at medium to low temperatures. See also FELDSPAR.

microcoquina see COQUINA.

microcrystalline *adj.* (of a rock texture) comprising crystals that are visible only with the aid of a microscope. Compare MACROCRYSTALLINE.

microgabbro see DOLERITE.

microlite *n.* a very small crystal not easily seen without the aid of a hand lens or microscope.

micrometeorite see METEORITE.

micropygous see PYGIDIUM.

microseism *n.* in the broadest sense, ground motion that is not caused by earthquakes or explosions; storms at sea are the main source. There is not yet a generally accepted theory of microseisms.

Mid-Atlantic Ridge *n.* that portion of the MID-OCEANIC RIDGE feature that lies within the limits of the Atlantic Ocean.

mid-oceanic ridge *n.* a submarine feature associated with SEA-FLOOR SPREADING typified by the MID-ATLANTIC RIDGE. Other ridges occur in the Indian, Antarctic and Pacific Oceans, the Norwegian Sea and the Arctic Basin. Altogether the oceanic ridges extend for a total distance of over 40 000 kilometres (25 000 miles).

According to the theory of PLATE TECTONICS, when two crustal plates separate, basaltic material wells up through the spreading centre and produces a ridge. Rapid separation of the plates builds ridges with gentle slopes and broad elevations, such as those of the East Pacific Rise. The steep flanks of the Mid-Atlantic Ridge were formed by slow spreading. Mid-oceanic ridges are characterised by linear MAGNETIC ANOMALIES that are symmetrically distributed parallel to the ridge axis. Earthquakes at these extensional margins are shallow and distributed along a submarine rift valley that extends along the crest of the ridge. The rift is also the locus of the volcanic activity associated with sea-floor spreading. Many sections of the ridge show above average heat flow, over 200 mW m–2, compared with the average heat flow for the ocean basin of about 60 mW m–2. Mid-ocean ridge basalts (MORB) are predominantly low-potassium tholeiitic basalts, although alkaline and transitional basalts occasionally occur, associated with SEAMOUNTS, aseismic ridges and fracture zones. See TRANS-FORM FAULTS.

migmatite *n.* an intimate mixture of apparently igneous material of granitic composition and high-grade metamorphic rocks characterised by a banded or veined appearance, often accompanied by evidence of plastic deformation. Theories for the origin of migmatites include the ideas that the granitic material was introduced as magma (see INJECTION GNEISS) or through the action of fluids permeating the host rock (see GRANITISATION). In most cases, however, migmatite appears to represent the most extreme case of metamorphism in which the less refractory components of gneiss start to melt, leading to the segregation of veins and patches of granitic liquid under conditions of elevated temperature, pressure and shearing stress. Often SCHLIEREN are developed in which streaked-out portions of the host rock (*paleosome*) enclose melted material (*neosome*). See also PTYGMATIC STRUCTURES.

migration *n*. **1.** the movement of gas and oil from their beds of origin into RESERVOIR ROCKS via permeable formations or materials.
2. the change in position of the crest of a divide, directed away from a strongly eroding stream on a steep slope towards a less active stream on a gentler slope.
3. the gradual shifting downstream of a system of MEANDERS, accompanied by widening of the meander course and enlargement of its curves.
4. the apparent movement of a DUNE because of the continuous transfer of sand from its windward to its leeward side.
5. the passage of plants and animals from one place to another over long periods of time.

Milankovitch hypothesis *n*. the hypothesis linking the periodicity of the changes from GLACIAL to INTERGLACIAL climates during the QUATERNARY with variations in the amount of solar radiation reaching the Earth caused by geometrical variations in the Earth's orbit.

A record of climatic fluctuations can be obtained from SEDIMENTS in deep ocean cores, either from the nature of the FOSSIL ASSEMBLAGES, which reflects the temperature of the surface water, or from the oxygen-isotope content of the sediments. A high proportion of ^{18}O in deep-ocean sediments is correlated with low sea levels and a cold glacial climate, since ^{16}O is preferentially evaporated from sea water and concentrated in ice sheets.

Milankovitch identified three mechanisms affecting the amount of solar radiation reaching the Earth: (a) changes in the eccentricity of the Earth's orbit occur with a period of about 100 000 years; (b) the angle of the Earth's rotation axis to the ecliptic plane varies over a period of 41 000 years; (c) the spinning Earth 'wobbles' on its axis like an unstable top. This movement has a period of 21 000 years.

Over the last 400 000 years there is a close correspondence between the calculated variations in solar radiation and fluctuations in sea level recorded by deep-ocean cores. It appears that Milankovitch mechanisms are adequate to explain the glacial/interglacial cycles during an ICE AGE but not why ice ages occur, since these orbital variations probably operated during much of geological time.

Miller indices *n*. a set of three or four symbols used to define the orientation of a crystal face or internal crystal plane.

millidarcy see DARCY.

milligal *n*. the normal unit used in geophysical gravity surveying. It is equal to 10^{-5} m s^{-2}.

mimetic growth *n*. the control of the growth of new minerals during METAMORPHISM by some pre-existing textural or compositional inhomogeneity in the rock. For example, platy minerals such as MICA tend to grow preferentially along the BEDDING or an original FOLIATION. The metamorphic texture thus mimics the original texture of the rock.

Mindel see CENOZOIC ERA, Fig. 13.

mineral *n*. **1.** an element or chemical

compound that is normally crystalline and that has been formed as the result of geological processes. 'Crystalline' is taken to mean a degree of atomic ordering that allows identification from the diffraction pattern produced when the substance is traversed by radiation such as X-rays or a beam of electrons. For practical purposes, 'geological' is taken to include extraterrestrial substances formed in ways analogous to those occurring on Earth (e.g. the silicate *tranquillityite*, $Fe_8(Zr,Y)_2Ti_3Si_{13}O_{24}$), found on the Moon). Minerals are usually formed by inorganic processes, and individual species can be identified by their physical properties, such as density, hardness, crystalline forms, chemical composition and the ways in which they interact with light and other forms of radiation. See MINERALOID.

2. (*economic geology*) a natural resource obtained from the Earth's crust. This can include minerals in the strict sense, building material or FOSSIL FUELS.

mineral assemblage *n.* the minerals that co-exist in equilibrium with each other in a metamorphic or IGNEOUS ROCK. See also PARAGENESIS, METAMORPHIC FACIES.

mineral deposit *n.* a naturally occurring accumulation of MINERALS that may be economically valuable. Compare ORE. The range of economic minerals is shown in Fig. 59 opposite.

Mineral deposits can be formed in a variety of ways:

(a) MAGMATIC DIFFERENTIATION processes; see MAGMATIC ORE DEPOSITS: for example, CHROMITE, MAGNETITE;

(b) PEGMATITES: rare lithophile elements such as uranium, lithium (see AFFINITY OF ELEMENTS);

(c) HYDROTHERMAL processes, including VOLCANIC-EXHALATIVE DEPOSITS, HYPOGENE mineralisation: mainly base metals,

lead-zinc, copper, tin and molybdenum;

(d) PYROMETASOMATIC deposits: magnetite, PYRRHOTITE, CHALCOPYRITE and GALENA;

(e) SEDIMENTARY processes: most construction materials (building stone, sands, gravels, etc), FOSSIL FUELS, PLACER deposits (gold, tin, gemstones), EVAPORITES, the sedimentary IRON FORMATIONS and IRONSTONES;

(f) residual deposits formed through leaching of soluble elements from rocks during WEATHERING: BAUXITE, lateritic nickel (see LATERITE);

(g) SECONDARY ENRICHMENT (supergene enrichment): sulphide deposits, principally copper.

mineral isochron see ISOCHRON DIAGRAM.

mineralisation *n.* **1.** the process by which valuable minerals are introduced into a rock, resulting in an ore deposit, either actual or potential. The term includes various types of this process, e.g. impregnation, fissure filling, REPLACEMENT.

2. the process of fossilisation in which inorganic materials replace organic features.

mineralogy *n.* the study of minerals, including their formation, composition, properties and classification.

mineraloid *n.* a solid substance produced by geological processes in which the degree of atomic ordering is insufficient to define it as a MINERAL. This includes so-called amorphous minerals such as LIMONITE and natural glasses.

mineral wax see OZOCERITE.

minette *n.* **1.** a variety of LAMPROPHYRE containing BIOTITE as PHENOCRYSTS and in the GROUNDMASS, together with ALKALI FELDSPAR (orthoclase), AUGITE, minor PLAGIOCLASE and accessory APATITE.

2. a type of IRONSTONE composed of CHAMOSITE and LIMONITE ooliths (see

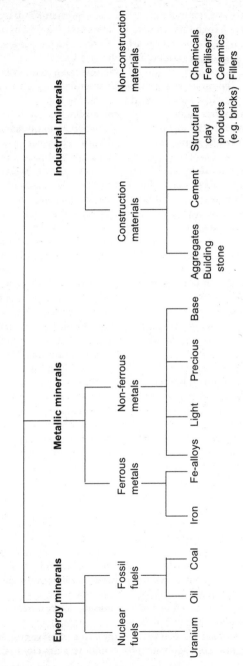

Fig. 59 **Mineral deposits**. The range of economic minerals

OOLITE), with some REPLACEMENT by SIDERITE. They have a high lime content so may be self-fluxing. Minette ores are common in the MESOZOIC of Europe. See also CLINTON ORE.

mining geology *n.* the geological aspects of MINERAL DEPOSITS, with special regard to problems related to mining.

Miocene *n.* the epoch of the Tertiary Period between the Oligocene and Pliocene Epochs. See CENOZOIC.

miogeosyncline see GEOSYNCLINE.

mirabilite or **Glauber's salt** *n.* a yellow monoclinic mineral, $Na_2SO_4 \cdot 10H_2O$, that occurs as a residue from playas and saline lakes; it is a source of sodium sulphate.

miscible *adj.* (of two or more liquids) capable, when introduced, of mixing and forming a single liquid. Compare IMMISCIBLE.

mispickel see ARSENOPYRITE.

Mississippian *n.* the Lower Carboniferous subperiod typical of the Mississippi Valley. Together with the term PENNSYLVANIAN for the Upper Carboniferous, it has been accepted by the United States Geological Survey since 1953, although not used outside North America.

Mississippian-Pennsylvanian boundary *n.* the boundary between the Mississippian and Pennsylvanian Periods in North America – the two American periods together being equivalent to the Carboniferous Period in Britain. The boundary is difficult to fix since, in both the Mississippi valley and in Arkansas, there are non-sequences (unconformities) between the Mississippian and Pennsylvanian, and where there are continuous sequences there are no INDEX FOSSILS present, thus preventing correlation. The age of the top of the Mississippian is *c.* 320 Ma. The boundary is probably equivalent to

the Alportian-Kinderscoutian boundary of the Namurian in the British Carboniferous. See also STRATIGRAPHICAL CORRELATION.

Missourian *n.* the lower part of the Upper Pennsylvanian of North America.

mixed-base crude *n.* a crude oil in which both naphthenic and paraffinic hydrocarbons are present in about equal proportions. Compare PARAFFIN-BASE CRUDE, ASPHALT-BASE CRUDE.

mixtite *n.* an unsorted, mixed-grain-size SEDIMENTARY ROCK. *Mixton* is the name given to the unconsolidated equivalent. Mixtites may be GLACIAL deposits (see TILL) or may result from DEBRIS FLOWS or MUDFLOWS. A mixtite can be identified as being of glacial origin only by association with such features as GLACIAL STRIATIONS and DROPSTONES.

mixton see MIXTITE.

mobile belt *n.* a long narrow belt within the CONTINENTAL CRUST where tectonic and metamorphic activity occurs, i.e. one of the younger fold-mountain belts.

mobilisation *n.* **1.** any process by means of which a solid rock becomes sufficiently plastic to permit the geochemical migration of particular mobile components. Compare RHEOMORPHISM.
2. any process that relocates and concentrates valuable ROCK components into an ORE deposit.

mode *n.* the actual MINERAL composition of a rock, expressed in terms of the weight or volume percentage of the constituent minerals. Compare NORM.

modified Mercalli scale see EARTHQUAKE MEASUREMENT.

modulus of compression see COMPRESSIBILITY.

modulus of elasticity or **modulus of volume elasticity** *n.* the ratio of STRESS to STRAIN within the range in which a material is deformed according to HOOKE'S LAW. See also YOUNG'S MODULUS, MODULUS OF RIGIDITY, BULK MODULUS.

modulus of incompressibility see BULK MODULUS.

modulus of rigidity *n.* an expression of the resistance to change in shape. It is given as $G = \tau/\gamma$, where G is rigidity modulus, τ is SHEAR stress and γ is shear strain.

modulus of volume elasticity see MODULUS OF ELASTICITY.

mofette *n.* a low-temperature FUMAROLE rich in carbon dioxide.

mogote *n.* a large residual hill of limestone associated with a KARST erosion landscape. Mogotes are found only in regions of tropical or subtropical rainfall, such as western Cuba. Gigantic examples are characteristic of South China and are the hills of fantasy that appear so frequently in Chinese paintings.

Moho *abbreviation for* MOHOROVICIC DISCONTINUITY.

Mohole project *n.* a United States scientific project with the research objective of studying the Earth as a planet, to be accomplished through sampling of all layers of the Earth's crust and mantle by core drilling near Hawaii to a total depth of approximately 9755 metres. Phase 1 of Project Mohole, conducted in 1961, consisted of drilling into several hundred metres of the Earth's crust off the California coast. Phase II was concerned with the problems created by drilling in deep oceans and how to solve them. This work progressed from 1962 to 1965 but in 1966 the project was terminated for lack of funding.

Mohorovičić discontinuity, Moho or **M-discontinuity** *n.* the boundary surface that separates the Earth's crust from the mantle. It marks the level at which velocities of seismic P waves abruptly change from 6.7 to 7.2 kilometres per second (in the lower crust) to 7.6 to 8.6 kilometres per second (top of upper mantle) as a result of the greater rigidity and elastic modulus of the mantle rock. The depth of the DISCONTINUITY ranges from about 5 to 10 kilometres beneath the ocean floor; its depth averages about 35 kilometres below the continents but increases to a maximum of about 90 kilometres beneath mountain ranges. Its thickness is variously judged to be between 0.2 and 3 kilometres. Seismic sounding has not detected its presence beneath the MID-ATLANTIC RIDGE, a possible indication of low rigidity because of the high heat flow associated with the injection of magma associated with ocean-floor spreading. The Mohorovičić discontinuity is named after its discoverer, Andrija Mohorovičić (1857–1936), a Croatian seismologist. Compare GUTENBERG DISCONTINUITY.

Mohr diagram *n.* a graphical construction used to represent states of STRESS or of STRAIN.

Mohs scale *n.* a scale of relative hardness of minerals proposed in 1922 by Friedrich Mohs, an Austrian mineralogist. The points on the scale are defined in terms of 10 standard minerals: 1 talc (softest), 2 gypsum, 3 calcite, 4 fluorite, 5 apatite, 6 orthoclase, 7 quartz, 8 topaz, 9 corundum, 10 diamond (hardest).

Moine Supergroup *n.* a thick monotonous succession of deformed and polymetamorphosed Middle to Upper PROTEROZOIC sedimentary rocks found in the north of Scotland. The supergroup crops out over much of the northern Highlands, mainly between the Moine Thrust Zone and the Great Glen Fault.

The rocks consist largely of psammites and semipelites in which sedimentary structures are often preserved. Moine rocks unconformably overlie the LEWISIAN GNEISS, and Moine sedimentation occurred after *c.* 1000 Ma. The Moine Supergroup is divided into the Morar, Glenfinnan and Loch Eil groups. The Grampian group, which occurs to the southeast of the Great Glen Fault, was formerly considered to be part of the Moine Supergroup but is now regarded by most authorities to be the lowest division of the DALRADIAN SUPERGROUP. See UNCONFORMITY.

molasse *n.* an association of shallow marine and non-marine conglomerates and sandstones, the non-marine sediments being deposited as ALLUVIAL FANS and LACUSTRINE deposits. Rocks of the facies are associated with erosion and deposition during the post-tectonic phase of mountain-building, in contrast to the syntectonic FLYSCH. Molasse was first described in the Alps.

mollusc *n.* an invertebrate of the phylum Mollusca, most of which secrete an external calcium carbonate shell.

The shell may be external UNIVALVE, external bivalve, internal or absent. The body is covered by two folds of tissue, the mantle, which secretes the shell and encloses the gill cavity. Molluscs have a muscular 'foot' that is used for movement and is modified in the various groups. The molluscs are a diverse group with a range of life-styles. Their habitat may be marine, freshwater or even terrestrial.

The principal FOSSIL classes are the BIVALVES (Bivalvia), GASTROPODS (Gastropoda) and the CEPHALOPODS (Cephalopoda). The Amphineura (AMPHINEURANS) and Scaphopoda (SCAPHOPODS) have minor fossil records. The earliest molluscs date from the lower CAMBRIAN.

molybdate *n.* a mineral compound in which the MoO_4 radical is a constituent. WULFENITE, $PbMoO_4$, is an example.

molybdenite *n.* a bluish-grey hexagonal mineral, MoS_2, the principal ore of molybdenum. Molybdenite usually occurs as bladed or foliated masses and is found in pegmatites and high-temperature pneumatolytic veins.

monadnock *n.* an isolated mountain or hill of temperate regions, rising above a lowland that has been levelled almost to the theoretical limit (BASE LEVEL) by fluvial EROSION. Such a lowland is called a PENEPLAIN and represents the almost complete advance of the levelling process. Monadnocks remain as residual features, either because they are composed of more resistant rock or are farther away from the base level. Mount Monadnock in New Hampshire, USA, is the classic example from which these hills derive their name. Compare UNAKA, INSELBERG.

monazite *n.* a rare-earth phosphate, $(Ce,La,Y,Th)PO_4$, yellow to brownish-red, generally occurring in disseminated granules. It is an accessory in granites and gneisses, and is the main ore for the radioactive element thorium. See also ACTINIDE SERIES.

monchiquite *n.* an alkaline variety of LAMPROPHYRE composed of titanaugite (see AUGITE), AMPHIBOLE (barkevikite and/or kaersutite) and BIOTITE-PHLOGOPITE, sometimes with OLIVINE, in a glassy groundmass containing NEPHELINE or ANALCIME.

monocline *n.* a local steepening of the dip of layered rocks in areas where the BEDDING is relatively flat. Compare HOMOCLINE.

monoclinic system see CRYSTAL SYSTEM.

monogenetic *adj*. **1.** derived from a single source or resulting from a single process of formation, e.g. a single eruption.
2. consisting of one element or material, e.g. a gravel composed of a single rock type. Compare POLYGENETIC.

monomictic *adj*. **1.** (of a clastic sedimentary rock) composed of a single mineral type.
2. (of a lake) having only a single yearly OVERTURN. Compare OLIGOMICTIC, POLYMICTIC.

monomyarian see BIVALVE.

monophyletic *adj*. a monophyletic group is a group of organisms that comprise all the descendants of a common ancestor.

monotropy *n*. the relationship between two different forms of the same substance where there is no transition point since only one of the forms is stable and the change from the unstable to the stable form is irreversible. Compare ENANTIOTROPY.

montmorillonite *n*. a monoclinic clay mineral of the smectite group, $(Al,Mg)_8$ $(Si_4O_{10})(OH)_8 \cdot 12H_2O$, the dominant clay mineral in BENTONITE. Two noteworthy characteristics are its capacity to expand by absorbing liquids and its potential for ion exchange when, under certain conditions, the cations sodium, potassium and calcium, present between the structural layers, are replaced by other cations present in solution.

monzodiorite *n*. a plutonic rock intermediate in composition between MONZONITE and DIORITE. Monzodiorite is the term recommended by the IUGS Subcommission on the Systematics of Igneous Rocks to replace *syenodiorite*.

monzogabbro *n*. a plutonic rock composed principally of PYROXENE and PLAGIOCLASE (like GABBRO) but with minor amounts of ORTHOCLASE in addition to the plagioclase. It was formerly known as *syenogabbro*.

monzonite *n*. a plutonic rock with a composition intermediate between SYENITE and DIORITE. It contains approximately equal amounts of PLAGIOCLASE and ALKALI FELDSPAR, and little or no quartz; AUGITE is often the main MAFIC mineral. Monzonite is the intrusive equivalent of LATITE.

moonstone *n*. a gemstone-quality ALKALI FELDSPAR that shows a silvery or bluish iridescence; an opalescent variety of adularia (a low-temperature K-feldspar related to orthoclase). Compare SUNSTONE.

moraine *n*. an accumulation of rock material that has been carried or deposited by a GLACIER. It ranges in size from BOULDERS to SAND and shows no BEDDING or SORTING. Different kinds of moraine are distinguished based on the nature of the landform:
(a) *ground moraine* – an irregular covering of TILL deposited under a glacier and composed chiefly of CLAY, SILT and sand. It is the most prevalent deposit of continental glaciers;
(b) *lateral moraine* – debris derived from EROSION and avalanches from the valley wall on to the glacier edge and finally deposited as a long ridge when the glacier recedes;
(c) *medial moraine* – an enlarged zone of debris formed when lateral moraines join at the intersection of two glaciers. It is deposited as a ridge running approximately parallel to the direction of ice movement;
(d) *terminal* or *end moraine* – a ridgelike mass of glacial debris formed by the foremost glacial snout and dumped at the outermost edge of a given ice advance. It shows convex curving down-valley and may form lateral

moraines up the side. It may take the appearance of hummocky ground, perhaps with KETTLES;
(e) *recessional moraine* – secondary end moraine deposited during a temporary halt in a glacial retreat. Series of such moraines thus show the history of glacial retreat.

MORB *abbreviation for* mid-ocean ridge basalts (see MID-OCEANIC RIDGE).

morphogenetic region *n*. a climatic zone in which prevailing geomorphic processes produce landscape features distinct from those of other regions that formed under different climatic conditions.

morphological unit *n*. 1. a rock STRATIGRAPHICAL UNIT specified by its topographical features.
2. a surface that is recognised by its topography.

morphology *n*. 1. GEOMORPHOLOGY.
2. the study of soil properties and distribution arrangements of SOIL HORIZONS.

Morrowan *n*. the Lower Pennsylvanian of North America.

mortar structure *n*. an apparently cataclastic structure resembling stones in mortar. It is found in gneisses and granites in which spaces between particles of feldspar and quartz are filled in with pulverised grains of the same minerals. See CATACLASIS.

morvan *n*. the intersection of two erosional surfaces. The name is derived from the Morvan Plateau of northeastern France, which exhibits such a relationship.

mosaic 1. *n*. DESERT PAVEMENT.
2. *n*. a granular or saccharoidal rock texture.
3. *adj*. (of a substructure within a CRYSTAL) in which small blocks of the crystal have slightly different orientations because of dislocations

within the CRYSTAL LATTICE. This substructure may be formed during crystallisation from a liquid or by deformation of the crystal.

moss agate *n*. CHALCEDONY with a moss- or fern-like INCLUSION. The inclusion is actually a dendrite of pyrolusite (MnO_2). Moss agate is often mistaken for a fossil. See also PSEUDOFOSSIL.

Mössbauer spectrometry see Appendix 4.

mould *n*. 1. an impression of a FOSSIL shell or other organic structure made in the encasing material. An EXTERNAL MOULD shows the surface form and markings of the outer hard parts of a structure. The encasing material the surface of which received these moulds is also called an external mould. An internal mould shows the form and markings of the inner surface of a structure. It is made on the surface of the material that fills the hollow interior of the structure. When a shell becomes filled with sediment and the shell material is later removed by solution or erosion, the remaining hard core is called a *steinkern* ('stone kernel'). Compare CAST.
2. a FLUTE MARK or other mark made on a sedimentary surface the filling of which produces a cast.

moulin or **glacial mill** *n*. a circular or subcircular cylindrical shaft extending down into a GLACIER; it is produced by the abrasive scouring action of swirling meltwater.

mountain belt *n*. a linear region of the Earth's crust characterised by rocks that are strongly deformed and usually have suffered extensive METAMORPHISM.

mountain-building *n*. orogenesis, see OROGENY.

moveout see NORMAL MOVEOUT.

mud *n*. 1. a mixture of water with CLAY- to SILT-sized particles, ranging in consistency from semifluid to soft and plastic.

2. a fine-grained red, blue or green marine sediment found mainly on the continental shelf; its colours are because of a predominance of various mineral substances. (See TERRIGENOUS DEPOSITS.)

3. the material of a MUDFLOW.

4. DRILLING MUD.

mud breccia see DESICCATION BRECCIA.

mud crack, shrinkage crack or **desiccation crack** *n.* an irregular polygonal pattern formed by the shrinkage of mud, silt or clay in the process of drying. Mud cracks develop generally under conditions of long exposure to a dry warm climate. Compare NONSORTED POLYGON.

mud-crack polygon see NONSORTED POLYGON.

mud flat *n.* a relatively flat expanse of fine silt and clay lying along a coastline, often in the shelter of an island. It is a muddy TIDAL FLAT devoid of vegetation and may be covered by shallow water or alternately covered and uncovered by the tide.

mudflow *n.* a flow of fine-grained, water-saturated sediment in a stream channel. It is typical of semi-arid to arid regions where heavy rains transport earth material collected on hill slopes into pre-existing stream channels. Compare EARTHFLOW. See also DEBRIS FLOW, LAHAR.

mud pot or **sulphur-mud pool** *n.* a variation of a hot spring in which the channelway is filled with mud that is broken down from the surrounding rock by acid volcanic gases dissolved in the water. Mud pots are frequently associated with GEYSERS. See THERMAL SPRING.

mudrock *n.* a fine-grained SEDIMENTARY ROCK composed chiefly of particles in the silt-clay size range. Mudrock is a general term that can be used to distinguish the finer-grained sedimentary rocks, from sandstones or limestones. Mudrocks can be further identified as SHALE, MUDSTONE, ARGILLITE, SILTSTONE, CLAYSTONE or MARL, depending on the dominant GRAIN size, composition and the presence of fissility or laminations.

mudstone *n.* **1.** a commonly used synonym for MUDROCK, in particular a massive or blocky non-fissile mudrock.

2. a mud-supported carbonate sedimentary rock containing less than 10 per cent particles of clay and fine silt size; the original components are not bound together during deposition. Compare WACKESTONE, PACKSTONE, GRAINSTONE. See DUNHAM CLASSIFICATION.

mud-supported *adj.* (of LIMESTONE) containing insufficient sand- and gravel-sized grains to form the supporting framework.

mugearite *n.* a dark-coloured fine-grained VOLCANIC ROCK composed of OLIGOCLASE and minor ANORTHOCLASE; OLIVINE and MAGNETITE are more abundant than PYROXENE. It is part of a series of rocks: ALKALI BASALT–HAWAIITE–mugearite–BENMOREITE–sodic TRACHYTE. Mugearite is defined chemically as the sodic variety of basaltic TRACHYANDESITE.

mullion structure *n.* a structure resembling a series of parallel columns or rods. Each unit may be several centimetres in diameter and several metres long. They are formed during the regional METAMORPHISM of micaceous SANDSTONE as the result of intense folding along an axis parallel to the rodding.

multi- *prefix meaning* many or much.

multiple intrusion *n.* an igneous INTRUSION formed by two or more successive injections of MAGMA of more or less identical chemical and mineral composition. (Compare COMPOSITE.) DYKES and SILLS are often multiple; when dykes occur as large numbers of separate intrusions of about the same age they are called DYKE SWARMS.

multiple reflection *n.* a SEISMIC WAVE that has been reflected more than once.

mural pore see TABULATA.

muscle scars *n.* the traces left by the attachment of muscles inside the shells of BIVALVES, Fig. 7, and BRACHIOPODS, Fig. 9(b).

muscovite *n.* a mineral of the MICA group, $KAl_2[AlSi_3O_{10}](OH)_2$, colourless and transparent in thin CLEAVAGE flakes but translucent silvery or pale shades of yellow, brown or green in thicker CRYSTALS. Muscovite is a widespread and common rock-forming mineral, especially in PEGMATITE, GRANITE and low- or medium- to high-grade METAMORPHIC ROCKS (GREENSCHIST and AMPHIBOLITE facies).

muskeg see BOG.

m.y. *abbreviation for* million years = 10^6 years. See MA.

mylonite *n.* a hard, fine-grained CHERT-like rock with banded or streaky structure, formed by the extreme granulation of rocks that have been pulverised during faulting or intense dynamic metamorphism. It is typically related to FAULTS (especially thrust faults) or intense shear zones. Lenses of undestroyed parent rock may persist in the re-formed granulated groundmass. There is debate about whether new crystals form in the groundmass. When extremely fine-grained, mylonite may be described as an *ultramylonite*.

mylonitisation *n.* the deformation of a rock by microbrecciation resulting from mechanical forces that are applied in a definite direction; there is no signifi-cant chemical reconstitution of the granulated minerals.

myrmekite *n.* an intergrowth of a plagioclase FELDSPAR and QUARTZ, generally replacing alkali feldspar; it is formed during the later stages of CONSOLIDATION in an IGNEOUS ROCK. The quartz occurs as blebs or as vermiform shapes within the feldspar. See Fig. 60 below.

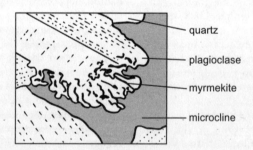

Fig. 60 **Myrmekite**. Vermiform intergrowth of quartz blebs (black) and plagioclase feldspar

N*n*

N see NEWTON.

nacreous *adj*. **1.** pearly; having the lustre of mother-of-pearl.

2. relating to or consisting of mother-of-pearl (nacre).

nailhead spar *n*. a common CRYSTAL HABIT of the mineral CALCITE, having six prominent prism faces and rhombohedral (three-faced) terminations. See FORM.

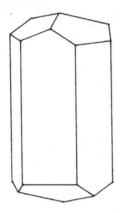

Fig. 61 **Nailhead spar**. The rhombohedral terminations are an indication that calcite crystallises in the trigonal system

Namurian *n*. the lowest series of the SILESIAN (CARBONIFEROUS).

nannofossil or **nanofossil** *n*. the smallest size of microfossil, normally including forms of less than 60 µm in largest dimension, such as coccolithophores. A more restrictive definition limits the name to those forms of less than 2µm in largest dimension. Members of the nannoplankton form nannofossils.

nappe or **decke** *n*. a large sheet-like body of solid rock that has moved a long distance (generally a kilometre or more) at low angles over the underlying rocks either by overthrusting or recumbent folding (in which the fold axes are approximately horizontal). After long EROSION, portions of a nappe may become isolated remnants called *klippen*. A *klippe* is a nappe outlier and is distinguished from outliers of CUESTAS or plateaux in having younger rocks surrounding it. If, in the early stages of dissection of a nappe, erosion breaks through the overthrust sheet, exposing the rocks beneath the fault, a *window* (or *fenster*) is formed; the name applies because it is possible to look through the upper sheet to the lower.

native element *n*. an element found in nature uncombined and in a non-gaseous state. Metals that occur in native form include silver, gold, copper, iron and mercury. Non-metallic examples are carbon, sulphur and selenium.

NATM *abbreviation for* New Austrian Tunnelling Method, a method of tunnel lining in which SHOTCRETE is applied directly after blasting, sometimes followed by a supplementary lining.

natrolite *n*. an orthorhombic tectosilicate mineral, $Na_2(Al_2Si_3O_{10})\cdot2H_2O$. This fibrous ZEOLITE often occurs as radiating crystal groups.

natural arches and bridges *n*. arch-like rock features formed by natural agents such as WEATHERING and EROSION. When LATERAL EROSION by a stream enables it to cut through a MEANDER neck, a *bridge* or *arch* will be formed if the upper level of the neck does not collapse. In KARST regions, characterised by tunnels created by the GROUNDWATER solution of LIMESTONE, the partial collapse of a tunnel roof may leave short un-collapsed portions in the form of a bridge. Caves that develop on either side of a headland, such that they ultimately join together because of wave erosion, give rise to a *sea arch*. Upon collapse of the arch, the head-land remains as a STACK.

natural gas *n*. HYDROCARBONS such as METHANE (CH_4) and ETHANE (C_2H_6) that exist as a gas or vapour at ordinary temperatures and pressures. Natural gas may occur alone or associated with OIL or COAL (coal-bed methane). Like oil and coal, it is an important FOSSIL FUEL. It is believed to have originated from the remains of microscopic life forms buried in sediments of prehistoric seas where they eventually underwent decomposition.

natural levee *n*. an embankment of silt and sand built up by a stream along both its sides. The process is particu-larly effective during flooding, when the water overflows its banks and deposits its coarsest sediment.

natural remanent magnetisation *n*. the permanent magnetism in rocks, resulting from the orientation of the Earth's magnetic field at the time the rock was formed. It provides the basic information for the palaeomagnetic studies of POLAR WANDERING and CONTINENTAL DRIFT. It can derive from several processes:
(a) *thermoremanent magnetisation* (TRM) is the most important. It is the magneti-sation acquired by the magnetic minerals in a rock as they cool through their CURIE POINTS. It records the

direction and strength of the Earth's magnetic field at that time;
(b) *detrital remanent magnetisation* (DRM), the result of the settling of small grains of magnetic minerals into a sedimentary matrix. It is thought that the grains arrange themselves in the direction of the geomagnetic field during deposition and before final consolidation of the rock.
(c) *chemical remanent magnetisation* (CRM), non-magnetic minerals that are chemically altered to magnetic forms can acquire remanent magnetisation in the presence of the geomagnetic field.
(d) *viscous magnetisation* may ultimately cause already cooled igneous rocks to acquire remanent magnetisation. The superimposition of later or earlier magnetisation must be considered when determining remanent magneti-sation. See also PALAEOMAGNETISM, GEOMAGNETIC POLARITY REVERSALS.

nautiloid *n*. any CEPHALOPOD of the subclass Nautiloidea, characterised by an external chambered shell with an approximately central SIPHUNCLE and retrochoanitic septal necks, see Fig. 62(a) opposite. The SUTURE LINES are straight or gently curved (Fig. 62(b)), unlike the AMMONOID suture lines. Nautiloid shells may be straight (orthocone), curved (cyrtocone) or planispirally coiled. Some of the terms used to describe planispirally coiled shells are given under ammonoid.

The nautiloids reached their peak in the ORDOVICIAN and SILURIAN. The last surviving genus of nautiloids is Nautilus. Range, CAMBRIAN to Recent.

Nebraskan *adj*. the first glacial stage of the Pleistocene Epoch in North America; it was succeeded by the Aftonian interglacial.

nebular hypothesis *n*. a theory of the origin of the planetary system, accord-

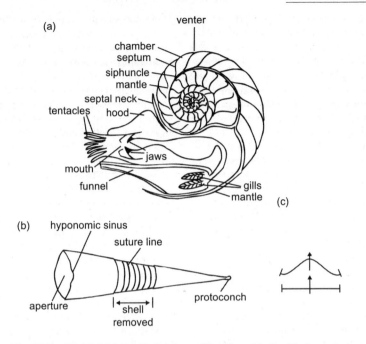

Fig. 62 **Nautiloid**. (a) internal structure of Nautilus; (b) simplified external
structure of a straight (orthocone) form; (c) nautiloid sutures

ing to which the system formed within
a gas cloud that was part of the
original Sun. In the nebular hypothesis
of Kant (1775), developed by La Place
(1796), a rotating cloud of gas contracts, the outer region of the cloud
becomes a cool gaseous disc in which
the planets condense and the innermost region contracts to form the Sun.
The problem of how such protoplanets
condense in a Laplacian disc of gas is
still unsolved. Moreover, according to
this theory, the Sun should rotate much
more rapidly than is observed. Compare JEANS-JEFFREY THEORY.

neck *n*. **1.** a narrow strip of land that
connects two larger areas, as an
isthmus or cape.
2. a volcanic neck (see VOLCANIC PLUG).

3. a mineral-bearing pipe.
4. the narrow band of water that forms
the part of a rip current where feeder
currents converge.

neck cut-off *n*. a term for the CUT-OFF that
takes place in the final stage of mean-
der-loop development when a MEANDER
neck is cut through. It results in
shortening of the river course and local
steepening of the stream gradient.
Compare CHUTE CUT-OFF.

needlestone see HAIRSTONE.

nekton *n*. any organism able to swim.
Seals, whales, tuna, herring and squid
are examples. See also PLANKTON,
PELAGIC.

nektonic see PELAGIC.

nema *n*. a thread-like extension of the
apex of a GRAPTOLITE SICULA (first-formed

structure) (Fig. 41). In some forms the nema is embedded in the wall of the graptolite colony (rhabdosome) and is known as the *virgula*.

nematoblastic *n.* a texture of METAMORPHIC ROCKS in which an abundance of slender prismatic crystals are oriented in linear fashion, thus imparting a LINEATION to the rock.

Neogene *n.* the youngest of the Tertiary periods (24 Ma–1.8 Ma), equivalent to the Miocene and Pliocene Epochs of the Tertiary. Compare PALAEOGENE.

neomagma *n.* a little used term for MAGMA formed by the fusion of pre-existing rock under conditions of plutonic METAMORPHISM. Compare ANATEXIS.

neosome see MIGMATITE.

neomineralisation see METAMORPHISM.

neoteny *n.* a mode of evolution in which the onset of sexual maturity in a juvenile animal removes the morphological characters of the adult animal. See HETEROCHRONY.

Neotethys *n.* a name used for the oceanic embayment of Panthalassa during the Permian and Triassic and more accurately for the Jurassic–Cretaceous ocean opened along the northern margin of that embayment of Panthalassa. Compare PALAEOTETHYS.

neotype see TYPE SPECIMEN.

nepheline *n.* a hexagonal mineral of the FELDSPATHOID group, $(Na,K)AlSiO_4$, generally occurring in compact granular aggregates of white, yellow, grey or green. It is typically found in silica-UNDERSATURATED, intrusive and extrusive IGNEOUS ROCKS such as NEPHELINE SYENITE, MONCHIQUITE.

nepheline syenite *n.* a plutonic IGNEOUS ROCK composed chiefly of ALKALI FELDSPAR and NEPHELINE with a small amount of MAFIC minerals (ALKALI AMPHIBOLE and/or alkali PYROXENE). It is

the coarse-grained intrusive equivalent of PHONOLITE. SODALITE and CANCRINITE are common accessories.

nephelinite *n.* an extrusive IGNEOUS ROCK, commonly light grey in colour, composed primarily of NEPHELINE and CLINOPYROXENE. It is often associated with CARBONATITE-IJOLITE complexes.

nephrite *n.* an extremely tough, compact tremolite or actinolite (see AMPHIBOLE). It is one of the two minerals referred to as JADE, JADEITE being the other. Nephrite normally ranges in colour from white to dark green.

neritic deposits *n.* sea-bottom deposits of the shallow-water marine zone, found chiefly at the edge of the CONTINENTAL SHELF but farther out than where TERRIGENOUS DEPOSITS would usually be found. They consist of material from the land and organic substances from shallow coastal waters, including the remains of crustacea, worm tubes, molluscs, etc. See OCEAN-BOTTOM DEPOSITS.

neritic zone *n.* **1.** (*sedimentology, stratigraphy*) the shallow-sea environment from low water down to 200 metres (100 fathoms).

2. (*oceanography, ecology*) the LITTORAL.

nesosilicate, orthosilicate *n.* a SILICATE in which the SiO_4 tetrahedra occur as isolated units, i.e. they do not share oxygen ions as do all other silicate types. In nesosilicates, the tetrahedra are bound together by other ions. The ratio of silicon to oxygen is 1:4. OLIVINE is an example of a nesosilicate.

net balance *n.* the change in glacial mass from the time of minimum mass during one year to the time of minimum mass in the year following. Compare BALANCE.

net slip *n.* the distance between two previously adjacent points on either side of a FAULT, measured on the fault surface or parallel to it. Net slip defines both the direction and relative magnitude of DISPLACEMENT. See Fig. 31, also DIP SLIP, STRIKE SLIP.

Neuropteris *n.* a typical seed fern (Pteridospermae) foliage fossil that is common in certain Carboniferous strata. Oval leaflets alternate on either side of the stem. The length of the pinnules is 0.6 to 1.3 centimetres. *Neuropteris* is the foliage of the plant *Medullosa*. See also FOSSIL PLANTS.

neutral shoreline *n.* a SHORELINE the basic character of which is independent of the SUBMERGENCE of a former land surface or the emergence of a surface that was formerly under water. It includes alluvial and OUTWASH planes, VOLCANOES and CORAL REEFS.

neutron activation analysis see Appendix 4.

neutron-gamma log see WELL LOGGING.

neutron logging see WELL LOGGING.

Nevadan orogeny *n.* a series of orogenic pulses on the western margin of North America. These pulses extended from Late Jurassic to Middle Cretaceous. The Nevadan OROGENY was the result of underthrusting of the western portion of the North American plate by oceanic crust and microcontinents along a former oceanic trench.

névé see FIRN, GLACIER.

New Austrian Tunnelling Method see NATM.

new catastrophism see CATASTROPHISM.

new global tectonics see PLATE TECTONICS.

New Red Sandstone *n.* the red SANDSTONE facies of the Permian and Triassic systems, particularly well developed in northwestern England. Compare OLD RED SANDSTONE.

newton (N) *n.* the SI system unit of force. The force required to produce an acceleration of $1m/s^2$ in a mass of 1 kg. A pressure of $1N/mm^2$ is equal to 1 mega-pascal (Mpa). See also Appendix 1.

Newton's scale *n.* the complete sequence of POLARISATION COLOURS. It is a repeating pattern, divided into orders, of sets of colours somewhat similar to the white-light spectrum. The first-order colours are an exception, being grey or white. The sequence may be seen if a quartz wedge is examined between CROSSED POLARISERS under a PETROLOGICAL MICROSCOPE.

Niagaran *n.* the Middle Silurian of North America.

niccolite *n.* redundant name for NICKELINE.

nickeliferous *adj.* containing nickel.

nickeline *n.* a copper-red nickel arsenide, NiAs, a minor ore of nickel. Nickeline occurs with other arsenides and sulphides in low-temperature hydrothermal vein deposits; it was formerly called *kupfernickel*.

nick point see KNICK POINT.

Nicol prism *n.* a device for obtaining plane-polarised light. It consists of a rhombohedron of optically clear CALCITE, cut and cemented together in such a way that the ordinary ray produced by double refraction in the calcite is totally reflected while the extraordinary plane-polarised ray is transmitted. It has largely been replaced by sheet polarisers. See POLARISER, BIREFRINGENCE.

niter see NITRE.

nitrate *n.* a compound characterised by the $(NO_3)^-$ radical. NITRATINE ($NaNO_3$) and NITRE (KNO_3) are examples. Compare CARBONATE, BORATE.

nitratine *n.* a trigonal NITRATE mineral, $NaNO_3$, isostructural with CALCITE; it is also known as *soda nitre*.

nitre, niter or **saltpetre** *n.* a white mineral, KNO_3, found in arid regions and as surface efflorescence in caves. Compare CHILE SALTPETRE, NITRATINE.

nivation *n*. processes acting on high mountain slopes beneath a persistent snow mass that alter the BEDROCK or REGOLITH. Nivation includes CREEP, FROST ACTION, SOLIFLUCTION and sheetwash, all of which take place on the snow-patch edge or beneath it. GLACIATION is initiated when the snow that accumulates in such nivation hollows becomes thick enough to move downslope as glacial ice.

noble gas see INERT GASES.

noble metal *n*. any of those metals that do not readily lose their metallic character when heated in air, e.g. gold, silver, platinum. Such metals are resistant to oxidation or to attack by corrosive agents. Compare BASE METALS.

node *n*. **1.** the point on a FAULT at which the apparent DISPLACEMENT changes.
2. that point on a standing wave at which there is no vibration.
3. a branching point in a CLADE.

nodular *adj*. composed of or having the form of NODULES.

nodule *n*. **1.** a rounded concretionary mass or lump; there is no connotation of size. It can be applied to a lump of mineral aggregate that is normally without internal structure and contrasts with its embedding matrix. See CONCRETION.
2. the concretionary lumps of metals found on the ocean floors, for instance MANGANESE NODULES.
3. a XENOLITH composed of earlier-formed minerals, found as an inclusion in an IGNEOUS ROCK.

noise *n*. all recorded energy not derived from the explosion of a seismic shot. See SEISMIC SHOOTING.

nonconformity see UNCONFORMITY.

nonflowing artesian well *n*. an ARTESIAN well in which the hydrostatic pressure is not great enough to elevate the water above the land surface; the water does, however, rise above the local WATER TABLE. Compare FLOWING ARTESIAN WELL.

non-penetrative fabric see PENETRATIVE FABRIC.

nonsorted see PATTERNED GROUND.

nonsorted polygon *n*. a form of PATTERNED GROUND showing a dominantly polygonal mesh the units of which are not bordered by stones, as are SORTED polygons. The structure typically results from the filling in of fissures that commonly mark its borders. The outlining fissures of a *desiccation polygon* (also called a *mud-crack polygon*) contain vegetation. A *fissure polygon* is marked by intersecting grooves that form a slightly convex pattern. It is typical of areas of the northwestern Canadian lowlands. When the outline is formed by intersecting frost cracks, the configuration is called a *frost-crack polygon*. An *ice-wedge polygon* is a large feature averaging 10 to 110 metres in diameter and typically bordered by intersecting ice wedges. It is found only in PERMAFROST regions and is formed by the contraction of frozen ground.

nontronite *n*. a CLAY MINERAL, $Fe_2(Al,Si)_4O_{10}(OH)_2Na_{0.3} \cdot nH_2O$, of the SMECTITE group.

Norian *n*. a STAGE of the Upper TRIASSIC.

norite *n*. a coarse-grained gabbroic IGNEOUS ROCK. Its essential components are calcic PLAGIOCLASE and ORTHOPYROXENE. See also GABBRO.

norm *n*. the theoretical mineral composition of a rock, given in terms of anhydrous NORMATIVE MINERAL molecules and calculated from the bulk chemical composition. Compare MODE. The norm calculation has been used for a variety of purposes in PETROLOGY, in particular to determine the extent of SILICA SATURATION and as a means of presenting information in some variation diagrams and PHASE DIAGRAMS. See CIPW NORMATIVE COMPOSITION.

normal fault see FAULT.

normal horizontal separation see OFFSET.

normal moveout n. the increase in ARRIVAL TIME of a seismic reflection incident as a result of an increase in the distance from the shot to the GEOPHONE or of the dip of the reflecting interface. *Moveout* is the difference in reflection times, as observed from adjacent traces of a seismic record, in particular when the difference is caused by dip of the reflecting interface.

normal polarity n. a NATURAL REMANENT MAGNETISATION consistent with the present orientation of the Earth's MAGNETIC FIELD. (See also GEOMAGNETIC POLARITY REVERSAL.)

normal stress n. that component of STRESS that is perpendicular to, i.e. normal to, a given plane. The stress may be either compressive or tensile.

normal zoning n. in PLAGIOCLASE, that change by which crystals become more sodic in their outer parts. Compare INVERSE ZONING.

normative mineral or **standard mineral** n. a MINERAL the presence of which in an IGNEOUS ROCK is theoretically possible according to the basis of the CIPW NORMATIVE calculation.

nosean n, a rare isometric FELDSPATHOID mineral, $Na_8(AlSiO_4)_6SO_4 \cdot H_2O$, related to HAÜYNE.

notothyrium n. a triangular opening in the BRACHIAL VALVE of some BRACHIOPODS. It lies opposite the DELTHYRIUM so that the PEDICLE opening is diamond-shaped. The notothyrium may be closed by two *chilidial plates* or a *chilidium*.

nucleation n. the initiation of CRYSTAL growth at one or more points in a melt or solution.

nuée ardente or **glowing cloud** n. a PYROCLASTIC FLOW produced by the explosive collapse of an actively growing LAVA DOME or flow erupted from a VOLCANO. A nuée ardente consists of a block-and-ash flow surmounted by a highly mobile, turbulent, sometimes incandescent gaseous cloud and forms a block-and-ash flow deposit. A famous example of a nuée ardente occurred at Mount Pelée in Martinique during the eruptions of 1902. The basal part of a pyroclastic flow heading for St Pierre was diverted by the topography but the hot cloud of gas and ash above it carried straight on to engulf the town; it annihilated 30 000 inhabitants. See IGNIMBRITE.

nummulite n. very large foraminiferans of the Superfamily Rotaliacea with discus-shaped or coin-shaped multichambered cells. Nummulites were extremely abundant in the seas of the Mediterranean region during Eocene and Oligocene time, and their tests contributed largely to the nummulite limestones that are widely distributed in southern Europe, northern Africa and the Himalayan region.

nunatak n. an isolated peak of BEDROCK that projects above an ice cap or GLACIER and is surrounded by ice. Nunataks can be seen along the coastal region of Greenland.

Oo

obduction *n.* the emplacement of part of the oceanic LITHOSPHERE on to the CONTINENTAL CRUST at a DESTRUCTIVE PLATE BOUNDARY. See OPHIOLITES.

oblique fault *n.* a FAULT that trends oblique to, rather than perpendicular or parallel to, the direction (STRIKE) of the dominant rock structure. See DIP FAULT, STRIKE FAULT. Compare OBLIQUE-SLIP FAULT.

oblique joint see JOINT.

oblique-slip fault *n.* a FAULT in which the NET SLIP has substantial DIP SLIP and STRIKE SLIP components, in comparison with those, such as normal faults, where the DISPLACEMENT is dominantly dip slip or with strike-slip faults. Compare OBLIQUE FAULT.

obsequent fault-line scarp see FAULT-LINE SCARP.

obsequent stream *n.* a watercourse that flows in a direction opposite that of the original inclination of the land, i.e contrariwise to the CONSEQUENT stream and usually, but not always, opposite the dip of the SEDIMENT.

obsidian *n.* VOLCANIC GLASS, usually black but sometimes brown or red, characterised by CONCHOIDAL fracture. Most obsidian is formed from rhyolitic or dacitic MAGMA. See GLASS.

obtuse bisectrix see BISECTRIX.

occipital ring see GLABELLA.

occult minerals *n.* minerals that, on the basis of the chemical analysis, should be present in a glassy IGNEOUS ROCK.

ocean basin *n.* the part of the sea floor that lies beyond the CONTINENTAL SHELF, including ABYSSAL PLAINS and HILLS, oceanic ridges, volcanic islands and trenches. Some oceanographers include the slope and rise of the continental shelf.

ocean-bottom deposits *n.*
(a) TERRIGENOUS DEPOSITS – found on the CONTINENTAL SHELF; these are mainly materials from the land, e.g. shallow-water sands and muds.
(b) NERITIC DEPOSITS – found farther out along the continental shelf; these consist of land (terrigenous) materials mixed with organic substances from shallow coastal waters.
(c) PELAGIC DEPOSITS – found beyond the direct influence of land. These comprise biogenic and nonbiogenic materials the origins of which are almost solely the sea.

Ocean Drilling Program see ODP.

oceanic crust *n.* the part of the Earth's surface that underlies the ocean basins. It varies in thickness from about 5 to 10 kilometres. Compare CONTINENTAL CRUST. See also CRUST.

oceanography *n.* the application of various disciplines to the study of the oceans and seas of the Earth. Such disciplines include marine biology, marine geology, chemistry, physics and meteorology.

oceanography, optical *n.* a branch of physical OCEANOGRAPHY that embraces the study of sea water's optical properties and the underwater field of natural light from the sky and Sun. Optical measurements are useful in work on oceanographic problems, such as diffusion mixing and the characterisation of water masses, as well as in marine biology.

oceanography, physical *n.* that science involved with describing the physical state of the sea and its variations in space and time, and with formulating the physical processes occurring in it.

ocean trench *n.* a long, narrow, submarine depression with relatively steeply inclined sides. The deepest ocean trenches exceed 10 000 metres in depth and are characterised by strongly negative GRAVITY ANOMALIES. Deep oceanic trenches occur on the outward (convex) side of ISLAND ARCS or at the margins of CORDILLERA mountain belts; they are associated with SUBDUCTION ZONES. Although they are located in the three major oceans, the largest number of trenches occurs in the Pacific. See also PLATE TECTONICS.

ocellar *adj.* (of the texture of an IGNEOUS ROCK) containing aggregates of small crystals arranged about larger crystals in such a fashion that eyelike structures are formed.

Ochoan *n.* the uppermost Permian of North America.

ochre *n.* a powdery red, brown or yellow iron oxide used as a pigment, e.g. brown or yellow ochre (LIMONITE) and red ochre (HEMATITE); also, any of various clays deeply coloured by iron oxides. Compare UMBER, SIENNA.

octahedral cleavage see CLEAVAGE.

octahedrite *n.* **I.** see ANATASE.

2. an IRON METEORITE containing between about 6 and 15 per cent nickel. Surfaces that have been polished and etched reveal WIDMANNSTÄTTEN patterns caused by the intergrowth of coarsely crystallised plates of the nickel-iron alloys KAMACITE and TAENITE.

octahedron *n.* (*pl.* **octahedra**) (*crystallography*) an eight-sided closed crystal FORM in which all the sides are equilateral triangles, found in the isometric system (see CRYSTAL SYSTEM). DIAMOND crystals in their natural state commonly exhibit this form.

ocular ridge *n.* a ridge between the eye and the GLABELLA normally found in Cambrian trilobites but rare in later forms. See CEPHALON.

ODP *abbreviation for* Ocean Drilling Program, the successor to DEEP SEA DRILLING PROGRAM and JOIDES.

offlap *n.* the successive contraction in the area covered by BEDS in a SEDIMENTARY sequence caused by deposition in a regressing sea. See Fig. 63 below. It is a term used to describe a conformable sequence of inclined strata lying above an UNCONFORMITY in which each STRATUM is succeeded laterally by a younger unit. Compare OVERLAP.

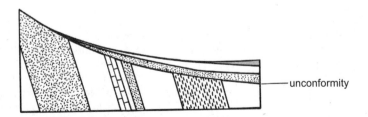

Fig. 63 **Offlap**. The area covered by beds above the unconformity is progressively reduced

offset, also called **horizontal normal separation** *n*. **1.** the horizontal component of DISPLACEMENT along a FAULT, measured perpendicular to the interrupted horizon.
2. in reflection shooting, the horizontal DISPLACEMENT of the reflection point from the vertical plane of the profile. The displacement is used as a correction for converting slant time to vertical time.

offshore bar *n*. **1.** see LONGSHORE BAR.
2. see BARRIER ISLAND.

offshore oil *n*. oil that resides in the largely marine sedimentary rocks that are the submerged extensions of the continents.

ogive *n*. **1.** a wave-like structure on the surface of a GLACIER that develops at the foot of an ice fall (a steep section of a glacier). Ogives are bowed downstream.
2. a ridge in a LAVA FLOW.

oikocryst *n*. a host crystal through which smaller crystals (*chadacrysts*) of different minerals are distributed as poikilitic inclusions.

oil field *n*. a group of HYDROCARBON reservoirs in a common geological setting.

oil pool *n*. a HYDROCARBON reservoir, a subsurface accumulation of petroleum that can yield crude oil in quantities that are economically favourable. The oil occurs in the pores of the RESERVOIR ROCK.

oil sand *n*. a general term for any porous sand or sandstone strata that contain, or are impregnated with, PETROLEUM hydrocarbons or BITUMEN. Such sands can be mined, as in the Athabasca tar sands of Canada. See also GAS SAND, TAR SAND.

oil shale *n*. a SHALE that contains PETROLEUM hydrocarbons. Oil shales exist because their burial was never deep enough to allow the formation of free petroleum. See OIL SOURCES.

oil sources *n*. oil is found in permeable rocks confined by virtually non-permeable ones and in various OIL TRAPS formed by folding, faulting, unconformities or salt movement. Many PETROLEUM deposits occupy the crests of elongated anticlinal folds and domes. Others form around SALT-DOME occurrences. Petroleum (oil and gas) originates exclusively in sediments and almost all petroleum deposits are located in sedimentary rocks. Some rare exceptions receive oil by migration from adjacent sedimentary rocks. Petroleum originates from organic materials accumulating in fine-grained sediments (MUDROCKS), usually in marine environments. Such mudrocks form oil sources or source rocks. The organic material may be derived from marine microfaunas, leading to light oils, or from macroplant debris, leading to heavy oils. The mudrocks contain high proportions of organic matter, in the form of complex carbon compounds, and were accumulated in oxygen-depleted basins. The pressure-temperature conditions resulting from their burial lead to the production of KEROGEN and, from kerogen, petroleum. The pressure-temperature conditions required for the production of petroleum define the 'oil window'. Little free oil has been found in sedimentary rocks of freshwater origin. Instead, such rocks commonly form oil shale (a shale containing no free oil but from which oil can be distilled) in rocks of Carboniferous age in the Midland Valley of Scotland and in Eocene lake deposits in Utah and Wyoming in North America. Oil and coal are not the respective liquid and solid alteration products of PEAT, even though both

olivine

sometimes occur in the same sedimentary sequences in mid-continent. Insofar as oil is absent from Holocene strata, although petroleum precursor compounds have been extracted from them, the oil-forming process seems to be extremely slow.

oil trap *n.* any barrier that impedes the upward movement of oil or gas and allows either or both to accumulate. A distinction is usually made between traps that result from some deformation, such as folding or faulting (e.g. an anticlinal trap), which are called *structural traps*, and those in which certain lithological peculiarities are accountable. These are called *stratigraphical traps* or *lithological traps*. An example would be the grading of a porous section into a non-porous section. See also SHOESTRING SAND.

oil-water contact *n.* the boundary layer or surface between an oil accumulation and the underlying 'bottom water'.

oil window see OIL SOURCES.

Old Red Sandstone *n.* a continental sedimentary facies and its associated volcanic rocks that accumulated in the Late Silurian and Devonian in northwest Europe. Large quantities of sediment, often sands and conglomerates, accumulated in basins between the ranges of the Caledonian Mountains (Caledonides). The sediments are often poorly sorted and quite variable, consisting of conglomerates, red, green and grey sandstones, and grey shales. The sediments are mainly water-deposited, but some aeolian sands also accumulated. A fossil fauna and flora also occur in the facies. Compare NEW RED SANDSTONE.

oligo- *prefix meaning* few or a little.

Oligocene *n.* the epoch of the Tertiary Period between the Eocene and Miocene Epochs. See CENOZOIC.

oligoclase *n.* a plagioclase feldspar ranging in composition from $Ab_{90}An_{10}$ to $Ab_{70}An_{30}$. It is common in IGNEOUS ROCKS of intermediate to high SILICA content and is present as a major constituent in GRANODIORITE and MONZONITE rocks. See FELDSPAR, Fig. 33.

oligomictic *adj.* 1. (of a lake) circulating only at times of inordinately cold weather.
2. (of a clastic SEDIMENTARY ROCK, particularly a CONGLOMERATE) composed almost entirely of a single type of fragment. Compare MONOMICTIC, POLYMICTIC.

oligotrophic lake *n.* a lake characterised by a deficiency in plant nutrients and generally by large amounts of dissolved oxygen in the bottom layers; the bottom deposits have only small amounts of organic matter.

olistholith see OLISTHOSTROME.

olisthostrome or **olistostrome** *n.* a SEDIMENTARY DEPOSIT containing a chaotic mixture of BLOCKS, some very large, in a muddy MATRIX that is believed to form by submarine GRAVITY SLIDING or DEBRIS FLOW in a CONTINENTAL SLOPE and rise environment. The CLASTS are known as olistholiths and are extra-formational in origin. Olisthostromes are sometimes called *sedimentary mélanges*. See also MÉLANGE.

olivine *n.* an olive-green to pale yellow (or brown through alteration) orthorhombic NESOSILICATE, $(Mg,Fe)_2SiO_4$. There is a continuous SOLID-SOLUTION series from FORSTERITE (Mg_2SiO_4) to fayalite (Fe_2SiO_4). Olivine is a common rock-forming mineral, typical of ULTRAMAFIC and MAFIC intrusive or VOLCANIC ROCKS; it is an essential mineral in PERIDOTITE and often occurs in GABBROS and BASALTS. In METAMORPHIC ROCKS, olivine occurs in high-temperature regional or contact-metamorphosed dolomitic limestones. See also PERIDOT, CHRYSOLITE, HORTONOLITE.

omphacite *n*. a sodium-rich INOSILICATE mineral of the PYROXENE group, usually occurring as light-green crystals. It is intermediate in composition between AUGITE and JADEITE and is found with GARNET as an essential mineral of ECLOGITES. It also occurs in KIMBERLITES.

oncolite *n*. a rounded calcareous accretionary body of algal origin. Most are between one and two centimetres in diameter; occasionally they grow to diameters of between 5 and 10 centimetres. Internally they have an irregular concentric laminated structure. Growth takes place on whichever side is uppermost, and the action of water currents rolling the oncolites over will result in a rounded shape. Compare PISOLITH.

onion-skin weathering see SPHEROIDAL WEATHERING.

onlap see OVERLAP.

ontogeny *n*. the development of an individual organism from fertilisation through to the adult form. Compare PHYLOGENY.

onyx *n*. a variety of CHALCEDONY similar to BANDED AGATE in that it consists of alternating white and black or brown layers; the layers of chalcedony, however, are straight and parallel, unlike those of banded agate. Onyx, especially that with alternating black and white layers, is used in making cameos. Compare AGATE, SARDONYX.

ooid see OOLITE.

oolite *n*. a SEDIMENTARY ROCK, usually a LIMESTONE, composed mainly of small accretionary bodies cemented together. These bodies, called *ooliths* or *ooids* from their resemblance to fish eggs, are usually one millimetre or less in diameter. They typically have a concentric or radial internal structure. Ooliths are mostly CALCAREOUS, formed by the precipitation of concentric layers of calcium carbonate around a nucleus such as a shell fragment or a sand grain as it is rolled about the sea floor by currents. (See Fig. 64 below). Oolitic IRONSTONES contain ooliths composed of iron minerals like CHAMOSITE and LIMONITE. They may be the result of primary precipitation, but other iron

Fig. 64 **Oolite**. Ooliths showing radial and concentric structures

minerals such as SIDERITE are the result of early diagenetic replacement (see DIAGENESIS, also MINETTE). See also PISOLITH.

oolith see OOLITE.

oomicrite *n*. a LIMESTONE containing ooliths in a MICRITE matrix (see OOLITE, FOLK CLASSIFICATION).

oosparite *n*. a LIMESTONE composed of ooliths in a matrix of sparry CALCITE (SPARITE). See OOLITE, FOLK CLASSIFICATION.

ooze see PELAGIC DEPOSITS.

opal *n*. a substance variously referred to as a MINERAL, cryptocrystalline, a MINERALOID (if classified as non-crystalline) or a mineral gel. Its chemical composition, $SiO_2 \cdot nH_2O$, indicates the content of an indefinite amount of water, usually 3 to 9 per cent. X-ray studies show that while much opal is amorphous, the precious varieties contain an ordered packing of silica spheres. Opal is commonly white and milky, translucent to opaque, but may also display pale shades of yellow, red, brown, green, grey and blue. Precious opal is distinguished by its often spectacular play of colours. *Fire opal*, a variety of precious opal, has hyacinth-red to yellow colours. Opal is commonly precipitated at low temperatures from silica-rich solutions. It also replaces the skeletons of many marine organisms. *Wood opal* replaces the fibres in fossilised wood without destroying texture or detail. See SILICIFIED WOOD.

opalescence *n*. a milky, lustrous appearance of a mineral, such as of a moonstone or OPAL. Compare PLAY OF COLOUR.

opalised wood see SILICIFIED WOOD.

opaque minerals *n*. MINERALS that, even in THIN SECTION, do not transmit light. They can be studied with a microscope using reflected light; see ORE

MICROSCOPY. Many sulphide and oxide minerals are opaque, e.g. HEMATITE, PYRITE.

opencast mining or **strip mining** *n*. surficial mining in which the valuable substance is exposed by removal of overlying material. Coal, iron and copper are often worked in this way.

open fold see FOLD.

open packing see PACKING.

operculum *n*. **1.** a plate that closes the aperture of a GASTROPOD shell when the animal is drawn up inside.
2. a hinged lid covering the CALICE of some RUGOSE CORALS.

ophicalcite *n*. a marble containing green serpentine.

ophiolite *n*. a suite of MAFIC and ULTRAMAFIC igneous rocks consisting of basaltic PILLOW LAVAS, DOLERITE DYKES, GABBROS and layered CUMULATE rocks, and PERIDOTITES, associated with PELAGIC sediments, which represent segments of oceanic LITHOSPHERE emplaced on the continent during plate collisions (see OBDUCTION). The Troodos ophiolite complex in Cyprus is an example.

ophitic *n*. a texture of an IGNEOUS ROCK in which EUHEDRAL or SUBHEDRAL plagioclase crystals are enclosed in crystals of pyroxene, commonly augite. Compare POIKILITIC.

ophiuroid or **brittle star** *n*. a member of the subclass Ophiuroidea, star-shaped ECHINODERMS with five or more thin flexible arms radiating from a central disc that carries the mouth on the lower surface. They are not common fossils except in the unusual 'starfish' beds, see ASTEROID. Range, Lower ORDOVICIAN to present.

opisthodetic see LIGAMENT.

opisthoparian see CEPHALON.

optical calcite see ICELAND SPAR.

optical emission spectroscopy see Appendix 4.

optic axial angle *n*. the acute angle between the optic axes of a BIAXIAL crystal, designated as 2*V*.

optic axial plane (**OAP**) see AXIAL PLANE.

optic axis *n*. a direction in an ANISOTROPIC crystal along which there is no double refraction.

optic normal see INDICATRIX.

opx *abbreviation for* orthopyroxene. See PYROXENE.

orbicule *n*. a large spherical onion-like mass with distinct concentric shells. The usual size range is between two and 15 centimetres, but some are three metres or more in diameter. They are variously distributed through coarse-grained rocks of silicic to basic composition, for example orbicular granite. Shells within individual masses are sharply defined (compare LITHOPHYSAE), and each differs from its immediate neighbours in texture or composition. The minerals of most orbicules are the same as those of the host rock. The concentric structure seems to be a result of rhythmic crystallisation around centres that sometimes hold a xenolithic nucleus. The formation of orbicules usually occurs at early stages of rock consolidation. See also SPHERULITE.

ordinary ray *n*. that one of the two rays transmitted through a UNIAXIAL mineral that vibrates at right angles to the OPTIC AXIS. It has a constant refractive index and constant velocity. See EXTRAORDINARY RAY, VIBRATION DIRECTIONS, INDICATRIX.

Ordovician *n*. a period of time extending for 70 Ma from 510 to 440 Ma. The name was proposed for rocks in the Arenig Mountains of North Wales, formerly inhabited by an ancient tribe, the Ordovices. The calibration of Ordovician time is based on the distinctive GRAPTOLITE fossils contained in the strata and by use of the abundant igneous rocks found in the system. The Ordovician system is divided into five series, Tremadoc, Arenig, Llanvirn, Caradoc and Ashgill, all defined by successions in Eastern Avalonia (specifically England and Wales). The base of the Ordovician is defined by the first appearance of the CONODONT *Iapetognathus*, and its top is defined by the base of the *Akidograptus acuminatus* BIOZONE of the overlying Silurian. In early Ordovician time Scotland was separated from Eastern Avalonia (southern Britain and southern Ireland) by the IAPETUS Ocean, as shown by the marked differences in the TRILOBITE and BRACHIOPOD faunas of the two areas. The history of the Ordovician was dominated by the closure of Iapetus by the successive collisions of the Scottish terranes and of Eastern Avalonia with Laurentia, the process ending in the earliest Silurian. Faunal differences between Eastern Avalonia and Laurentia had almost disappeared by the end of the Ordovician. subduction-related magmatism and the resulting extensive volcanism from the destruction of Iapetus characterises much of the Ordovician. Volcanic activity was particularly intensive in the Lake District, Wales and southeast Ireland, with eruptions of andesites, rhyolites and ignimbrites.

The Ordovician rocks of Britain and Ireland can be divided into a series of terranes that formerly lay on either side of Iapetus. In Scotland, in the Hebridean terrane, the Lower Ordovician is represented by the limestones and dolomites of the Durness Limestone Formation. In the Grampian terrane, mudstones of the Dalradian Supergroup around Banff may be

Ordovician, and the limestones, shales and sandstones of the Highland Border Complex extend, often in fault-bounded units, through much of the Ordovician. The Midland Valley terrane consists of an older ophiolite and a younger proximal fore-arc sequence, and the Southern Upland terrane is an accretionary prism. Ordovician sedimentary rocks in the Eastern Avalonian terrane of southern Britain and southern Ireland were deposited on shallow marine platforms, such as the Midland Platform and the Irish Sea Platform, and in deep basins, such as the Welsh and Leinster-Lake District Basins. Sediments characteristic of the platforms are calcareous sandstones, mudstones and limestones, and of the basins are deepwater graptolitic shales and greywackes and cherts.

Ordovician faunas, in contrast to those of the cambrian, contain more evolutionarily developed trilobites and brachiopods. Ostracods, crinoids and the bryozoan groups become abundant in the Lower Ordovician. Graptolites, the most important index fossil for this system, are found in all types of sedimentary rock throughout the Ordovician. rugose and tabulate corals appear in the Ordovician, and nautiloid cephalopods diversify and spread widely. Fragmented fossils of jawless freshwater fish (ostracoderms) appear in North America. Spores thought to be from vascular land plants have been found in the mid-Ordovician, but undisputed land plants are not known until the early Silurian. Red, brown and green algae are known from the Ordovician seas, and it is likely that algae and possibly higher plants may be present in fresh water.

ore *n.* a naturally occurring material from which a MINERAL of economic value can be extracted. The word is commonly used of metalliferous minerals but may be expanded to include any natural material used as a source of a nonmetallic substance. A mass of rock or other material housing enough ore to make extraction economically feasible is called an *ore body*.

ore body see ORE.

ore control *n.* any tectonic, geochemical or lithological feature thought to have influenced the formation and placement of ORE.

ore microscopy *n.* the study of polished sections of ore minerals using reflected light. Properties that can be observed in reflected light include colour, PLEOCHROISM, shape, TWINNING and reflectance, which varies from low (dark grey) to high (bright white). In ANISOTROPIC minerals a variation in reflectance can be observed when the stage is rotated. The difference between the maximum and minimum values for reflectance is known as the *bireflectance*.

organic reef *n.* a REEF or BIOHERM.

organic rock *n.* a SEDIMENTARY ROCK consisting essentially of plant or animal remains. Compare BIOGENIC ROCK.

original dip see INITIAL DIP.

orogenic belt *n.* a linear or arcuate zone in the Earth's crust, characterised by deformed and metamorphosed rocks, frequently associated with large plutonic INTRUSIONS in the deeper levels of the belt. Orogenic belts are formed in zones of convergence of plates, especially intercontinental collisions. They are widely represented in the geological record as recorded in the continental crust and often provide evidence of the former existence of mountain ranges now worn down by erosion. See also MOBILE BELT, OROGENY.

orogeny or **tectogenesis** *n*. **1.** mountain-building. The formation of fold-belt mountains.

2. the process by which features such as thrusting, folding and faulting within fold-belt mountainous regions are formed. It is now widely accepted that orogeny is a result of compressive forces acting on crustal segments. See also PLATE TECTONICS.

orpiment *n*. a rare, bright, lemon-yellow to orange monoclinic arsenic mineral, As_2S_3, that occurs in veins of lead, sulphur and gold ores associated with REALGAR. Orpiment is deposited in hot springs and also found in metamorphic dolomites.

ortho- *prefix denoting* upright, straight, genuine, right or rectangular. In metamorphic petrology it indicates derivation from an IGNEOUS ROCK, for example orthogneiss.

orthoclase *n*. a white or pink monoclinic mineral of the potassium feldspar group, $KAlSi_3O_8$, found as prismatic crystals, often twinned, or as compact granular masses. Orthoclase is a common rock-forming mineral and an essential component of many acid and alkaline intrusive rocks that cooled slowly: granites, granodiorites, syenites. It is also common in sedimentary rocks such as arkoses and greywackes. See FELDSPAR; compare SANIDINE, MICROCLINE.

orthoconglomerate *n*. a CONGLOMERATE in which the matrix forms less than 15 per cent of the rock. The larger CLASTS are in contact with each other, forming a framework, and held together by a mineral cement. Orthoconglomerates are commonly well SORTED, with a bimodal particle size distribution. Compare PARACONGLOMERATE.

orthocumulate see CUMULATE.

orthomagmatic (stage) *n*. the main stage of CRYSTALLISATION of silicates from a typical MAGMA during which as much as 90 per cent of the magma may crystallise.

orthopyroxene *n*. orthorhombic PYROXENE.

orthoquartzite or **quartz arenite** *n*. a clastic SEDIMENTARY ROCK composed almost totally of silica-cemented quartz sand; it is a quartzite of sedimentary origin. The rock is distinguished by a scarcity of heavy minerals, lack of fossils and prominent cross-beds. Compare METAQUARTZITE.

orthorhombic system see CRYSTAL SYSTEM.

orthosilicate see NESOSILICATE.

orthotectic see ORTHOMAGMATIC.

Osagean *n*. the upper part of the Lower Mississippian of North America.

oscillation ripple mark see RIPPLE MARK.

ossicle *n*. an ECHINODERM plate.

ostracod *n*. a small bivalved crustacean of the subclass Ostracoda of the phylum Arthropoda, found mostly in oceans but also in fresh water. The carapace has pits, lobes and spines, is hinged at the upper margin and in many species shows sexually dimorphic forms. Ostracods are divided according to features of the carapace, which is shed regularly. They range from Cambrian to present and are usually of microscopic size although larger forms are known. Ostracods are significant as INDEX FOSSILS in limestone, marine marls and shales. Certain types are often used as indicators in subsurface petroleum exploration.

outcrop *n*. an area where rocks occur at the surface. They may or may not be covered by ALLUVIUM and vegetation. Compare EXPOSURE.

outgassing *n*. the removal of occluded gases from MAGMA. Outgassing, in which gases and water vapour are released during volcanic processes, is responsible for the formation and maintenance of the Earth's atmosphere.

outlier *n.* an area of rock surrounded by rocks of older age. Compare INLIER.

outwash *n.* GRAVEL and SAND deposited by meltwater streams on land. *Outwash plains* are produced by the merging of a series of *outwash fans* or *aprons*. See also SANDUR, VALLEY TRAIN.

overburden *n.* **1.** the rocks and/or sediments overlying a certain horizon. **2.** the economically useless material overlying a valuable deposit that is removed in OPENCAST MINING. This material may be unconsolidated or consolidated (see CONSOLIDATION).

overconsolidated clay *n.* **1.** a CLAY that at some stage has been subjected to greater OVERBURDEN pressure than now applies and is therefore stronger than expected. See CONSOLIDATION. **2.** a CLAY that has hardened more rapidly than normal because of desiccation or CEMENTATION.

overland flow *n.* above-ground RUNOFF of water that contributes to STREAMFLOW.

overlap or **onlap** *n.* the successive increase in the area covered by BEDS in a SEDIMENTARY sequence, perhaps caused by a marine TRANSGRESSION. See Fig. 65 below. The term 'overlap' describes the relationship between beds in a se-quence above an unconformity. The term 'onlap' refers to the process by

which overlap is produced. Compare OFFLAP, see also OVERSTEP.

overprinting *n.* the superposition of a set of metamorphic or TECTONIC structures on earlier ones. See METAMOR-PHISM.

oversaturated *adj.* **1.** SILICIC. **2.** silica-oversaturated. Compare UNDERSATURATED. See SILICA SATURATION.

overstep *n.* the truncation of beds lying below an angular UNCONFORMITY by the surface of unconformity. It is a term that indicates the relationship between the beds beneath the unconformity and the surface of unconformity itself, formerly a surface of EROSION. Compare OVERLAP.

overthrust see FAULT.

overturn *n.* the circulation, especially in spring and autumn, of waters of a lake or sea whereby surface water sinks and mixes with bottom water; it is the result of density differences resulting from changes in temperature. See also TURNOVER, CIRCULATION.

overturned fold see FOLD.

oxbow see OXBOW LAKE.

oxbow lake *n.* an arc-shaped body of standing water located in river bends (MEANDERS) that are abandoned when-ever a stream crosses the neck between bends, thus shortening its course. The abandoned meander itself is referred to as an *oxbow*. After the channel is bypassed, the ends of the channel are

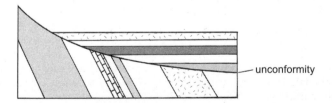

Fig. 65 **Overlap**. Beds above the unconformity cover a progressively wider area

rapidly silted and eventually become a crescent-shaped lake.

Oxfordian *n*. the lowest STAGE of the Upper JURASSIC.

oxidates *n*. sediments composed of the oxides and hydroxides of iron and manganese that have crystallised from an aqueous solution.

oxidation *n*. the process of combining with oxygen.

oxide *n*. a mineral compound of oxygen and one or more metallic elements.

oxidised zone *n*. the upper portion of an ORE body that has been modified by a solution of surface waters bearing oxygen, soil acids and carbon dioxide. It extends from the surface to the WATER TABLE. Primary ore minerals in this zone react with this solution to produce various secondary minerals containing oxygen. Chemical alteration of sulphide minerals within the zone of oxidation results in the formation of acids that then enable METEORIC WATER to act as a solvent. For example, the weathering of PYRITE, FeS_2, produces ferric sulphate, $Fe_2(SO_4)_3$, and sulphuric acid, H_2SO_4. The ferric sulphate reacts with sulphides of copper, lead and zinc to form soluble sulphates of these metals. The sulphuric acid dissolves carbonates such as the GANGUE minerals, CALCITE and DOLOMITE. As a result of these processes, buried deposits of copper and copper sulphide will change to malachite, $Cu_2 CO_3(OH)_2$, or azurite, $Cu_3(CO_3)_2(OH)_2$, under increasingly oxidising conditions, i.e. towards the surface of the Earth. See SECONDARY ENRICHMENT, PROTORE.

oxycone see AMMONOID.

ozocerite or **mineral wax** *n*. a dark-coloured paraffin wax that occurs in irregular veins. Many varieties occur in association with coal and bitumen.

ozone layer *n*. a region of the stratosphere that contains most of the atmospheric ozone (O_3) where it is produced by photochemical reactions. The ozone layer absorbs much of the ultraviolet radiation, with wavelengths between about 200 and 300 nm, which is harmful to organisms. For much of geological time, life was confined to shallow water; this provided a shield from ionising radiation. As atmospheric oxygen levels increased (as a result of photosynthesis) and the ozone layer developed, organisms could eventually begin to colonise the land. The existence of spores of land plants of SILURIAN age is evidence that this process had begun by about 440 Ma ago.

The destruction of ozone by the reaction with CFCs (halocarbons) from AEROSOL propellants and refrigerants, etc, leads to an increase in the amount of UV radiation reaching the Earth's surface. See also ATMOSPHERE.

Pp

Pacific-type coastline *n*. a coastline that runs generally parallel to the dominant trend of the land structures, such as mountain chains. The margins of the Pacific Coast CORDILLERAS of North America and the island area of the western Pacific are ideal representations. See also ATLANTIC-TYPE COASTLINE.

packing *n*. the spatial arrangement of grains or crystals in a rock. It is a function of packing density and packing proximity. If uniform, solid spheres are packed as loosely as possible as porosity is at a maximum; this arrangement is known as *open packing*. If the spheres are packed as closely as possible so that porosity is at a minimum, the arrangement is known as *close packing*. An aggregate with *cubic packing* has a porosity of about 47 per cent. This is the 'loosest' systematic arrangement of uniform spheres in a CLASTIC sediment or CRYSTAL LATTICE. The unit cell is a cube the eight corners of which are the centres of the spheres. *Rhombohedral packing* is the 'tightest' systematic arrangement of uniform solid spheres in a crystal lattice or clastic sediment. The unit cell is composed of six planes that pass through the centres of eight spheres positioned at the corners of a regular rhombohedron. An aggregate with such packing has the lowest porosity possible (25.95 per cent) without grain distortion. The average porosity of a *chance-packed* aggregate of uniform spheres is slightly under 40 per cent. This is a random combination of grains packed in ordered fashion and grains packed haphazardly. See Fig. 66 below.

packstone *n*. a clastic limestone supported by its own grains with a calcareous mud matrix. Compare MUDSTONE, GRAINSTONE, WACKESTONE. See DUNHAM CLASSIFICATION.

paedomorphosis *n*. see HETEROCHRONY.

pahoehoe *n*. a basaltic flow with a smooth undulating surface formed from lava of low viscosity. The surface often displays *ropy* structures and bulbous protrusions (*pahoehoe toes*). Pahoehoe commonly forms thin flow units of up to about two metres in thickness that overlap and build up in multiple layers during individual

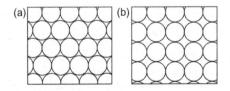

Fig. 66 **Packing**. (a) rhombohedral or close packing;
(b) cubic or open packing

eruptions. Internally within each layer there is an upper zone rich in spheroidal VESICLES trapped below the skin formed on the upper surface as it cools, and a lower zone largely devoid of vesicles. The term 'pahoehoe' originated in Hawaii. Compare AA. See also LAVA FLOWS.

palaeo- *prefix denoting* old or ancient, as in PALAEOCURRENTS.

palaeobotany see FOSSIL PLANTS.

Palaeocene *n.* the lowermost epoch of the Tertiary Period. See CENOZOIC ERA.

palaeoclimatology *n.* the study of climate through geological time, extending back to the Precambrian. Climatic data prior to modern instrumental measurement is obtained by *proxies*, geological evidence from which palaeotemperatures and other climatic indicators can be derived. These proxies include:

(a) *glaciological evidence* derived from ice – geochemistry, gas content of bubbles;

(b) *geological evidence*, marine – biogenic sediments, aeolian dust, clay mineralogy;

(c) *terrestrial* – glacial features and deposits, SHORELINES (eustatic and glacio-eustatic features), PALAEOSOLS, SPELEOTHEMS;

(d) *biological evidence* – DENDROCHRONOLOGY, PALYNOLOGY, plant macrofossils, coral, GEOCHEMISTRY, microfossils;

(e) *isotopic evidence* – oxygen isotopic compositions of macro- and micro-fossils;

(f) *sedimentological evidence* – RAIN PRINTS, DUNES, Heinrich cycles (ice melt-out cycles), ice-rafted debris.

Such proxies can be used to produce weather maps and other climatic interpretations for any point within the later Precambrian and Phanerozoic. See also PALAEOCURRENTS, PALAEOECOLOGY, PALAEOGEOGRAPHICAL MAP.

palaeocurrents *n.* currents of water or wind, active in the geological past, the direction of which can be inferred from sedimentary structures such as CROSS-BEDDING, RIPPLE MARKS and STRIATIONS. DUNE type and heavy mineral distribution are also used to determine palaeocurrent direction.

palaeoecology *n.* the study of all aspects of relationships between FOSSIL organisms and between fossil organisms and their ancient life environments. These relationships are accessed by the use of PALAEONTOLOGY, SEDIMENTOLOGY and the application of modern ecological laws. Much palaeoecological work depends on comparative studies of modern and ancient forms.

Palaeogene *n.* the oldest of the Tertiary periods, the Palaeocene, Eocene and Oligocene epochs of the Tertiary, when grouped together. Compare NEOGENE.

palaeogeographical map *n.* a map showing reconstructed physical GEOGRAPHY at a specified time in the geological past; it includes the distribution of land and seas, depth of the seas, GEOMORPHOLOGY of the land and climatic belts.

palaeogeological map *n.* a map that shows the areal GEOLOGY of a land surface at some time in the geological past, e.g. showing an unconformity as it was before it was covered with deposits.

palaeomagnetism *n.* the study of changes of the geomagnetic field during geological time. The basic assumption is that some rocks, at the time of their formation, acquire a permanent record of the Earth's magnetic field because of the presence of magnetic iron-bearing minerals. (See NATURAL REMANENT MAGNETISATION.) Palaeomagnetism has provided substantial evidence to support aspects of the theory of PLATE TECTONICS from studies of apparent POLAR WANDERING. It has been used to locate the past positions of landmasses in relation to each other. Studies of alternating magnetic patterns on either side of mid-ocean ridges (see GEOMAGNETIC POLARITY REVERSALS) have provided evidence for the hypothesis of SEA-FLOOR SPREADING.

palaeontology *n.* the study of life forms that existed in past geological periods, as represented by FOSSIL PLANTS and animals. See also HISTORICAL GEOLOGY.

palaeosols *n.* soils of past geological ages, usually buried by more recent materials.

palaeotectonic map *n.* a map intended to represent geological and tectonic features as they were at some time in the geological past rather than the totality of all tectonics of a region.

palaeotemperature *n.* surface tempera-ture during the geological past, as determined indirectly from the evidence of different minerals, rocks, fossils and geochemical isotope ratios. For example, certain corals can live only within a particular temperature range. The presence of iceberg-transported glaciated boulders in water the present temperature of which does not fall below 10°C indicates much lower palaeotemperatures. Because the isotope ratio of ^{16}O to ^{18}O is partly temperature-dependent, its measure-ment in carbonate shell matter of marine organisms indicates the ambient water temperature at the time of shell secretion.

Palaeotethys *n.* a name variously used to indicate:
(a) the equatorial ocean that existed during the Devonian between GONDWANA and a group of northern equatorial continental units;

(b) as an embayment of the ocean Panthalassa (see PANGAEA) during the Permian and Triassic. The term NEOTETHYS has been used for the same embayment of Panthalassa and more accurately for the Jurassic-Cretaceous ocean opened along the northern margin of the embayment Palaeotethys.

Palaeozoic *n.* the ERA extending from about 544 Ma to 245 Ma. The Lower Palaeozoic includes the Cambrian, Ordovician and Silurian periods. The Upper Palaeozoic includes the Devonian, Carboniferous and Permian periods; it was preceded by the Precambrian and followed by the Mesozoic.

palaeozoology *n.* that branch of PALAE-ONTOLOGY in which both vertebrate and invertebrate animals are studied.

palagonite *n.* a yellow to orange-brown MINERALOID that is optically isotropic and often concentrically banded. It is produced by the hydration of basaltic glass fragments formed during the submarine or subglacial eruption of lavas. See also PALAGONITE TUFF.

palagonite tuff *n.* a consolidated deposit of glassy basaltic ash in which most of the constituent particles have been altered to PALAGONITE.

paleosome see MIGMATITE.

palimpsest *n.* the fabric or texture of a METAMORPHIC ROCK in which remnants of a pre-existing fabric or texture are still preserved. Compare RELICT.

palingenesis *n.* the formation of a new magma by *in situ* melting of pre-existing rocks. Compare ANATEXIS, NEOMAGMA.

palinspastic map *n.* a type of GEOLOGICAL MAP in which the effect of various Earth movements is removed in order to re-store the rocks to their original posi-tions, for example by unfolding FOLDS.

pallasites *n.* STONY IRON METEORITES consisting of crystalline OLIVINE embedded in a matrix of nickel-iron (10 to 20 per cent Ni).

pallial line *n*. a curved line joining the MUSCLE SCARS on the inner surface of a BIVALVE shell (Fig. 7). It marks the attachment of the animal to the VALVE. In forms with a retractable siphon, there is an indentation in the line, the *pallial sinus*.

pallial sinus see PALLIAL LINE.

palpebral lobe *n*. a flat area of the fixed cheek above the eye on the CEPHALON of a TRILOBITE, see Fig. 89.

paludal *adj*. relating to a MARSH.

palustrine *n*. pertaining to substances growing or deposited in a marsh or PALUDAL environment.

palygorskite *n*. a fibrous CLAY MINERAL, $(Mg,Al)_2Si_4O_{10}(OH)\cdot4H_2O$, with lath-like crystals, that forms a series with SEPIOLITE. It occurs in sediments from PLAYA LAKES and DESERT soils and can be a component of FULLER'S EARTH.

palynology *n*. the study of pollen and spores, and their dissemination. Ancient pollens are used in stratigraphical work, especially in the petroleum industry.

panfan see PEDIPLAIN.

Pangaea *n*. a supercontinent that existed about 300 to 200 Ma ago. This landmass included most of the continental crust of the Earth. The assembly of Pangaea took place as a series of continental collisions during the later Palaeozoic and earlier Mesozoic. About 320 Ma ago, during the Carboniferous, the continents GONDWANA and LAURASIA converged, forming the Hercynian-Appalachian fold belt. During the Permian, the continents of Siberia and Kazakhstania (Central Asia) collided, the resulting unit itself colliding with Laurasia along the line of the Ural fold belt. Finally, in the Triassic the continental units of China collided with

eastern Pangaea. By about 225 Ma, Pangaea was a single landmass that stretched from pole to pole. Pangaea was surrounded by a universal ocean called Panthalassa. The present continents are derived from the fragmentation of Pangaea, which began during the later Triassic, about 200 Ma ago. The concept of the supercontinent Pangaea was part of the theory of CONTINENTAL DRIFT proposed in 1912 by Alfred Wegener, a German climatologist. The theory remained controversial until the 1960s, when supporting evidence led to its increased acceptance. See PLATE TECTONICS.

panning *n*. a technique of prospecting for heavy metal by swirling PLACER or crushed vein material in a pan. Lighter material is washed away, and the heavy metals remain in the pan.

Pannotia *n*. a short-lived supercontinent that existed between about 570 Ma and 550 Ma. Pannotia was produced from the destruction of the preceding supercontinent, RODINIA.

pantellerite *n*. a PERALKALINE RHYOLITE containing PHENOCRYSTS of ANORTHOCLASE and aenigmatite (a sodium/titanium silicate mineral), sometimes with sodic PYROXENE, in a GROUNDMASS that is mainly QUARTZ and FELDSPAR. See also COMENDITE.

Panthalassa see PANGAEA.

para- *prefix denoting* parallel to, beside or resembling. For example paraconformity, see UNCONFORMITY. Para- as a prefix to the name of a metamorphic rock indicates derivation from a sedimentary rock, for example paragneiss.

parabolic dune see DUNE.

paraconformity see UNCONFORMITY.

paraconglomerate *n*. a CONGLOMERATE containing more than 15 per cent matrix. The CLASTS are not in contact but are matrix-supported. Paraconglomerates are commonly poorly SORTED, with a very variable particle size distribution. Compare ORTHOCONGLOMERATE.

paraffin-base crude *n*. CRUDE OIL that will yield large quantities of paraffin during distillation. Compare ASPHALTIC-BASE CRUDE, MIXED-BASE CRUDE.

paraffin series *n*. the methane, CH_4, series; its general formula is C_nH_{2n+2}

paragenesis *n*. of minerals, the characteristic association of a mineral or mineral assemblage in a particular geological setting.

paralic *adj*. **1.** relating to the environments of marine margins such as littoral basins and lagoons.
2. (of coal deposits) formed along sea margins, as contrasted with LIMNIC deposits.

parallel cleavage see BEDDING-PLANE CLEAVAGE.

parallel drainage pattern see DRAINAGE PATTERN.

parallel fold see FOLD.

paramagnetic *adj*. having a magnetic permeability slightly greater than 1 and a small positive magnetic susceptibility. Platinum has a permeability value of 1.00002. A paramagnetic mineral placed in a magnetic field shows a magnetisation in direct proportion to the field strength; magnetic movements of the atoms have random directions. Compare DIAMAGNETIC, FERROMAGNETIC.

parametral plane *n*. a CRYSTAL FACE, selected as a standard to which all other faces may be referred, that cuts all three CRYSTAL AXES. See MILLER INDICES.

paramorph *n*. a PSEUDOMORPH having the same composition as the original CRYSTAL.

paraphyletic *adj*. a paraphyletic group is a group of organisms that does not include all the descendants of a common ancestor. Compare MONOPHYLETIC. See CLADISTICS.

parasitic cone see VOLCANIC ERUPTION, SITES OF.

parasitic folds *n*. small FOLDS developed within larger folds. They usually show S- or Z-shaped forms on the limbs of the larger fold and M forms in the

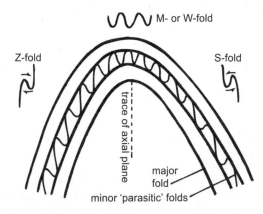

Fig. 67 **Parasitic folds**. The sense of asymmetry or vergence of the minor folds is directed towards the axial trace of the major fold and is shown by an apparent anticlockwise rotation of the S-fold and an apparent clockwise rotation of the Z-fold

HINGE. See Fig. 67 on previous page. See also VERGENCE.

parautochthonous *adj.* (of TECTONIC structures such as NAPPES) having characteristics intermediate between allochthonous and autochthonous. It can be difficult, however, to establish how far the structures have moved. See AUTOCHTHON, ALLOCHTHON.

parent element *n.* the radioactive element from which daughter products are formed by radioactive decay.

parental magma *n.* the MAGMA from which a particular IGNEOUS ROCK was formed or from which a MINERAL crystallised; it may also be the source of another magma. Compare PRIMARY MAGMA. See also MAGMATIC DIFFERENTIATION.

particle size *n.* the general dimensions, such as average diameter or volume, of particles or grains that make up a sediment or rock. The gauge of particle size is based on the premise that particles are spheres. The size range of particles in a sedimentary deposit can be determined by measurement of pebbles and larger fragments, sieving (for sands and gravels) and ELUTRIATION of silt and finer material. Sedimentologists commonly use the Udden-Wentworth grade scale (commonly called the Wentworth scale) or the PHI GRADE SCALE to classify particle size. See WENTWORTH GRADE SCALE and Fig. 94.

particle velocity *n.* in wave motion, the velocity with which an individual particle moves in a medium.

parting *n.* a plane along which a mineral will split fairly readily but which is not a CLEAVAGE plane. The weakness may have been produced by deformation.

parting lineation see PRIMARY CURRENT LINEATION.

passive continental margin see CONTINENTAL MARGIN.

patch reef see CORAL REEF.

paternoster lakes *n.* a chain of small lakes occupying a rock basin in a glacial valley. Their name derives from the similarity of their arrangement to beads on a rosary.

patterned ground *n.* various forms, more or less symmetrical, that are characteristic of but not restricted to surficial material subject to extensive frost action. *Striped ground*, a pattern of alternating stripes, is produced on a sloping surface by FROST ACTION. If the pattern is oriented down the steepest available slope, the alternating bands of fine and coarser material comprising the pattern is called a *sorted stripe*. The finer component is a *soil stripe* and the coarser a *stone stripe*; if the material is very coarse, it is a *block stripe*. A sorted stripe never forms singly; an individual stripe may range from a few centimetres to two metres wide and exceed 100 metres in length. *Sorted circles* and *sorted polygons* are dominantly circular or polygonal shapes with a border of stones surrounding finer material. The diameters of both range from a few centimetres to 10 metres. A *sorted net* is a form whose mesh is intermediate between that of a sorted circle and a sorted polygon. A steplike form with a downslope border of stones containing an area of finer material is called a *sorted step*. Forms of patterned ground in which the border of stones is absent and in which vegetation characteristically outlines the pattern, are described as *nonsorted*. Thus there are nonsorted forms of stripe, step, circle, polygon and net. See also NONSORTED POLYGON.

peacock coal *n.* an iridescent BITUMINOUS or ANTHRACITE COAL. It occurs in the upper levels of mines where water seepage through rock fractures deposits a film of iron oxide.

peacock ore *n*. an informal name for a copper mineral, BORNITE, with a lustrous, tarnished, iridescent surface.

peak zone see ACME ZONE.

peat *n*. a deposit of semi-carbonised plant remains in a water-saturated environment such as a BOG or a meadow with spongy soil. As dead plants fall into the bog or swamp waters, anaerobic conditions permit partial but not total decomposition of the remains. Successive accumulations result in the compaction of material at the bog bottom, and eventually the entire bog may be covered. A vertical section of a peat bog shows the stages of development: living moss, dead moss, laminated peat and solid peat.

Peat that has been cut into blocks and dried is used as fuel in some European countries. Although its carbon content (50 to 60 per cent) and calorific value are fairly high, it is an inferior fuel because of its smoke and heavy ash. Peat is considered to be an early stage in the development of COAL. The climatic conditions of the CARBONIFER-OUS, over land surfaces north and south of the Equator, fostered the growth of the great tree ferns and created vast swamp areas; this led to the accumulation of the first beds of forest peat. Under pressure, peat dries and hardens to become LIGNITE.

pebble *n*. a rock fragment, generally rounded, with a diameter of 4 to 64 millimetres. Its size is between that of a GRANULE and a COBBLE.

pedestal rock see ROCK PEDESTAL.

pedicle *n*. the stalk by which most BRACHIOPODS are fixed to the sea floor (see Fig. 9). See also DELTHYRIUM.

pedicle foramen see DELTHYRIUM.

pedicle valve *n*. the VALVE from which the PEDICLE emerges in a BRACHIOPOD shell (see Fig. 9). Usually, it is the larger of the two valves and is ventral in position.

pediment or **rock pediment** *n*. a very slightly inclined erosion surface that slopes away from a mountain front; it is typically formed by running water. The bedrock may be exposed or thinly covered with alluvium. Pediments are sometimes mistaken for bajadas; they are most conspicuous in BASIN-AND-RANGE-type arid regions. Compare BAJADA. See also PEDIPLAIN.

pediplain *n*. any degradational PIEDMONT surface produced in arid climates and either exposed or covered with a thin blanket of contemporary ALLUVIUM.

pedogenesis *n*. SOIL formation. The study of the MORPHOLOGY, origin and classification of soils is *pedology*.

pedology see PEDOGENESIS.

pegmatite *n*. a very coarse-grained rock (with grain size larger than 1 to 2 centimetres), typically found in veins or lenticular or podlike bodies around the margins of large deep-seated PLUTONS, usually extending from the pluton itself into the surrounding country rocks. Although diorite and gabbro pegmatites occur, SYENITE and GRANITE ones are the most common pegmatites. Many pegmatites have an internal zonation of fabric and composition. *Simple pegmatites* are generally very coarse-grained equivalents of common plutonic rocks, such as granite, when they would comprise mainly quartz and potassium feldspar, with lesser amounts of muscovite and small amount of tourmaline. Of greater mineralogical and economic interest are *complex pegmatites*, which contain concentrations of rare minerals rich in REE, Li, Be, etc. Complex pegmatites contain the Earth's largest recorded crystals, such as SPODUMENE, up to about 15 metres in length and individual

TOPAZ crystals weighing hundreds of kilograms. Pegmatites represent a volatile-rich, late stage in the CRYSTALLI-SATION of MAGMA. Coarse-grained veins and patches in GNEISSES and SCHISTS, produced under metamorphic conditions, are often referred to as pegmatites but are normally of the simple type.

pelagic *adj.* **1.** pertaining to the water of the ocean environment.

2. (of marine organisms) living without direct dependence on the sea bottom or shore. Among pelagic organisms are the free-swimming (*nektonic*) animals, such as fish, and floating forms (*planktonic*), such as jellyfish and sargassum weed, that are carried by the ocean currents.

3. (of sediments) covering the floor of the open sea, as distinct from the TERRIGENOUS DEPOSITS of coastal areas. See also CONTINENTAL SHELF.

pelagic deposits *n.* deep OCEAN-BOTTOM DEPOSITS beyond the direct influence of the land, composed of materials that originate almost exclusively in the sea and are fine enough to be carried in suspension. Pelagic deposits consist of four primary biogenic *oozes* and non-biogenic RED CLAY. Great quantities of manganese oxide nodules are occasionally found in association with the oozes:

(a) *calcareous deposits*: *pteropod ooze* – occurs principally in tropical regions at depths of 1000 to 2500 metres. The shells are delicate and readily dissolve at greater depths; *globigerina ooze* is found at depths of 2000 to 4000 metres. About two-thirds of the Atlantic sea floor and more than one-third of the total sea floor are covered with globigerina ooze. These oozes are unusual at depths exceeding 4000 metres, since the $CaCO_3$ of the shells dissolves in sea water under great pressure. See GLOBIGERINA, PTEROPOD.

(b) *siliceous deposits*: *diatom ooze* – found at depths of 1100 to 4000 metres in Antarctic regions and in the southern and far northern Pacific. Although DIATOMS themselves are not restricted to these regions, their shells are so abundant here that other skeletal remains are obscured; *radiolarian ooze* is found at depths greater than 5000 metres in isolated areas of the tropical Pacific and Indian Oceans. At depths greater than 4000 metres, $CaCO_3$ is more soluble than silica, hence only siliceous skeletal remains survive in sediments deposited below these depths. Siliceous oozes cover about 38 x 10^6 square kilometres (15 x 10^6 sq miles) of ocean bottom.

(c) *red clay*: mostly inorganic material deposited anywhere on the sea bottom but more notable in the great depths where it is not obscured by organic deposits. It consists of ultrafine particles of volcanic ash, both submarine and terrestrial, desert dust and some meteoritic material; oxides of iron and manganese give it its red colour. It constitutes more than half the Pacific floor and about one-third the combined area of all sea floors. There is no red clay on the land that corresponds to the red clay of the ocean floor. See also OCEAN-BOTTOM DEPOSITS.

pelecypod see BIVALVE.

Pele's hair *n.* golden brown to dark brown glassy filaments of solidified basaltic lava, sometimes exceeding two metres in length, resembling human hair. When drops of liquid lava of low viscosity are ejected during fountaining at a LAVA LAKE, they draw out a thread of liquid behind them. The threads solidify, break off from the parent blobs and may drift with the wind for distances up to several kilometres. They are often observed during eruptions of the Hawaiian volcanoes Kilauea and Mauna Loa. See PELE'S TEARS

Pele's tears *n.* cylindrical, spherical or pear-shaped drops of glassy basaltic lava from 6 to 13 millimetres long, named after the Hawaiian goddess of volcanoes and fire. They are ejected as small blobs of liquid lava and solidify during flight. See PYROCLASTIC MATERIAL.

pelite *n.* a MUDSTONE or LUTITE, or the metamorphic derivative of a lutite. See also PSAMMITE, PSEPHITE.

pelitic *adj.* pertaining to or derived from PELITE. It is most commonly used now for metamorphosed argillaceous rocks. Compare ARGILLACEOUS, LUTACEOUS.

pellets *n.* small ovoid or spherical particles of MICROCRYSTALLINE CALCITE without any internal structure (unlike ooliths, see OOLITE). They can range in size from 0.03 to 0.15 millimetres in diameter, but in any one rock the pellets tend to be all the same size and shape. Pellets may be faecal or of algal origin: if the origin is not clear they may be called *peloids*. See ALLOCHEMS; see also LIMESTONE.

pelmatozoan *n.* any ECHINODERM living attached to a substrate, either with or without a stem. The term relates to life habit and is not a true taxonomic identification. See CRINOID, CYSTOID.

pelmicrite *n.* LIMESTONE composed of a variable proportion of carbonate mud (MICRITE) and small rounded aggregates of sedimentary material. See FOLK CLASSIFICATION.

peloids see PELLETS.

pelsparite *n.* a LIMESTONE composed of PELLETS in a matrix that consists mainly of SPARITE. See FOLK CLASSIFICATION.

pendent see GRAPTOLITE.

Pendleian *n.* the lowest STAGE of the NAMURIAN (CARBONIFEROUS) in Britain and Western Europe.

penecontemporaneous *adj.* formed during or shortly after the deposition of the containing rock stratum.

peneplain *n.* a hypothetical surface to which landscape features are reduced through long-continued MASS WASTING, stream EROSION and sheet wash (*peneplanation*). The concept of such a surface is based on deductions from stream behaviour that suggest a reduction to base level of even resistant rocks by continuous stream erosion.

penetration twin see TWINNING.

penetrative fabric *n.* a structure or texture in a METAMORPHIC ROCK that affects all the mineral grains, for example SLATY CLEAVAGE. A *non-penetrative* fabric is one affecting only some parts of the rock, for example FRACTURE CLEAVAGE.

Pennsylvanian *n.* the Upper Carboniferous subperiod in North America, compare the Lower Carboniferous subperiod MISSISSIPPIAN. The term has been formally recognised by the United States Geological Survey since 1953. It is used extensively in North America but not elsewhere. See MISSISSIPPIAN-PENNSYLVANIAN BOUNDARY, STRATIGRAPHICAL CORRELATION.

pentane *n.* any of three paraffin hydrocarbons, formula C_5H_{12}, found in NATURAL GAS and PETROLEUM.

pentlandite *n.* a bronze to yellow isometric mineral, $(FeNi)_9S_8$, the principal ore of nickel. It occurs in massive form, commonly intergrown with PYRRHOTITE.

peperite *n.* a BRECCIA composed of a mixture of glassy LAVA fragments and SEDIMENTARY material formed when MAGMA flows over or is intruded through wet sediments.

peralkaline *adj.* (of igneous rocks)

having $Na_2O + K_2O > Al_2O_3$ (in molecular proportions). Peralkaline rocks typically contain alkali PYROXENES and/or alkali AMPHIBOLES, for example AEGERINE, RIEBECKITE. The CIPW norm will contain acmite and perhaps sodium metasilicate. See CIPW NORMATIVE COMPOSITION.

peraluminous *adj.* (of igneous rocks) having $Al_2O_3 > CaO + Na_2O + K_2O$ (in molecular proportions). The characteristic minerals include MUSCOVITE, TOPAZ, TOURMALINE, SPESSARTINE-ALMANDINE and CORUNDUM. Peraluminous rocks are much less common than PERALKALINE ones.

peramorphosis see HETEROCHRONY.

perched boulder *n.* a large detached rock lying stably or unstably on a hillside.

perched water table *n.* a local WATER TABLE above an IMPERMEABLE layer of very limited extent, such as a lens of CLAY within a SANDSTONE bed.

percolation *n.* the slow laminar movement of water through any small openings within a porous material. See also INFILTRATION.

percussion mark *n.* a lunate scar formed on a hard, compact PEBBLE by a sharp blow, possibly indicating high-velocity flow.

percussion drilling see CABLE-TOOL DRILLING.

perennially frozen ground see PERMAFROST.

pergelation *n.* the formation of permanently frozen ground in either the past or the present.

pergelisol see PERMAFROST.

peri- *prefix denoting* near or around.

periclase *n.* an isometric mineral, MgO. It may be found in LIMESTONE affected by contact METAMORPHISM. It readily alters to BRUCITE, $Mg(OH)_2$.

pericline *n.* a FOLD in which the DIPS are inclined in all directions about a central area.

peridot *n.* an old name for OLIVINE. The name peridot is now confined to the olivine-green variety when used as a GEMSTONE.

peridotite *n.* a coarse-grained plutonic igneous rock; its essential minerals are OLIVINE, CLINOPYROXENE and/or ORTHOPYROXENE, with little or no FELDSPAR. It is often found in OPHIOLITE complexes and LAYERED INTRUSIONS. Many peridotitic XENOLITHS found in basalt lava are thought to be samples of the Earth's upper MANTLE brought to the surface by mantle-derived MAGMA. Peridotite is often associated with deposits of nickel-bearing minerals. There are many varieties of peridotite depending on the type of pyroxene and accessory minerals, e.g. WEHRLITE (olivine + cpx), LHERZOLITE (olivine + opx + cpx) and HARZBURGITE (olivine + opx).

periglacial *adj.* (of areas) having locations, conditions, processes and topographical features that are adjacent to the borders of a GLACIER. A periglacial climate, for instance, is characterised by low temperatures, many fluctuations about the freezing point and strong wind action, at least during certain times of the year.

perignathic girdle *n.* an extension of the PERISTOME margin inside the TEST of an ECHINOID, which forms a support for the jaws (see Fig. 27(d)).

period see GEOCHRONOLOGICAL UNIT.

periproct *n.* a membrane studded with small plates and containing the anus, which lies at the centre of the APICAL DISC of an ECHINOID TEST (see Fig. 27(c)).

peristerite *n.* a fine intergrowth of ALKALI FELDSPAR (ALBITE) and a more Ca-rich PLAGIOCLASE in approximately equal amounts. It is a type of ANTIPERTHITE.

peristome *n.* a membrane studded with

small plates surrounding the mouth on the oral surface of an ECHINOID (Fig. 27(c)).

perlite *n.* lustrous pearly grey glass with perlitic texture.

perlitic *adj.* (of glassy and devitrified rocks) having a texture consisting of clusters of concentric curved fractures. A small kernel of black uncracked OBSIDIAN is sometimes found in the core of a cluster of cracks. Perlitic glasses contain up to 10 per cent, by weight, of water while obsidian without cracks usually has less than 1 per cent water. Contraction during cooling is one theory for the development of perlitic texture. Another is that hydration of a thin outer layer of the original obsidian causes this part to expand and crack away from the non-hydrated section. As hydration continues, perlitic cracks develop inwards.

permafrost *n.* SOIL or subsoil that is permanently frozen. It occurs in those regions (arctic, subarctic, alpine) where the mean annual temperature of the soil remains below freezing. Permafrost underlies almost 20 per cent of the Earth's land area, where its thickness ranges from 30 centimetres to more than 1000 metres. Exploration during the past decade has revealed the existence of sub-sea permafrost off the Alaskan shore. See also PINGO.

permanently frozen ground see PERMA-FROST.

permeability *n.* 1. the measure of the ability of earth materials to transmit a fluid. It depends largely on the size of pore spaces and their connectedness; it is less dependent on the actual porosity. For example, a clay or shale with a porosity of 30 to 50 per cent may be far less PERMEABLE to water than gravel with a porosity of 20 to 40 per cent, because the pore interstices of the clay

are so small that flow is inhibited. *Intrinsic permeability* is the property of a rock that controls the permeability (HYDRAULIC CONDUCTIVITY). It is a property of the rock itself and remains constant whatever fluid flows through the rock. The customary unit of measurement of intrinsic permeability is the millidarcy (see DARCY). Compare ABSOLUTE PERMEABILITY, EFFECTIVE PERME-ABILITY, RELATIVE PERMEABILITY.

2. the ratio of magnetic flux density B to the inducing field intensity H, expressed as $\mu = B/H$.

permeable *adj.* (of a rock or sediment) allowing a gas or fluid to move through it at an appreciable rate via large capillary openings.

Permian *n.* the period at the end of the Palaeozoic Era, from 286 to 245 Ma, which takes its name from the district of Perm in Russia, the site where this sequence was first recognised. Permian rocks are subdivided into a Lower Permian series, the Cisuralian, and two Upper Permian Series, the Guadalu-pian and the Lopingian, the boundary between Lower and Upper Permian being placed at the base of the Roadian stage of the Guadalupian. The stages of the Lower Permian are the Asselian, Sakmarian, Artinskian and Kungurian. The stages of the Upper Permian are, from the base, the Roadian, Wordian, Capitanian, Wuchiapingian and Changhsingian. Correlation in the dominantly continental facies of the Permian is difficult but can be done using spores. In the marine facies of the Permian, FUSULINIDS and AMMONOIDS are the most useful INDEX FOSSILS. Following the Variscan orogeny in the Late Carboniferous/Early Permian, Laur-asia and Gondwana were joined to form PANGAEA. Britain lay just north of the Equator at this time, and the Lower

Permian sediments were mainly continental: desert sandstone, red marls and some fluvial breccias and sandstones. Sediments deposited during the Upper Permian included marine dolomitic limestones and EVAPORITES formed during repeated incursions of the Zechstein Sea (which stretched across northern and central Germany and extended into northern England). The evaporites are important economic deposits and also form traps for PETROLEUM accumulation in the North Sea.

The end of the Permian saw the greatest of the mass EXTINCTIONS, many fossil groups, including the TRILOBITES, tabulate and RUGOSE CORALS, productid BRACHIOPODS, most crinoids, goniatite cephalopods and fusulinid Foraminifera, becoming extinct. One of the reasons for these extinctions is thought to be the reduction in the area of shelf seas with the formation of Pangaea. Amphibians and some fish decline, although labyrinthodonts still thrive. Freshwater fish are known in much greater variety than marine forms. The first major reptile evolution took place, with the emergence of three principal groups: cotylosaurus, pelycosaurus and therapsids, although considerable extinctions of terrestrial faunas also took place. Many new insects appear, including beetles and true dragonflies. In the Lower Permian, the descendants of Carboniferous ferns (PTERIDOPHYTES) and seed ferns (pteridosperms) predominate. During the transition from Early to Late Permian these primitive forms are replaced by *Gigantoperis* in eastern Asia and western North America, and by *Glossopteris* in the southern continents (Gondwanaland). Most horsetails (sphenopsids) become extinct. The great SCALE TREES *Lepidodendron* and *Sigillaria* diminished greatly during the Late Carboniferous/Permian glaciation of Gondwanaland but reappeared to a limited extent during the Late Permian.

The VARISCAN orogeny of Europe and the Allegheny orogeny of the Appalachian fold belt of North America were diminishing in intensity. All the major continents were locked together to form Pangaea. The Gondwanaland continents were in high southern latitude, and the glaciations that began in the Early Carboniferous continued into the Permian.

permineralisation or **petrification** *n.* the preservation of hard parts of many organisms by mineral-bearing solutions after burial in sediment. Percolating groundwater may infiltrate porous shells and bones, and deposit its mineral content in pores and the open spaces of skeletal parts. These added minerals tend to increase the hardness of the original hard parts of the organisms. SILICA, CALCITE and various iron compounds are the usual *permineralisers*, but others are known. Petrified wood (see SILICIFIED WOOD) is the most common example of this type of preservation. The term 'petrified' should be applied only to remains that have been permineralised. See also FOSSIL, REPLACEMENT, MINERALISATION.

permitted intrusion *n.* **1.** the magmatic emplacement wherein the spaces occupied by the MAGMA have been created by forces other than its own, for example by CAULDRON subsidence or under conditions of crustal tension. **2.** the MAGMA or rock body emplaced in this manner. Compare FORCEFUL INTRUSION.

perovskite *n.* **1.** a pseudocubic mineral, $CaTiO_3$, found in nepheline syenites and carbonatites.

2. a high-pressure mineral, (Mg,Fe) SiO₃, with a similar crystalline structure, that occurs as inclusions in diamonds. The high-pressure form is thought to make up more than 70 per cent of the Earth's lower mantle.

perthite *n*. a variety of sodium-rich MICROCLINE in which there is an intergrowth of Na-rich PLAGIOCLASE FELDSPAR in a K-rich ALKALI FELDSPAR host, formed as the result of unmixing of initially homogeneous alkali feldspar during slow cooling. A *perthitic intergrowth* that can be distinguished without the use of a microscope is a *macroperthite*; *microperthites* can be detected only under a microscope while *cryptoperthites* can be detected only by X-ray methods. Compare ANTIPERTHITE; see FELDSPAR.

pervious *adj*. (of a rock) able to transmit fluid or gas via interconnected JOINTS and FISSURES. See also PERMEABLE.

petrification see PERMINERALISATION.

petrified wood see SILICIFIED WOOD.

petro- *prefix denoting* rock.

petrochemistry *n*. **1.** an aspect of GEOCHEMISTRY in which the chemical composition of rocks is studied.
2. PETROLEUM chemistry.

petrofabric analysis see STRUCTURAL PETROLOGY.

petrofabric diagram see FABRIC DIAGRAM.

petrofacies see PETROGRAPHIC FACIES.

petrogenesis *n*. the study of the mode of origin and evolution of rocks.

petrographic facies *n*. FACIES distinguished and grouped on the basis of their composition or appearance, without regard to boundaries or natural relations, e.g. 'red-bed facies'.

petrographic province *n*. a wide area in which most or all of the igneous rocks are believed to have been formed during the same interval of magmatic activity. See also COMAGMATIC.

petrography *n*. that branch of GEOLOGY involved with the description and systematic classification of rocks, especially by means of microscopic examination. Petrography is more restricted in its scope than is PETROLOGY.

petroleum *n*. a naturally occurring complex liquid HYDROCARBON that, after distillation and the removal of impurities such as nitrogen, oxygen and sulphur, yields a variety of combustible fuels. The term 'petroleum' may also be used to cover solid hydrocarbons, such as bitumen, and gaseous hydrocarbons. See also OIL SOURCES.

petroleum geology *n*. the branch of GEOLOGY concerned with the origin, migration and accumulation of oil and gas, and the search for and location of commercial deposits of these materials.

petrological microscope *n*. a transmitted-light microscope used in the study of THIN SECTIONS of minerals and rocks. It is fitted with two polarising filters, one above and one below the microscope stage (ANALYSER and POLARISER). Thin sections can be examined either in PLANE-POLARISED LIGHT or between crossed polars (with analyser in position) and various optical properties observed. See PLEOCHROISM, EXTINCTION, VIBRATION DIRECTIONS, POLARISATION COLOURS, BIREFRINGENCE, INTERFERENCE FIGURE. See also ACCESSORY PLATES.

petrology *n*. a branch of GEOLOGY that is broader in scope than PETROGRAPHY. It deals with the origin, distribution, structure and history of rocks.

PGM *abbreviation for* PLATINUM GROUP METALS.

pH *symbol for* hydrogen-ion concentration. It is used to express the degree of alkalinity or acidity of a substance. Pure water has pH of 7, acidic solutions have pH less than 7 and alkaline solutions greater than 7, e.g. acid mine waters are commonly 2 to 4.5 and alkaline soils about pH 11.

phaceloid see COLONIAL CORAL.

phacolith *n.* a concordant lenticular igneous INTRUSION. It is generally assumed that the igneous material was intruded during or before folding of the host rock and was deformed along with it. Phacoliths sometimes occur in vertical groups in the domes of anti-clinal folds. Compare LACCOLITH.

phaneritic or **coarse-grained** *adj.* (of an IGNEOUS ROCK) having a texture of which the individual components are discernible with the naked eye. Compare APHANITIC.

Phanerozoic *n.* that EON during which sediments containing obvious plant and animal remains accumulated. It includes the periods from the CAMBRIAN Period to the present (Quaternary) period. See also PRECAMBRIAN, CRYPTOZOIC.

phase change *n.* an alteration in the internal structure of a substance so that the physical properties are different but the composition remains the same. One example is the change from ice (solid) to water (liquid) to steam (gas), all of which are H_2O; another is the change from one MINERAL polymorph to another under different conditions of temperature and pressure (see POLYMOR-PHISM).

phase diagram *n.* a graphic method for showing the boundaries of equilibrium between different phases in a chemical system, the parameters usually being temperature, pressure and composi-tion. Solubilities and sequences of precipitation can be determined from such diagrams; the point the coordi-nates of which represent the chemical composition of a phase is the *composi-tion point*. A *phase* is a physically and chemically homogeneous part of a system that is distinct and separable from other parts of the system. Several phases can co-exist in equilibrium in a system. A *system* is any part of the Universe that is isolated for the purpose of study. Interpretation of phase diagrams and phase equilibria is facilitated by the use of *Gibb's phase rule*, which states that $P + F = C + 2$, where P equals the number of phases, F equals DEGREES OF FREEDOM and C equals the number of *components*. The number of components comprising a system is the minimum number of chemical constituents required to define completely the composition of every phase of the system. A system is defined by the number of its compo-nents: a two-component system is a *binary system*; the albite-anorthite-potassium feldspar system is *ternary*; if water is involved in the ternary, the system becomes *quaternary*.

phase velocity and group velocity *n.* *phase velocity* is the velocity with which an individual wave or wave crest is propagated through a medium, i.e. the velocity of constant phase. If a line is drawn connecting a point of constant phase in each of several SEISMOGRAMS, such as the crest of a certain wave, its slope gives the phase velocity of waves of a certain period. If the centres of groups of waves of about the same frequency are connected, starting from the same point as the phase-velocity line, the slope of the second line drawn is the *group velocity* for waves of the same period. Both kinds of velocity are greater for longer waves than for shorter waves. See also DISPERSION.

phenocryst *n.* a relatively large crystal in an IGNEOUS ROCK, contrasting sharply in size from crystals of the matrix and formed during the early stages of CRYSTALLISATION of a MAGMA. Its size is a function of both cooling rate and chemical composition. Phenocrysts will be EUHEDRAL if the magma system in which they formed was near equilibrium when solidification took place, but during conditions of rapid cooling skeletal crystals may be formed. Differing rates of nucleation and crystal growth of various mineral families in a cooling magma may produce phenocrysts of mineral composition different from the groundmass minerals and compositional ZONING. See PORPHYRY, PORPHYRITIC.

phenotype *n.* the morphologically expressed characteristics of an organism produced by the GENOTYPE.

phi grade scale *n.* a logarithmic transformation of the size divisions of the WENTWORTH GRADE SCALE such that

$$\phi = -\log_2 \left(\frac{d}{d_0} \right)$$

where d is a class division and d_0 is the 'standard' grain diameter of 1 millimetre. See Fig. 94.

phlogopite *n.* a magnesium mica mineral, $KMg_3(AlSi_3O_{10})(OH)_2$, light brown or yellowish, often in foliated aggregates. Phlogopite is found in metamorphosed magnesian LIMESTONES and some ULTRAMAFIC igneous rocks. It is common in KIMBERLITE. See MICA.

phonolite *n.* a fine-grained UNDERSATURATED extrusive igneous rock related to TRACHYTE, composed primarily of potassic feldspar (especially sanidine), nepheline, clinopyroxene or sodic amphibole. It commonly has a trachytic texture and emits a characteristic ringing sound when hammered. It is the extrusive equivalent of nepheline syenite.

phosphate *n.* a chemical compound containing the phosphate ion radical $(PO_4)^{3-}$. Phosphates comprise a very large group of minerals; however, all

but a few are rare. The most abundant phosphate mineral is APATITE.

phosphatic nodule *n.* a rounded mass consisting of organic and inorganic debris including corals, shells, coprolites, mica flakes and sand grains, all covered by and within a matrix of calcium phosphate. Such nodules range in size from a few millimetres to greater than 30 centimetres. Phosphatic nodules occur in marine strata, e.g. the Permian beds of the western United States and the Cretaceous chalk of England.

phosphorescence *n.* a luminescence in which the emission of visible light from a substance continues after termination of the radiation that activated the luminescence. Compare FLUORESCENCE.

phosphorite *n.* a sedimentary rock that contains at least 20 per cent of phosphate minerals. In ancient phosphorites, the phosphorus is present in fluorapatite in the form of pellets, laminae, oolites and bone fragments.

photic *adj.* allowing the penetration of light, specifically sunlight. The photic zone in the sea extends down to 200 metres. Compare APHOTIC.

photogeology *n.* the geological interpretation of aerial and satellite photographs. The study of satellite photographs is useful in the investigation of large-scale geological phenomena.

phragmocone *n.* the chambered part of a CEPHALOPOD shell. See AMMONOID Fig. 1(a) and BELEMNITE Fig. 5.

phreatic cycle *n.* a daily, annual or other time period during which the WATER TABLE rises and falls.

phreatic explosion *n.* a volcanic explosion or eruption of steam, mud or other non-incandescent material caused by the rapid conversion of GROUNDWATER to steam as a result of contact between the water and an underlying igneous heat source. Phreatic explosions often occur as precursors to major explosive volcanic eruptions including SUBGLACIAL activity.

phreatic water *n.* originally applied only to water that occurs in the upper zone of saturation under WATER-TABLE conditions but now applied to all water in the zone of saturation. Thus, it is synonymous with GROUNDWATER.

phreatic zone see WATER TABLE.

phreatomagmatic eruption *n.* the explosive eruption of magmatic material resulting from the interaction between water and hot MAGMA. By contrast, a PHREATIC EXPLOSION produces very little magmatic material but instead mainly steam and fragments of the COUNTRY ROCK.

phyllite *n.* a regional metamorphic rock characterised by lustrous, often undulating, well-developed CLEAVAGE surfaces, light silvery-grey in colour, intermediate in metamorphic grade between slate and mica schist. Phyllites are derived from ARGILLACEOUS sedimentary rocks. The main mineral constituents are muscovite, chlorite and quartz.

phyllonite *n.* a rock similar in appearance to PHYLLITE, formed from a coarser-grained rock by cataclastic RETROGRADE METAMORPHISM (see CATACLASIS).

phyllosilicate *n.* a silicate mineral such as MICA the tetrahedral silicate groups of which are linked in sheets, each group containing four oxygen atoms, three of which are shared with other groups, so that the ratio is two of silicon to five of oxygen. See SILICATES.

phylogenetic systematics see CLADISTICS.

phylogeny *n.* the patterns of relationship by descent, or the relationship of a group or groups of organisms as seen in their evolutionary history.

-phyro, -phyric *suffix denoting* PORPHYRITIC.

physical geology *n.* a division of GEOLOGY in which the processes and forces associated with the evolution and MORPHOLOGY of the Earth are studied along with its constituent minerals and rocks and the materials of its interior.

physical oceanography see OCEANOGRAPHY.

physical weathering see MECHANICAL WEATHERING.

phyteral *n.* vegetal material in coal the morphological forms of which are still discernible, e.g. cuticle or spore casts. Phyterals are distinguished from MACERALS, the organic compositional units that form the COAL mass.

phytolith *n.* a part of a plant that is mineralised, such as the silica phytoliths in the cells of grasses.

phytoplankton *n.* **1.** marine or freshwater microscopic unicellular plant cells that include many algal groups. **2.** drifting and free-floating plants of the sea (such as sargassum weed), the more important microscopic diatoms, dinoflagellates and many groups of photosynthetic bacteria that are part of the PELAGIC region. See also ZOOPLANKTON.

Piacenzian *n.* the top STAGE of the PLIOCENE.

picrite *n.* an ULTRABASIC igneous rock belonging either to the tholeiitic or the alkali basaltic rock association, rich in magnesian OLIVINE along with CLINOPYROXENE and or ORTHOPYROXENE and AMPHIBOLE and subordinate calcium-rich PLAGIOCLASE feldspar (see THOLEIITIC BASALT, ALKALI BASALT). Some rocks classified as picrites are CUMULATES, while others have crystallised from a picritic MAGMA, such as Tertiary volcanic rocks from west Greenland.

Picrites are defined chemically as a group of rocks containing MgO > 18 per cent by weight, SiO_2 < 47 per cent and $Na_2O + K_2O < 2$ per cent. Also, the name is used specifically for those rocks within the group that have total alkalis > 1 per cent. KOMATIITE with $Na_2O + K2O < 1$ per cent and $TiO_2 < 1$ per cent and meimechite with $Na_2O + K_2O < 1$ per cent and $TiO_2 > 1$ per cent are the other subdivisions.

piecemeal stoping see STOPING.

piedmont 1. *adj.* (of a terrace, etc) situated or formed at the base of a mountain.
2. *n.* a feature at the base of a mountain.

piedmont alluvial plain see BAJADA.

piedmont glacier see GLACIER.

piezoelectric effect *n.* the development, in some CRYSTALS, of an electrical potential when pressure is exerted parallel to certain crystallographic directions. Quartz and tourmaline are examples of piezoelectric crystals.

pillow lava *n.* spherical or ellipsoidal structures usually composed of basaltic LAVA but which can also be andesitic; they are generally about one metre in diameter. These formations are the result of the rapid cooling of hot fluid MAGMA that comes in contact with water, such as occurs when lava flows into the sea, into water-saturated sediments or beneath a glacier. Lava pillows are extruded from openings as glassy-skinned lava tongues in much the same way as toothpaste is squeezed from a tube. The margins of pillow lavas are composed of GLASS, which progressively alters to PALAGONITE. The interiors are composed of more crystalline material. See Fig. 68 below. See also LAVA FLOW.

pilotaxitic *adj.* (of groundmasses of holocrystalline volcanic rock) having a texture in which lath-shaped microlites (usually plagioclase) manifest a feltlike, interwoven pattern, common in flow lines. See also HYALOPILTIC, TRACHYTIC.

pinacoid *n.* a crystal FORM consisting of two parallel faces cutting only one CRYSTAL AXIS; if it cuts the c-axis it is called a *basal pinacoid*. Pinacoids can occur among crystals of all CRYSTAL SYSTEMS except the isometric system.

pinch-and-swell see BOUDINAGE.

pinch-out *n.* the end of a stratum or vein that thins progressively until it disappears and the rocks it had separated are in contact with one another.

pingo *n.* a mound or hillock found in areas affected by PERMAFROST in Arctic regions. Such mounds may be round or ovoid in outline, with diameters ranging from a few metres to about 1200 metres in diameter and a maximum height of about 100 metres,

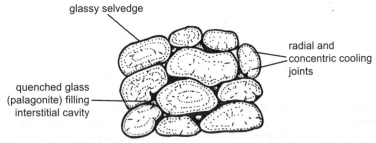

glassy selvedge

radial and concentric cooling joints

quenched glass (palagonite) filling interstitial cavity

Fig. 68 **Pillow lava**. Pillows show a radial and concentric internal structure because of cooling joints and sag into the underlying spaces

although most are much smaller than the maximum. Two types of pingo have been described. The *Mackenzie type* is formed in a closed hydrological system in areas of thick continuous permafrost, in flat terrain associated with former lake beds or river courses. The freezing of water at the base of the permafrost causes doming of the overlying ground at an annual rate of up to 1.5 metres in the initial stages. The *East Greenland type*, formed in an open hydrological system, occurs in areas of thin discontinuous permafrost at the foot of gentle slopes and valley flanks where they are associated with seasonal seepage of sub-permafrost groundwater that forms a body of INJECTION ICE.

pinnate drainage pattern see DRAINAGE PATTERN.

pinnate fractures see FEATHER JOINTS.

pinnules *n*. rows of slender branchlets on the arms of a CRINOID (see Fig. 21). Like the arms, the pinnules carry food grooves on the oral side. The skeletal plates supporting the pinnules are called *pinnular plates*. See also BRACHIALS.

pipe clay see BALL CLAY.

piracy see STREAM CAPTURE.

pisolite *n*. a SEDIMENTARY carbonate rock, composed mainly of PISOLITHS, large coarse-grained spherical to sub-spherical grains with an internal concentric structure. Some pisoliths may form in the same way as ooids (see OOLITE).

pisolith *n*. an accretionary formation in a sedimentary rock; it resembles a pea in size and shape, and is one of the units making up a PISOLITE. Pisoliths have the same radial and concentric internal structure as ooliths (see OOLITE) but are larger and less regular. They are formed of calcium carbonate. Pisoliths should not be confused with oncolites,

the latter being of algal origin. Compare ONCOLITE.

pitch *n*. the angle between a linear feature and the horizontal, measured in the inclined plane that contains the two (see FAULT, Fig. 30). Compare this with PLUNGE.

pitchblende *n*. a brown to black variety of the radioactive uranium mineral URANINITE, UO_2, found in pegmatites and high-temperature hydrothermal veins. It is massive or BOTRYOIDAL with a banded structure.

pitch lake see TAR PIT.

pitchstone *n*. a black volcanic glass with the lustre of resin or pitch, usually associated with shallow igneous intrusions; the chemical composition can vary between that of andesite and rhyolite. It contains more water, tends to contain a greater proportion of CRYPTOCRYSTALLINE material and generally has a less well-developed conchoidal fracture than OBSIDIAN.

piton *n*. a term used for VOLCANIC SPINES, especially in the West Indies, e.g. Petit Piton (750 metres) and Gros Piton (800 metres) of Santa Lucia.

pivot fault see FAULT.

placer *n*. a surficial DEPOSIT containing economic quantities of valuable MINERALS. The minerals found in placer deposits are those that have a high density and are resistant to chemical WEATHERING and physical ABRASION, such as gold, platinum, magnetite, chromite, ruby, diamonds and so on. They are concentrated by sedimentary processes. Gravity separation by flowing water produces *alluvial placers* and BEACH PLACERS. *Aeolian placers* are less common, but important deposits have been formed because of the reworking of beach placers by wind action. Weathering *in situ* can produce a *residual placer* immediately above a bedrock source by removal of less resistant rock materials. Residual placers form only on flat surfaces; on hillslopes the resistant and denser minerals will accumulate downhill of the source, forming an *eluvial placer* (also known as FLOAT). Placer deposits are loose and easily worked, and mining usually takes the form of dredging, with various processes to concentrate the heavy minerals. Fossil placer deposits (see BANKET) are more expensive to exploit.

placoderm *n.* a member of the class Placodermi, an extinct class of fishes; it was enveloped in armour-like plates and had well-developed articulating jaws. Range, Early to Late Devonian.

plagioclase *n.* a group of triclinic feldspars that forms a SOLID-SOLUTION series, from pure ALBITE ($NaAlSi_3O_8$) to ANORTHITE ($CaAl_2Si_2O_8$). The series is divided according to the increasing percentage of anorthite (An) as follows: albite (An_{0-10}); oligoclase (An_{10-30}); andesine (An_{30-50}); labradorite (An_{50-70}); bytownite (An_{70-90}); anorthite (An_{90-100}). See FELDSPAR.

planar flow structure see PLATY FLOW TEXTURE.

planation *n.* the process of levelling a surface by erosional agents.

plane-polarised light *n.* light that has been constrained (e.g. by a polarising filter or Nicol prism) to vibrate in a single plane; ordinary light vibrates in all directions perpendicular to its direction of propagation.

plane table *n.* an instrument for obtaining and plotting survey data direct from field observations. It consists of a drawing board mounted on a tripod; the board is fitted with a sighting device, e.g. an ALIDADE.

planetology *n.* study of the condensed matter of the solar system, including planets and their satellites, asteroids, meteors and interplanetary material. It should be distinguished from ASTROGEOLOGY.

planimetry *n.* the determination of angles, areas and horizontal distances by measurements on a map.

plankton *n.* free-floating and drifting organisms that have little or no effective swimming ability. Compare NEKTON. See also PELAGIC, PHYTOPLANKTON, ZOOPLANKTON.

planulate see AMMONOID.

plastic *adj.* capable of permanent DEFORMATION without rupture.

plastic deformation *n.* a DEFORMATION that is not recoverable or is only partly recoverable but does not involve failure by rupture.

plasticity index *n.* the range of moisture-content values in which a sediment is PLASTIC. It is the difference between the LIQUID LIMITS and PLASTIC LIMITS of the sediment.

plastic limit *n.* the moisture-content boundary between the PLASTIC and semi-solid states of a sediment, e.g. soil. It is one of the ATTERBERG LIMITS. Compare LIQUID LIMIT.

plastron *n.* a ridge forming an extension of the posterior interambulacrum towards the mouth on the oral side of some irregular ECHINOIDS. The end of the interambulacrum forms a projecting lip, called the *labrum*, below the mouth (Fig. 28). See also AMBULACRUM.

plate see PLATE TECTONICS.

plateau basalt see FLOOD BASALT.

plate tectonics *n.* a synthesis of geological and geophysical observations in which the Earth's LITHOSPHERE is thought to be divided into seven large rigid *plates*, and several smaller ones, that are moving relative to each other. The plates move over the weak ASTHENOSPHERE and interact with each

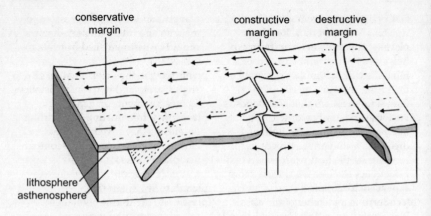

Fig. 69 **Plate tectonics**. The types of plate boundary: constructive (mid-oceanic ridge), destructive (oceanic trench), and conservative (transform fault)

other along relatively narrow zones of volcanic and seismic activity. Internally, the plates are aseismic and may contain both oceanic and continental CRUST. The theory of plate tectonics, formulated during the late 1960s, unified and expanded the earlier hypotheses of CONTINENTAL DRIFT and SEA-FLOOR SPREADING. When earthquake epicentres are plotted on a world map, it is found that they are mainly confined to narrow continuous zones that coincide with MID-OCEANIC RIDGES, OCEAN TRENCHES and TRANSFORM FAULTS. These define the *plate boundaries*.

The oceanic ridges are *constructive plate margins*, where the plates separate and MAGMA wells up to form new oceanic crust.

The oceanic trenches are the sites of DESTRUCTIVE PLATE BOUNDARIES, where plates converge and oceanic lithosphere is consumed in a SUBDUCTION ZONE. Where the oceanic parts of two plates converge, an ISLAND ARC is formed on the over-riding plate, for example the Japanese islands. Where

the oceanic part of one plate converges with the continental part of another, the oceanic plate is subducted beneath the continental plate, and the edge of the continental plate is compressed and folded to form a mountain range such as the Andes. Continents may collide if the oceanic lithosphere between them is subducted. The more buoyant continental crust cannot be subducted, so subduction will cease. The collision of two continents results in under-thrusting and crustal thickening, leading to the formation of mountain ranges such as the Alpine-Himalayan chain.

The third type of plate boundary is the CONSERVATIVE boundary, where material is neither created nor destroyed. These boundaries are transform faults, where plates slide past each other.

platform *n*. **1.** any level surface ranging in area from a bench or terrace to a plateau or peneplain.
2. a WAVE-CUT PLATFORM.
3. the part of a continent that consists of BASEMENT rocks covered by essen-

tially flat-lying beds of sedimentary rock; it is part of the CRATON.

platinum group metals (PGM) *n.* the rare, closely related elements ruthenium (Ru), osmium (Os), rhodium (Rh), iridium (Ir), palladium (Pd) and platinum (Pt). Platinum has an average abundance in the Earth's crust at about 0.001 grams per ton; the others are generally less abundant. The highest concentrations of PGM in the Earth's crust occur in ultrabasic igneous intrusions and related PLACER deposits. Relatively high concentrations also occur in chondritic and iron METEORITES.

platy flow texture or **planar flow structure** *n.* a texture found in IGNEOUS ROCK, resulting from the common alignment of platy and PRISMATIC minerals by the flow of the MAGMA. See also TRACHYTIC.

playa *n.* a flat dry barren plain at the bottom of a desert basin, underlain by silt, clay and EVAPORITES. It is often the bed of an ephemeral lake and may be covered with white salts.

playa lake *n.* a shallow recurring lake that covers a PLAYA after rains but disappears during a dry period.

Playfair's law *n.* a generalisation, proposed in 1802 by John Playfair, professor of mathematics at Edinburgh University, relating to river systems. His proposals were that:
(a) most rivers cut their own valleys;
(b) stream junctions in the river system are accordant, i.e. the streams are at the same level at the point of junction; and
(c) for each river the angle of slope, speed of flow and amount of material carried tend to adjust towards equilibrium.

play of colour *n.* an optical phenomenon observed in certain minerals and mineraloids, e.g. opal, in which PRISMATIC colours flash in rapid succession as it is rotated. It is caused by a dispersion of light that varies according to the ANGLE OF INCIDENCE, and the size of the oriented spherical particles of which the mineral is composed and which function as a DIFFRACTION grating. See IRIDESCENCE, LABRADORESCENCE, SCHILLER.

Pleistocene *n.* the lower epoch of the Quaternary Period. See CENOZOIC.

pleochroic halo see HALO.

pleochroism *n.* the ability of some crystals to exhibit different colours when viewed from different directions under transmitted PLANE-POLARISED LIGHT. It is a property only of ANISOTROPIC crystals and results from the selective absorption of certain wavelengths of light along different crystallographic directions. Tetragonal and hexagonal crystals exhibit *dichroism*, since there are only two directions in which there is differential absorption. In orthorhombic, monoclinic and triclinic systems there are three directions along which light may be transmitted to yield three different colours; this is called *trichroism*. See also DICHROISCOPE.

pleurae see THORAX.

Pliensbachian *n.* a STAGE of the Lower JURASSIC.

Plinian-type eruption see VOLCANIC ERUPTION, TYPES OF.

Pliocene *n.* the uppermost epoch of the Tertiary Period. It is followed by the Quaternary Period. See CENOZOIC.

plucking see GLACIAL PLUCKING.

plug see VOLCANIC PLUG.

plug dome *n.* a type of VOLCANIC DOME.

plumbago *n.* a common name for GRAPHITE, sometimes used of GALENA and MOLYBDENITE; use of the term is best avoided.

plunge *n.* the inclination of a FOLD axis measured in the vertical plane.

Together with the strike of the horizontal projection, the plunge defines the attitude of the HINGE of a fold.

pluton *n.* any major intrusive body of IGNEOUS ROCK formed beneath the surface of the Earth by the consolidation of MAGMA. See BATHOLITH, DYKE, LACCOLITH, SILL.

plutonic *adj.* **1.** (usually of igneous rocks) of deep-seated origin.
2. (of a rock) formed by any process at great depth. Compare VOLCANIC.

pluvial *adj.* pertaining to rain or precipitation.

pneumatolysis *n.* the alteration of rock or crystallisation of minerals by gaseous emanations from the late stages of a solidifying MAGMA. Both the COUNTRY ROCK and earlier-formed parts of an INTRUSION can be affected. The main volatiles include compounds of boron, fluorine, carbon dioxide and sulphur dioxide. There is not a clear distinction between pneumatolysis and HYDROTHERMAL processes, although pneumatolysis is regarded as a higher temperature process (above 600°C) while hydrothermal processes are caused by the action of hot aqueous fluids at temperatures from about 600°C down. Minerals formed as a result of pneumatolytic action include TOURMALINE, TOPAZ and FLUORITE (see LUXULLIANITE, GREISEN). ORE minerals such as CASSITERITE and WOLFRAMITE (a tungsten mineral) occur as pneumatolytic deposits.

pod *n.* an ore body or intrusion of elongate or lenticular form.

podzol *n.* a group of zonal soils of which the surface material is organic matter; beneath this there is an ashen-grey layer and then a zone in which iron and aluminium minerals accumulate. Podzol develops in a cool to temperate and moist climate.

poikilitic *adj.* **1.** (of certain minerals within igneous rock) having a speckled appearance. Usually, the host completely encloses the inclusions.
2. (of small granular crystals) variously oriented within larger host crystals (OIKOCRYSTS) of a different mineral. The normal CRYSTALLISATION of MAGMA, in which the INCLUSIONS are earlier-formed minerals, can produce poikilitic texture. It can also result from mineral REPLACEMENT during postmagmatic processes; in this case the crystal grains indicate remnants of the replaced mineral. Compare POIKILOBLASTIC.

poikilo- *prefix denoting* spotted.

poikiloblast *n.* a CRYSTAL in a metamorphic rock that encloses crystals of other minerals. Poikiloblasts may be larger than most other crystals in the rock and thus also PORPHYROBLASTS. If the poikiloblast is spongy with many randomly arranged small inclusions, this is known as *sieve texture*. The inclusions may form a pattern related to a RELICT structure (see HELICITIC TEXTURE), to the growth of the crystal (see SNOWBALL TEXTURE) or to the internal structure of the poikiloblast, as in CHIASTOLITE.

poikiloblastic *adj.* (of metamorphic rock) having a texture caused by the development during recrystallisation of a new mineral around remnants of the original minerals, simulating the POIKILITIC texture of igneous rocks.

point bar see FLOODPLAIN.

point group *n.* any one of the 32 CRYSTAL CLASSES.

Poisson's ratio *n.* the ratio of unit elongation, c, to unit lateral contraction, s, in a body stressed longitudinally within its elastic limits. The ratio c/s is constant for a given material.

polarisation colours or **interference colours** *n.* the colours seen when THIN SECTIONS of ANISOTROPIC minerals are examined between crossed POLARISERS.

The two light rays transmitted through a crystal travel with different velocities, so most wavelengths in white light are out of phase when the rays are recombined into the same plane at the ANALYSER. Their interference produces the colours seen, as certain wavelengths are cut out and others are reinforced.

The phase difference for any particular wavelength of light depends on the relative *retardation* of the two rays so that

$$P = \frac{t(n_1 - n_2)}{\lambda}$$

where P is the phase difference, t is the thickness of the section, $n_1 - n_2$ is the difference in refractive index between the two rays and λ is the wavelength of light. Retardation is equal to $t(n_1 - n_2)$.

The crystallographic orientation of a mineral section affects the polarisation colour. Sections cut at right angles to the OPTIC AXIS show no colour because all light transmitted parallel to the optic axis has the same velocity and there is no phase difference. Sections cut parallel to the optic axis show the maximum relative retardation and therefore the maximum phase difference. The polarisation colour will be the highest for that particular mineral. This is a characteristic property that can be used to help identify minerals in thin section. See also BIREFRINGENCE, NEWTON'S SCALE.

polarised light see PLANE-POLARISED LIGHT.

polariser *n*. a device for producing PLANE-POLARISED LIGHT, fitted below the stage of a PETROLOGICAL MICROSCOPE. See also ANALYSER.

polarity epoch *n*. corresponds to the *polarity chron*, which is the fundamental time interval in the magnetostratigraphical time scale. Polarity chrons have durations greater than 100 000 years and correspond to an interval, e.g. the *Brunhes normal epoch*, during which the Earth's magnetic field was mainly of a single polarity. See also GEOMAGNETIC POLARITY REVERSAL.

polarity interval *n*. the fundamental unit of worldwide polarity-chrono-stratigraphic classification.

polarity reversal see GEOMAGNETIC POLARITY REVERSAL.

polarity zone *n*. a unit of rock characterised by its particular magnetic polarity; it is the basic unit of polarity-lithostratigraphic classification.

polar migration see POLAR WANDERING.

polar symmetry *n*. a type of CRYSTAL SYMMETRY in which the two ends of a central CRYSTAL AXIS are not symmetrical. Such a crystal is HEMIMORPHIC.

polar wandering *n*. the concept that the Earth's magnetic poles appear to have migrated over the surface through geological time. The idea is implied in directional changes of the geomagnetic field as determined from the NATURAL REMANENT MAGNETISATION of rocks. Pole locations calculated from rocks younger than 20 Ma show only minor discrepancies from present pole locations. Increasingly substantial deviations occur in older rocks. The fact that apparent polar-wandering curves are different for different locations can be explained only by migration of the continents relative to the geomagnetic pole and to each other, and provides important supporting evidence for CONTINENTAL DRIFT. See also GEOMAGNETIC POLARITY REVERSAL, PALAEOMAGNETISM.

polje *n*. a steep-sided enclosed basin with a flat floor occurring in a KARST region. Poljes can range in area from two square kilometres up to 400 square kilometres. The floor may be covered with IMPERMEABLE ALLUVIUM. The location and orientation of poljes appear to be controlled by structural features like FAULTS and FOLD HINGES or contact with an impermeable horizon. It is suggested that these features guide and enhance EROSION, resulting in the formation of poljes.

pollen analysis see PALYNOLOGY.

poly- *prefix denoting* many.

polygenetic *adj*. I. derived from more than one source or resulting from more than one process of formation, e.g. a polygenetic mountain range is one that has been formed by many orogenic incidents.
2. consisting of more than one type of material or of materials from several sources. Compare MONOGENETIC.

polygonal ground see PATTERNED GROUND.

polymetamorphism *n*. polyphase or multiple metamorphism, whereby evidence of two or more successive metamorphic events occurs in the same rocks.

polymictic *adj*. I. (of a clastic sedimentary rock, particularly a conglomerate) composed of a variety of fragment types.
2. (of a lake) in which the water is continuously circulating and shows no persistent thermal stratification. Compare MONOMICTIC, OLIGOMICTIC.

polymorphism *n*. the characteristic of having two or more distinct forms.
I. a crystal that has different crystal structures. Iron sulphide crystallises as two distinct forms: marcasite and pyrite.
2. The presence of several forms in a species population. See also DIMORPHISM.

polyphyletic *n*. a grouping or clustering of organisms that do not possess a common ancestor. The polyphyletic grouping may be the result of convergent or parallel evolution causing life

forms to be placed in a single, often higher, TAXON. In CLADISTIC taxonomy, polyphyletic groupings can never be valid biological entities.

polytype *n*. a mineral that has the same chemical formula and the same structural sub-units as another but with a different stacking sequence of the sub-units. Sheet SILICATES, such as MICAS, or the CLAY MINERALS are examples as the silicate sheets and linking layers can be stacked in different ways.

polyzoan see BRYOZOAN.

polzenite see LAMPROPHYRE.

pontic *adj*. (of sediments) deposited in comparatively deep and still water.

porcellanite *n*. any compact dense siliceous rock having the general appearance of unglazed porcelain; the term is applied to various siliceous rocks or minerals of such description.

pore pairs see AMBULACRA.

poriferan see SPONGE.

porosity *n*. the ratio of the collected volume of interstices (VOIDS) in a soil or rock to the total volume, usually stated as a percentage. Compare EFFECTIVE POROSITY. See PERMEABILITY.

porphyritic *adj*. (of any igneous rock) containing large crystals (PHENOCRYSTS) in a groundmass (matrix) of smaller crystals or glass. Porphyritic texture is generally interpreted in terms of a two-stage cooling procedure: an initial period of slow cooling during which large crystals are formed, followed by a period of rapid cooling and the subsequent formation of a finer-grained groundmass. An explanation of porphyritic texture that does not entail two-stage cooling is that crystallisation occurs at a fixed degree of undercooling, but different minerals have different rates of nucleation and growth. See Fig. 70 on the opposite page.

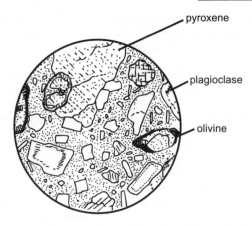

pyroxene

plagioclase

olivine

Fig. 70 **Porphyritic** texture. A thin section of olivine basalt viewed in plane-polarised light shows phenocrysts of olivine (altered), pyroxene and plagioclase in a fine-grained groundmass

porphyroblast *n.* a large crystal in METAMORPHIC ROCK, formed during METAMORPHISM.

porphyroblastic texture *n.* a METAMORPHIC ROCK texture in which large crystals are embedded in a finer-grained matrix.

porphyroclast *n.* a large rounded or angular rock or mineral fragment in the finer crushed matrix of a rock produced by CATACLASIS. See MORTAR STRUCTURE, also MYLONITE.

porphyry *n.* an IGNEOUS ROCK in which PHENOCRYSTS constitute a significant proportion of the volume; the ground-mass should be fine-grained or even APHANITIC. The term may be used in conjunction with a mineral name, for example quartz porphyry for a micro-granite containing phenocrysts of quartz and orthoclase.

Portlandian *n.* a STAGE of the Upper JURASSIC in Britain.

positive element *n.* a sizeable structural feature or area that has had an ex-tended history of continuing uplift.

potassium-argon method see DATING METHODS.

potassium bentonite see BENTONITE.

potassium feldspar or **K-feldspar** *n.* an alkali feldspar containing the molecule $KAlSi_3O_8$, e.g. orthoclase, microcline, adularia or sanidine. See FELDSPAR.

potentiometric surface *n.* an imaginary surface that represents the total head of groundwater; it is defined by the level to which water will rise in a well.

Pratt hypothesis *n.* a model that explains ISOSTACY by lateral variations in the DENSITY of the CRUST or MANTLE. Thus the density of the crust and/or mantle under topographically higher areas such as mountains would be lower than the density of the crust and/or mantle under topographically lower areas. Pratt considered that at a certain depth under the Earth's surface the pressure resulting from the weight of the overlying topographical masses became the same, regardless of their height. This depth was called the *depth of compensation*. See also AIRY HYPOTHESIS.

Precambrian *n.* the period of time from the formation of the Earth, 4530 Ma to about 544 Ma ago, during which the Earth's crust was formed and the first organisms appeared. The term Precambrian is now used informally to describe the HADEAN (4530–3900 Ma), ARCHAEAN (3900–2500 Ma) and PROTEROZOIC (2500–544 Ma) eons. The name PRISCOAN has also been used for pre-Archaean time, from the formation of the Earth to 3900 Ma ago. Precambrian rocks are covered by younger formations in approximately four-fifths of the world's land areas and are best seen in the Earth's SHIELD areas, such as the CANADIAN SHIELD. During the Hadean eon the Earth itself was formed, heated up and melted, following which the core, mantle and early crust differentiated and an atmosphere and ocean evolved. During the Archaean eon the first tectonic plates formed and protocontinents accumulated into CRATONS, with the first surviving sediments being produced in the oceans. These sediments are found (now in metamorphosed form) at Isua in west Greenland and date to 3800 Ma. Also by 3800 Ma, life existed in the oceans, which were anoxic (lacking in oxygen), as was the atmosphere. During the Proterozoic eon new continental crust continued to accumulate, and by the middle of the eon about 90 per cent of the earth's current continental crust had formed. Photosynthetic life forms produced sufficient oxygen to convert atmosphere and ocean to an oxic mode, and the first steps towards multicellular life were taken around 2000 Ma. Late in the Proterozoic eon macrofossils of multicellular plants began to appear in the stratigraphical record. Continental collisions late in the Proterozoic eon produced a single supercontinent, RODINIA, which by about 800 Ma had broken up and was reconstructed as a further supercontinent, PANNOTIA. Precambrian rocks contain rich deposits of iron, copper, nickel, silver and gold. A number of orogenies are known to have occurred during this time, and most Precambrian rocks are therefore strongly deformed and metamorphosed. However, some relatively unmetamorphosed Precambrian sediments are known. See ISUAN SUPRACRUSTALS.

precious metal *n.* gold, silver or any metal of the PLATINUM GROUP.

precious stone *n.* a relatively rare, hence sought-after, gemstone, such as diamond, ruby, emerald or sapphire.

precision depth recorder *n.* an echo sounder with an accuracy of better than 1 in 3000. See ECHO SOUNDING.

preferred orientation *n.* the alignment of certain elements of a rock in a particular direction. This may be displayed by the alignment of mineral grains or CLASTIC fragments (e.g. PEBBLES or even FOSSILS) of a particular shape (see DIMENSION-PREFERRED ORIENTATION) or by the alignment of the crystallographic directions of particular minerals (see LATTICE-PREFERRED ORIENTATION). Preferred orientation may be a primary feature resulting from the alignment of particles during the formation of the rock, for example by magmatic flow in igneous rocks (see FLOW LAYERING), or the settling of platy or elongate mineral grains in a sediment (for example MICA lying parallel to the BEDDING surface of a sedimentary rock).

During deformation and METAMORPHISM preferred orientation may be produced by rotation of existing mineral grains or by the nucleation and growth of new minerals aligned in directions related to the STRESS system. The pattern of preferred orientation in rocks can be studied using petrofabric diagrams (see FABRIC DIAGRAM) as part of STRUCTURAL PETROLOGY.

prehnite-pumpellyite facies see METAMORPHIC FACIES, Fig. 58.

pressure altimeter see ALTIMETER.

pressure gradient *n*. **1.** the rate of pressure variation in a given direction and at a fixed time, e.g. variation of pressure with ocean depth.
2. informally, the magnitude of the pressure gradient. Compare HYDRAULIC GRADIENT.

pressure solution *n*. the selective solution and re-deposition of minerals along the grain boundaries under STRESS. STYLOLITES are produced by this process.

pressure wave see SEISMIC WAVES.

Priabonian *n*. the top STAGE of the EOCENE.

Přídolí *n*. the top SERIES of the SILURIAN.

primary cosmic rays see COSMIC RADIATION.

primary current lineation or **parting lineation** *n*. a LINEATION formed by a series of faint, closely spaced ridges and hollows on the bedding surfaces of flat-bedded SANDSTONES. The planar BEDDING and primary current lineation are an indication of high current velocities.

primary magma *n*. a MAGMA of which the composition has not changed since it was generated. Compare PARENTAL MAGMA.

primary mineral *n*. a mineral, deposited or formed at the same time as the rock containing it, that retains its original composition and form. Compare SECONDARY MINERAL.

primary phase *n*. the only crystalline phase that can exist in equilibrium with a given liquid; it is the first to appear upon chilling from a liquid and

the last to disappear upon heating to the point of fusion.

primary wave see SEISMIC WAVES.

primordial dust see COSMIC DUST.

principal axes of stress see STRESS.

principal planes of stress see STRESS.

Priscoan *n*. the EON spanning the time from the formation of the Earth to 3900 Ma, an alternative name for HADEAN.

prism *n*. (*crystallography*) a FORM consisting of three, four, six or more faces parallel to the vertical axis (c-axis).

prismatic *adj*. **1.** (of a crystal) bounded by PRISM faces of varying dimensions but elongated in one of them.
2. (of metamorphic texture) characterised by such crystals.

prismatic cleavage see CLEAVAGE.

procaryotic *adj*. of cells that lack a true membrane-enclosed nucleus. Bacteria are procaryotic.

prod marks see TOOL MARK.

profile *n*. a graph or chart that shows the variation of one property with respect to another, e.g. gravity, seismic-wave velocity and geothermal energy may all be plotted against depth below the surface to obtain desired profiles for seismic or geophysical exploration. See also SOIL PROFILE.

profile of equilibrium or **graded profile** *n*. **1.** the longitudinal (or long) profile of a graded stream or stream whose gradient at every point is just great enough to enable it to transport its load of sediment.
2. the marine profile of equilibrium is a smooth sweeping curve, concave upwards. It is steep in the breaker zone and flattens sharply seawards. The slope is just sufficient to keep the material deposited by waves and currents in slow seaward transit.

proglacial *adj*. immediately in front of or just outside the limits of a GLACIER or ice sheet.

progradation *n*. the forward or outward building from a SHORELINE of a sedimentary rock unit, such as the advance of a delta.

pro-ostracum see BELEMNITE and Fig. 5.

propane *n*. an inflammable gaseous saturated HYDROCARBON, C_3H_8. It occurs naturally in natural gas and crude petroleum.

proparian suture see CEPHALON.

protaspis *n*. the earliest stage in the development of a TRILOBITE from larva to adult. It is a tiny disc (about 0.75 millimetres diameter) with a segmented central portion that later becomes the GLABELLA. The eyes are small and located on the anterior margin; later they move inwards. Many protaspids carry spines that disappear in the adult. The development of the larva continues with the formation of a transverse furrow separating the 'CEPHALON' from the 'PYGIDIUM' (the *meraspid* stage). Thoracic segments begin to form at and separate from the anterior margin of the 'pygidium' until the final adult number is

reached (the *holaspid* stage). See Fig. 71 below.

protegulum *n*. the first-formed part of each VALVE in BRACHIOPODS. It becomes part of the UMBO.

Proterozoic *n*. the latest of the three major subdivisions of the PRECAMBRIAN. The Proterozoic has been divided into three eras; the Palaeoproterozoic (2500–1600 Ma), Mesoproterozoic (1600–900 Ma) and the Neoproterozoic (900–544 Ma). See Appendix 2.

protists *n*. an informal name for all single-celled eucaryans. See EUCARYA.

proto- *prefix meaning* first or foremost.

protoconch *n*. the first-formed part of the shell in CEPHALOPODS and GASTROPODS.

protore *n*. an ore having material too low in concentration or too poor in grade to be mined profitably. See also SECONDARY ENRICHMENT.

protozoa *n*. an informal name for minute aquatic or parasitic PROTISTS that consist of a single cell or a colonial aggregate of cells, without differentiation of function, that are often highly motile and feed heterotrophically and

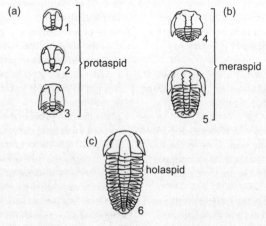

Fig. 71 **Protaspis**. Stages in the development of a trilobite: (a) the protaspid stage; (b) the meraspid stage; (c) the holaspid stage

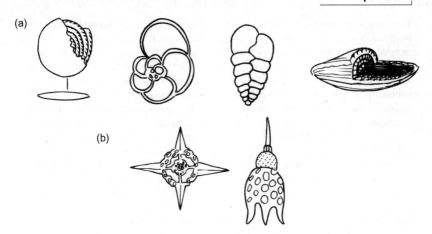

Fig. 72 **Protozoa**. (a) Foraminifera; (b) Radiolaria

can thus be considered informally as 'animals'. The former designation of Protozoa as a subkingdom of Animalia cannot be sustained since the 'Proto- zoa' are POLYPHYLETIC. The typical protozoan is microscopic, but some are as much as six centimetres long. Protozoa may or may not secrete a skeleton. They are divided into a number of groups, but the only two with major geological importance are the foraminifera and the radiolaria.

Foraminifera, or *forams*, are predomi- nantly marine benthic or planktonic protozoans. The TEST, which may be simple or multi-chambered, is gener- ally composed of calcium carbonate but may consist of silica or chitin. Certain species build arenaceous tests. Forams are important as age indicators, as rock-building agents and in sea-floor deposits. Range, Cambrian to Recent. (See GLOBIGERINID).

Radiolarians are marine planktonic protozoans. The test is siliceous, unchambered and often netlike in appearance. *Radiolarian ooze* is the accumulation on the deep sea bottom of the siliceous deposits composed of vast numbers of these tests. These siliceous exoskeletons may be partially responsible for some flint and CHERT deposits. (Certain cherts are referred to as *radiolarite*.) Their fragility and minute size somewhat limit the radiolarian test in fossil use; however, they have been recorded in rocks from Cambrian to Recent time. See PELAGIC DEPOSITS.

proustite, a mineral ore, Ag_3AsS_3, of silver. See RUBY SILVER.

provenance *n*. a place of origin, in particular the area or region from which the constituents of a sedimen- tary rock or facies were derived. Compare DISTRIBUTIVE PROVINCE.

province *n*. 1. part of a region defined by certain peculiarities of climate, geology, flora, fauna, etc.
2. an ecological unit defined by climate and geography and characterised by clusters of families or other higher taxonomic groupings that are endemic to the area.

proxies see PALAEOCLIMATOLOGY.

proximal *adj*. close to the source. The *proximal end* of a LAVA FLOW is the part of it nearest the vent from which it was erupted. See DISTAL.

psammite *n.* a SANDSTONE or ARENITE, or the metamorphic derivative of an arenite. See also PELITE, PSEPHITE.

psephite *n.* **1.** a coarse sediment such as gravel, breccia or conglomerate; compare RUDITE.
2. the metamorphic derivative of a RUDITE. See also PELITE, PSAMMITE.

pseudo- *prefix meaning* false. It generally denotes a deceptive resemblance to the object or substance to whose name it is prefixed.

pseudocraters *n.* small CRATERS that have no connection with a VOLCANIC pipe, produced as a result of the explosion of steam trapped at the base of a LAVA FLOW. They may be found where lava flowed into water or over wet sediment. See also LITTORAL CONE.

pseudofossil *n.* an object of inorganic origin that resembles one of organic origin, often found in sedimentary or metamorphic rock. Common false fossils are DENDRITES, CONCRETIONS, CONE-IN-CONE STRUCTURES and SEPTARIA. Compare FOSSIL, DUBIOFOSSIL.

pseudomorph *n.* a mineral that takes the outer form of another. This may occur in different ways:
(a) by REPLACEMENT;
(b) the original mineral may be dissolved, leaving a cavity in which the new mineral is deposited (INFILTRATION);
(c) a coating of one mineral may be deposited on another (*incrustation*);
(d) one mineral may alter to another, but the original crystal form is retained, for example SERPENTINE forms pseudomorphs after OLIVINE.

pseudonodules see LOAD CASTS.

pseudotachylyte *n.* a dense, glassy, dark grey or black CATACLASTIC rock associated with intense FAULT movement involving extreme MYLONITISATION and/

or partial melting typically occurring in erratically branching veins. Pseudotachylite may also occur in the subcrater basement below major meteorite impact sites. See IMPACTITE.

psilomelane *n.* a mixture of hydrated manganese oxide minerals found in SECONDARY deposits. It occurs as hard BOTRYOIDAL masses. It is an important ore of manganese.

pteridophyte *n.* a spore-bearing vascular plant, belonging to the division Pteridophytina, that appeared in the Devonian and includes horsetails, club mosses and ferns. Compare BRYOPHYTE, SPERMATOPHYTE.

pteropod *n.* a marine nektonic GASTROPOD of the order Pteropoda. Pteropods have thin, calcareous shells that, in different species, may be snail-like, slender, conical, globular and spinous, straight or curved. Their skeletal remains accumulate in the ocean bottom to form *pteropod ooze*, one of the constituents of PELAGIC DEPOSITS. Range, Cretaceous to present.

pteropod ooze see PELAGIC DEPOSITS, PTEROPOD.

ptygmatic structures *n.* disharmonically folded VEINS developed in a COMPETENT layer surrounded by incompetent material. Contorted quartzo-feldspathic veins in high-grade METAMORPHIC ROCKS such as MIGMATITES commonly show this structure. See Fig. 73 on the opposite page.

pudding stone see CONGLOMERATE.

pull-apart basin see RHOMBOCHASM.

pumice *n.* white or grey highly vesiculated pyroclastic material, normally of rhyolitic or dacitic composition. When release of pressure at the time of eruption causes effervescence of dissolved gases, the high viscosity of the magma traps the bubbles so that they expand to form a froth. Its specific

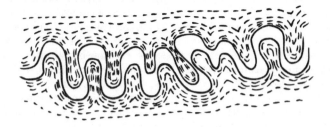

Fig. 73 **Ptygmatic structure**. Highly contorted folds

gravity is usually less than 1 g/cm^3. It could have formed OBSIDIAN had it cooled under greater pressure. Basaltic and andesitic magma is normally of sufficiently low viscosity to enable the gases to escape more easily and form volcanic cinder (scoria) rather than pumice. An unusual basaltic pumice is sometimes found in Hawaii, however, both as pyroclastic material and as froth on lava flows; it forms the lowest density rock known, with a specific gravity of 0·3. Compare SCORIA. See PYROCLASTIC MATERIAL.

Purbeckian *n*. an obsolete name for the stratigraphical STAGE spanning the top of the JURASSIC/base of the CRETACEOUS in Britain. It is equivalent to Tithonian (Jurassic) and Ryazanian (Cretaceous). The Purbeck Group is the final Jurassic group in the south of England and consists of limestones and shales deposited in evaporitic lagoons.

pure gold see CARAT.

pure shear *n*. a strain in which a body is lengthened in one direction and shortened at right angles to that direction.

P wave see SEISMIC WAVES.

pycnometer or **specific-gravity bottle** *n*. a small glass bottle containing a known volume of a HEAVY LIQUID of known density, used for determining the specific gravity (and hence the density) of mineral grains.

pygidium *n*. a triangular or rounded plate of fused segments forming the tail section of a TRILOBITE exoskeleton (Fig. 89). If the pygidium is small it is described as *micropygous*; if smaller than the cephalon it is *heteropygous*. An *isopygous* pygidium is the same size as the cephalon while a *macropygous* pygidium is larger than the cephalon (but this is rare).

pyramid see FORM.

pyralspite *n*. a little-used acronym for garnets intermediate in composition between pyrope, almandine and spessartine. These three end members of the GARNET group form SOLID SOLUTIONS with each other but not with members of the UGRANDITE group.

pyrargyrite *n*. a sulphide mineral, Ag_3SbS_3, an ore of silver. See RUBY SILVER.

pyrite *n*. a common yellow isometric sulphide mineral, FeS_2, dimorphous with MARCASITE. Pyrite forms under a wide range of pressure-temperature conditions, so is found in many geological environments. It occurs in masses, associated with chalcopyrite, among magmatic segregation deposits in MAFIC rocks. Pyritised concretions are formed by chemical deposition under water. It is the most abundant and widespread of the sulphide minerals and an important ore of sulphur. Its yellow-gold colour has often led to its being mistaken for gold. See also FOOL'S GOLD, PYRITISATION.

pyrites *n.* a redundant name for PYRITE.

pyritisation *n.* **1.** the introduction of iron disulphide, FeS_2, into any type of rock. **2.** a type of fossilisation wherein the original hard parts of an organism are replaced by PYRITE, FeS_2. The iron disulphide is probably formed from the sulphur of the decomposing organisms and the iron from the encasing sediments. Remains of BRACHIOPODS and some molluscs are often found in a pyritised state. See FOSSIL, REPLACEMENT.

pyroclast *n.* an individual fragment of any size ejected during a volcanic eruption.

pyroclastic *adj.* **1.** pertaining to fragmental rock materials formed by volcanic explosion or aerial ejection from a volcanic vent. It is not synonymous with VOLCANICLASTIC (fragmental volcanic) insofar as the latter applies to any fragmental aggregate of volcanic parentage. (See PYROCLASTIC MATERIAL.) **2.** a textural term describing the fabric of a rock the fragmented appearance of which is the result of explosive volcanic processes, e.g. TUFF and AGGLOMERATE. Compare VOLCANICLASTIC.

pyroclastic fall *n.* a shower of PYROCLASTIC MATERIAL that has been blown out of the VOLCANIC VENT and falls at varying distances from the source, depending on fragment size, wind direction and the height of the ERUPTION COLUMN. The deposits are better sorted than most pyroclastic deposits (because of AEOLIAN processes) and mantle the topography. Some fine-grained fall deposits are derived from the ash cloud that forms above a PYROCLASTIC FLOW.

pyroclastic flow *n.* a hot mixture of PYROCLASTIC material and gas, with a high concentration of particles, that moves as a surface flow away from the vent. The flow is controlled by gravity but may be partly fluidised (see FLUIDISATION). Pyroclastic flows are formed by the violent disintegration of a growing VOLCANIC DOME or LAVA FLOWS in the summit area of andesitic to rhyolitic volcanoes. Flow deposits are topographically controlled and tend to be confined to valleys and depressions, unlike the PYROCLASTIC FALL deposits. They are usually poorly SORTED, although larger fragments may show some grading. The three main types of deposit formed are BLOCK-and VOLCANIC ASH-flow deposits, SCORIA-flow deposits and PUMICE-flow deposits or IGNIMBRITES. See also NUÉE ARDENTE, BASE SURGE.

pyroclastic material or **ejecta** *n.* those clastic rock materials formed by volcanic explosion or aerial ejection from a VOLCANIC VENT. Although in the full geological record of the Earth, the weight of material from LAVA FLOWS exceeds the weight of pyroclastic products, the latter are spread over a greater surface area during most VOLCANIC ERUPTIONS that are observed. Pyroclastic material may be classified according to size, origin or petrographic composition:
(a) VOLCANIC BOMBS and BLOCKS: fragments between several metres and 64 millimetres in diameter.
(b) LAPILLI: fragments between 64 and 2 millimetres in diameter.
(c) VOLCANIC ASH: fragments between 2 and 0.25 millimetres in diameter.
(d) *volcanic dust:* fragments less than 0.25 millimetre in diameter.
(e) PUMICE: a light-coloured highly vesicular glassy material, commonly dacitic or rhyolitic in composition.

(f) SCORIA: a vesicular material that is usually heavier, darker and more crystalline than pumice. Also known as *cinder*, it may be BASIC or INTERMEDIATE in composition. Both pumice and scoria may exist as pyroclastic material or as froth on lava flows.

Ashes, pumice, scoriae and bombs are formed from ejected fragments of fluid lava. Other ejecta, including lapilli and blocks, consist of fragments of all sizes thrown out in a completely solid or semi-solid state. Pyroclastic material, having a rhyolitic composition, is generally white or light grey in colour; basaltic material is commonly black or red, and andesitic material intermediate in colour. See also PYROCLASTIC ROCKS.

pyroclastic rocks *n.* rocks consisting of consolidated volcanic ejecta. Such rocks are generally classified according to the dominant particle size (see Fig. 74 below).

Rocks composed entirely of lapilli or of volcanic blocks are rare, since finer particles usually fill in the spaces. Consolidated mixtures of ash and lapilli or of ash and blocks or bombs are common. See IGNIMBRITE, PYROCLASTIC MATERIAL.

pyroclastic surge *n.* a turbulent, low-particle concentration flow that transports PYROCLASTIC MATERIAL along the ground surface. Compare PYROCLASTIC FALL, PYROCLASTIC FLOW. Pyroclastic surge deposits mantle the topography but are thickest in the hollows. Unidirectional SEDIMENTARY STRUCTURES such as DUNES and CROSS-BEDDING are common. See also BASE SURGE.

pyroelectric *adj.* (of minerals) developing an electric charge in response to a change in temperature. TOURMALINE is a pyroelectric mineral.

pyrogenesis *n.* the INTRUSION and EXTRUSION of MAGMA and products derived from it.

pyrolite *n.* a hypothetical peridotitic rock composition proposed by Ringwood to represent upper MANTLE material from which a basaltic magma could be produced by partial melting. See PERIDOTITE, BASALT.

pyrolusite *n.* a soft, black tetragonal mineral, MnO_2, the most important ore of manganese. It usually occurs as fibrous, dendritic or concretionary aggregates in secondary deposits.

pyrometamorphism *n.* the recrystallisation of rocks under extremely high

Clast size (mm)	Pyroclasts	Consolidated deposit (pyroclastic rock)
	bombs	agglomerate (bomb > block)
	blocks	pyroclastic breccia (block > bomb)
64	lapilli	lapilli tuff
2	coarse ash	coarse ash tuff
$1/16$	fine ash (dust)	fine ash tuff (dust tuff)

Fig. 74 **Pyroclastic rocks**. Nomenclature of pyroclastic rocks according to particle size

temperatures, most commonly seen in XENOLITHS included in igneous rocks or in the wall rocks of some INTRUSIONS. Partial melting may occur, with the production of BUCHITES.

pyrometasomatism see METASOMATISM.

pyrope *n.* the magnesium-aluminium end member of the GARNET series, $Mg_3Al_2Si_3O_{12}$; some Ca and Fe^{2+} is usually present. Its colour varies from deep red to almost black. It is found in detrital sediments, in garnet peridotite and in KIMBERLITE.

pyrophyllite *n.* a yellowish white, grey or pale-green mineral, $Al_2Si_4O_{10}(OH)_2$, resembling TALC. It usually occurs in lamellar or radiating foliated aggregates in schists.

pyroxene-hornfels facies *n.* a rock facies characteristic of high-grade CONTACT METAMORPHISM characteristic of the AURIOLE close to GRANITIC, DIORIOTIC or GABBROIC bodies. In pelitic rocks, minerals such as quartz, orthoclase, cordierite, hypersthene and plagioclase occur. Minerals found in calcareous rocks include plagioclase, diopside and grossularite. See also METAMORPHIC FACIES.

pyroxenes *n.* a group of common rock-forming minerals with the general formula $X_{1-n}Y_{1+n}Z_2O_6$, where X = Ca or Na, Y = Mg, Fe^{2+}, Ni, Li, Fe^{3+}, Cr or Ti, and Z = Si or Al. The pyroxenes are single-chain silicates (compare the AMPHIBOLES) and are characterised by two PRISMATIC cleavages in the basal plane that intersect almost at right angles. The minerals crystallise in both the orthorhombic and monoclinic crystal systems.

The principal ORTHOPYROXENE minerals are ENSTATITE and HYPERSTHENE. The principal monoclinic pyroxenes (CLINOPYROXENES) are DIOPSIDE, *hedenbergite* (the iron-rich member of the diopside solid solution series, $CaFe^{2+}Si_2O_6$), AUGITE, AEGERINE, JADEITE and SPODUMENE. Pyroxenes occur in a wide variety of BASIC, ultrabasic and ALKALINE IGNEOUS ROCKS and high-grade METAMORPHIC ROCKS and SKARNS.

pyroxenite *n.* an ULTRAMAFIC intrusive IGNEOUS ROCK composed essentially of PYROXENE with accessory olivine, hornblende or chromite.

pyroxenoid *n.* a mineral, such as RHODONITE or WOLLASTONITE, that is chemically analogous to pyroxene but the SiO_4 tetrahedra of which are joined in chains with a repeat unit of 3, 5, 7 or 9.

pyrrhotite *n.* a yellowish-brown mineral, $Fe_{1-x}S$ (x=0–0.2). Its composition is close to that of PYRITE but it is deficient in iron. Some pyrrhotite is magnetic, and for this reason it has sometimes been referred to as *magnetic pyrites*, but the intensity decreases with increasing iron content. A source of sulphur and iron, it occurs in association with nickel, platinum and copper ores.

Qq

Q *n.* a measure of dissipation in energy-storing systems; it is proportional to the energy stored in the system divided by the energy dissipated during a cycle. Among the many ways in which Q can be characterised is as the degree of perfection of elasticity; one measure of it in the Earth is the width of spectral peaks in free oscillations. The decay of Love waves and Rayleigh waves (see SEISMIC WAVES) after repeated circling of the Earth is an alternative method. Using both methods, it is found that the Q of rocks increases with pressure, decreases with temperature and depends upon the rock's composition and physical state.

QAPF diagram see IGNEOUS ROCK, Appendix 3.

QAPFM classification see IGNEOUS ROCK.

quaking bog see BOG.

quartz *n.* a trigonal mineral, one of the crystalline polymorphs of SILICA, SiO_2. It commonly occurs as six-sided crystals with pointed terminations. For cryptocrystalline varieties of silica see CHALCEDONY. Colourless and transparent quartz, if found in good crystals, is known as *rock crystal*. Varieties that are coloured because of the presence of impurities may be used as gemstones: AMETHYST, purple to blue-violet; *rose quartz*, pink; CITRINE, orange-brown; *smoky quartz*, pale yellow to deep brown.

Quartz has no CLEAVAGE but a CONCHOIDAL fracture. It is one of the commonest rock-forming minerals and is an essential constituent of acid IGNEOUS ROCKS, for example granite. It occurs in pegmatites and in hydrothermal veins in association with ore minerals. Its hardness (7 on MOHS SCALE) and lack of cleavage make it resistant to ABRASION, so it is an important constituent of SEDIMENTARY ROCKS and occurs in many METAMORPHIC ROCKS, especially psammites and pelites. See also ALPHA QUARTZ, BETA QUARTZ, TRIDYMITE and CRISTOBALITE.

quartzarenite see ORTHOQUARTZITE.

quartz diorite *n.* a coarse-grained plutonic rock with similar composition to DIORITE but containing subordinate amounts of quartz (less than 5 per cent) and alkali feldspar. As the quartz and alkali feldspar content increases, quartz diorite grades into GRANODIORITE.

quartz index *n.* the mineralogical maturity of a SANDSTONE, expressed as the ratio of quartz plus chert to the combined percentage of feldspar, rock fragments and clay matrix.

quartzite *n.* **1.** a METAMORPHIC ROCK consisting primarily of QUARTZ grains, formed by the recrystallisation of sandstone by thermal or regional metamorphism; a METAQUARTZITE. **2.** a sandstone composed of quartz grains cemented by silica; an ORTHOQUARTZITE.

quartz monzonite *n.* a granitic rock in which 5 to 20 per cent of the felsic constituents comprise QUARTZ and in which the ratio of ALKALI FELDSPAR to total FELDSPAR is between 35 and 65 per cent. It is the approximate plutonic equivalent of LATITE, grading into GRANODIORITE with an increase in plagioclase and quartz, and into GRANITE with more potassium-rich alkali feldspar.

quartzose *n*. a sediment or rock (especially sands and sandstones) containing QUARTZ as a principal constituent.

quartz wedge see ACCESSORY PLATES.

Quaternary *n*. the most recent period of geological time, a division of the CENOZOIC.

quaternary system see PHASE DIAGRAM.

quick clay see SENSITIVE CLAY.

quicksand *n*. a mass or bed of fine sand comprising smooth grains that do not adhere to each other. Water usually flows up through the pores, resulting in the formation of an extremely unstable mass that yields readily to pressure.

Rr

radial drainage pattern see DRAINAGE PATTERN.

radial symmetry *n.* a pattern of symmetry in which similar parts of an organism are regularly arranged about an axis or central point, e.g. a starfish. Compare BILATERAL SYMMETRY.

radioactive age determination see DATING METHODS.

radioactive decay *n.* the spontaneous transformation of a nucleus into one or more different nuclei by the emission of a particle or photon, by fissioning or by the capture of an orbital electron. See also ALPHA DECAY, BETA DECAY, GAMMA DECAY.

radioactive series *n.* a series of isotopes, each of which becomes the next in sequence by some type of decay until a stable element is reached. The major radioactive series are the actinium, thorium and uranium series. See also PARENT ELEMENT, DAUGHTER ELEMENT, END PRODUCT.

radioactivity *n.* a property of certain elements that spontaneously and constantly emit ionising and penetrating radiation. See also RADIOACTIVE DECAY.

radioactivity logging see WELL LOGGING.

radio altimeter see ALTIMETER.

radiocarbon *n.* either of the radioactive isotopes of carbon; ^{14}C is the one normally used in radiometric dating. See CARBON-14, DATING METHODS.

radiocarbon dating see CARBON-14, DATING METHODS.

radiogenic isotope *n.* an ISOTOPE produced by RADIOACTIVE DECAY; it is not necessarily radioactive itself. Compare RADIOISOTOPE.

radioisotope *n.* a radioactive isotope of an element. Compare RADIOGENIC ISOTOPE.

radiolarians see PROTOZOA.

radiolarian ooze see PELAGIC DEPOSITS, PROTOZOA.

radiometric dating see DATING METHODS.

radula *n.* a flexible belt with rows of teeth found in many MOLLUSCS and used for rasping food.

rain prints *n.* circular or elliptical shallow pits with a raised rim, up to one centimetre in diameter, found on the surface of some SILTSTONES or MUDROCKS. They are often associated with DESICCATION CRACKS, indicating that the surface dried out, and are formed by the impact of raindrops or hailstones on the wet sediment surface. Sandy surfaces do not cohere when dried, so do not preserve these marks. Rain prints may however be found as CASTS on the base of a SANDSTONE bed, see SOLE MARK. They are useful as WAY-UP INDICATORS.

raised beach *n.* a BENCH or terrace with beach DEPOSITS marking the position of a former SHORELINE above the present sea level. It may be backed by inland cliffs. Many raised beaches found around the Scottish coastline and the coast of Scandinavia result from the isostatic uplift of the land following the melting of the ice sheet at the end of the last GLACIAL period, some 10 000 years ago. (See ISOSTACY.) However, some raised beaches (or *coastal terraces*) result from TECTONIC rather than isostatic uplift, for example terrace sequences in New Guinea or Bermuda.

range see STRATIGRAPHICAL RANGE.

range finder *n.* an instrument for determining the distance from a single point of observation to other points at which no instruments are located.

range zone or **range biozone** *n.* a body of strata that represents the total extent of occurrence, horizontal and vertical, of any particular fossil form (*biozonal index*) selected from the collection of fossil forms in a stratigraphical sequence. The biozone thus defined is a *total range biozone*. Compare BIOZONE, ACME ZONE.

rank *n.* the position of a COAL in the transition series from PEAT to ANTHRACITE, reflecting the degree of COALIFICATION. The rank of coal is largely determined by the amount of fixed carbon, volatiles and water it contains. Compare GRADE, COAL TYPE.

rapakivi texture *n.* an overgrowth of sodic PLAGIOCLASE on large, usually ovoid, potassium FELDSPAR crystals originally found in Proterozoic granites in Finland. See also REACTION RIM, CORONA.

rare earth elements (REE) *n.* a series of metallic elements with closely similar chemical properties. The series consists of the lanthanide elements (atomic numbers 57, lanthenum to 71, lutetium). Sometimes yttrium (atomic number 39) is included as it has an ionic radius similar to the rare earth holmium. In most rocks and minerals they occur only in trace amounts, with the exception of a few minerals such as MONAZITE. See INCOMPATIBLE ELEMENTS.

ray *n.* a vector normal to a wave surface, indicating direction and sometimes velocity of propagation.

Rayleigh number *n.* a parameter used in the mathematical description of fluid CONVECTION. It is proportional to the ratio of the time required to heat a fluid layer by conduction and the time required for fluid particles to circulate once around the CONVECTION CELL. Rather large Rayleigh numbers occur in the mantle and have a lower limit of about 10^6. For small values (Ra < 10^3), thermal convection cannot occur; for large values, the system becomes unstable and develops disturbances that grow into convection cells.

Rayleigh wave see SEISMIC WAVES.

raypath *n.* the imaginary line along which wave energy travels. It is always normal to the WAVEFRONT in isotropic media.

reaction pair *n.* in geology, two minerals that exhibit the *reaction principle*, i.e. one mineral species is converted into the other by the reaction of the crystal phase with the liquid magma containing it. Thus, forsterite, which is stable at high temperatures, can be converted into enstatite at a lower temperature if there is sufficient silica in the magma containing it. See also REACTION SERIES.

reaction point *n.* a point on a melting relations diagram in which it is impossible to state the composition of the liquid in terms of all the solid phases in equilibrium at this point.

reaction rim *n.* a zone of contrasting composition at the periphery of a CRYSTAL where the rim is composed of another mineral species and represents the reaction of the earlier solidified mineral with the surrounding MAGMA. The crystal outlines of the original mineral characteristically have a corroded appearance.

reaction series *n.* a sequence in which early-formed minerals within a cooling melt react progressively with the melt to modify their composition or to form new minerals as the temperature falls. This concept, originally proposed by

N. L. Bowen, suggests two series. The *discontinuous reaction series*, in which ferromagnesian minerals may react with the melt; early-formed crystals of olivine react to form PYROXENE, which later reacts to form AMPHIBOLE, which in turn reacts with the residual liquid to form BIOTITE. The *continuous reaction series* is seen in the crystallising behaviour of PLAGIOCLASE feldspar; early-formed plagioclase, rich in anorthite, reacts continuously with the liquid, exchanging calcium for sodium as the temperature falls, and as a result the composition of the crystals is progressively enriched in ALBITE. If crystallisation takes place slowly under equilibrium conditions, the plagioclase is completely altered, remains homogeneous and will end up with a composition that reflects that of the sodium and calcium content of the original melt. If the cooling is too rapid for equilibrium to be established, the reaction will be incomplete and cores of earlier composition will be preserved within the plagioclase crystals as concentric compositional zones. See also REACTION PAIR, ZONING.

realgar *n.* a monoclinic mineral, arsenic sulphide, AsS, usually found in compact aggregates and brilliant red-orange films. It is deposited in hot springs and low temperature hydrothermal veins. If exposed to light and air, it gradually converts to ORPIMENT.

Recent see HOLOCENE.

recession *n.* **1.** GLACIAL recession.
2. the continuing landward progression of a SHORELINE as a result of EROSION.
3. the net landward movement during a specific time interval. Compare ADVANCE.
4. the backward movement of an eroded ESCARPMENT or the retreat of a slope from a former position without a change in its angle. Compare REGRESSION.

recessional moraine see MORAINE.

recharge or **intake** *n.* the refilling of an AQUIFER either by natural or artificial recharge; the sum of the processes involved in the downward movement of water to the zone of saturation or the amount of water added.

recrystallisation *n.* in METAMORPHISM, the formation, essentially while in the solid state, of new crystalline mineral grains in a rock. The new grains may be larger than the primary grains, as in PORPHYROBLASTS, and their mineral composition may differ or be the same.

rectangular drainage pattern see DRAINAGE PATTERN.

recumbent fold see FOLD.

red beds *n.* predominantly red sedimentary strata composed primarily of SANDSTONE, SILTSTONE and SHALE; their colour is the result of the presence of a coating of HEMATITE (ferric oxide) on the sedimentary grains. They are generally considered to indicate sedimentation in an arid continental environment, with the hematite originating from the *in situ* oxidation of detrital iron minerals. Examples of red beds include the OLD RED SANDSTONE facies of the European Devonian and the NEW RED SANDSTONE of the Permian and Triassic.

red clay *n.* a fine-grained PELAGIC DEPOSIT, reddish-brown or chocolate-coloured, formed by the accumulation of material at depths generally greater than 3500 metres and at a far distance from the continents. Red clay contains rather large amounts of meteoric and volcanic dust, pumice, shark teeth, manganese concretions and ice-transported debris. Compare red mud (see TERRIGENOUS DEPOSITS).

red mud see TERRIGENOUS DEPOSITS.

red ochre *n.* a red, earthy HEMATITE used as a pigment. See also OCHRE.

redox potential *n.* **1.** the oxidation-reduction potential, with the symbol E_H. **2.** the oxidation-reduction potential of an environment, i.e. the voltage obtainable between a normal hydrogen electrode and an inert electrode placed in the environment.

reduction spot or **bleach spot** *n.* a spherical feature common in RED BEDS, especially red SANDSTONE, where the colour has been bleached by local chemical reduction of the ferric iron compounds in the red cementing material to the ferrous state. This appears as a white or pale-green circular area on the rock surface, generally with a diameter up to about 10 centimetres.

redusates *n.* sediments that have accumulated under reducing conditions and are therefore typically rich in organic carbon and iron sulphide; black shale and coal are examples. Compare RESISTATES, EVAPORITES, HYDROLY-SATES, OXIDATES.

REE see RARE EARTH ELEMENTS.

reef *n.* **1.** a build-up of organic skeletal material and organic debris formed by the growth of calcareous ALGAE and sessile calcareous animals, especially CORALS, and composed largely of their remains. The resulting structure may be patch-like, mound-like, wall-like or a flat surface (*reef flat*). It may also be such a structure built in the geological past and now enclosed in rock, usually of dissimilar LITHOLOGY. See also BARRIER REEF, CORAL REEF, ATOLL, BIOHERM. **2.** a gold-bearing QUARTZ deposit.

reef rock *n.* a resistant massive unstratified rock composed of the calcareous remains of reef-building organisms, often interspersed with carbonate sand; the entire substance is cemented by calcium carbonate.

reference locality *n.* a place containing a reference section designated as a supplement of the TYPE LOCALITY.

reference plane see DATUM PLANE.

reference section or **hypostratotype** *n.* a rock section specified as a supplement of the TYPE SECTION (*holostratotype*) to extend the knowledge of the unit to other geographical areas or facies. See also STRATOTYPE.

reflection *n.* the return of a wave that is incident upon a surface to the original medium. A SEISMIC WAVE that is reflected at an interface between different media is called the *reflection wave*.

reflection shooting see SEISMIC SHOOTING.

reflux *n.* a process in which heavy, concentrated brines move downwards through the floor of an EVAPORITE basin. Because such brines may be rich in magnesium, reflux is thought to contribute to the DOLOMITISATION of carbonate rocks in some sequences. The depth of dolomitisation in a SABKHA may depend on the extent to which reflux occurs.

refraction *n.* (*physics*) the directional change of a ray of light, heat or sound, or an energy wave, such as a SEISMIC WAVE, upon passing obliquely from one medium into another in which its speed is different. See LAW OF REFRACTION.

refraction shooting see SEISMIC SHOOTING.

refractive index (**RI**) see INDEX OF REFRACTION, LAW OF REFRACTION.

refractory clay see FIRECLAY.

refractory ore *n.* ore from which it is difficult to recover valuable substances.

reg or **serir** *n.* a stony desert, a stone-covered desert surface, especially in North Africa, from which FINES have been removed by DEFLATION and other EROSION processes. Compare ERG, HAMMADA.

regelation *n.* a twofold process involving the melting of ice that is subjected to extreme pressure and the refreezing of the resultant meltwater upon removal of that pressure. Regelation is one of the factors involved in the BASAL SLIDING of glaciers. See GLACIER.

regime *n*. **1.** in a stream channel, the existence of a balance between deposition and erosion over several years.
2. the condition of a stream with respect to the rate of its average flow, as determined by the volume of water flowing past different cross-sections within a defined time period.
3. a regular pattern of occurrence or a condition showing widespread influence, e.g. a sedimentary regime.
4. (*glaciology*) see BALANCE.

regimen *n*. **1.** the flow characteristics of a stream, such as volume, velocity and sediment-transport capacity. Compare REGIME.
2. the total quantity of water associated with a DRAINAGE BASIN and its behaviour as determined from such quantities as rainfall, surface and subsurface flow and storage.
3. an analysis of the total quantity of water associated with a lake over a specified time interval.
4. glacial BALANCE.

regional dip *n*. a nearly uniform inclination, usually at a small angle, of strata extending over a wide region.

regional metamorphism see METAMORPHISM.

regolith *n*. unconsolidated rock material resting on bedrock, found at and near the surface of the Earth. *Residual regolith* is formed by the mechanical and chemical weathering of bedrock; *transported regolith* is moved and deposited by processes acting at or near the Earth's surface. Regolith material includes soils, alluvium, till, loess, dune sand and volcanic dust.

regression *n*. **1.** the retreat of the sea from land areas.

2. any change that moves the boundary between marine and non-marine deposition towards the centre of a marine basin or converts deep-water conditions to near-shore, shallow-water conditions. Compare TRANSGRESSION, REGRESSION.

rejuvenation *n*. the reversal or renewal of some landscape feature or geological form to a former degree of effectiveness, or to a condition leading to a new cycle of EROSION. A stream is said to be *rejuvenated* when, after having developed to maturity, it has its erosive ability renewed as a result of regional uplift. Other factors commonly related to rejuvenation are the lowering of sea level and climate change.

rejuvenation head see KNICK POINT.

relative age *n*. the geological age of a rock, fossil or event, as defined relative to other rocks, fossils or events rather than in terms of years. Compare ABSOLUTE AGE.

relative dating *n*. the chronological arrangement of fossils, rocks or events with respect to the geological time scale, without reference to their ABSOLUTE AGES.

relative humidity *n*. the ratio, expressed as a percentage, of the actual amount of water vapour in a given volume of air to the amount that the same volume would hold if the air were saturated at the same temperature.

relative permeability *n*. the ratio between effective and absolute PERMEABILITY in a rock, where EFFECTIVE PERMEABILITY is the permeability to a given fluid at partial saturation and ABSOLUTE PERMEABILITY is the permeability at 100 per cent saturation.

relic *n*. **1.** a landform, such as an erosion remnant, that has survived disintegration or decay.
2. a metamorphic RELICT.

relict *adj.* **1.** (of a feature or fabric structure in a previous rock) persisting in a later rock despite METAMORPHISM. **2.** (of topographic features) remaining after other parts have disappeared, for example a VOLCANIC PLUG.

relief *n.* the apparent roughness a mineral shows in THIN SECTION. It is an indication of the refractive index (RI) of the mineral, as a mineral with an RI very different from that of the mounting medium will appear 'rough' or obvious, whereas one with an RI similar to the cement will appear smooth. The BECKE TEST or the SHADOW TEST can be used to determine whether the RI of the mineral is higher or lower than that of the cement.

remanent magnetisation see NATURAL REMANENT MAGNETISATION.

remnant arc or **dead arc** *n.* an inactive arc located behind the active volcanic arc and separated from it by an inter-arc basin. See ISLAND ARC and Fig. 53.

remote sensing *n.* the collection of information or data without the direct contact of a recording instrument. Devices such as cameras, infrared detectors, microwave frequency receivers and radar systems are used in remote sensing.

reniform *adj.* (of a crystal structure) consisting of radiating clusters of crystals that terminate in rounded kidney-shaped masses. Compare BOTRYOIDAL, COLLOFORM.

replacement *n.* **1.** the progressive substitution of one mineral for another by the action of capillary solutions, for example DOLOMITISATION. See also METASOMATISM. **2.** a process of fossilisation also referred to as MINERALISATION. There is progressive substitution of new minerals, such as calcite, silica, pyrite or limonite, for the original hard parts of the organism. Compare PERMINERALISATION; see also FOSSIL.

reptile *n.* a tetrapod vertebrate of the class Reptilia. They evolved from the AMPHIBIANS in the early CARBONIFEROUS and were able to colonise new habitats on land because of the development of a shelled egg that was protected from DESICCATION. Reptiles spread widely during PERMIAN times and gave rise to the mammals in the Upper TRIASSIC and the birds in the JURASSIC. The MESOZOIC was dominated by spectacular reptiles: of the subclass Archosauria, the DINOSAURS (terrestrial), ichthyosaurs and plesiosaurs (marine reptiles) and pterosaurs (flying reptiles). These became extinct at the end of the CRETACEOUS. Forms that have survived to the present day, such as lizards, snakes, crocodiles, turtles and tortoises, were relatively insignificant in the MESOZOIC.

reservoir rock *n.* any permeable and porous rock that yields gas or oil. SANDSTONE, DOLOMITE and LIMESTONE are the most prevalent types.

residual clay *n.* clay material formed in place by the weathering of rock; it is derived either from chemical decomposition of FELDSPAR or by the removal of non-clay mineral constituents from clay-bearing rock.

residual liquid *n.* the still-molten portion of a magma that remains in the MAGMA CHAMBER after some crystallisation has occurred.

residual placer see PLACER.

reserves *n.* that quantity of an ore that is located and that can be exploited economically using existing technologies. Compare RESOURCES.

resistates *n.* sediments composed of chemically resistant minerals, such as quartz or muscovite, with abundant weathering residues, e.g. highly quartzose sediments. Compare HYDRO-LYSATES, OXIDATES, REDUSATES, EVAPORITES.

resistivity n. **1.** (*electrical*) also called **specific resistance**, the resistance per unit length of a unit cross-sectional area of a material. It is the reciprocal of the electrical conductivity and is measured in ohm-m. $\rho = RA/l$ = ohm metres, where ρ is the resistivity, R the conductor resistance, A and l the cross-sectional area and length of the conductor, respectively.
2. (*thermal*) the reciprocal of thermal CONDUCTIVITY.

resistivity log see WELL LOGGING.

resistivity method n. an electrical method of GEOPHYSICAL EXPLORATION in which a voltage is applied between electrodes implanted in the ground and the resulting current is measured. Measurements are usually made between pairs of electrodes placed at different separations along a line crossing the area under study. Variations in RESISTIVITY, as noted by a change in current, may reveal a resistivity anomaly indicating some obscured feature, such as an ore body or fault or other geological structure.

resorbtion n. the remelting of a previously formed crystal by the enclosing MAGMA. This produces rounded or embayed corroded PHENOCRYSTS that may be surrounded by a REACTION RIM of other minerals.

resources n. known or inferred deposits of ores that cannot yet be economically exploited with current technologies. Compare RESERVES.

restite n. residual rock remaining after partial melting.

retardation see POLARISATION COLOURS.

reticulated adj. (of a vein or a lode) having a netlike structure or in which partially altered crystals form a network.

retrograde metamorphism n. the response of MINERAL ASSEMBLAGES to decreasing temperature and pressure. The minerals may form new combinations that are capable of stability within the adjusted conditions. See METAMORPHISM.

reverse fault n. a FAULT along which the HANGING WALL is displaced upwards with respect to the FOOTWALL.

reversed polarity n. a NATURAL REMANENT MAGNETISATION opposite the present geomagnetic field direction. Compare NORMAL POLARITY. See also GEOMAGNETIC POLARITY REVERSAL.

reversed zoning see INVERSE ZONING.

Reynolds number n. a dimensionless number that expresses the ratio of inertial to viscous forces during flow. For flow in a pore space, as happens in GROUNDWATER flow, the Reynolds number $Re = pvd/\eta$, where p is the fluid density, v the fluid velocity, d the mean pore dimension and η the viscosity. DARCY'S LAW for groundwater flow is valid for values of Re between about 1 and 10. For flow in a pipe, the Reynolds number $Re = plU/\mu = lU/v$, where p is the fluid density, l is the diameter of the pipe, U is the flow velocity and v is the kinematic viscosity of the fluid. The Reynolds number is the parameter that determines whether LAMINAR FLOW or TURBULENT FLOW will occur in a given condition. For Reynolds numbers below 2000, pipe flow is laminar and for values above 2000 it is turbulent. Reynolds numbers can also be used to indicate the drag experienced by self-propelling organisms in a fluid, where the higher the kinematic viscosity the lower the Reynolds number. For example, a large organism such as a dolphin swimming in sea water at c. 10 metres per second experiences low drag and has a high

Reynolds number (c. 3×10^6), whereas a bacterium swimming at 0.01 metres per second experiences high drag and has a low Reynolds number (0.00001). At low Reynolds numbers a fluid is very viscous.

rhabdosome n. the GRAPTOLITE colony.

Rhaetian n. the top STAGE of the UPPER TRIASSIC.

rheomorphism n. the process by means of which a rock becomes mobile and flows. The term is used in a number of contexts, including the formation of MIGMATITE in high-temperature regional METAMORPHISM, and in VOLCANISM to the process when PYROCLASTIC FLOWS, during or immediately following deposition, sometimes start to flow in a manner similar to lava. Compare MOBILISATION.

rheophilic adj. used to describe those CRINOIDS that seek water currents to feed from, turning their arms towards the oncoming current to collect food particles.

rheophobic adj. used to describe those CRINOIDS that seek current-free waters, spreading their arms horizontally to collect food particles.

rhodocrosite n. a banded pink rhombohedral carbonate mineral, $MnCO_3$, commonly occurring in granular, concretionary or stalactitic masses. It forms in medium-temperature hydrothermal veins in association with copper, silver and lead sulphides and other manganese minerals. It is used as an ore of manganese when quantities are sufficient; it is used to a minor extent for ornamental purposes.

rhodonite n. a deep-pink to brown manganese silicate mineral of the triclinic system, $MnSiO_3$. A limited amount of substitution of calcium for manganese is normal. Iron and zinc may also be present. It occurs as compact granular masses with black veins of manganese dioxide. Rhodonite is a mineral typical of the METAMORPHISM of impure LIMESTONE.

rhombochasm or **pull-apart basin** n. a rhomb-shaped major crustal depression that forms in regions where crustal extension accompanies STRIKE-SLIP fault movement. See TRANSTENSION.

rhombohedral cleavage see CLEAVAGE.

rhombohedral packing see PACKING.

rhourds see DRAAS.

rhyodacite n. an extrusive IGNEOUS ROCK, usually porphyritic, with PHENOCRYSTS of plagioclase, sanidine (an alkali feldspar) and quartz in a glassy to microcrystalline groundmass. Mafic phenocrysts of any type may be present but are usually sparse or absent. It is intermediate in composition between RHYOLITE and DACITE.

rhyolite n. one of a group of extrusive rocks commonly showing flow texture, and typically porphyritic, with phenocrysts of quartz and potassium feldspar in a glassy to microcrystalline groundmass. Rhyolite is the extrusive equivalent of GRANITE. See also RHYODACITE, DACITE

rhythmite n. any sediment with a distinctive couplet arrangement developed by cyclical sedimentation. Fine-textured dark laminae may alternate with coarser-grained lighter layers, or a distinct sediment type such as mudrock may alternate with another such as limestone. An individual pair of such laminae is a *couplet*. Rhythmite refers to those couplets that do not represent a one-year period; *varve* refers to those couplets that represent a single year. Varves comprise a pair of laminae, one fine and dark, the other coarser and lighter, that have been deposited within a one-year period and can be used as a means of dating. Great care must be exercised, however, to ascertain the annularity of a particular rhythmite before it is used for this purpose. In the case of glacial lake laminations that are true varves, the fine and coarse layers are deposited in winter and summer respectively. There are many conditions that may produce non-annual rhythmites, e.g. warm and cold spells of a non-annual nature. Repeated sequences of three or more sedimentary types are called *sedimentary cycles*.

RI *abbreviation for* refractive index. See LAW OF REFRACTION.

ria *n.* **1.** a long, narrow arm of the sea produced by the partial submergence of an area dissected by subaerial erosion; it is shallower and shorter than a FJORD.

2. a long, narrow river inlet that gradually decreases in depth from mouth to head.

Richter scale see EARTHQUAKE MEASUREMENT.

ridge see MID-OCEANIC RIDGE.

riebeckite *n.* a dark-blue or black monoclinic mineral, a sodium-rich AMPHIBOLE. It occurs as slender acicular crystals or bluish fibrous crystals in asbestiform aggregates (CROCIDOLITE) and has a characteristic CHATOYANCY.

rift valley *n.* an elongate topographical depression bounded by steep-dipping parallel or sub-parallel FAULTS that have a large DIP-SLIP component (i.e. a GRABEN bounded by normal or step-faults). Rift valleys result from crustal extension, with a component perpendicular to the rift zone, and commonly occur along the crest of broad domes in the Earth's CRUST. The uplift may be caused by the rise of hot MANTLE material beneath the crust, resulting in thinning and stretching of the crust. The East African rift system and the Rhine Valley graben in Germany are well-known examples of continental rift valleys.

Rifts are also found along the crest of many MID-OCEANIC RIDGES, particularly the MID-ATLANTIC RIDGE. Crustal extension here is linked with plate separation, and it is thought that the evolution of an ocean basin begins with the formation of a continental rift system. However, not all continental rift valleys evolve into ocean basins, see AULACOGEN. See also PLATE TECTONICS.

rigidity modulus see MODULUS OF RIGIDITY.

rille *n.* one of many variously shaped trench-like valleys on the lunar surface. Rilles are thought to represent collapsed LAVA TUBES.

ring complex *n.* a complex of igneous INTRUSIONS the individual components of which appear as circular or arcuate outcrops. RING DYKES AND CONE SHEETS are the two main members of ring complexes. Compare CAULDRON SUBSIDENCE. Both components can be created from the upsurge of MAGMA and its migration into subsequent tensile fractures in the host rock. Ring complexes are considered to be formed as part of the substructure of major central VOLCANOES where their evolution is linked to formation of CALDERAS. Ring complexes are well known in the Inner Hebrides and adjacent areas of Scotland and northeast Ireland, associated with centres of TERTIARY (early Palaeogene) igneous activity.

ring dykes and cone sheets *n.* curved dyke forms commonly found together. *Ring dykes* are annular or arcuate in plan, with a more or less vertical axis. The diameter varies considerably: some measure a few hundred metres, others reach several kilometres. The outcrop of individual intrusions may extend uninterrupted for the larger part of a circle. *Cone sheets* have the form of a conical surface converging to

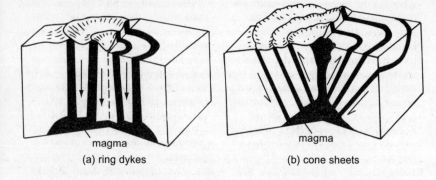

magma

(a) ring dykes

magma

(b) cone sheets

Fig. 75 **Ring dykes and cone sheets**. (a) cauldron subsidence of the central
block allows magma to fill annular spaces; (b) the upward pressure caused by
magma intrusion opens conical fractures along which magma can penetrate

a buried focus (presumably a magmatic
centre). Outcrops of these dyke forms,
often occurring in concentric sets, are
generally thin and interrupted. The
formation of cone sheets is ascribed to
the injection of a body of magma into
the Earth's crust; magmatic pressure
produces cracks in the overlying crust
into which magma from the primary
body is injected. As the magma
subsides, relaxation of pressure allows
the structure to slump (CAULDRON
SUBSIDENCE) into the primary magma
body and causes magma to well up
along vertical fractures to form ring
dykes. Cone sheets and ring dykes may
be seen in the Ardnamurchan penin-
sula on the west coast of Scotland. See
also RING COMPLEX.

Ring of Fire n. the volcanic chain that
occurs around most of the perimeter of
the Pacific Ocean. See VOLCANO.

ring structure see RING COMPLEX.

ringwoodite n. a high-pressure polymorph of OLIVINE with a SPINEL structure, found in meteorites.

rip-rap n. broken rock of weights from 7
to 70 kilograms, used as armour stone

to protect embankments or constructions from the action of water.

Riphean n. a division of the Russian
PROTEROZOIC, below VENDIAN.

ripple drift cross-lamination see CROSS-
BEDDING.

ripple marks n. small-scale ridges and
troughs formed by the flow of wind or
water over loose sand-grade sediment.
They are usually less than three
centimetres high, with spacing distances less than 30 centimetres. *Current
ripples* are asymmetrical in PROFILE, with
gentle upcurrent (STOSS) slopes and
steeper concave-upward lee slopes and
result from current flow in one
direction. They may range from
straight to sinuous crested or even
linguoid (tongue-shaped) forms.
Climbing ripples are formed where
sediment is deposited rapidly. *Oscillation ripple marks* are symmetrical in
outline and are produced by the
oscillating or lapping movement of
water. Two sets of differently oriented
ripple marks on the same surface
sometimes create a crossed or cell-like
pattern called an *interference ripple*

mark. Cross-lamination structures are formed within the sediment by the migration of ripples. See CROSS-BEDDING.

Riss see CENOZOIC, Fig. 13.

river bar *n*. an accumulation of ALLUVIUM in the channel, at the mouth or along the banks of a river or stream. At low water, it is commonly exposed and constitutes a navigational obstruction. *Channel bars* are located within the course of a stream and seem to be most characteristic of BRAIDED STREAMS although not restricted to them.

river capture see STREAM CAPTURE.

river terrace see KNICK POINT.

roche moutonnée *n*. an asymmetrical mound on a glaciated bedrock surface. The up-ice side of a roche moutonnée has been glacially scoured and is smoothly abraded; the down-ice side has a steeper, jagged slope as a result of GLACIAL PLUCKING. Thus the long axes of the roches moutonées are aligned parallel to the general flow of the former ice masses. The term originated from the resemblance of roches moutonées to sheeps' backs in form and texture. A CRAG AND TAIL differs from a roche moutonnée in scale and by the presence of a long, tapered sediment ridge extending down-ice. See also GLACIAL SCOURING.

roche rock see LUXULLIANITE.

rock *n*. **1.** any aggregate of MINERALS or organic matter, whether consolidated or not (see CONSOLIDATION). A rock may consist of only one type of mineral (a *monomineralic* rock) but more commonly contains a variety of minerals (a *polymineralic* rock). The geological definition includes materials such as SANDS and GRAVELS as well as the consolidated SEDIMENTARY, IGNEOUS or METAMORPHIC ROCKS, but in ordinary usage such materials are more commonly referred to as ALLUVIAL deposits, DRIFT, etc.

2. (*engineering*) a hard, ELASTIC material not significantly affected by immersion in water. Compare SOIL.

rock association *n*. a group of IGNEOUS ROCKS within a petrographic PROVINCE that are related both chemically and petrographically.

rock crystal see QUARTZ.

rock cycle see GEOCHEMICAL CYCLE.

rock flour *n*. finely ground rock particles resulting from glacial erosion.

rock-forming *adj*. (of minerals) comprising part of the composition of rocks and determining their classification. Among the more important rock-forming minerals are quartz, feldspars, micas, amphiboles and pyroxenes.

rock glacier see TALUS.

rock head *n*. the boundary between unconsolidated superficial DEPOSITS and BEDROCK. See CONSOLIDATION.

rock mantle see REGOLITH.

rock mechanics *n*. the science that covers the characteristics of ROCKS as materials and the ways in which rocks are tested for stress, strain, elasticity, fracture, etc.

rock pedestal *n*. a land feature (most common on desert surfaces) sculpted by wind EROSION. The more resistant parts of a rocky mass, formed of alternate layers of hard and soft rock, are worn away more slowly; because maximum ABRASION occurs at ground level, the mass may eventually resemble an object atop a pedestal.

rock pediment *n*. a PEDIMENT formed on a bedrock surface.

rock salt *n*. HALITE when it occurs as a massive or granular aggregate.

rock stratigraphic unit see STRATIGRAPHICAL UNIT.

rock stream see BLOCK STREAM.

rock unit see STRATIGRAPHICAL UNIT.

rock waste see DEBRIS.

Rodinia *n*. the supercontinent that formed in the PROTEROZOIC around 1000 Ma and broke up around 800 Ma.

rollover fold *n*. a fold that forms in the HANGING WALL of a LISTRIC FAULT. See Fig. 55.

roof *n*. **1.** the rock lying above a COAL bed.
2. the COUNTRY ROCK bordering the upper surface of an igneous INTRUSION.

roof pendant *n*. a mass of COUNTRY ROCK projecting down into the top part of an igneous INTRUSION such as a BATHOLITH or STOCK. It may not be possible to tell whether an OUTCROP of country rock within an intrusion is a roof pendant or a large XENOLITH unless the structures within the mass are now discordant to the structures of the rocks surrounding the intrusion.

room-and-pillar *n*. a system of mining in which ore is mined in chambers supported by columns of undisturbed rock that have been left intact for the purpose of roof support. It is also known as STOOP-AND-ROOM.

root zone *n*. **1.** that zone in the Earth's CRUST from which thrust FAULTS emerge.
2. the source of original attachment of the basal part of a fold NAPPE. See also ISOSTASY.

ropy lava see PAHOEHOE.

rose quartz see QUARTZ.

Rossi-Forel scale see EARTHQUAKE MEASUREMENT.

rostrum *n*. a ventral projection of the aperture of an AMMONOID shell, Fig. 1.

rotary drilling *n*. the principal method of drilling deep wells, especially for gas and oil. Rotary drilling may be adapted for any angle and is suitable for underground mining. In this method the drill bit rotates while bearing down. A great advantage of rotary drilling over CABLE-TOOL (percus-sion) DRILLING is that the borehole is kept full of drilling mud during the process. DRILLING MUD is a weighted fluid that is pumped down the drill pipe, out through openings in the drill bit and back up to a surface pit where it re-enters the mud pump. In addition to cooling and lubricating the drill bit, this fluid, by its hydrostatic pressure, inhibits the entry of formation materials into the well, thus preventing blowouts and gushers. The drilling mud also carries the crushed rock to the surface. See also AIR DRILLING.

rotation axis see CRYSTAL SYMMETRY.

rotational fault see FAULT.

rotational movement *n*. an apparent FAULT-BLOCK displacement in which the blocks have rotated in relation to each other; the alignment of previously parallel features is disturbed.

rotational wave see SEISMIC WAVES.

Rotliegendes *n*. the name formerly used for the European Lower Permian RED BEDS underlying the Zechstein salt deposits. The name is still used in the petroleum industry.

rounded *adj* (of a SEDIMENTARY particle) of which the original edges and faces have been almost removed by ABRASION although the original shape is still clearly seen. Compare ANGULAR.

roundness *n*. (in SEDIMENTARY particles) the form of a CLAST related to the sharpness of curvature of edges and corners following ABRASION. Roundness is calculated as average radius of corners and edges/radius of maximum inscribed circle, i.e. that circle which will fit within the edges of the clast. Roundness is independent of sphericity. See Fig. 76 opposite. Compare SPHERICITY.

rubellite *n*. a pale, rose-red to deep-red lithian variety of TOURMALINE that is used as a gemstone.

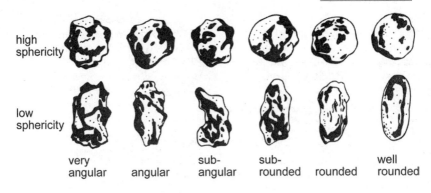

sphericity

low
sphericity

very
angular   angular   sub-angular   sub-rounded   rounded   well rounded

Fig. 76 **Roundness**. A chart of roundness and sphericity for clastic fragments

rubidium-strontium method see DATING
   METHODS.
ruby *n.* the red gem variety of CORUNDUM;
   its colour is attributed to the presence
   of a small amount of chromium.
ruby silver *n.* the red silver-sulphide
   minerals PROUSTITE, Ag_3AsS_3, (light ruby
   silver) and PYRARGYRITE (dark ruby
   silver).
rudaceous *adj.* (of a SEDIMENTARY ROCK or
   its texture) composed of a significant
   proportion of fragments larger than
   sand grains, such as PEBBLES or GRAVEL.
rudite *n.* any SEDIMENTARY ROCK composed
   of coarse fragments larger than sand
   grains (i.e. 2 millimetres in diameter).
   Included are both CONGLOMERATE
   (rounded pebbles and grains) and
   BRECCIA (angular pieces). A clastic limestone consisting of detrital carbonate
   particles that are greater than sand size
   (i.e. more than 2 millimetres) is referred
   to as a *calcirudite*; it may be a consolidated calcareous gravel or a limestone
   breccia or conglomerate. See PSEPHITE.
   See also LUTITE, ARENITE.
rugose coral *n.* any ANTHOZOAN of the
   order Rugosa. The outer skin of the
   CORALLUM, the EPITHECA, may be transversely wrinkled, hence the name
   'rugose'. Rugose corals range in form

from simple solitary to complex
colonial types.
   The CORALLITES are divided vertically
by six primary septa (walls), with later
septa being inserted at four points in
the corallite wall, giving a bilateral
symmetry, see Fig. 77 overleaf. (Compare SCLERACTINIA.) Various axial
structures are found in the centre of
each corallite, and these, together with
the septa, the horizontal TABULAE and
the DISSEPIMENTS, are used in classification.
   Solitary forms are usually cone-
shaped and may be straight or curved
(HORN CORAL) although the later-formed
part may be cylindrical. The corallum
of colonial forms depends on the
relationships of the individual
corallites – see COLONIAL CORAL for a
description of the different types.
   Rugose corals were not cemented to
the sea floor and the colonial forms
were not really REEF builders, unlike the
TABULATA, which existed at the same
time. They were most abundant in the
Lower CARBONIFEROUS and some have
been used as zone fossils. Range,
Middle Ordovician to Upper Permian.
runoff *n.* that part of precipitation that
   appears in or makes its way to surface

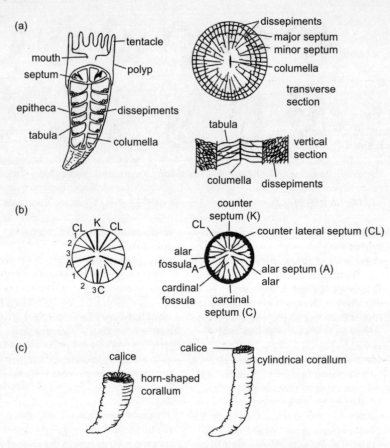

Fig. 77 **Rugose corals**. (a) general morphology of a rugose coral;
(b) transverse section showing the six major septa with the minor
septa inserted in groups of four; (c) forms of the solitary corallum

streams. The runoff that reaches stream channels immediately after a rainfall or the melting of snow is *direct runoff*. Soil and rock permeability, vegetation, temperature and ground slope all affect runoff. Ground that is water-saturated or filled with ice favours runoff.

Rupelian *n.* the lower STAGE of the OLIGOCENE.

rutile *n.* a reddish, brown or black tetragonal mineral, TiO_2, an important ore of titanium. Rutile occurs as a very common ACCESSORY MINERAL in intrusive igneous rocks or dispersed through quartz veins.

rutilated quartz *n.* QUARTZ in which needle-like crystals of RUTILE are enclosed. See also SAGENITE.

R wave (Rayleigh wave) see SEISMIC WAVES.

Ss

Saalian see CENOZOIC ERA, Fig. 13.

sabkha *n*. the Arabic term *sabkha* refers to HALITE-encrusted coastal flats or salt flats (see ALKALI FLAT). In the geological sense the term includes both coastal and continental salt flats, the latter known as *playas* (North America), *salars* (South America) and *schotts* (North Africa). Sabkhas are either flat or slope very gently and are developed where AEOLIAN sand is in limited supply. In a sabkha the WATER TABLE lies 1–2 metres below the surface. Coastal sabkhas run parallel to the coastline and consist of carbonate sediment, mainly aragonite, produced by depositional OFFLAP of nearshore marine sediments. During times of onshore winds and spring tides, extensive areas of sabkha are flooded and subsequent intense evaporation causes the upward movement of concentrated brine, with the resultant precipitation of halite, gypsum and occasionally celestite and magnesite. Underlying, dominantly aragonitic sediment is dolomitised by brines with a relatively high Mg to Ca ratio. (See also REFLUX.) Sabkhas are found on many modern coastlines, e.g. the Persian Gulf, the Gulf of California. The facies may be indicated by EVAPO- RITES, the absence of FOSSILS, PEBBLE conglomerates, algal mats and DOLOMI- TISATION. Continental sabkhas differ from coastal sabkhas, being evaporite- dominated.

saccharoidal *adj*. having a granular texture resembling that of a loaf of sugar. The term is used to describe some MARBLES and SANDSTONES.

saddle reef *n*. a MINERAL DEPOSIT found in the crest of an ANTICLINE following the BEDDING PLANES. These deposits are usually found in vertically stacked succession, e.g. in Bendigo, Australia, where gold-bearing deposits occur in the domes of anticlinal folds of Lower Ordovician quartzites and slates. In Nova Scotia, gold-bearing saddle reefs are found in association with folded beds of SLATE.

Sakmarian *n*. a STAGE of the Lower PERMIAN.

sagenite *n*. a variety of RUTILE that occurs in crossed needle-like crystals, often enclosed in QUARTZ. Similar crystals of tourmaline, goethite and other miner- als that penetrate quartz have also been referred to as sagenite. The quartz is said to be *sagenitic*. See also RUTILATED QUARTZ

salic *n*. a mnemonic term derived from silica and aluminium and applied to the group of standard normative minerals (CIPW classification) in which one or both of these elements occur in large amounts. These include quartz, feldspars and feldspathoids. The corresponding term for the silicic and aluminous minerals actually present in a rock is FELSIC. See CIPW NORMATIVE COMPOSITION. Compare FEMIC, MAFIC.

salina *n*. a small evaporitic basin or lagoon.

saltation *n*. a type of SEDIMENT TRANSPORT in air or water in which particles are moved forwards in short, abrupt leaps. The process is intermediate between SUSPENSION and traction.

salt dome *n.* a circular piercement structure (DIAPIR) produced by the upward movement of a pipelike plug of salt. Salt domes occur in certain areas underlain by thick layers of SEDIMENTARY ROCK and are derived from deposits of deeply buried HALITE. A CAP ROCK of gypsum, anhydrate and varying amounts of sulphur and calcite is found on top of many salt plugs, and some have associated PETROLEUM deposits. Oil and gas may accumulate in traps created by the deformation of the rocks surrounding the salt plug. Salt walls are diapirs where the salt mass is linear rather than domal.

salt flat see ALKALI FLAT, SABKHA.

salt lake *n.* a body of water in an arid or semi-arid region having no outlet to the sea and containing a high concentration of salt (NaCl), e.g. Great Salt Lake in Utah, USA, or the Dead Sea in the Middle East. See also ALKALI LAKE.

salt marsh see MARSH.

saltpetre *n.* **1.** a naturally occurring potassium nitrate, NITRE.
2. a general name for cave deposits of nitrate minerals.

salt plug see SALT DOME.

salt tectonics *n.* the study of the formation and mechanics of emplacement of SALT DOMES and other salt-related structures.

Samfrau fold belt *n.* an extensive OROGENY the remnants of which are found from northeast Australia to Tasmania, in the Ellsworth mountains of Antarctica, the Cape fold belt in South Africa and in Brazil. It formed along the edge of GONDWANA during the Lower PALAEOZOIC.

sand *n.* **1.** non-cohesive granular material the individual grains of which have a

diameter in the range of 0.062 to 2 millimetres.
2. a CLASTIC sediment of such particles.
3. a driller's term for any productive porous SEDIMENTARY ROCK in a well. See also OIL SAND.

sandar see SANDUR.

sand crystal *n.* a large CRYSTAL with irregular external form, consisting of as much as 60 per cent sand inclusions. Calcite, barite or gypsum sand crystals develop by growing within a deposit of unconsolidated sand during cementation. See DESERT ROSE.

sand dune see DUNE.

sand-shale ratio *n.* the ratio in a stratigraphical section of the amount of sandstone and conglomerate to that of shale, without considering the non-clastic material. Compare CLASTIC RATIO.

sand sheet see BLANKET SAND.

sandstone *n.* a sedimentary rock composed of SAND-sized particles with varying amounts of a fine-grained MATRIX of clay or silt. The grains in a sandstone are usually cemented by materials such as calcium carbonate or silica. Quartz is the dominant detrital fraction of sandstone, together with feldspars, mica and LITHIC fragments. Particle shapes in sandstone vary from ANGULAR to ROUNDED, the shape depending on the nature of the transporting medium and the duration of transport. A particle size close to the average gives a sandstone with well-sorted grains, and a particle size including many larger or smaller than the average gives a sandstone with poorly sorted grains. Sandstones may be classified according to their relative content of feldspar, quartz, rock fragments and matrix. See Fig. 78 opposite. Such a classification defines four major groups: *quartzarenites*, made up almost entirely of quartz grains;

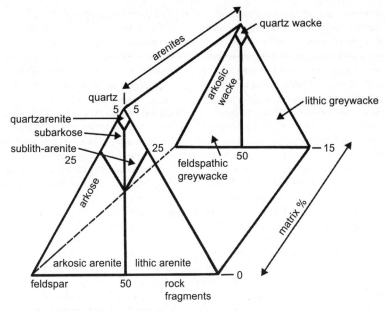

Fig. 78 **Sandstone**. Classification of sandstones based on the relative proportions of quartz, feldspar, rock fragments and matrix. The wacke group can contain up to 75% matrix

arkoses, with more than 25 per cent feldspar; LITHIC ARENITES, in which are fragments derived from pre-existing rock types; and GREYWACKES, heterogeneous mixtures of quartz, rock fragments and feldspar in a fine-grained clay matrix.

The colour of sandstones is commonly determined by the cementing agents or by AUTHIGENIC minerals such as GLAUCONITE.

Sandstones originate by various means, some as water-deposited materials in marine and fluvial systems and others as aeolian deposits (sand DUNES).

sandstone dyke *n.* a vertical body of SANDSTONE, commonly TABULAR in shape, cutting across other rocks in a manner similar to an igneous DYKE. It was formed by the injection of liquefied sand from an under lying bed (or rarely from an overlying bed), into the host layer (see LIQUEFACTION). The host layer may be anything from MUDROCK to a coarse CONGLOMERATE.

sandur *n.* (*pl* **sandar**) an Icelandic term for a low-angle sheet of OUTWASH material beyond the terminal MORAINE of a GLACIER. *Valley sandar* form the flat floors of U-shaped valleys downstream from glacier snouts, and *sandur plains* occur in coastal or other flat areas. Compare VALLEY TRAIN.

sand volcano *n.* a conical or dome-like structure on the upper surface of a SANDSTONE bed, sometimes with a crater-like depression in the middle and radially arranged sand lobes on the flanks. They range in size from 10 centimetres in diameter to several metres. Sand volcanoes are produced as a result of the localised upward escape of water from a liquefied bed, which extrudes sand at the surface (see LIQUEFACTION). The underlying bed typically shows evidence of post-depositional disturbance, such as CONVOLUTE LAMINATION, and may be associated with SANDSTONE DYKES.

sandwaves *n*. large, asymmetric marine or FLUVIAL bed forms, commonly with straight or sinuous crests and lacking localised scours on the lee-side. In a river bed they may be up to 200 metres long and 2 metres high, but BED FORMS interpreted as sandwaves on some areas of the CONTINENTAL SHELF are up to 15 metres high, with wavelengths of up to one kilometre. Internally, sandwaves show tabular CROSS-BEDDING.

sanidine *n*. a colourless or whitish silicate mineral, $KAlSi_3O_8$, a high-temperature potassium FELDSPAR. It occurs in VOLCANIC ROCKS of rhyolitic, trachytic or high-potassium type that have been cooled rapidly.

sanidinite facies *n*. a FACIES represented by small fragments of XENOLITHS or the wall rocks immediately adjacent to basic intrusions. The ASSEMBLAGE of minerals indicates maximum temperatures, often at very low pressures. Pelitic rocks contain minerals such as sillimanite, sanidine, hypersthene and anorthite. Calcareous rocks tend to lose almost all carbon dioxide, but pure calcite may survive. Typical metamorphic minerals are wollastonite, anorthite and diopside.

sannaite *n*. an ALKALINE variety of LAMPROPHYRE essentially composed of PHENOCRYSTS of barkevikite/kaersutite AMPHIBOLE, Ti-AUGITE, OLIVINE and/or BIOTITE in a fine-grained GROUNDMASS of ALKALI FELDSPAR, ACMITE, CHLORITE, CALCITE and pseudomorphs of MICA after NEPHELINE.

Santonian *n*. a STAGE of the Upper CRETACEOUS.

saponite *n*. a trioctahedral magnesium-rich clay mineral of the SMECTITE group. See CLAY MINERALS.

sappharine *n*. a term that has been used for blue quartz that resembles sapphire. It should not be confused with *sapphirine*, a rare sapphire-blue magnesium-aluminium silicate, $(Mg, Al)_8(Al,Si)_6O_{20}$, found in west Greenland and Labrador.

sapphire *n*. any pure, gem-quality CORUNDUM other than RUBY. Varieties other than the well-known blue sapphire are green sapphires and yellow sapphires. The colour of blue sapphire is attributed to the presence of small amounts of oxides of cobalt, chromium and titanium. See STAR RUBY, STAR SAPPHIRE.

sapphirine see SAPPHARINE.

saprolite *n*. a soft, earthy, red or brown clay-rich and totally decomposed rock, formed in place by chemical WEATHERING of IGNEOUS or METAMORPHIC ROCKS, particularly in humid climates. Structures that were in the unweathered rock are preserved in saprolite. Compare LATERITE.

sapropel *n*. an unconsolidated deposit composed chiefly of the remains of certain ALGAE with mineral grains and spores as minor constituents; these remains accumulate and decompose in anaerobic conditions on the shallow bottoms of lakes and seas. When consolidated, sapropel becomes OIL SHALE, bituminous shale or BOGHEAD COAL. It is distinguished from PEAT by its high content of fatty and waxy substances (because of algal remains) and small amount of cellulose material.

sardonyx *n*. a gem variety of CHALCEDONY

that has straight parallel bands of sard (a reddish-brown chalcedony) alternating with chalcedony of another colour, usually white.

sastrugi features *n*. wave formations caused by persistent winds blowing on a snow surface. Sastrugi features vary in size according to the duration and force of the wind and the condition of the snow surface in which they are formed.

satin spar *n*. a white, translucent, fibrous GYPSUM with a silky lustre.

saturated *adj*. **1.** (of pores) filled with water or another liquid.
2. (of a rock) having QUARTZ as part of its norm. See SILICA SATURATION.
3. (of a mineral) capable of forming in the presence of free SILICA. See SILICA SATURATION.

saturated zone see WATER TABLE.

saturation *n*. **1.** the degree to which the pores in a rock hold water, oil or gas; it is generally expressed as a percentage of the total pore volume.
2. the degree of SILICA SATURATION in an igneous rock.
3. the maximum possible amount of water vapour in the Earth's atmosphere for a specific temperature.

saturation line *n*. **1.** a boundary line on a compositional VARIATION DIAGRAM for IGNEOUS ROCKS. Rocks to the right of it are oversaturated with respect to silica, those to the left are undersaturated. (See SILICA SATURATION.)
2. the boundary on a GLACIER between the zone of partial melting (where the FIRN layer is not fully soaked) and that where the firn layer is saturated with meltwater.

saussurite *n*. a compact mineral aggregate consisting of albite and zoisite or epidote; it is an alteration product of calcic plagioclase.

savanna or **savannah** *n*. tropical or subtropical grassland characterised by scattered shrubs or trees and a pronounced dry season. Savanna areas occur primarily in Africa (*veld* or *bushveld*) and South America (*llano*), with similar areas in Australia and Madagascar.

scale trees *n*. Palaeozoic *Lycopsid* species the bark pattern of which has a scale-like appearance. Some attained heights of 30 metres. Two Late Palaeozoic species, *Lepidodendron* and *Sigillaria*, are commonly found in Carboniferous COAL deposits. These are the 'coal plants'. *Lepidodendron* is tall and branching, with slender leaves and diamond-shaped leaf scars. *Sigillaria* has a stout trunk, bladed leaves and vertical leaf scars. See FOSSIL PLANTS.

scandent see GRAPTOLITE and Fig. 42.

scaphopod *n*. a marine UNIVALVE of the class Scaphopoda, having an elongated bilaterally symmetrical tusk-like shell that is open at both ends. The maximum length, rarely attained, is 13 centimetres. Scaphopods inhabit shallow shelf waters. They range from the Devonian to the present but are of minor importance as fossils.

scapolite *n*. a group of sodium-calcium, aluminium silicate, metamorphic minerals somewhat analogous to the plagioclase FELDSPARS. Scapolite minerals occur in schists, gneisses and crystalline limestone.

scarp see ESCARPMENT.

scheelite *n*. calcium tungstate, a yellow, green or brownish tetragonal mineral, $CaWO_4$, an ore of tungsten. It is found in granite PEGMATITES, contact metamorphic AUREOLES and high-temperature HYDROTHERMAL veins.

schiller *n*. the bronze metallic lustre seen in orthopyroxenes that is caused by the interference of light reflecting from exsolved mineral plates within the crystal. However, the term has also been used to refer to iridescent effects (see IRIDESCENCE)in plagioclase FELDSPAR. See also LABRADORESCENCE.

schist *n.* a METAMORPHIC ROCK that is not defined by mineral composition but by the well-developed parallel orientation of more than 50 per cent of the minerals present, in particular those minerals of lamellar or elongate prismatic habit, such as mica and hornblende. Compare GNEISS, PHYLLITE.

schistosity *n.* the oriented or planar structure, mainly seen in SCHISTS and sometimes PHYLLITES, caused by the parallel alignment of platy or prismatic mineral grains. Compare FOLIATION.

schizochroal eyes *n.* compound eyes in which the lenses are large and separated from each other by interstitial material. Each lens has a separate corneal covering. Compare HOLOCHROAL EYES. This type of eye is unique and occurs in only one suborder of TRILOBITES.

schlieren *n.* pencil-like, discoidal or blade-like mineral aggregates, varying greatly in size, that occur in plutonic rocks. They may have the same general mineralogy as the host rock but, because of differences in mineral proportions, the colour of schlieren is either lighter or darker. Some schlieren may be elongated inclusions.

schorl *n.* a black iron-rich variety of TOURMALINE, often occurring in radiating clusters.

scintillation counter *n.* an instrument that measures radiation by counting the individual scintillations emitted by the substance or object being investigated. The counter consists of a phosphor and a photomultiplier. The phosphor emits visible light when struck by high-speed particles, and the photomultiplier tube registers the phosphor flashes.

scissors fault see FAULT.

Scleractinia or **Hexacorallia** *n.* an order of solitary or COLONIAL CORALS of the subclass Zoantharia. They have aragonite skeletons and the septa are inserted in multiples of six, unlike the RUGOSE CORALS. See Fig. 79 below. They appeared in the Middle TRIASSIC in small patch reefs and became widespread during the JURASSIC. The earliest scleractinians were HERMATYPIC CORALS, but ahermatypic types developed as some forms spread into colder, deeper water. See AHERMATYPIC CORALS. Scleractinians are important in modern CORAL REEFS.

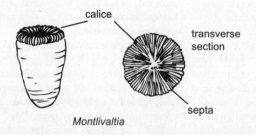

Fig. 79 **Scleractinia**. Solitary and colonial scleractinian corals; septa inserted in multiples of six

scoria *n.* (*pl.* **scoriae**) a dark-coloured vesiculated fragment of PYROCLASTIC MATERIAL of basaltic to andesitic composition produced from the spray of particles from LAVA FOUNTAINS. The degree of vesiculation, hence density, is highly variable. The larger fragments may be drawn out and contorted in flight. Basaltic scoriae when fresh often have a bluish surface IRIDESCENCE because of the presence of a very thin layer of volcanic glass. Scoriae are commonly deposited in *scoria cones* surrounding active VOLCANIC VENTS.

scour-and-fill *n.* a SEDIMENTARY STRUCTURE produced by the alternate scouring out and refilling of a channel or depression by water that fluctuates in current speed and volume. The scours are concave and filled with coarser material.

Scourian complex see LEWISIAN GNEISS.

scour mark *n.* a variety of smooth, streamlined marks produced by current action eroding a cohesive sediment surface such as SILT or MUD. TURBULENT FLOWS at large REYNOLDS NUMBERS are required to produce scour marks, using sediment grains as the erosive material. Coarser-grained sediments do not develop these structures because they move as BEDLOAD. Typically, scour marks are found on the base of a coarse-grained horizon overlying a finer-grained one, so they are a type of SOLE MARK. FLUTE MARKS, obstacle scours, longitudinal scours and *gutter casts* are all varieties of scour marks. *Obstacle scours* are crescentic ridges, dying out downstream, which filled the trough formed round a PEBBLE or similar CLAST. The clast itself may not remain. *Longitudinal scours* are closely spaced parallel ridges and furrows 0.5 to 1 centimetre apart. The ridges may die out or coalesce, and

broader ones may have a blunt end somewhat similar to flutes. These structures can show the general current direction as well as being WAY-UP INDICATORS.

scree *n.* a heap of rock debris produced by WEATHERING at the base of a cliff or a sheet of coarse waste covering a mountain slope. Scree is frequently considered to be a synonym of TALUS but is a more inclusive term. Whereas talus is an accumulation of debris at a cliff base, scree also includes loose debris lying on slopes without cliffs. The term 'scree' is more commonly used in Great Britain, whereas talus is more commonly, but often incorrectly, used in the United States.

screw dislocation see DISLOCATION.

Scythian *n.* a name formerly used for a STAGE of the Lower TRIASSIC.

sea arch see NATURAL BRIDGES AND ARCHES.

sea cave *n.* a marine EROSION feature that forms along a line of weakness at the base of a cliff that has been subjected to prolonged wave action. It is a cylindrical tunnel extending into the cliff. If a BLOWHOLE is opened on the cliff top, continued wave action ultimately causes the roof of the cave to collapse, and a narrow sea inlet forms.

sea-floor spreading *n.* a hypothesis, proposed by the American geophysicist Harry Hess at Princeton University in 1960, that OCEANIC CRUST forms along the MID-OCEANIC RIDGE system and spreads out laterally away from it. The concept was pivotal in the development of PLATE TECTONICS. Until the 1950s, it was believed that the ocean floor represented the oldest parts of the Earth's crust. Discovery in the late 1940s of the mid-oceanic ridge system and new knowledge that the crust forming the ocean floor is thinner and younger than the CONTINENTAL CRUST provided the basis for the Hess hypothesis: MAGMA wells up continuously along the mid-oceanic ridges, forming new oceanic crust as the older crust moves away on either side.

The sea-floor spreading hypothesis has been supported by a great deal of substantial evidence. Thermal anomalies over the mid-oceanic ridges (three to four times normal values) are considered to reflect the INTRUSION of molten material near the ridge crests. Anomalously low SEISMIC-WAVE velocities have also been observed at the ridge crests; thermal expansion and fracturing associated with the up-welling magma are thought to be the responsible factors. Sea-floor spreading was further corroborated when investigation of oceanic MAGNETIC ANOMALIES revealed a pattern of magnetic stripes running parallel with the ridges. The geomagnetic field is alternately high and low with increasing distance from the axis of the mid-oceanic ridge system. These linear trends seem to be underlain by alternating bands of normally and reversely magnetised basaltic rocks generated at the ridge axis where they acquired a thermoremanent magnetisation as they cooled; their magnetic polarity is determined by the polarity of the geomagnetic field at the time of solidification. (See GEOMAGNETIC POLARITY REVERSAL.) The farther distant from the ridge axis a band is located, the older the rock. The oldest marine bottom sediments, as determined from core samples, date only to the Jurassic (190 Ma ago).

The concept of sea-floor spreading played a key role in establishing the theory of PLATE TECTONICS. It is believed that the spreading out of the sea floor and the continuous upward flow of molten material are linked to the migration of the continents. See SUBDUCTION.

seamount *n.* a submarine mountain rising about one kilometre or more above the deep ocean floor but not reaching the surface. Seamounts vary in shape and are found singly as well as in groups or chains. They may be relatively small conical peaks or more massive structures. Evidence from gravity and seismic measurements and from comparisons with the Hawaiian volcanic chain indicate that seamounts are of volcanic origin and are probably composed of PILLOW LAVA and associated CLASTIC debris. See also GUYOT.

sea stack see STACK.

seat-earth *n.* a PALAEOSOL, often containing fossilised plant rootlets, underlying a COAL seam. There are different types of seat-earth. *Ganister* is a fine-grained silica-rich SANDSTONE or SILTSTONE that can be used in the manufacture of silica bricks. Clay-rich seat-earths are known as *underclays* (see also FIRECLAY).

secondary *adj.* 1. (of rocks and minerals) formed by the alteration of pre-existing minerals (PRIMARY MINERALS). Secondary minerals may be found at the site as PARAMORPHS or PSEUDOMORPHS or may be deposited from solution in rock interstices through which the solution is percolating, as, for example, AMYGDALES. 2. formed of material derived from the disintegration or erosion of other rocks, as, for example, CLASTIC ROCKS.

secondary cosmic rays see COSMIC RADIATION.

secondary enrichment or **supergene enrichment** *n.* a near-surface process of mineral deposition by which a primary ore body or vein is later enriched to a higher grade. The

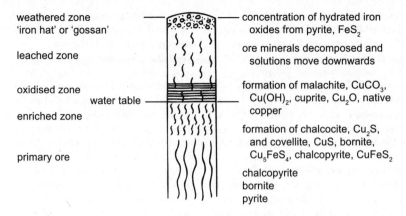

weathered zone 'iron hat' or 'gossan' — concentration of hydrated iron oxides from pyrite, FeS_2

leached zone — ore minerals decomposed and solutions move downwards

oxidised zone

water table — formation of malachite, $CuCO_3$, $Cu(OH)_2$, cuprite, Cu_2O, native copper

enriched zone — formation of chalcocite, Cu_2S, and covellite, CuS, bornite, Cu_5FeS_4, chalcopyrite, $CuFeS_2$

primary ore

chalcopyrite
bornite
pyrite

Fig. 80 **Secondary enrichment**. The processes of secondary enrichment in a copper deposit

enrichment process often consists of two phases:

(a) oxidation of overlying ore masses above the WATER TABLE (*zone of oxidation*); oxidation may then produce acidic solutions that leach out metals and carry them downwards towards the water table;

(b) below the zone of oxidation, reaction of the solutions with the primary material by coating or replacement with an ore of more valuable content. The water table thus marks the lower limit of oxidation, near which there may be a zone of secondary enrichment. This process has been significant in the formation of many copper deposits. See also GOSSAN, OXIDISED ZONE.

secondary mineral *n*. a mineral formed after the rock enclosing it has been formed; the process usually involves the alteration of a PRIMARY MINERAL by METAMORPHISM, WEATHERING, SEISMIC WAVES or solution.

secondary wave see SEISMIC WAVES.

second law of thermodynamics see ENTROPY.

sectile see TENACITY.

sector collapse *n*. the structural failure of a large part of the slopes of a VOLCANO. This can result in a catastrophic avalanche and may trigger a major explosive eruption. A horseshoe-shaped valley or open-sided CRATER is formed on the side of the volcano.

secular variation *n*. a slow, steady, progressive change in part of the Earth's magnetic field caused by the internal state of the planet. These changes are immediately apparent from yearly averages of the values of magnetic elements as reported by worldwide magnetic observatories. By taking period averages, transient disturbances caused by field interaction with solar-particle emissions are eliminated and what remains are slow variations of internal origin.

SEDEX see SEDIMENTARY-EXHALATIVE DEPOSITS.

sediment *n*. solid material, organic or inorganic in origin, that has settled out from a state of SUSPENSION in a fluid and has been transported and deposited by wind, water or ice. It may consist of fragmented rock material, products derived from chemical action or from the secretions of organisms. Loose sediment such as sand, mud and till may become consolidated and/or cemented to form coherent sedimentary rock. Compare DEPOSIT.

sedimentary *adj.* containing SEDIMENT or formed by its deposition.

sedimentary cycle see CYCLE OF SEDIMENTATION, RHYTHMITE.

sedimentary-exhalative deposits (SEDEX) *n.* produced by the upwelling of mineralising fluids into marine subsurface and surface rocks, especially at MID-OCEAN RIDGES. BLACK SMOKERS are associated with SEDEX mineral deposits. See VOLCANIC-EXHALATIVE DEPOSITS.

sedimentary facies *n.* an areally confined part of an assigned STRATIGRAPHICAL UNIT that shows characteristics clearly distinct from those of other parts of the unit.

sedimentary mélange see OLISTHOSTROME.

sedimentary rock *n.* a rock formed by the CONSOLIDATION of SEDIMENT settled out of water, ice or air and accumulated on the Earth's surface, either on dry land or under water. CLASTIC or DETRITAL sedimentary rocks originate with the accumulation of discrete mineral or rock particles derived from WEATHERING and the EROSION of preexisting rocks. Nonclastic varieties originate from chemical precipitation or from the biogenic action of organisms (e.g. coal, limestone). Sediments are consolidated into a rock mass by LITHIFICATION. Sedimentary rock is typically stratified or bedded; beds can vary greatly in thickness.

TEXTURE and *composition* are the two principal properties used in identifying sedimentary rocks. The texture in this case usually refers to PARTICLE SIZE; the particle reflects the nature of weathering and of transportation of particles and the conditions at the site of deposition. Texture of sedimentary rock is described, according to the classification of Grabau, as RUDACEOUS (coarse), ARENACEOUS (medium) or LUTACEOUS (fine); rocks characterised by these textures are referred to as RUDITE, ARENITE and LUTITE respectively. These are purely textural terms, independent of composition. Rocks showing the presence or absence of a wide range of particle sizes are distinguished as *poorly sorted* or *well sorted*. The texture of rocks consisting of chemical or biogenic precipitates is called MASSIVE, regardless of crystal size.

The *composition* of sedimentary rocks is determined by conditions prevailing at the source of the sediments and at the site of deposition as well as by the nature of transport. (See ENVIRONMENT OF DEPOSITION.) The rocks may contain fragments of any pre-existing rock or mineral. For example, a CONGLOMERATE may include fragments of quartz, limestone, chert, etc. Rocks resulting from biogenic or chemical accumulation are usually composed of calcium or magnesium carbonate, silica or EVAPORITE salts. Primary structures exhibited in sedimentary rock, such as bedding (stratification) and fossil inclusion, reflect environmental and climatological conditions at the time of deposition.

The major types of sedimentary rocks are mudrocks (65 per cent), sandstones (20 to 25 per cent) and carbonate rocks (10 to 15 per cent).

sedimentary structure *n.* any of a variety of features in a SEDIMENTARY horizon produced by sedimentary processes. Structures formed during or shortly after deposition and those produced during DIAGENESIS may be included under this heading. The approximately planar bottom and top surfaces of most sedimentary units (i.e. BEDDING) are the commonest feature of sedimentary successions.

It is convenient to group sedimentary structures according to whether they are observed on the bedding surface or within the bed. In the second group may be found LAMINATION, CROSS-BEDDING, FLASER BEDDING, GRADED BEDDING and soft-sediment deformation structures like CONVOLUTE LAMINATION, FLAME STRUCTURES and SLUMP structures. There are also structures of biogenic origin such as various types of burrows, BIOTURBATION and STROMATOLITES. Structures of chemical origin (mostly diagenetic) found within beds include CONCRETIONS and NODULES. Bedding surface structures may be divided into those typically observed on the bottom surface (SOLE MARKS): such as TOOL MARKS, SCOUR MARKS and LOAD CASTS; and those found on the top surface: RIPPLE MARKS, DUNES, SANDWAVES, PRIMARY CURRENT LINEATION, DESICCATION CRACKS, RAIN PRINTS, SAND VOLCANOES and TRACE FOSSILS.

The structures in a particular horizon are indicative of the processes (physical, chemical and biological) affecting the sediment during and after deposition. See also ENVIRONMENT OF DEPOSITION.

sedimentation *n*. the process of depositing sediments.

sedimentology *n*. **1.** the scientific study of SEDIMENTARY ROCKS and the processes responsible for their formation. **2.** the origin, description and classification of SEDIMENTS.

sediment transport *n*. the movement of DETRITAL particles by air, water, ice or gravity. In the last case, sedimentary material moves downslope when the SHEAR STRESS exerted by gravity exceeds the strength of the sediment. Mass gravity transport of material may occur sub-aerially or in the subaqueous environment, with processes ranging from rock falls and SLIDES to DEBRIS FLOWS and TURBIDITES.

Grains transported by air and water (*fluid transport*) move as BED LOAD (by rolling and sliding and by SALTATION) or in SUSPENSION, when grains are kept up by turbulence. The different size/density populations of moving grains promote SORTING of the material, and mechanical ABRASION during transport removes corners and edges so the grains become more ROUNDED. The lower density and viscosity of air means that air transports a smaller range of grain sizes than does water, and generally only the finest dusts move in suspension. As a result, AEOLIAN deposits are usually much better SORTED than water-laid sediments. Grain impacts are more vigorous during wind transport, so particles become rounded more quickly and commonly have a frosted or etched appearance. Ice transports a very wide range of material, from CLAY GRADE to BOULDERS. GLACIAL deposits are characteristically very poorly sorted, with ANGULAR fragments.

segregation *n*. a secondary feature, such as a NODULE of iron sulphide, that is formed by the chemical rearrangement of minor components within a SEDIMENT subsequent to its deposition.

seiche *n*. periodic oscillation of a moderate-sized enclosed body of fluid, characteristic of the system only and independent of the exciting force except as to initial magnitude. Where water is the medium, the oscillating system is characterised by the shape of the basin and the depth of the water.

The restoring force is provided by gravity. Earthquakes commonly effect seiches in ponds and lakes. Sudden barometric pressure changes plus wind velocity and direction are also important. See also TSUNAMI.

seif see DUNE.

seism see EARTHQUAKE.

seismic *adj.* pertaining to a naturally or artifically induced EARTHQUAKE or earth vibration.

seismic activity see SEISMICITY.

seismic area *n.* an EARTHQUAKE ZONE or the region affected by a particular earthquake.

seismic detector *n.* an instrument, such as a seismometer or GEOPHONE, that receives seismic impulses and converts them into readable signals.

seismic discontinuity or **discontinuity** *n.* a boundary between rocks of differing DENSITY and/or different elastic properties at which the velocities of SEISMIC WAVES change abruptly. For most waves incident on such a boundary, part of the wave energy is reflected and part refracted, and the laws relating to refraction and reflection of light also apply to seismic waves (see LAW OF REFRACTION). The MOHOROVICIC DISCONTINUITY between CRUST and MANTLE and the GUTENBERG DISCONTINUITY between mantle and CORE are two major seismic discontinuities within the Earth. (See also ELASTIC CONSTANT.)

seismic event *n.* an EARTHQUAKE or a similar transient earth motion caused by an explosion.

seismic exploration or **seismic prospecting** *n.* the use of seismic surveying and explosion SEISMOLOGY in prospecting for oil, gas or other mineral sources. The use of such methods in seismic prospecting has led to important mathematical findings about the transmission of elastic waves from particular source types and through media of various complexities. The two basic practical methods commonly used are reflection shooting and refraction shooting (see SEISMIC SHOOTING).

seismic focus or **hypocentre** *n.* in SEISMOLOGY, the point within the Earth that is the centre of an EARTHQUAKE. It is the initial rupture point of an earthquake, where strain energy is transformed into elastic wave energy. Compare EPICENTRE.

seismicity or **seismic activity** *n.* the phenomenon of Earth movements and their geographical distribution.

seismic magnitude see EARTHQUAKE MEASUREMENT.

seismic map *n.* a contour map constructed from seismic data. The data values may be in units of depth or of time and may be plotted with respect to the observing (receiving) station at the surface or with respect to a series of subsurface reflecting or refracting points.

seismic moment *n.* a relatively new method of measuring the strength of an EARTHQUAKE and a more physical measure of its size than that provided by earlier methods. It is defined as the product of the rigidity of the rock, the area of faulting and the average amount of slippage. The seismic moment is determined by the Fourier analysis of SEISMIC WAVES of such long period that the details of the rupture are smoothed out, with the effect that the entire fault may be considered to be a point-source. The periods at which the seismic moment is determined increase with the size of the FAULT. If the fault is based on such long-period waves, the slip from unruptured to ruptured state appears to be instantaneous. The actual pattern of the seismic radiation emitted by the instantaneous rupture is mathematically equivalent to the theoretical radiation pattern emitted by two hypothetical torque couples embedded in an unruptured elastic medium. The torque couples, rotating in opposite directions, deform the medium, thus radiating elastic waves in a pattern identical with that in which an earthquake source radiates seismic waves. The moment can be calculated from this model See also EARTHQUAKE MEASUREMENT.

seismic prospecting see SEISMIC EXPLORATION.

seismic regions *n.* a division of the Earth's interior into seven regions (A to G), according to depth based on seismic velocity distributions. See Fig. 81 below. See SHADOW ZONE.

seismic sea wave see TSUNAMI.

seismic shooting *n.* a method of GEOPHYSICAL EXPLORATION using tech-niques of explosion SEISMOLOGY in which elastic waves are produced in the Earth by firing explosives. Of several methods used, reflection shooting and refraction shooting have been the most successful. In *reflection shooting*, the travel times of SEISMIC WAVES from small explosions in shallow holes are recorded on portable SEISMOGRAPHS, thus permitting identification of various strata by their differing elastic properties. A comparison of travel times to different points indicates the structure of buried strata.

Refraction shooting is designed to determine seismic velocities and configurations of layers in those regions where velocity does not decrease with depth. Recording instruments are set up at distances from an explosion source that are several times the depth of the most remote layer being investigated. This

Shell	Depth (km)	Shells and discontinuities	State
A	Variable	Crust	solid
	———————	*Mohorovičić discontinuity (Moho)*	———————
B	Moho to 410	Upper Mantle	mostly solid
C	410–660	Transition zone (phase changes)	solid
D'	660–2600	Lower Mantle	solid
D"	2600–2900	Transition zone	solid (lower velocity)
	———————	*Gutenberg discontinuity*	———————
E	2900–4590	Outer Core	liquid
F	4590–5120	Transition zone	liquid
G	5120–6370	Inner Core	solid

Fig. 81 **Seismic regions**. The main shells of the Earth. The average thicknesses of the continental and oceanic crust are 36 km and 6 km respectively. Between about 100–200 km in the upper mantle seismic properties of the low velocity zone (asthenosphere) indicate the presence of small amounts of molten material

contrasts with reflection shooting, where the recording instruments are close to the source, the object being to record near-vertical reflections from the boundaries of the layer being investigated. Various filtering procedures are used to isolate the desired phases from background effects such as scattered waves and surface waves.

seismic velocity see SEISMIC WAVES.

seismic waves *n*. the manifestation of energy that is released when rocks within the Earth fracture or slip abruptly along FAULT planes during an EARTHQUAKE. Seismic waves are of two main types: body waves and surface waves. *Body waves*, appropriately, travel through the body of the Earth; *primary (P) waves*, which are compression (longitudinal) waves, are the fastest of this type and can travel through both solid and fluid matter; *secondary (S) waves*, the slower body waves, are *transverse shear waves* that can propagate through solids but not fluids. Primary waves, which are similar to sound waves, emerge from the earthquake focus and alternately compress and stretch the medium through which they travel. As secondary waves travel through a medium, individual particles vibrate at right angles to the direction of wave propagation. The velocities of both P and S waves vary in different media. The ratio of their velocities, V_p / V_s, is almost always a constant but decreases some weeks, months or even years before an earthquake and increases to almost normal level just prior to the earthquake. See EARTHQUAKE PREDICTION.

When body waves travel outwards through the Earth to the surface, some are transformed into *surface waves*, two types of which predominate: *Rayleigh waves*, which have a rotating motion in the vertical plane, aligned in the direction of travel, and *Love waves*, which have only horizontal shear motion. Rayleigh and Love waves have lower frequencies than the body waves from which they originate and usually travel more slowly, but they persist for great distances, sometimes circling the Earth many times before subsiding.

The use of seismic waves has been invaluable in determining the physical properties of the Earth and the structure of its interior. ARRIVAL TIMES are used to locate reflecting discontinuities within the Earth, and the velocities of P and S waves give information about the ratio of any two of the parameters incompressibility, elastic rigidity and density of the material through which the waves pass. Seismological evidence for a distinct core of the Earth, and for a clearly delineated inner core, has been derived from reflected P waves.

seismogram *n*. the record made by a SEISMOGRAPH.

seismograph *n*. an instrument that detects and records seismic vibration. The basic principle of operation for all seismographs is essentially the same. A frame that is rigidly fastened to a body of rock capable of transmitting Earth movements will move in conjunction with the rock support. Another element of the seismograph is a free-swinging pendulum to which is attached some scribing device that records the motion of the pendulum and, thus, of the Earth as well. Since SEISMIC WAVES can cause the earth to move vertically as well as horizontally, seismographs are designed accordingly.

seismology *n*. the study of EARTHQUAKES, including their origin, propagation, energy manifestations and possible techniques of prediction.

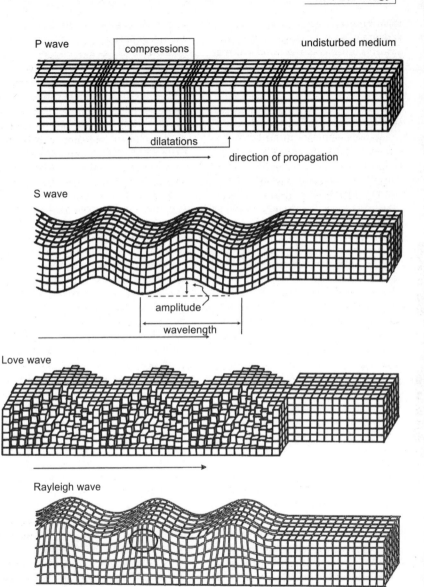

Fig. 82 **Seismic waves**. The motion of the different types of seismic waves

seismometer see SEISMIC DETECTOR.

selenite *n.* a variety of gypsum occurring as broad colourless and transparent monoclinic crystals or large crystalline masses that yield such folia upon cleavage.

selenizone see EXHALANT SLIT.

selenology *n.* a division of astronomy that deals with various aspects of lunar science, including lunar geology.

self-potential see SPONTANEOUS POTENTIAL.

selvage *n.* a marginal zone of a rock form showing some distinguishing peculiarity, such as in composition or fabric, for example the chilled margin of an INTRUSION.

Senecan *n.* a SERIES forming the lower Upper Devonian of North America.

sensitive clay or **quick clay** *n.* a CLAY that loses most of its strength after being remoulded. When the clay is disturbed, repacking of the grains results in an increase in pore-water pressure and the clay becomes mobile, i.e. it liquefies (see LIQUEFACTION, also PACKING). Many sensitive clays are marine clays that have lost their original flocculated structure because the ionic species responsible for the strength of the bonds have been removed by fresh water passing through (see FLOCCULATION).

sensitive tint see ACCESSORY PLATES.

separation *n.* a measure of the distance between the cut-off traces of a particular horizon on either side of a FAULT plane. Separation can be measured in any direction within the fault plane, so the direction must be specified. See DIP SEPARATION, STRIKE SEPARATION.

sepiolite or **meerschaum** *n.* a phyllosilicate CLAY MINERAL, $Mg_4(Si_2O_5)_3(OH)_2$. $6H_2O$, occurring as white or yellowish concretions or porcellaneous masses. Sepiolite is found as a surface alteration product of magnesite or serpentine and is used in the manufacture of tobacco pipes.

septarian *adj.* **1.** (of a SEPTARIUM) having an irregular pattern of internal cracks resembling the polygonal shrinkage cracks developed in mud during desiccation.
2. (of mineral deposits) that may occur as fillings of, e.g. calcite, in these cracks.

septarian nodule see SEPTARIUM.

septarium (*pl.* **septaria**) or **septarian nodule** *n.* a large, roughly spheroidal concretion, generally of limestone or clay ironstone, characterised by a network of radiating and intersecting mineral-filled cracks; the mineral is usually calcite. Internally, a septarium consists of irregular polygonal blocks cemented together by the mineral deposits within these cracks. Their formation is thought to be caused by desiccation and shrinkage of a gel-like mass of clay minerals. The cracks thus formed are later filled with precipitated minerals.

septum *n.* (*pl.* **septa**) any thin plate or wall that partitions a cavity or divides structures.

serac *n.* a jagged pinnacle of ice that is formed when several glacial crevasses intersect each other at the terminus of a GLACIER.

seriate texture *n.* a continuous range in the sizes of the CRYSTALS of the principal MINERALS in an IGNEOUS ROCK. It can be difficult to identify correctly because minerals in a THIN SECTION represent random slices through three-dimensional crystals.

sericite *n.* fine-grained MUSCOVITE mica occurring in small flakes. It is particularly common in SCHISTS where it imparts a silky lustre to FOLIATION planes but is also found as an alteration of FELDSPARS and other silicate minerals.

series *n.* **1.** a STRATIGRAPHICAL UNIT above STAGE and below SYSTEM; it includes the rocks formed during an EPOCH of geological time. Some series are recognised worldwide whereas others are regional.
2. IGNEOUS ROCK SERIES.
3. RADIOACTIVE SERIES.

serir see DESERT PAVEMENT, REG.

serpenticone see AMMONOID, Fig. 2.

serpentine *n.* a group of common rock-forming minerals often showing variegated shades of light and dark green, with a waxlight lustre when massive and silky when fibrous. They include the three common polymorphs: antigorite, $(Mg,Fe)_3Si_2O_5(OH)_4$, lizardide, $Mg_3Si_2O_5(OH)_4$, and CHRYSOTILE, $Mg_3Si_2O_5(OH)_4$. Serpentine is widely distributed, usually as a product of the alteration of magnesium silicates, in particular olivine, pyroxene and amphibole. It is often associated with magnesite, chromite and magnetite. Chrysotile is the principal ASBESTOS mineral and as such is known as white asbestos.

serpentine asbestos see CHRYSOTILE.

serpentinite *n.* a METAMORPHIC ROCK composed almost completely of serpentine-group minerals, e.g. antigorite, lizardite, chrysotile. Its colour varies from dark green to black, it may be streaked with red and its texture may be lamellar or felt-like with zonations. It is formed by the late-stage HYDROTHERMAL alteration of ULTRA-MAFIC rocks such as PERIDOTITES and PICRITES and is commonly found in OPHIOLITE complexes. It alters readily to TALC.

Serravalian *n.* a STAGE of the Middle-MIOCENE.

shadow test *n.* a method of determining the relative refractive index (RI) of a mineral using oblique illumination. A piece of card is placed below the microscope stage to darken one side of the field of view. At the contact between the mineral and the mounting medium, a shadow visible on the same side as the card lies within the higher RI material, whereas a shadow on the opposite side from the card lies within the lower RI material. See also BECKE TEST.

shadow zone *n.* a region, 103° to 142° of arc from the focus of an EARTHQUAKE, in which there is no direct penetration of SEISMIC WAVES. It is a 4350-kilometre belt circling the Earth. Primary waves are not observed in this region because they are refracted below the boundary between the Earth's core and mantle. Secondary waves are not observed beyond 103° from their course because they cannot travel through a liquid. The non-penetration of secondary waves within this zone was used as evidence that the Earth's core is part liquid. See Fig. 83 overleaf. See also GUTENBERG DISCONTINUITY.

shale *n.* a fine-grained fissile SEDIMENTARY ROCK formed by the compaction of clay or silt. It is the most abundant of all sedimentary rocks. See MUDROCK. Compare ARGILLACEOUS.

shallow-focus earthquake *n.* an EARTHQUAKE with a focus at a depth of less than 70 kilometres. Most earthquakes occur in this depth range. Compare INTERMEDIATE-FOCUS EARTHQUAKE, DEEP-FOCUS EARTHQUAKE.

shard *n.* a small (less than 2 millimetres), ANGULAR fragment of VOLCANIC GLASS.

shatter cone *n.* a striated conical rock fragment along which fracturing has occurred. It varies from a centimetre to several metres in length and is usually found in nested or composite structures that bear a resemblance to CONE-IN-CONE formations in sedimentary rocks. Shatter cones are found in

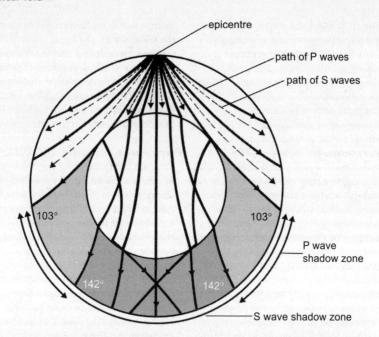

Fig. 83 **Shadow zone**. S waves are not received more than 103° of arc from the epicentre of an earthquake; P waves are not received between 103° and 142°

limestone, dolomite, sandstone, shale and granite; they are believed to have been formed by shock waves created by meteorite impact.

shear fold or **slip fold** *n*. a FOLD that results from minute displacements along closely spaced FRACTURES each of which is a minute fault. If the fractures are very close, and the beds tend to run parallel to the fractures, the resulting structure is a major fold with several associated minor folds.

shear fracture *n*. a FRACTURE caused by stresses that tend to slide one part of a rock past the adjacent part.

shear modulus see MODULUS OF RIGIDITY.

shear strain or **angular shear strain** *n*. an angular change for any specific direction measured in a body between two lines that were originally perpendicular, arising from displacements that act tangentially to a particular surface. The resulting angle is referred to as ψ (the Greek letter 'psi'). The change in shape of the body, for example, could be the change in shape from a square to a rhombus in response to the strain. Movement may be clockwise or anticlockwise, depending on which way the perpendicular is rotated relative to the specific direction. This may be imagined as being a displacement along closely spaced parallel planes within the body, that is, simple shear. The vertical sides of the square are rotated through the angle ψ, the tangent of which is the *unit shear*, a measure of the shear strain.

shear strength *n*. the internal resistance of a body to SHEAR STRESS. The total shear strength of an isotropic substance is the sum of the internal friction and the cohesive strength.

shear stress *n*. the component of STRESS that is tangential (i.e. acts parallel) to a plane passing through any point in the body.

shear wave see SEISMIC WAVES.

shear zone *n*. a tabular zone of rock showing evidence of SHEAR STRESS in the form of crushing and brecciation by many parallel fractures. In a ductile shear zone, no discontinuities are visible, although there is evidence of deformation in the displacement of stratigraphical markers.

sheeted-zone deposit *n*. a mineral deposit comprising veins or lodes that occupy a SHEAR ZONE.

sheet erosion *n*. the removal of surface material from a wide area of gently sloping or graded land by broad continuous sheets of running water rather than by streams. Compare GULLY EROSION.

sheet flood *n*. a broad sheet-like area of moving water. It is characteristic of desert fans in which loose silt and sand become water-loaded so quickly that the water is unable to scour into the surface. If it is diverted by cobbles and clumps of vegetation, a sheet flood forms small braided channels; if unobstructed, it may cover the entire surface. The corresponding geomor-phological process is called *sheetwash*.

sheeting *n*. a form of rupture, similar to jointing, that occurs in massive rock bodies. It is characterised by tabular surfaces that are somewhat curved and essentially parallel to the topographical surface except in regions of recent rapid EROSION. Sheeting is best exposed in openings such as quarries. At the surface of the Earth, the FRACTURES are close together; the interval increases with depth and, at a point, the visible sheeting disappears. At greater depths, the planes of weakness parallel to the sheeting are utilised in working the quarry. Sheeting is a result of the release of pressure as overlying material is removed. The rock mass, which was formed under conditions of hydrostatic stress, will now tend to expand as its distance from the surface decreases.

sheet sand see BLANKET SAND.

sheetwash see SHEET FOLD.

shelf break see CONTINENTAL SHELF.

shelf ice *n*. the ice comprising an ICE SHELF.

shield *n*. a large area of the Earth's CRUST, consisting mostly of Precambrian rocks, that is generally shield-like in form, i.e. broad and gently sloping from the centre. Shields are the exposed areas of the larger structural units known as CRATONS. A Precambrian shield is a mass of rocks made up of OROGENIC BELTS the active lives of which have been completed within the Precambrian and which have therefore been relatively stable over a long period of time. Shields were once the sites of Precambrian mountain belts and consist predominantly of IGNEOUS ROCK such as GRANITE, metamorphosed basic lavas (see GREENSTONE BELT) or of highly metamorphosed rock such as gneiss. Each continent has at least one shield, one of the best known being the CANADIAN SHIELD in North America.

shield volcano *n*. a central VOLCANO, approximately circular in outline, with sides that slope at angles generally between 2º and 10º around a central vent. They are named for their fancied resemblance to the round shields of ancient Nordic warriors. The *simple* or

Icelandic type of shield volcano is built by a single eruptive cycle of activity when very fluid basaltic lava flows outwards in a series of pulses that distribute themselves more or less evenly around the vent. Because of the EFFUSIVE nature of the activity, the proportion of PYROCLASTIC products is small, and this prevents the development of a steep cone. Most Icelandic shields range in diameter up to about 10 kilometres and in height up to about 1000 metres above their surroundings. POLYGENETIC shield volcanoes go through many cycles of activity over periods of tens of thousands of years. They are typified by the *Hawaiian-type* in which the central vent takes the form of a CALDERA from which one to three fissures, or *rift zones*, radiate outwards. During the build-up to eruptions, magma from the MANTLE is fed into a shallow MAGMA CHAMBER, causing the whole edifice to swell detectably. Initially eruptions tend to take place in the summit caldera, but the magma then migrates into one or other of the rift zones where it breaks out as a FISSURE ERUPTION. The Hawaiian-type shields are the largest volcanoes on Earth; Mauna Loa stands over 9000 metres (30 000 feet) above the sea floor although only 4172 metres (13 680 feet) of it are above sea level.

shock metamorphism *n*. metamorphic alteration caused by METEORITE impact or underground nuclear explosions. It is a type of dynamic METAMORPHISM in which very high temperatures and pressures are produced in the rocks over a very short period of time by the passage of high-energy SHOCK WAVES. It can be detected by the presence of IMPACTITES such as SUEVITE and the high-temperature polymorphs of quartz, COESITE and STISHOVITE.

shock wave *n*. a compressional wave that is formed whenever the speed of an object, relative to the medium in which it is travelling, exceeds the speed of sound transmission for that medium. Its amplitude is greater than the elastic limit of the medium through which it is propagating, and it is characterised by a region within which there occur abrupt and drastic changes in the temperature, pressure and density of the medium. If the medium is a rock, a shock wave is capable of melting, vaporising, mineralogically altering or profoundly deforming the rock materials. See also HYPERVELOCITY IMPACT, SHOCK METAMORPHISM.

shoestring sand *n*. an elongate, buried sand body that may be an LONGSHORE BAR, a buried coastal BEACH or BAR, a stream channel filling or some other geomorphic feature. Particular shoestring sands may hold oil, but the specific origin of the formation must first be recognised. For example, an longshore bar differs considerably from a stream channel filling in cross-section, pattern and trend with respect to the ancient SHORELINE. Longshore bars are likely to be discontinuous or segmented; bar segments may be oil-productive, whereas the intervening gaps will not. See also CHANNEL SAND.

shore *n*. **1.** the narrow strip of land along any body of water.
2. the zone ranging from the low-tide limit to the landward limit of wave action.
3. SHORELINE. See also BACKSHORE, FORESHORE.

shoreline *n*. the line along which water of a sea or lake meets the shore or beach. It marks the position of the water level at any given time and on marine margins varies between high-tide and low-tide shoreline positions. A shoreline the features of which result from the dominant relative submergence of a landmass is a *shoreline of submergence*. Such a shoreline is usually very irregular and is either of RIA or FJORD type. There is no implication in the term 'submergence', as used here, as to whether it is land or sea that has moved. If an absolute subsidence is implied in a submerged shoreline, it is called a *shoreline of depression*.

A *shoreline of emergence* is one the features of which are the result of the dominant relative emergence of a lake or ocean floor. Such shorelines are generally less irregular than submergent shorelines. The most pronounced change along an emergent shoreline involves the migration of its LONGSHORE BAR. Whether it is the land or the sea that has moved cannot be inferred from the term. The succession of changes through which a shore PROFILE passes during its development is called the *shoreline cycle*.

shore reef see FRINGING REEF.

shoshonite *n*. a potassium-rich trachy-andesitic rock containing essential K-feldspar (ORTHOCLASE) in the GROUNDMASS in addition to OLIVINE, AUGITE, LABRADORITE and sometimes BIOTITE. Absarokite is a similar but more MAFIC rock. Shoshonite has been defined chemically as the potassic variety of basaltic TRACHYANDESITE. In ISLAND ARCS lavas of shoshonitic affinity occur in islands located on the concave side of the arc.

shot break *n*. a record of the moment of seismic wave inception, as by explosion, during SEISMIC EXPLORATION.

shotcrete, *n*. sprayed concrete containing aggregate particles of greater than 10 millimetres. It is used in stabilising tunnel walls. In the New Austrian Tunnelling Method (NATM) shotcrete is applied immediately after blasting.

shot depth *n*. the vertical distance from the surface to an underground man-made explosive charge.

shot point *n*. that point at which an explosive charge is detonated in order to generate seismic energy. In field work, the point includes the hole in which the charge is planted and its immediate surrounding area.

shrinkage crack *n*. a crack, such as a mud crack, formed in fine-grained sediment by moisture loss during drying.

sial *acronym for si*lica and *al*umina. It is a name proposed by Edouard Suess for the part of the Earth's crust composed of rocks rich in silica and alumina. Compare SIMA.

Sicilian *n*. a STAGE of the PLEISTOCENE in southern Europe.

sicula *n*. (*pl*. **siculae**) a cone-shaped cup with the aperture pointing downwards, the first-formed part of a GRAPTOLITE colony (Fig. 42). The sicula has two parts, an upper *prosicula*, which is marked with longitudinal and spiral lines, and a lower *metasicula*, which is patterned with growth bands, or *fusellae*.

siderite *n*. **1.** a hexagonal iron mineral, $FeCO_3$, belonging to the calcite group of CARBONATES.

2. see IRON METEORITES.

siderolites see STONY IRON METEORITES.

sideromelane *n*. pale brown to reddish-brown basaltic glass. Compare TACHYLITE.

siderophile see AFFINITY OF ELEMENTS.

Siegenian *n*. a STAGE of the Lower DEVONIAN.

sienna *n*. any brownish-yellow, earthy limonitic pigment for paints and oil stains. When burnt, it becomes dark orange-red to reddish brown. See LIMONITE.

sieve texture see POIKILOBLAST.

Sigillaria see SCALE TREES.

silcrete *n.* **1.** a hard surface deposit composed of sand and gravel cemented by opal, chert and quartz, formed by chemical WEATHERING and water evaporation in a semi-arid climate. Extensive deposits of silcrete are found in South Africa and Australia. Silcrete is a siliceous DURICRUST. Compare CALCRETE, FERRICRETE.
2. siliceous DURICRUST. Compare CALCRETE, FERRICRETE.

Silesian *n.* the upper sub-SYSTEM of the European Carboniferous.

silica *n.* silicon dioxide, SiO_2, which occurs in a great variety of crystalline and cryptocrystalline forms depending on physical and chemical conditions. See CHALCEDONY, CHERT, ALPHA QUARTZ, BETA QUARTZ, TRIDYMITE, CRYSTOBALITE, COESITE and STISHOVITE.

silica concentration *n.* the percentage by weight of SiO_2 in an IGNEOUS ROCK. SiO_2 is, with few exceptions, the principal constituent of igneous rocks, so the SiO_2 content is often used as a basis for classification. The categories below have no direct correlation with the modal quantity of quartz in a rock, although, generally speaking, acidic rocks contain quartz and ultrabasic ones do not. Two rocks with equal concentrations of silica may have very different quantities of quartz, and two rocks of similar quartz content may have different silica concentrations.

Silica concentration (wt. %)	Description
63% or more	acidic or silicic
52 to 63%	intermediate
45 to 52%	basic
45% or less	ultrabasic

silica-oversaturated see SILICIC.

silica sand *n.* a SAND containing a high percentage of quartz. It is a raw material for glass and other products.

silica saturation *n.* the concentration of SILICA relative to the concentrations of other chemical components in a rock that combine with it to form SILICATE minerals. By calculating the NORM of a rock the following classifications can be made:
1. (a) *silica-saturated minerals*, which can exist at equilibrium with quartz; these minerals include feldspar, amphiboles, micas and many others; (b) *silica-unsaturated minerals*, which are never found with quartz; these include all feldspathoids, Mg-olivine and corundum;
2. (a) *silica-oversaturated rocks*, such as granite, which contain quartz or its polymorphs; (b) *silica-saturated rocks*, which contain neither unsaturated minerals nor quartz, for example diorite; (c) *silica-unsaturated rocks*, which contain unsaturated minerals, for example nepheline syenite. Compare SILICA CONCENTRATION. See NORM.

silicate *n.* a compound the basic structural unit of which is the silicate tetrahedron, SiO_4, which is either isolated or joined by one or more of its oxygen atoms to form configurations of independent or double groups, rings, chains, sheets or three-dimensional frameworks. Silicates are classified according to the manner in which the basic tetrahedral groups are joined. See NESOSILICATE, SOROSILICATE, CYCLOSILICATE, INOSILICATE, PHYLLOSILICATE, TECTOSILICATE.

silicate tetrahedron *n.* the fundamental structural unit in the SILICATE minerals. It consists of a silicon atom attached to four oxygen atoms by COVALENT BONDS. See Fig. 84 on the opposite page.

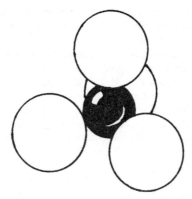

Fig. 84 **Silicate tetrahedron**. A single silicon atom is attached to the four surrounding oxygens by covalent bonds to form an SiO_4 ion. The oxygens are positioned at the corners of a regular tetrahedron

silication *n.* the process of replacing by or converting to SILICATES, such as the formation of SKARN minerals in carbonate rocks. Compare SILICIFICATION.

siliceous *adj.* containing abundant SILICA, particularly as free silica.

siliceous ooze see PELAGIC DEPOSITS.

silicic, acidic or **silica-oversaturated** *adj.* (of an igneous rock or magma) rich in silica. The amount of silica generally recognised as minimum is at least 63 per cent of the rock. In addition to combined silica, silicic rocks usually contain free silica in the form of QUARTZ. Granite and rhyolite are typical examples. See SILICA CONCENTRATION.

silicification *n.* **1.** the introduction of, or REPLACEMENT by, silica, particularly in the form of fine-grained quartz, chalcedony or opal, which may fill cavities and pores and replace existing minerals. (Compare SILICATION.) Silicification may result from the filling of cavities with silica-saturated groundwater. The silica is often deposited in concentric rings around a central nucleus.
2. the process of fossilisation wherein the original hard parts of an organism are replaced by quartz, chalcedony or opal. See FOSSIL, REPLACEMENT.

silicified wood, petrified wood or **opalised wood** *n.* a material formed by the SILICA permineralisation of wood in such a manner that the original shape and structural detail (grain, growth rings, etc) are preserved. The silica is generally in the form of CHALCEDONY or OPAL.

sill *n.* a tabular igneous INTRUSION with boundaries conformable with the planar structure of the surrounding rock. Sills with a small area may be restricted to a single plane; larger ones may be transgressive, i.e. they cut the bedding or foliation of the surrounding rock. Sills may be multiple or composite and are generally medium-grained, although larger ones may be coarse-grained.

silled basin or **barred basin** *n.* an area on the ocean floor set off by a rise that prevents maximum water circulation over it, a condition often resulting in oxygen depletion.

sillimanite *n.* an orthorhombic nesosilicate, Al_2SiO_5, trimorphic with ANDALU-

SITE and KYANITE. Its colours are grey, brown or pale green, and it often occurs in silky, fibrous aggregates. Sillimanite is widespread in high-temperature ARGILLACEOUS metamorphic rocks and also occurs in contact metamorphic types; it is an indicator of the temperature and pressure at which the host rock formed. One of its main uses is in the manufacture of refractories and high-temperature minerals.

silt *n*. **1**. DETRITAL particles, finer than very fine SAND and coarser than CLAY, in the range of 0.004 to 0.062 millimetres. **2**. an unconsolidated CLASTIC sediment of silt-sized rock or mineral particles. **3**. fine earth material in SUSPENSION in water.

siltstone *n*. a fine-grained SEDIMENTARY ROCK principally composed of SILT-grade material. Siltstones are intermediate between sandstones and clays but less common than either. Most siltstones are composed of clastic quartz, together with some feldspar and mica. See also MUDROCK.

Silurian *n*. a division of the Palaeozoic Era extending from 440 to 408 Ma ago, named in 1835 by Murchison after the Silures, an ancient Celtic tribe of the Welsh borderland. Here the rock system was studied and subdivided according to its graptolitic and shelly fauna into the series Llandovery, Wenlock, Ludlow and Přidoli. The base of the Silurian is the base of the *Akidograptus acuminatus* BIOZONE in the Dob's Linn section, Scotland, and its top is the base of the zone of *Monograptus uniformis* in the Klonk section of the Czech Republic. During the Silurian the IAPETUS Ocean, which separated Scotland as a part of Laurentia from Eastern Avalonia (England, Wales, southern Ireland), was closing, with continental collision occurring about

420 Ma ago in the later stages of the CALEDONIAN OROGENY. In Britain the volcanic activity of the Early Silurian took place in the east-west zone, from southeast Ireland through southwest Wales and towards the southern edge of the Midland Platform. Volcanic activity drew to a close by the Wenlock. In mid-Silurian southern Britain, the area was dominated by a series of platforms and basins, including the Midland Platform and the Welsh, Leinster and Lakes basins, all of which accumulated thick turbidite deposits and graptolitic shales. On the Midland Platform, mudstones and carbonates, including reef limestones, were deposited, the latter dominating in Wenlock times. Marine deposition in Eastern Avalonia came to an end during late Ludlow and Přidoli times. Silurian sedimentation in Scotland and northern Ireland, in the Midland Valley TERRANE, was that of the marginal deposits of Laurentia, including the sediments of the Girvan inlier where marine sequences terminate in Wenlock times. Elsewhere in the Midland Valley terrane early marine turbidite sequences pass up into shallow marine and then fluvial sequences. In the Southern Uplands terrane the Silurian succession is interpreted as the deposits of an accretionary prism developed above a northerly dipping SUBDUCTION ZONE, although other interpretations have been suggested. Sedimentation ended with the collision of Eastern Avalonia with the Laurentian margin about 420 Ma.

In relation to life, a mass EXTINCTION at the end of the Ordovician was followed by the rediversification and recovery of many groups, including the ECHINODERMS (including CRINOIDS), BRACHIOPODS, GRAPTOLITES, BRYOZOANS, BIVALVES,

TABULATE and RUGOSE CORALS. TRILOBITES continued to decline. Other arthropods, however, such as the eurypterids, became important members of brackish and freshwater faunas. Ostracoderms (jawless freshwater fish) become abundant towards the end of the period. The first fish with jaws (Acanthodians) appear in fresh water in the Silurian. On land, Middle and Upper Silurian fossils indicate that psilopsids and early lycopsids were present, together with terrestrial arachnids such as scorpions. The earliest members of the Insecta were also present. Widespread REEF formation by Silurian corals suggests warm shallow seas and little seasonal temperature variation. Britain lay in tropical or subtropical latitudes south of the Equator. See also CONTINENTAL DRIFT.

silver *n.* an isometric mineral, the native element Ag, that normally occurs as malformed crystals in branching or reticulated forms. Native silver is widely distributed in small amounts as a secondary mineral in the OXIDISED ZONE of ore deposits. It is found in larger amounts deposited from hydrothermal solutions. Native silver is found in deposits associated with zeolites, calcite, quartz, arsenides and sulphides of cobalt, nickel and silver.

sima *n. acronym for si*lica and *ma*gnesia. It is the name suggested by Edouard Suess for that part of the Earth's crust composed of rocks rich in silica and magnesium. Compare SIAL.

similar folding see FOLD.

simple shear *n.* a deformation consisting of a displacement in one direction of all straight lines that are initially parallel to that direction; it is closely similar to shearing a deck of cards. Compare PURE SHEAR.

Sinemurian *n.* a STAGE of the Lower JURASSIC.

sinistral fault see FAULT.

sink *n.* **1.** see DOLINE.
2. a depression on the flank of a VOLCANO resulting from collapse.
3. a generally depressed area where a desert stream ends or disappears by evaporation.

sinter *n.* a white, porous opaline variety of SILICA, deposited as an incrustation by precipitation from geyser and hot-spring waters (*siliceous sinter*).

siphonal canal *n.* a grooved projection at the anterior of the aperture of some GASTROPOD shells that supports the inhalent siphon. See Fig. 38.

siphuncle *n.* a tube containing body tissue that extends from the apex of the shell to the mantle of CEPHALOPODS. Gas and liquid are added to or extracted from the chambers of the shell via the siphuncle, thus controlling buoyancy and attitude. See AMMONOID, Fig. 1, NAUTILOID, Fig. 62 and BELEMNITE, Fig. 5.

skarn *n.* a poorly defined term with a usage that includes contact rock containing calcium, magnesium and iron silicates derived from nearly pure limestone or dolomite into which abundant amounts of silicon, iron, aluminium and magnesium were metasomatically introduced during contact metamorphism. Such skarns are often host rocks for deposits of MAGNETITE and many calcium-magnesium silicates, such as GROSSULAR garnet. Skarn has also been used for any calc-silicate metamorphic rock with unusual mineralogy, especially if it has resulted in the concentration of ore minerals. See also METAMORPHISM.

skip marks see TOOL MARKS.

slag *n.* **1.** the fused waste product from the smelting of ores.

2. an informal, loosely defined term sometimes used for volcanic SCORIA.

slaking *n.* the crumbling and disintegration of fine-grained rocks on exposure to moisture.

slant drilling see DIRECTIONAL DRILLING.

slate *n.* a fine-grained METAMORPHIC ROCK derived mostly from SHALE. It is characterised by SLATY CLEAVAGE, i.e. the ability to be split into large, thin, flat sheets. Such cleavage is caused by the alignment of platy minerals into parallel planes, which occurs during METAMORPHISM.

slaty cleavage or **flow cleavage** *n.* well-developed, closely spaced parallel planes along which a rock may be easily split. It is caused by the parallel alignment of small plate-like minerals and develops in SLATE or other homogeneous rock by deformation and low-grade METAMORPHISM. See Fig. 85 below.

slickensides *n.* the characteristic texture on a rock surface that has become polished and striated from the grinding or sliding motion of an adjacent rock mass, commonly found along FAULT planes or zones. The striations or flutings may indicate the direction of relative motion; they tend to end abruptly in the form of small steps in the direction of movement.

slide *n.* **1.** a type of MASS WASTING in which a mass of rock or rock debris moves downhill along a planar or concave-up surface, usually accompanied by break-up of the debris. Compare SLUMP.

2. a low-angle FAULT.

slip *n.* the relative displacement of formerly contiguous points on opposite sides of a FAULT, as measured on the fault surface.

slip fold see SHEAR FOLD.

slit band see EXHALANT SLIT.

slope stability *n.* the resistance of a slope, natural or artificial, to landsliding. See also FAILURE.

slope wash *n.* **1.** earth material moved down a slope, principally by the action of gravity aided by non-channelled running water. Compare COLLUVIUM.

2. the process itself by which such material is moved.

sluicing *n.* the concentration of HEAVY MINERALS by flushing unconsolidated material through boxes (*sluices*) equipped with grooves or slats of

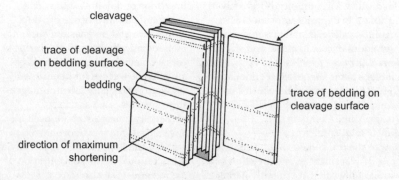

cleavage

trace of cleavage on bedding surface

bedding

trace of bedding on cleavage surface

direction of maximum shortening

Fig. 85 **Slaty cleavage**. The cleavage forms at right angles to the direction of maximum compression. Traces of the sedimentary bedding may still be visible

wood (*riffles*) that trap the heavier minerals on the bottom of the box.

slump *n*. **1.** the downward sliding of rock DEBRIS as a single mass, usually with a backward rotation relative to the slope along which movement takes place. Compare DEBRIS SLIDE.
2. also called **subaqueous gliding**, the slipping or sliding down of a mass of sediment relatively soon after its deposition on a subaqueous slope.

slushball Earth see SNOWBALL EARTH.

smectite *n*. a group of clay minerals characterised by a three-layer CRYSTAL LATTICE, by deficiencies of charge in the tetrahedral and octahedral positions, and by swelling upon wetting. The smectite minerals are the primary constituents of FULLER'S EARTH and BENTONITE. The smectite group includes the dioctahedral members, MONTMORIL-LONITE, BEIDELLITE and NONTRONITE; HEC-TORITE and SAPONITE are trioctahedral. See EXPANSIVE CLAYS.

smithsonite *n*. an ore of zinc, $ZnCO_3$, variously coloured white, pale blue or green, yellow or pink, depending on the impurities present. It is found in MAMILLATED, BOTRYOIDAL, RENIFORM or STALACTITIC aggregates, and is of secondary origin produced by the reaction of carbonated waters on zinc sulphide. Smithsonite is characteristi-cally found in the oxidation zone of sulphide deposits.

smoky quartz see QUARTZ.

SMOW *acronym of* Standard Mean Ocean Water, the international stand-ard for $^{18}O/^{16}O$ ratios.

Snell's law see LAW OF REFRACTION.

snowball Earth *n*. a hypothesis that suggests that in late PROTEROZOIC times the polar ice sheets spread south and north to cover the entire surface of the Earth. It has been devised to explain the widespread distribution of glacial deposits of late Proterozoic age and in particular the apparent occurrence of glacial deposits in equatorial regions. According to the hypothesis, a build-up of greenhouse gases, such as carbon dioxide, from volcanic activity and other sources led to an abrupt end to this worldwide GLACIATION and the deposition of dolomitic limestones (CAP CARBONATES) that overlie late Protero-zoic glacial deposits in many parts of the Earth. A less extreme variant of this hypothesis, which suggests that parts of the equatorial regions of the Earth were not completely ice-covered, is the *slushball Earth*.

snowball texture *n*. S-shaped trails of inclusions in a POIKILOBLAST produced by rotation of the crystal as it grows, (see Fig. 86 below). If the apparent rotation of the crystal is less than 90°, the structure is called *rotational* and *snowball* if the apparent rotation is greater than 90°. These structures are syntectonic, compared with the post-tectonic HELICITIC TEXTURES. It can be difficult to distinguish the two, but all snowball crystals in the same THIN

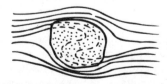

Fig. 86 **Snowball texture**. The 'S' shape of the inclusions is caused by rotation of the crystals as it grows.

SECTION should contain inclusion trails of the same shape and sense of rotation, whereas the inclusion trails in helicitic crystals will have different shapes.

soapstone *n*. **1.** a massive METAMORPHIC ROCK composed mostly of talc, with varying quantities of mica, chlorite and other minerals. It can be cut, sawn and carved.
2. STEATITE.

soda lake *n*. an ALKALI LAKE containing large quantities of dissolved sodium salts, especially sodium carbonate, together with sodium chloride and sulphate. Examples are found in Mexico and Nevada, USA.

sodalite *n*. an isometric mineral of the FELDSPATHOID group, $Na_8(AlSiO_4)_6Cl_2$. It commonly occurs as compact masses in bright blue, white or grey, with green tints in undersaturated plutonic igneous rocks. Sodalite is widely used in jewellery and for carved ornaments.

soda nitre see CHILE SALTPETRE, NITRATINE.

sodium bentonite see BENTONITE.

soft coal see BITUMINOUS COAL.

soil *n*. **1.** a layer of organic and inorganic weathered material that accumulates at the Earth's surface. The soil plus sub-soil together make up the REGOLITH. See SOIL PROFILE. Soils reflect the influence of climate, relief and nature of the parent material as well as the physical, chemical and biological processes involved. See also CHERNOZEM, PODZOL, AZONAL SOILS, GLEY SOILS.
2. (*engineering*) a naturally occurring loose or soft deposit that either disintegrates or softens on immersion in water. Compare ROCK.

soil blister see FROST MOUND.

soil horizon *n*. a layer of a soil that can be distinguished from adjacent layers by physical properties (structure, texture, colour) or chemical composition. See SOIL PROFILE.

soil mechanics *n*. the application of the concepts of mechanics and hydraulics to engineering problems associated with the nature and behaviour of soils, sediments and other unconsolidated accumulations.

soil profile *n*. a vertical arrangement of the various discrete horizontal layers, called *horizons*, that make up any soil from the surface downwards to the unaltered parent material or BEDROCK. As commonly defined, the ideal soil profile is divided into A, B and C horizons and their subdivisions.
(a) *A horizon* or TOPSOIL: the layer from which soluble substances have been removed by leaching or eluviation and carried downwards to the B horizon;
(b) *B horizon* or SUBSOIL: a zone of accumulation enriched by the clay minerals leached from the A horizon;.
(c) *C horizon*: the layer of weathered bedrock at the profile base, which has undergone little alteration by organisms.

soil stripe see PATTERNED GROUND.

sol *n*. that component of a colloidal system that is more fluid than a GEL.

sole *n*. **1.** the undersurface of a STRATUM or VEIN.
2. the FAULT plane underlying a thrust sheet.
3. the lower parts of the shear surface of a LANDSLIDE.
4. the basal ice of a GLACIER.

sole mark *n*. a SEDIMENTARY STRUCTURE found on the underside (*sole*) of a SANDSTONE bed. Some sole marks are produced as a result of the infilling by coarser sediment of a variety of depressions and other irregularities on the surface of a MUD or other FINE-GRAINED sediment. After LITHIFICATION the sandstone is more resistant to WEATHERING than the MUDROCK, so the surface texture of the underlying fine-grained sediment is most often preserved in counterpart on the base of the sandstone bed. Marks caused by current action, such as SCOUR MARKS and TOOL MARKS, are characteristically found as sole marks. LOAD CASTS, caused by soft sediment deformation, are another type of sole mark. Many structures may be preserved either on the top of a bed or moulded on to the sole of the overlying bed. These include RIPPLE MARKS, DESICCATION CRACKS, RAIN PRINTS and marks left by organisms, such as footprints, worm casts and various kinds of tracks. Sole marks can be used as WAY-UP INDICATORS.

solfatara *n.* a FUMAROLE in which sulphur gases are the most abundant after water (steam).

solid earth geophysics see GEOPHYSICS.

solid state flow *n.* flow in a solid by means of rearrangement among or within the component particles. See INTERGRANULAR MOVEMENT and INTRAGRANULAR MOVEMENT. Compare VISCOUS FLOW.

solid solution *n.* the substitution of one ION for another in a CRYSTAL LATTICE without any major change in structure. The composition of minerals that are members of a solid-solution series is thus not fixed but varies between certain limits defined by the END MEMBERS of the series. For example, the composition of OLIVINE varies between Mg_2SiO_4 (forsterite) and Fe_2SiO_4 (fayalite), as either Mg^{2+} or Fe^{2+} may enter the lattice. Slight adjustments of the lattice to accommodate ions of different size are reflected in variations in physical properties such as refractive index.

solidus see LIQUIDUS-SOLIDUS.

solifluction *n.* the slow downslope movement of water-logged earth materials (REGOLITH). Solifluction can occur in the tropics but is particularly characteristic of cold regions where

PERMAFROST is found. In areas of frozen terrain, the term *gelifluction* is used. Because permafrost is impermeable to water, the overlying soil may become oversaturated and will slide down under the force of gravity. Soil that has been weakened by FROST ACTION is most susceptible to this process. Original SOIL PROFILES become greatly altered or destroyed. Solifluction deposits are usually poorly SORTED but can contain a preferred orientation of rock fragments.

solution breccia *n.* a COLLAPSE BRECCIA resulting from the removal of soluble material by solution, thus permitting fragmentation of overlying rock.

solution load see DISSOLVED LOAD.

solution mining *n.* 1. the mining of soluble rock material, especially salt, from underground deposits by pumping water downwards into contact with the deposit, removing the brine and then processing it. 2. the in-place dissolution of mineral constituents of an ore deposit by trickling down a leaching solution through the broken ore to collection chambers farther down.

Solvan *n.* the lower STAGE of the Middle CAMBRIAN.

solvus *n.* the curve of surface in a PHASE DIAGRAM that separates a domain of stable SOLID SOLUTION from a domain of two or more phases that may form from the stable one by unmixing.

sonic log see WELL LOGGING.

sorosilicate *n.* those silicates each of whose tetrahedral silicate groups shares one of its oxygen atoms with a neighbouring silicate group. The ratio of silicon to oxygen is 2:7, S_2O_7. EPIDOTE and the MELILITE group are examples of sorosilicates.

sorted, also called **graded**, *adj.* 1. (of a SEDIMENT) composed of particles of uniform size.

2. (of PATTERNED GROUND) describing features that show a border of stones surrounding finer material such as SAND, SILT or CLAY.

sorted polygon see PATTERNED GROUND.

sorting *n.* the process by which sedimentary particles are naturally separated according to size, shape and specific gravity from associated but dissimilar particles.

soufrière *n.* a French term for a VOLCANO giving off sulphurous gases, or a SOLFATARA. The village of Soufrière on St Lucia in the Caribbean volcanic arc derives its name from a nearby solfatara. The active volcanoes on the islands of St Vincent and Guadeloupe in the same island arc are both called La Soufrière.

sounding *n.* **1.** the vertical distance from the surface of a body of water to its floor or some specific point beneath the water surface.

2. the procedure by which the depth of a body of water is determined. This is accomplished either by

(a) direct physical measurement using a lead line, or

(b) indirectly, by measuring the time it takes for a sound pulse, generated at the surface, to reflect from the bottom and return to the surface as an echo (ECHO SOUNDING). Sounding is performed for navigational safety, scientific investigation and economic or engineering purposes. See SEISMIC EXPLORATION, ECHOGRAM.

source rock *n.* SEDIMENT that forms the source of HYDROCARBONS.

SP see SPONTANEOUS POTENTIAL.

spaced cleavage see FRACTURE CLEAVAGE.

spall see EXFOLIATION.

spar *n.* any transparent or translucent light-coloured mineral that is somewhat lustrous and usually shows easy CLEAVAGE.

sparite *n.* **1.** sparry calcite, the coarsely crystalline interstitial cement of LIMESTONE. It may be transparent or translucent and consists of calcite or aragonite that precipitated during deposition or was later introduced as a cement. Its crystalline structure is much coarser than that of MICRITE.
2. a limestone with more sparite cement than micrite matrix.

sparker *n.* a device for generating underwater seismic energy by means of a high-voltage electrical discharge.

spatter *n.* agglutinised masses of primary magmatic ejecta erupted in a plastic or fluid state. It often accumulates around vents as mounds or cones, or along fissures, and is produced chiefly from the spray of erupting magma in lava fountains. See LAVA, SPATTER CONE.

spatter cone *n.* a low, steep-sided mound or hill formed by LAVA FOUNTAINS along a fissure or about a central vent. It consists mainly of SPATTER, the glassy skin of which adheres to form AGGLUTINATE. VOLCANIC BOMBS may also be present. Spatter cones may exceed 30 metres in height and sometimes give rise to short 'rootless' LAVA FLOWS. They are characteristic of basaltic eruptions. Compare HORNITO. See LAVA.

spatter rampart *n.* a wall of welded SPATTER that may accumulate along one or both sides of a volcanic fissure during the fountaining of lava of low viscosity (generally basaltic). Although usually less than three metres in height, they occasionally exceed six metres.

species *n.* a group of organisms the members of which are capable of breeding with one another to produce fertile offspring but not with members of other groups.

specific conductance see CONDUCTIVITY.

specific gravity *n*. the ratio of the weight of a given volume of a substance to the weight of an equal volume of water.

specific heat capacity *n*. the quantity of heat required to raise the temperature of a unit mass by 1° ($J\ K^{-1}\ kg^{-1}$).

specific resistance SEE RESISTIVITY.

specific retention *n*. the ratio of the volume of water retained by the pores in the rock of an unconfined AQUIFER when water is removed from that aquifer to the volume of the aquifer itself. The water is retained by capillary forces and cannot be removed by pumping or natural drainage. See SPECIFIC YIELD.

specific yield or **unconfined storativity** *n*. 1. the volume of water that an unconfined AQUIFER releases from storage per unit surface area of aquifer per unit decline in the WATER TABLE.
2. the ratio of the volume of water drained from a rock to the total volume of the pore space.

spectrographic analysis *n*. analysis of a substance by matching the lines of its spectrum with lines in the spectra of elements the wavelengths of which are known.

spectrometry see Appendix 4.

specularite or **specular hematite** *n*. a grey or black variety of HEMATITE that occurs in aggregates of tabular crystals or foliated masses; it has a brilliant metallic lustre.

speleology *n*. the exploration and scientific study of caves and caverns. As a science, it involves exploration, surveying, biology, geology and mineralogy.

speleothem *n*. any one of the variously shaped MINERAL DEPOSITS formed in a cave by the action of water. Almost all speleothems are made of calcium carbonate crystals in the form of calcite. See CAVE ONYX, DRIPSTONE, STALACTITE AND STALAGMITE.

spermatophyte *n*. a vascular seed-bearing plant, i.e. ANGIOSPERM, GYMNOSPERM or pteridosperm. Spermatophytes range from the Devonian to the present. Compare PTERIDOPHYTE.

spessartine *n*. the manganese-aluminium END MEMBER of the GARNET series, $Mn_3Al_2Si_3O_{12}$. It ranges in colour from pale orange-yellow to deep red. It occurs in SKARN deposits and in association with RHODONITE and other manganese minerals.

spessartite *n*. 1. a LAMPROPHYRE composed of phenocrysts of clinopyroxene or green hornblende in a matrix of sodic plagioclase with accessory biotite, olivine and apatite.
2. a redundant synonym for SPESSARTINE

sp. gr. *abbreviation for* SPECIFIC GRAVITY.

sphaerocone see AMMONOID.

sphalerite, blende, zinc blende or **blackjack** *n*. an isometric sulphide mineral, $(Zn,Fe)S$; its colour varies from yellow or reddish brown to blackish. It occurs in aggregates of distorted crystals and banded concretionary masses. Sphalerite is found in massive sulphide deposits and hydrothermal veins associated with GALENA and other sulphides. It is the main ore for zinc and generally also provides cadmium and other minerals as byproducts.

sphene see TITANITE.

Sphenopteris *n*. the name by which the leaves of *Lyginopteris*, a fossil seed fern, were originally described. The leaves consist of small symmetrical lobed leaflets that show radiating veins. Range, Devonian to Upper Carboniferous. See FOSSIL PLANTS.

sphenopsid *n*. any of a subdivision of the Pteridophytina (Devonian to Recent) characterised by grooved stems that are jointed at the nodes from which come whorls of leaves or branches. Sphenopsids have fertile and sterile shoots, the fertile shoots being terminated by cones (*sporangia*) that produce spores. Sphenopsids evolved in the Devonian, and arborescent sphenopsids, such as *Calamites*, with heights of up to 10 metres, were abundant in the Carboniferous, with herbaceous sphenopsids common throughout the history of the group.

sphericity *n*. the degree to which the form of a sedimentary particle approaches that of a sphere. Compare ROUNDNESS.

spheroid *n*. (*geodesy*) a mathematical figure closely approximating to the GEOID in size and form and used as a reference surface for geodetic surveys (see GEODESY).

spheroidal weathering *n*. concentric shells of WEATHERING products resulting from the penetration of chemical weathering processes along JOINT surfaces. Jointed IGNEOUS ROCKS such as GRANITE or BASALT are particularly susceptible to this form of weathering, which results in rounding of the edges of joint blocks and the production of CORESTONES.

spherulite *n*. generally spherical aggregates, commonly consisting of alkali feldspar fibres radially arranged about some nucleus, such as a PHENOCRYST. The usual size is between one millimetre and 2 to 3 centimetres but some may be as great as 3 metres in diameter. They occur mainly in acid glasses, such as volcanic rhyolites, but are present in some partly or wholly crystalline rocks that include shallow intrusive types. They can form as the result of rapid crystallisation in a quickly cooled magma, or by DEVITRIFICATION. See also LITHOPHYSAE, VARIOLITIC.

spilite *n*. an altered basaltic rock

consisting of low-metamorphic-grade minerals, including albite, chlorite, actinolite, sphene and calcite. It is thought that spilites form from sea-floor basalts near oceanic rifts as a result of the metasomatic interchange of calcium from the lava with sodium from heated sea water. Ancient PILLOW LAVAS are commonly spilitic.

spine *n*. **1.** a VOLCANIC SPINE.
2. a pointed projection found on the shell surface of certain invertebrates such as the ECHINOIDS.

spinel *n*. **1.** an isometric mineral, $MgAl_2O_4$, that occurs as crystals, aggregates or rounded grains. They can be white, pink, lavender, light blue, green, black or colourless. Spinels are found in contact metamorphic rocks, especially in many Mg-rich dolomites. Spinel is an ACCESSORY MINERAL in some basic or ultrabasic igneous rocks. Some varieties are used as gemstones.
2. spinels are a group of minerals with the general formula $A^{2+}B_2^{3+}O_4$. There are three series: the spinel series in which Al^{3+} is the dominant trivalent cation, the MAGNETITE series with Fe^{3+} as dominant trivalent cation, and the CHROMITE series in which Cr^{3+} is dominant.

spinifex texture *n*. a parallel or radiating arrangement of elongate ACICULAR OLIVINE or PYROXENE PHENOCRYSTS frequently observed in KOMATIITES. It is caused by extremely rapid cooling (quenching) of the LAVA.

spirit-level structure see GEOPETAL STRUCTURE.

SP interval *n*. (*seismology*) the time interval between the first arrivals of transverse and longitudinal waves, which is a measure of the distance from the EARTHQUAKE source.

spiralia see BRACHIDIUM.

spire *n*. **1.** all whorls except the BODY

WHORL in a helically coiled shell; see GASTROPOD, Fig. 38.

2. see HORNITO.

spit *n.* a narrow ridge or embankment of SEDIMENT forming a finger-like projection from the shore into open water. Spits are formed by the action of waves and currents, and are frequently found on irregular coastlines. Good examples are found along the east coast of England – Spurn Head on the Humber estuary and Orford Ness in Suffolk. See also TOMBOLO.

splay faults *n.* minor faults diverging from the main FAULT at angles of less than 45°. They are commonly found near the ends of major faults.

splendent lustre *n.* mineral lustre of the highest intensity.

spodumene *n.* a mineral of the PYROXENE group, $LiAlSi_2O_6$, occurring as prismatic crystals, sometimes as large as 16 metres, or rod-like aggregates of compact cryptocrystalline masses; its colours include whitish, yellow, grey, pink (kunzite) and emerald green. Spodumene is found almost exclusively in lithium-bearing pegmatites associated with quartz, feldspars, beryl and tourmaline. It is a source of lithium and its salts.

spondylium *n.* a vertical plate, Y-shaped in cross-section, to which muscles were attached inside the PEDICLE VALVE of BRACHIOPODS of the order Pentamerida. Sometimes it lies opposite a pair of vertical plates, the *cruralium*, inside the DORSAL VALVE, and together they enclose the muscles. See also BRACHIOPOD, Fig. 10.

sponge or **poriferan** *n.* the most primitive multicellular aquatic invertebrate, with an internal skeleton consisting of interlocking spicules of calcite or silica and chitinous fibres. Of the thousands of species known, all but about 20 are marine. The division of fossil sponges is based on: (a) the nature of the spicules (spikes) and (b) the way they interlock. The complexity of the water current system of a particular sponge connotes the degree of its advancement. Range, Precambrian to present.

spontaneous potential or **self-potential** (**SP**) *n.* the natural potential difference that exists between any two points on or under the ground. It is related to naturally occurring electric currents that flow within the ground resulting from electrochemical reactions in the rocks (see EARTH CURRENT). The *SP log* is one of the electrical methods of WELL LOGGING and can be used for stratigraphical correlation between boreholes, as particular geological formations within an area tend to have characteristic values of SP.

spring *n.* a natural flow of water from underground, occurring where the WATER TABLE intersects the ground surface. The line of intersection is called the *spring line*. The spring line, and therefore the position of the springs, may migrate as the position of the water table varies depending on the rate of RECHARGE.

stabile *adj.* resistant to chemical change or to decomposition. Compare LABILE.

stable *adj.* **1.** (of a constituent of a sedimentary rock) able to resist mineralogical change and representing a terminal product of sedimentation, e.g. quartz.
2. (of a part of the Earth's crust) manifesting neither subsidence nor uplift.
3. (of a substance) not spontaneously radioactive.

stack *n.* an erosional remnant in the form of a small bedrock island a short distance offshore. Stacks are pillar-like formations detached from a headland by marine EROSION. During the erosion of a coast, the more resistant parts will usually form headlands; progressive erosion forms caves, then arches or bridges. If the seaward part of the headland remains after collapse, a stack is formed. Further erosion reduces the stack to a WAVE-CUT PLATFORM. Stacks are found off coasts around the world. The Old Man of Hoy in Orkney is a well-known example. See MARINE ABRASION, NATURAL BRIDGES AND ARCHES.

stadia *n*. a surveying technique used to determine distances and elevation differences. The direction of a point is established by sighting the point through a telescope, after which a direction line is drawn along a straight edge and its distance scaled. Distances are read by noting the interval on a graduated rod (*stadia rod*) intercepted by two parallel hairs (*stadia hairs*) mounted in the telescope of a surveying instrument. The rod is placed at one end of the distance to be measured and the surveying instrument at the other.

stadia hairs see STADIA.

stadial *n*. a very cold period during a GLACIAL. Compare INTERSTADIAL.

stadial moraine see MORAINE.

stadia rod see STADIA.

stadia tables *n*. mathematical tables from which may be determined, without calculation, the vertical and horizontal components of a reading made with a STADIA rod and an ALIDADE.

stage *n*. **1.** a STRATIGRAPHICAL UNIT ranking above CHRONOZONE and below SERIES; it is based on BIOZONES considered to approximate to deposits that are equivalent in time and includes the rocks formed during an age of geological time.
2. a major subdivision of a glacial epoch, which includes GLACIAL and INTERGLACIAL stages.

staining test *n*. a method of identifying MINERAL compositions by staining the surface of a slab of rock or uncovered THIN SECTION with dyes. Initial preparation of the surface usually involves etching with an acid. The dye may be applied in an alkaline or acid solution and reacts to produce a coloured precipitate on the surface of a particular mineral. Many of the reagents are poisonous or corrosive and need great care in use. Details of staining methods are given by J. Miller in *Techniques in Sedimentology*, edited by M. Tucker (Blackwell Scientific Publications, Oxford, 1988).

Staining techniques are most useful for CARBONATE minerals. A method combining acidified solutions of Alizarin red S and potassium ferricyanide can differentiate CALCITE (pale pink to red), ferroan calcite (mauve to royal blue with increasing Fe content), DOLOMITE (unstained) and ferroan dolomite (shades of turquoise, deepening with increasing Fe content). Magnesian varieties of calcite stain pink or red with an alkaline solution of titan yellow, the depth of colour increasing with the content of $MgCO_3$. ARAGONITE can be distinguished from calcite using a solution of Ag_2SO_4 and $MnSO_4$, which stains aragonite black but does not affect the calcite. A variety of staining methods can be used to identify GYPSUM and ANHYDRITE, differentiate FELDSPAR compositions, identify CLAY MINERALS and even enhance structures formed by burrowing organisms in SANDSTONES and SILTSTONES.

stalactite and stalagmite *n*. *stalactites* are conical or cylindrical mineral deposits, usually calcite, that hang from ceilings of LIMESTONE caves and range in length from a fraction of a centimetre to several metres. They are chemical precipitates deposited from water that is supersaturated with calcium bicarbonate and enters the caves through cracks and joints in the ceiling. Depending on the cave temperature when deposition occurred, the calcium carbonate precipitate may be calcite or aragonite. As the water drips through the ceiling, successive rings of crystals form a minute tube; if the tube is blocked, the water flows down the outside and the deposits thicken the tube into a stalactite.

Water dripping on to the cave floor builds *stalagmites*, which are usually blunter than stalactites. A stalagmite and stalactite may eventually join to form a column. The growth of stalactites is limited by the tensile strength of calcite, whereas the height of stalagmites is limited by the massiveness of the calcite. The limiting effect of the second factor is greater, so stalactites are able to grow to a greater length than stalagmites. Both types of deposit may be white, translucent or, if impurities are contained, shades of brown, yellow or grey.

Growth rates for both forms are variable, since they are affected by many factors; thus the age of either type cannot be determined accurately by size, current growth rate or weight. Stalactites may also form in subways, tunnels, under bridges and in mines. Little mounds of LAVA that rise from the floor of a LAVA TUBE are also referred to as stalagmites, but lava stalactites, formed by drips from the ceiling, are more common.

Standard Mean Ocean Water see SMOW.

standard mineral see NORMATIVE MINERAL.

standard section see STRATOTYPE.

standing wave or **stationary wave** *n.* a wave the form of which oscillates vertically between two fixed points or nodes without progressive lateral movement.

stanniferous *adj.* containing or yielding tin.

stationary wave see STANDING WAVE.

star ruby, star sapphire *n.* the gem CORRUNDUM with a star-like OPALESCENCE when viewed along the c CRYSTAL AXIS. Stones are cut specially to take advantage of this property.

staurolite *n.* a red-brown to black orthorhombic silicate mineral, $(Fe^{2+}, Mg)_2(Al,Fe^{3+})_9O_6(SiO_4)_4(O,OH)_2$. It occurs as stubby prismatic crystals or frequently in cruciform twins; it is very rarely massive. Staurolite is a metamorphic mineral typical of medium-temperature/pressure conditions associated with GARNET and KYANITE; it is used as a means of defining the metamorphic grade of the COUNTRY ROCK.

steatite *n.* 1. a compact, massive rock composed mainly of talc. See SOAPSTONE.
2. TALC, especially the grey-green or brown massive variety that can be easily carved.

steatite talc, *n.* a high-grade variety of TALC, the purest commercial form.

Stebinger drum see GALE ALIDADE.

S-tectonite see TECTONITE.

steinkern see MOULD.

Stephanian *n.* the top SERIES of the Upper CARBONIFEROUS in Britain and western Europe.

steppe *n.* a broad, treeless, generally flat area of an arid region on which scattered bushes and short-lived grasses grow and furnish scant pasturage. Although by some definitions many deserts of North America would be steppes, the term is usually restricted to the semi-arid mid-latitudes of southeast Europe and Asia.

stereozone *n*. a zone where the septa (walls) are so thickened as to be contiguous around the outer margin of some RUGOSE CORALLITES.

stibnite *n*. a steel-grey orthorhombic mineral, Sb_2S_3, the main ore of antimony. It is opaque, with a bright metallic lustre, and occurs as prismatic or acicular crystals. Stibnite is found in low-temperature hydrothermal veins with silver, lead and mercury minerals.

Stigmaria *n*. root stocks of lycopod trees, e.g. *Sigillaria* and *Lepidodendron* (see SCALE TREES). The root system of these trees consisted of horizontally spreading bifurcating main trunk roots without a tap root. The real rootlets sprang directly from the sides of these trunk roots and extended radially to distances of several metres. Examples of Stigmaria are often found in the FIRECLAYS under coal beds and sometimes in the COAL itself.

stilbite *n*. a white, grey or reddish-brown monoclinic ZEOLITE mineral, $NaCa_2Al_5Si_{13}O_{36} \cdot 14H_2O$, occurring usually in sheaf-like aggregates. Stilbite is a hydrothermal mineral found in cavities of basaltic rocks and is associated with calcite and other zeolites.

stipe *n*. one branch of a GRAPTOLITE rhabdosome.

stishovite *n*. a high-pressure, dense polymorph of quartz, SiO_2, that is produced under static conditions at pressures greater than 100 kb (10 GPa). Stishovite is found only in shock-metamorphosed quartz-bearing rocks; its presence provides a criterion for METEORITE impact.

stock *n*. an intrusive body of deep-seated igneous rock, usually discordant and resembling a BATHOLITH except for its size. It covers less than 100 square kilometres in surface exposure. Contacts with the country rock generally dip outwards and may vary from a steep to a shallow angle. A stock is more or less elliptical or circular in cross-section. See INTRUSION.

stockwork *n*. a MINERAL DEPOSIT in the form of a branching network of veinlets associated with a PLUTONIC INTRUSION, particularly one of ACID to INTERMEDIATE composition. There may be extensive HYDROTHERMAL ALTERATION of the host rock.

Stokes' law *n*. an expression for the rate of settling or rising of spherical particles in a fluid, e.g. gravitational separation of crystals or settling of sediments in water.

$$v = \frac{2r^2 g \Delta\rho}{9\eta}$$

where v is the velocity of the crystals, g the acceleration caused by gravity, $\Delta\rho$ the difference of densities between crystal and melt and η the coefficient of viscosity of the melt.

stolon *n*. an internal rod-like system running from the SICULA (first-formed structure) through all the branches of a dendroid GRAPTOLITE. See Fig. 42.

stolotheca see GRAPTOLITE and Fig. 42.

stomach stones see GASTROLITH.

stone circle see SORTED 2.

stone field see BLOCK FIELD.

stone kernel see MOULD.

stone net, stone polygon see PATTERNED GROUND.

stone ring see PATTERNED GROUND.

stone stream see BLOCK STREAM.

stone stripe see PATTERNED GROUND.

stony iron meteorites or **siderolites** *n*. one of three major divisions of METEORITE types. Stony irons contain substantial amounts of silicate minerals (olivine and pyroxene) and metal (Fe and Ni) intermixed, and represent an intermediate type between stony and IRON METEORITES. In some stony irons, the Ni-Fe is a coherent mass containing discrete silicate minerals. Others are similar to stony meteorites but contain large quantities (up to 40 per cent) of interspersed metal. See STONY METEORITE. See also PALLASITE, MESOSIDERITE.

stony meteorite n. one of three major divisions of METEORITE types. Stony meteorites consist largely of silicate minerals, olivine, pyroxene and plagioclase with some Ni-Fe and Fe-S, and are the most abundant of meteorites seen to fall (more than 80 per cent). This group is divided into two sub-groups: CHONDRITES (containing CHONDRULES) and ACHONDRITES (lacking chondrules).

stoop-and-room n. a Scottish term for a method of mining in which the roof of the excavation is supported by pillars of rock (the *stoops* or *stoups*); the space where the rock has been removed is the *room*.

stoping n. **I.** a process by means of which a body of MAGMA intruding upwards through the Earth's crust makes room for itself by displacing the COUNTRY ROCK. The country rock is enveloped or sinks into the intruding magma where it may be preserved en masse, assimilated or remain as XENOLITHS. Where the country rock is broken into a large number of individual fragments the process is referred to as *piecemeal stoping*. See CAULDRON SUBSIDENCE.

2. in ore mining, any excavation made to remove the ore made accessible by shafts or ADITS.

stoss adj. the side of a hill facing the direction from which a GLACIER moves. The *stoss side* of hills (also called *stoss-seite*) is most exposed to the glacier's abrasive action, so will have rounded edges and gentle slopes. Compare LEE.

stoss-and-lee topography n. a grouping of hills or prominences in a glaciated area with gradual slopes on the STOSS side and sharper slopes with a rougher profile on the LEE side. Compare CRAG AND TAIL.

stoss-seite see STOSS.

straight extinction see EXTINCTION ANGLE

strain or **deformation** n. the change in the shape or volume of a body in response to STRESS or a change in relative position of the particles of a substance.

strain ellipsoid or **deformation ellipsoid** n. an imaginary distorted sphere to represent STRAIN in rock. Deformation, or STRAIN, can be conceptualised by visualising the change in shape of an imaginary sphere in the rocks. Such a sphere, in a body of granite, would become deformed into an oblate spheroid if the granite were compressed from top to bottom. The most general solid resulting from the deformation is an ellipsoid called the strain ellipsoid. The sphere is considered to have unit radius, and the ellipsoid has principal semi-axes of length proportional to the magnitude of the principal strains. See STRESS ELLIPSOID.

strain hardening n. the gradual increase in the STRESS required to produce further deformation after the ELASTIC LIMIT has been exceeded. It is a process that occurs at low temperatures and is observed in the cold working of metals.

strain-slip cleavage n. a structure found in low-grade semi-pelitic regionally metamorphosed rocks in which mechanical failure under STRESS leads to the development of closely spaced SHEAR planes a few millimetres apart.

strata-bound ore body n. a MINERAL DEPOSIT of any form (CONCORDANT or

DISCORDANT) that is restricted to a single stratigraphical unit (see GEOLOGICAL COLUMN). Compare STRATIFORM.

stratification *n.* a bedded or layered arrangement of materials. The term is used especially in reference to SEDIMENTARY ROCK. Stratification can also be seen in LAVA FLOWS, METAMORPHIC ROCKS, masses of snow, FIRN or ice, while the waters of a lake may be arranged in layers of differing temperature or density.

stratified drift see GLACIAL DRIFT.

stratiform *adj.* having the configuration of a layer or bed but not necessarily being BEDDED.

stratigraphical correlation *n.* the procedure by which the mutual correspondence of STRATIGRAPHICAL UNITS in two or more removed locations is shown. It is based on FOSSIL content, geological age, lithographic features or some other property. The term usually implies the identification of rocks of equivalent age in different places.

stratigraphical geology see STRATIGRAPHY.

stratigraphical range, range or **geological range** *n.* the duration and distribution of any taxonomic group of organisms through geological time, as indicated by their presence in strata of known geological age.

stratigraphical trap see OIL TRAP.

stratigraphical unit or **rock stratigraphical unit** *n.* a stratum or body of strata, recognisable as a unit, that may be used for mapping, description or correlation; it does not constitute a time-rock unit (see below). A *lithostratigraphical unit* is a body of rock strata with certain unifying lithological features. Although it may be sedimentary, igneous or metamorphic, it must meet the critical requirement of an appreciable degree of overall homogeneity. A lithostratigraphical unit has a

three-part designation consisting of a locality name with the lithology and unit term, e.g. Appin Quartzite Formation, Snowdon Volcanic Group. Lithostratigraphical units, in descending order, are: *supergroup*, GROUP, FORMATION, MEMBER, BED. A body of strata that is identified by particular fossil content is a *biostratigraphical unit*. BIOZONE (*biostratigraphical zone*) is the term used for any biostratigraphical unit. A *chronostratigraphical unit*, also called *chronolithological unit*, *time-rock unit* or *chronolith*, is one that was formed during a defined interval of geological time; it represents all, and only, those rocks formed during a designated time span of geological history. Chronostratigraphical units are, in descending order: *eonothem*, ERATHEM, SYSTEM, SERIES, STAGE, CHRONOZONE.

stratigraphy *n.* **1.** also called **stratigraphical geology**, the branch of GEOLOGY concerned with all characteristics and attributes of rocks as they are in strata, and the interpretation of strata in terms of derivation and geological background. **2.** that aspect of the GEOLOGY of an area that pertains to the character of its stratified rock.

stratotype *n.* the original or later designated type representative of a named STRATIGRAPHICAL UNIT or of a stratigraphical boundary, established as a point in a distinctive sequence of rock strata. The stratotype is the standard for the definition and recognition of a particular stratigraphic unit or boundary. See also TYPE SECTION, BOUNDARY STRATOTYPE.

strato-volcano *n.* a composite cone. See VOLCANIC CONE.

stratum *n.* (*pl.* **strata**) a defined layer of SEDIMENTARY ROCK that is usually separable from other layers above and below; a BED. Compare LAMINA.

streak *n*. the mark left by a mineral when it is rubbed on a piece of unglazed porcelain (*streak plate*). The colour of the streak is important in mineral identification as it often differs markedly from the colour of the mineral itself, for example GALENA, although silvery in appearance, gives a black streak.

stream capture, capture or **piracy** *n*. the diversion of one stream into another as a result of the erosional encroachment of one upon the drainage of the second. The stream whose waters are intercepted is known as the *beheaded stream*. There are three main ways of stream capture: ABSTRACTION, HEADWARD EROSION and subterranean diversion. *Abstraction* occurs when two streams join as a result of the EROSION of the divide between them. It takes place in ravines and gullies at the higher end of drainage lines. Stream capture by *headward erosion* takes place where two streams oppose each other on a ridge; one lowers its valley in the upper areas at a faster rate than the other, because there is a steeper gradient at its head or because it cuts through softer bedrock. *Subterranean capture* occurs where water from a higher stream percolates down to a lower stream through soluble rock and subsequently forms a diversion tunnel.

stream flow *n*. above-ground water flow via watercourses.

streamline flow see LAMINAR FLOW.

stream load *n*. all the material that is transported by a stream. The DISSOLVED LOAD consists of material in ionic solution; the ions are part of the fluid and move with it. They are derived from groundwater that has filtered through weathering soil and rocks. The *suspended load* or *wash load* comprises clay, silt and sand particles so fine that they remain in SUSPENSION almost indefinitely. Material that is too coarse to be lifted by the stream water is bounced or rolled along the bottom; this is the BED LOAD.

stress *n*. the force per unit area that acts on or within a body. It is measured in newtons per square metre or pascals ($1Nm^{-2} = 1Pa$). For *homogeneous stress*, the stress has the same magnitude and direction at every point in the body.

At any point within a homogeneous stress field there are three planes at right angles to each other on which the SHEAR STRESSES are zero. These are the *principal planes of stress*. The normals to these planes are the *principal axes of stress* along which act the *principal stresses*. The state of stress at any point within the body can be defined with reference to the three principal stresses.

The *effective stress* acting on a rock is a result of the interaction of the stress state within the rock with the hydrostatic stress exerted by the pore fluid (*Terzaghi's principle*).

stress difference *n*. the algebraic difference between the greatest and least principal STRESSES.

stress ellipsoid *n*. a geometric representation of a state of STRESS as defined by three mutually perpendicular principal stresses and their magnitudes.

stress relaxation *n*. the decrease with time in the STRESS required to sustain STRAIN. This kind of behaviour is shown by materials that have both viscous and ELASTIC characteristics, for example many waxes and pitches. See also VISCOUS FLOW.

strewn field *n*. an area of the Earth's surface within which TEKTITES are found. These include the central European field, where the tektites are thought to originate in the Ries crater in Bavaria; the North American field, possibly originating from a site in the Atlantic; the Ivory Coast field, where the source may be the Lake Bosumtwi crater in Ghana; and the immense Australasian strewn field covering much of southeast Asia, Australia and part of the Indian Ocean, where the tektites are from an unknown source.

striation *n.* one of many thin lines or scratches, generally parallel, incised on a rock by some geological agent such as a glacier or a stream.

strike 1. *n.* the direction taken by a structural surface such as a FAULT or BEDDING PLANE as it intersects the horizontal; it is the compass direction of the horizontal line in an inclined plane. Compare TREND, TRACE.
2. *v.* to be aligned in a direction at right angles to the direction of DIP.

strike fault *n.* a FAULT that strikes essentially parallel to the STRIKE of the adjacent rocks. Compare DIP FAULT, OBLIQUE FAULT.

strike separation *n.* the distance between two previously adjacent BEDS on either side of a FAULT surface, measured parallel to the STRIKE of the fault. Compare DIP SEPARATION.

strike shift *n.* the relative displacement of rock units parallel to the STRIKE of a FAULT but beyond the fault zone itself. Compare STRIKE SLIP.

strike slip *n.* the component of movement parallel with the STRIKE of a fault. See FAULT, Fig. 31. Compare DIP SLIP, STRIKE SEPARATION, STRIKE SHIFT.

strike-slip fault see FAULT.

stripe see PATTERNED GROUND.

striped ground see PATTERNED GROUND.

strip mining see OPENCAST MINING.

stromatoporoid *n.* any of a group of sessile calcareous benthic marine organisms considered to be extinct members of the phylum Porifera. They were sponge-like colonial structures of calcium carbonate, tabular, dome-like or bulbous in form. The external view shows a stomatal texture while the internal structure is that of a dense laminated mass. Stromatoporoids were especially abundant in reefs of the Ordovician-Devonian. Range, Cambrian to Cretaceous.

stromatolite *n.* laminated calcareous sedimentary formations produced by CYANOBACTERIA. Living stromatolites (Hamelin Pool, Shark Bay, Australia) and FOSSIL forms are in the shape of stony cushions or massive columns. Fossilised (silicified) stromatolites dating well back into the Precambrian are found in the Gunflint Chert of Lake Superior, in CHERTS of Africa and Australia, and in many calcareous sediments. The oldest known stromatolites are from the Warrawoona Group of Western Australia, dated at 3550 Ma. Until the presence of organic material in these deposits was ascertained, the forms embodied therein were called *cryptozoa*. Compare MICROBIALITE.

Strombolian-type eruption see VOLCANIC ERUPTION, TYPES OF.

strontianite *n.* a pale green, yellow or white mineral, $SrCO_3$. It occurs in VEINS, NODULES and GEODES in LIMESTONES. It is also found in association with lead MINERALISATION, as at Strontian in Scotland, where it occurs in veins with GALENA and BARITE.

structural *adj.* of or relating to rock deformation or to features that result from it.

structural feature *n.* a feature produced by the displacement or deformation of ROCKS, such as a FAULT or FOLD.

structural geology *n.* the branch of GEOLOGY concerned with the description, spatial representation and analysis of structural features, ranging from microscopic to the structure of regions. As such, it includes studies of the forces that produce rock deformation and of the origin and distribution of these forces. Structural features may be primary, i.e. those acquired in the genesis of a rock mass (e.g. horizontal layering), or secondary, i.e. resulting from later deformation of primary structures (e.g. folding or fracturing). Compare TECTONICS.

structural petrology or **petrofabric analysis** *n.* the study of the geometrical relationship between the FABRIC of a rock and major and minor geological STRUCTURES by which the deformational history may be determined. See also FABRIC DIAGRAM.

structural relief *n.* the difference in elevation between the lowest and highest points of a BED or stratigraphical horizon in a specified region.

structural trap see OIL TRAP.

structure *n.* **1.** the attitude and positions of rock masses of an area; the sum total of structural features resulting from processes such as folding and faulting. **2.** (*geomorphology*) a general term for underlying rocks of a landscape. **3.** a feature of a rock that embodies mutual relations of aggregates of grains such as bedding or foliation. Compare TEXTURE. **4.** (*petroleum geology*) any physical arrangement of rocks that may hold oil or gas accumulations. **5.** the form assumed by a mineral as, for example, cruciform.

structure contours *n.* lines of equal height, above or below a reference level, that show the three-dimensional form of a geological surface, e.g. between FORMATIONS, FAULT surfaces.

stylolite *n.* an irregular surface within a BED, usually of carbonate rocks,

characterised by pits and tooth-like projections on one side that fit into counterpart negatives on the other. *Stylolite seams* are formed of material less soluble than limestone, such as clay or organic matter. A diagenetic formation of stylolites is indicated by their cross-cutting of allochems, such as FOSSILS, in which the lower or upper part has apparently dissolved away. The generally parallel disposition of the stylolite seams implies that dissolution was caused by an interaction between overburden pressure and pore waters of the same type that forms the pressure-solution surfaces in some quartzarenites.

sub- *prefix* **1.** under or beneath. **2.** indicating a condition that is slightly outside the lower limit of a given condition, e.g. subtropical, subangular. **3.** indicating a division of a larger category, e.g. subgenus, subgroup.

subalkaline *adj.* (of igneous rocks) belonging to the tholeiitic and calc-alkaline series, or suites, containing no alkali minerals other than FELDSPARS.

subalkaline basalt *n.* a group of BASALTS that do not contain normative NEPHELINE. See NORM.

subaqueous gliding *n.* SOLIFLUCTION or SLUMP under water.

subarkose *n.* **1.** a SANDSTONE that contains between 5 and 25 per cent FELDSPAR. Compare ARKOSE. **2.** a SANDSTONE that is intermediate in composition between ARKOSE and pure quartz sandstone. See ARKOSIC SANDSTONE, FELDSPATHIC.

subautomorphic see SUBHEDRAL.

sub-bituminous *n.* the RANK between woody LIGNITE and BITUMINOUS COAL. It is characterised by a higher carbon content than lignite and a lower moisture content. The calorific value ranges between about 10 and 14 MJkg^{-1}.

subduction *n*. the movement of one crustal plate (lithospheric plate) under another so that the descending plate is 'consumed'. Subduction refers to the process, not the site. See LITHOSPHERE, SUBDUCTION ZONE.

subduction zone *n*. the boundary between two converging lithospheric plates where a slab of oceanic LITHO-SPHERE descends at an angle into the mantle. Where two oceanic plates collide, SUBDUCTION is associated with the development of a volcanic ISLAND ARC; where an oceanic plate is in collision with a continental plate, a CORDILLERA mountain belt is formed, with associated volcanic activity. In both cases an OCEANIC TRENCH occurs above the locus where the lithosphere bends downwards. The angle of dip of subduction zones varies greatly and appears to be related to the rate of CONVERGENCE: the slower the convergence, the steeper the dip. The stresses set up by subduction lead to the generation of EARTHQUAKES. It is by plotting the earthquake foci that subduction zones can be detected. The deepest known earthquakes are associated with subduction zones. Subduction zones are one of three kinds of boundary of lithospheric plates. The other two boundaries are MID-OCEANIC RIDGES and TRANSFORM FAULTS. See also PLATE TECTONICS.

subglacial or **infraglacial** *adj*. relating to the processes occurring and product formed at the base of a GLACIER or ice sheet.

subgroup see GROUP.

subhedral, subautomorphic or **hypidio-morphic** *adj*. (of a mineral grain) bounded only partially by geometric outlines. The crystal form is recognis-able but imperfect. Compare EUHEDRAL, ANHEDRAL.

sublimate *n*. a substance that can pass directly from a solid state to a vapour, and vice versa, without forming a liquid.

submarine canyon *n*. a V-shaped submarine valley, resembling a land CANYON, that has been cut by streams. It is characterised by high, steep walls and an irregular floor that slopes continuously outwards; some are found opposite the mouths of large rivers. All submarine canyons head on the continental shelf and most debouch at the base of the CONTINENTAL SLOPE. ALLUVIAL FANS are often found at the mouths. These valleys vary consider-ably in wall height, length and slope of the valley floor; their average length is about 55.5 kilometres and average wall height about 915 metres, but the Great Bahama Canyon has a wall height of 4285 metres (compare the Grand Canyon – 1700 metres). Several mechanisms have been proposed for the origin and maintenance of submarine canyons. There is also some question about whether these formations all share a common origin. At present, most geologists seem to subscribe to the following concepts:
(a) most canyons have been cut by the action (either subaerial or submarine) of downslope sediment movement;
(b) many of those canyons that were originally formed by subaerial erosion have since been drowned and are now actually kept free by submarine processes such as TURBIDITES, slumping or gravity flow. See SUBMARINE FANS.

submarine delta see SUBMARINE FANS.

submarine eruption *n*. an eruption from a volcanic vent on the oceanic or sea floor. Most of these eruptions are of basaltic lavas from fissure vents along the MID-OCEANIC RIDGES or from central vents on SEAMOUNTS. Pillow structures are characteristic (see PILLOW LAVAS) and associated glassy fragmentary deposits. Small amounts of submarine silicic lava have been erupted at oceanic islands. Lava can be erupted under water without explosive fragmentation, partly because of the *Leidenfrost effect*, which results in the formation of a film of water vapour that acts as a thermal barrier between the hot lava and the relatively cold water. The hydrostatic pressure of the overlying water is, however, the crucial factor as it inhibits the rapid exsolution of the dissolved fluids that leads to explosive PHREATOMAGMATIC activity when lava is erupted into shallow water. See also VOLCANIC ERUPTION, VOLCANIC ERUPTION, TYPES OF.

submarine fans *n.* fan- or cone-shaped submarine features that are accumulations of terrigenous sediment. They are also called *abyssal cones*, *deep-sea cones*, *submarine deltas*, etc. These fans are found offshore from most of the world's great rivers and extend downwards to abyssal depths. Periodic TURBIDITES that flush sediments through submarine canyons are responsible for their build-up.

submergence *n.* a rise of the water level relative to the land so that areas that were formerly dry land become inundated; it is the result either of the sinking of the land or of a net rise in sea level. Compare EMERGENCE.

subsequent *adj.* (of geological or topographical features) the development of which is governed by differences in erodibility of the underlying rocks. For example, a *subsequent valley* or a *subsequent stream* is one that has shifted from its original CONSEQUENT course to belts of more readily erodible rock.

subsidence *n.* **1.** the downward settling of material with little horizontal movement. The most common cause is the slow removal, either by geological processes or human activity, of material from beneath the subsiding mass. Processes such as compaction, solution, withdrawal of fluid lava from beneath a solid crust and subsurface mining can all be responsible for subsidence. KETTLES and sinkholes of limestone regions are subsidence landforms.
2. the downwarping of a large part of the Earth's crust relative to the surroundings, e.g. the formation of a RIFT VALLEY. See also CAULDRON SUBSIDENCE.

subsoil see SOIL PROFILE.

subsolidus *n.* a chemical system that is below its melting point and in which reactions may take place in the solid state.

substage *n.* **1.** a subdivision of a STAGE that includes the rocks formed during a sub-age of geological time.
2. the subdivision of a glacial stage during which there occurred a secondary fluctuation in glacial advance and retreat.

subterranean capture see STREAM CAPTURE.

subterranean diversion see STREAM CAPTURE.

subterranean water see GROUNDWATER.

suevite *n.* a BRECCIA formed by SHOCK METAMORPHISM in which the angular fragments are set in a matrix of glass (IMPACTITE GLASS).

sulcus *n.* **1.** a groove in the VENTER of an AMMONOID, see Fig. 1.
2. a concave fold in the anterior margin of the BRACHIAL VALVE that fits into a corresponding upfold in the PEDICLE VALVE. See BRACHIOPOD, Fig. 10.

sulfur see SULPHUR.

sulphate *n.* a chemical compound containing the sulphate radical $(SO_4)^{-2}$.

sulphide *n.* one of a group of compounds in which sulphur is combined with one or more metals; a mineral example is CINNABAR, HgS.

sulphide enrichment see SECONDARY ENRICHMENT.

sulphide zone see SECONDARY ENRICHMENT.

sulphur or **sulfur** *n.* an orthorhombic mineral, the native element sulphur (S). It occurs in bright yellow crystals and granular aggregates, often at or near hot springs and FUMAROLES in areas of active volcanicity. Sulphur is associated with sedimentary deposits of the EVAPORITE type and with oil-bearing deposits. See SULPHUR BACTERIA.

sulphur bacteria *n.* aerobic and anaerobic eubacteria, belonging to the Proteobacteria, that carry out the oxidation or reduction of sulphur. Colourless sulphur bacteria (of the β Proteobacteria) are aerobic and oxidise sulphur to sulphate. Some purple sulphur bacteria (of the γ Proteobacteria) are strict photosynthetic anaerobes and oxidise hydrogen sulphide to elemental sulphur, depositing the sulphur as granules within their cells. Compare IRON BACTERIA. See BACTERIOGENIC.

sulphur-mud pool see MUDPOT.

summit accordance *n.* the equal or nearly equal elevation of mountain summits or hilltops over a certain region. Such coincidence of elevation is generally thought to suggest that the summits are the preserved high features of a former erosion plain. See also ACCORDANT SUMMIT LEVEL, PENEPLAIN.

summit eruption see VOLCANIC ERUPTION, SITES OF.

sunstone *n.* an AVENTURINE feldspar, usually a translucent variety of OLIGOCLASE, that emits a golden shimmer or reddish glow. Its appearance is the result of minute, plate-like inclusions of HEMATITE oriented parallel to one another. Compare MOONSTONE.

super- *prefix meaning* above or over, either in position or condition.

supergene *n.* a term applied to ores or ore minerals formed by processes that almost always involve water, with or without dissolved material, descending from the surface. The word connotes an origin from above. Typical supergene processes are hydration, solution, oxidation and deposition from solution. Compare HYPOGENE.

supergene enrichment see SECONDARY ENRICHMENT.

supergroup see GROUP, STRATIGRAPHICAL UNIT.

superimposed drainage *n.* a drainage pattern inherited from an overlying structural system and not related to the structure of the rocks now exposed. For example, the river system may have developed on strata above an UNCONFORMITY and maintained the original pattern, perhaps with slight modification, as it cut down into the rocks below the unconformity.

superposition see LAW OF SUPERPOSITION.

superstructure *n.* the upper structural layer of an OROGENIC BELT that is subjected to relatively shallow or near-surface deformation processes, as distinct from an underlying and more profoundly deformed INFRASTRUCTURE.

supervolcano *n.* one of a small number of extremely large volcanic centres, including Lake Taupo caldera (New Zealand), Yellowstone (Wyoming, USA) and Lake Toba (Sumatra, Indonesia). There is clear evidence that all these centres have given rise to enormous volcanic eruptions in prehistoric times, on a scale much greater than any recorded in human history. There is reason to believe that further major activity at any of these centres is possible and that the environmental consequences of such an eruption would constitute a global catastrophy. The injection of vast amounts of gas and volcanic dust into the stratosphere would lead to lowering of average temperatures worldwide for several years, with consequent crop failures and famine.

surface of no strain or **neutral surface** *n.* a surface along which the original configuration remains unchanged after deformation of the body in which the surface is contained. For example, if a sheet is bent, the convex side is subject to tension whereas the concave side is subject to compression; between the two loci of points there is a surface of no strain.

surface wave see SEISMIC WAVES.

surging glacier or **glacier surge** *n.* a GLACIER that exhibits a period (usually brief) of very rapid flow, perhaps 100 or more times its normal velocity. Usually, surging glaciers have been stable or even retreating beforehand. The phenomenon occurs when ice that would normally move downglacier by normal flow mechanisms accumulates in an ice-reservoir area and surges downwards when a critical threshold is reached. It appears that most glaciers that have surged have done so in cycles of from 15 to 100 years.

susceptibility see MAGNETIC SUSCEPTIBILITY.

suspect terrane *n.* see TERRANE the origins of which are uncertain.

suspension *n.* a type of SEDIMENT TRANSPORT in which sediment particles are held in the surrounding water by the upward component of eddy currents associated with turbulent flow.

suspension current see TURBIDITES.

suspended load see STREAM LOAD.

suture see GEOSUTURE.

suture line *n.* the trace of each dividing wall (SEPTUM) on the inner wall of a CEPHALOPOD shell. It can be seen only on an internal MOULD. NAUTILOIDS and AMMONOIDS have characteristic suture lines, see Figs. 62, 1(b).

swale *n.* a long, narrow depression between BEACH RIDGES.

swallow hole *n.* a depression, closed or open, into which all or part of a river or stream disappears underground. See DOLINE.

swamp *n.* a type of wetland characterised by mineral soils with poor drainage and by plant life dominated by trees such as gums, willows and maples in temperate swamps, and by mangroves and palms in tropical regions. Swamps can develop in any area with poor drainage and a water supply sufficient to keep the ground waterlogged. Floodplains, abandoned river channels and oxbows all have conditions that support swamps and marshes. Salt swamps are formed when flat intertidal land is subjected to the flooding and draining of sea water. In tropical and subtropical regions, regularly flooded protected areas develop mangrove swamps. The mineral supply in swamp water is sufficient to stimulate decay of organisms and prevent accumulation of organic materials. If the rate of oxygen supply becomes too low, however, decay will be incomplete and organic matter will accumulate to form PEAT. A slow, continuous rise of the WATER TABLE and protection of the swamp area from major sea inundations help to promote thick peat deposits and the consequent formation of COAL seams. Compare BOG, MARSH.

swamp theory see IN-SITU THEORY.

swash see BREAKER.

swash mark see BREAKER.

S wave see SEISMIC WAVES.

syenite *n.* a coarse- to medium-grained

INTERMEDIATE igneous rock composed mainly of ALKALI FELDSPAR, such as microcline and orthoclase, with small amounts of plagioclase, hornblende, and/or biotite and little or no quartz. Syenite is the intrusive equivalent of TRACHYTE; with an increase in quartz, it grades into GRANITE.

syenodiorite see MONZODIORITE.

syenogabbro see MONZOGABBRO.

symmetrical fold see FOLD.

symmetry axis see CRYSTAL SYMMETRY.

symmetry elements see CRYSTAL SYMMETRY.

sympatry *n.* species occupying the same area. *Sympatric evolution* describes the evolution of a new species from an ancestral TAXON within the same area.

symplectic intergrowth or **symplectite** *n.* an intimate intergrowth of two minerals. This includes GRAPHIC and vermicular (worm-like) textures. MYRMEKITE is a variety of symplectite. Compare EPITAXY.

symplesiomorphies see CLADISTICS.

syn- *prefix denoting* with, at the same time.

synapomorphies see CLADISTICS.

synchronous *adj.* formed or occurring at the same time; contemporary, such as a rock surface on which every point is of the same geological age.

synclinal axis see SYNCLINE.

syncline *n.* a generally U-shaped fold or structure in stratified rock containing stratigraphically younger rocks towards the centre of curvature unless it has been overturned. A *synclinorium* is a large compound syncline composed of several minor folds. See also GEOSYNCLINE. The *synclinal axis* is that which, if displaced parallel to itself, generates the form of a syncline. A mountain the underlying structure of which is that of a syncline is called a *synclinal mountain*. Compare ANTICLINE. See also SYNFORM, FOLD.

synclinorium see SYNCLINE.

synclise *n.* an approximately circular depression of great areal extent (at least 1000 kilometres across) caused by crustal flexure in intraplate regions. See also ANTECLISE.

synform *n.* a U-shaped fold in strata of unknown stratigraphic sequence. Compare SYNCLINE, ANTIFORM.

syngenetic *adj.* **1.** (of mineral deposits) formed at the same time as the enclosing rock and by the same process. Compare EPIGENETIC.
2. (of small markings in sediments, such as ridges) formed at the same time as the deposition of the sediment.

syntectite *n.* a rock formed by SYNTEXIS. See also ANATEXITE.

syntexis *n.* **1.** ASSIMILATION.
2. the generation of MAGMA by the melting of two or more rock types. Compare ANATEXIS.

synthem *n.* an UNCONFORMITY-bounded STRATIGRAPHICAL UNIT.

synthetic fault *n.* a minor normal FAULT that is oriented in the same direction as the major fault with which it is associated. Compare ANTITHETIC FAULT.

system *n.* **1.** a major unit of rank in CHRONOSTRATIGRAPHY that serves as a worldwide reference unit. A system is based on a type area and includes all the rocks formed during a period of geological time. In rank, system is above SERIES and below ERATHEM.
2. a group of related natural features, e.g. a FAULT system.
3. a CRYSTAL SYSTEM.
4. a chemical system; see PHASE DIAGRAM.

Tt

tabulae *n.* (*sing.* **tabula**) flat or curved horizontal plates, usually in the central part of a CORALLITE, that represent successive floors of the CALICE. Towards the margin of the corallite the tabulae normally terminate in a zone of DISSEPIMENTS or a STEREOZONE.

tabular *adj.* **1.** (of a form) having two dimensions much greater than the third. For example, any igneous dyke is a tabular intrusion. Compare MASSIVE.
2. (of the shape of a sedimentary body) having a ratio of width to thickness greater than 50:1 but less than 1000:1. Compare BLANKET DEPOSIT.
3. (of a crystal shape) having one dimension markedly smaller than the other two. Compare PRISMATIC.
4. (of a metamorphic texture) characterised by a large proportion of tabular grains disposed in approximately parallel orientation.

tabular cross-bedding see CROSS-BEDDING.

Tabulata *n.* an extinct group of COLONIAL CORALS characterised by slender CORAL-LITES with prominent TABULAE and reduced or absent septa (dividing-walls). The corallite walls may be perforated by mural pores. In some advanced forms the corallites have no walls but are embedded in a complex mass of tissue, the *coenenchyme*. The tabulate corals are not important stratigraphically, although they were quite important in reef formation during the SILURIAN and DEVONIAN. Range, Lower ORDOVICIAN to PERMIAN. See Fig. 87 below.

tachylyte *n.* a black volcanic GLASS of basaltic composition. The colour is caused by the presence of minute crystallites of pyroxene. Compare SIDEROMELANE. See also PSEUDOTACHYLYTE.

Taconian (Taconic) orogeny *n.* a Mid-Ordovician OROGENY taking its name from the Taconic Range of the north-eastern United States. This disturbance was caused by the thrusting of major ophiolite NAPPES on to the Laurentian margin and culminated in a chain of

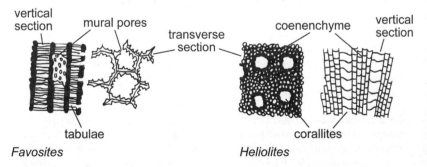

Fig. 87. **Tabulata**. Transverse and vertical sections of two tabulate corals

fold mountains extending from Newfoundland through the Canadian Maritime Provinces and New England and reaching as far south as Alabama. In northwest Europe, the CALEDONIAN OROGENY corresponds to the Taconian plus ACADIAN OROGENIES of North America. The Taconian may be considered an early phase of the Caledonian. It is now generally thought to have consisted of several pulses, extending from the Mid-Ordovician to Early Silurian.

taconite *n*. an iron-rich CHERT used as a low-grade iron ore. It contains about 25 per cent iron, mainly as MAGNETITE, and is widely distributed in the Precambrian crust. After crushing, the ore is concentrated by magnetic separation. See GREENSTONE BELTS.

tactite *n*. an American term for iron-rich SKARN found in the contact zones of calc-alkaline PLUTONS.

taenite *n*. an iron-nickel alloy containing about 27 to 65 per cent nickel, occurring in association with KAMACITE in IRON METEORITES. See WIDMANSTÄTTEN PATTERNS.

talc *n*. a very soft mineral, $Mg_3Si_4O_{10}(OH)_2$, with a hardness of 1 on the MOHS SCALE, that feels soapy when handled. It occasionally occurs as pale apple green, grey or white triclinic crystal but more commonly is compact, forming foliated, fibrous or granular masses; in SOAPSTONE it is dark grey or dark green. It is an alteration product of magnesium silicates or ultramafic rocks and is also formed by metasomatism in impure dolomitic marbles. See STEATITE.

talus *n*. a heap of coarse DEBRIS, a result of WEATHERING (FROST ACTION), at the foot of a cliff. Compare SCREE. The slow downslope movement of talus or scree produces *talus creep*.

talus cone *n*. a steep-sided pile of rock fragments lying at the base of a cliff from which they have been derived. Talus cones are formed primarily by the movement of materials aided by gravity. See COLLUVIUM.

tantalite *n*. a black mineral, $(Fe,Mn)Ta_2O_6$, the principal ore of tantalum. It is isomorphous with COLUMBITE and occurs in pegmatites. Because of its resistance to corrosion, tantalum is used in surgery for skull plates and in special alloys.

tarn see CIRQUE.

tarnish *n*. a discoloured, sometimes iridescent, surface film on a mineral, resulting from oxidation or reaction with atmospheric sulphur.

tar pit or **pitch lake** *n*. an accumulation of natural BITUMEN that is exposed at the land surface and becomes a trap into which animals sink; the hard parts of the carcasses are preserved. Examples are La Brea tar pits, Los Angeles, California, and the Pitch Lake, Trinidad.

tar sand *n*. a sedimentary formation that contains a commercial store of ASPHALT; it may be an OIL SAND from which more volatile materials have escaped.

TAS *abbreviation for* total alkali/silica diagram. See IGENOUS ROCK, Appendix 3.

Tatarian *n*. former name for the top STAGE of the PERMIAN.

taxon (*pl.* **taxa**) *n*. a grouping of organisms. The hierarchy of taxa contains all groupings from SPECIES to kingdoms. Taxa form the basis of *taxonomy*, the classification of organisms in ordered and natural groups.

TD *abbrev. for* time-distance.

TD curve see TIME-DISTANCE CURVE.

TΔT analysis *n*. a method of detecting such lateral variations of velocity as might distort seismic interpretation when calculating the average velocity to a reflecting horizon. In the TΔT method, records are picked from shots distributed over such a wide area that the net dip of the reflecting surface can be reasonably assumed to be zero. Velocities calculated by the TΔT method have been shown to be of considerably lower magnitude than well-shooting velocities at all depths.

tectogene *n*. a narrow, elongate unit of downfolding of sialic crust believed to be related to mountain-building processes. See SIAL.

tectogenesis see OROGENY.

tectonic or **geotectonic** *adj*. relating to structures of, or forces associated with, TECTONICS.

tectonic axis see FABRIC AXIS.

tectonic breccia *n*. an aggregation of angular rock fragments that have formed as a result of tectonic movement; a CRUSH BRECCIA.

tectonic conglomerate see CRUSH CONGLOMERATE.

tectonic fabric see DEFORMATION FABRIC.

tectonic facies *n*. rocks that owe their present features chiefly to tectonic activity.

tectonics *n*. a branch of GEOLOGY that is closely related to, but not synonymous with, STRUCTURAL GEOLOGY. Whereas structural geology is concerned primarily with the geometry of rocks, tectonics deals not only with larger features of the Earth but also with the forces and movements that produced them. See also PLATE TECTONICS.

tectonism see DIASTROPHISM.

tectonite *n*. a deformed rock the fabric of which reflects the coordination of the componental movements responsible for the deformation. Two principal types of tectonite are referred to in the literature: B-tectonite and S-tectonite. In *B-tectonite*, frequently called *L-tectonite*, lineation is prominent. In an *S-tectonite*, the fabric is characterised by planar elements caused by deformation, e.g. foliation.

tectonosphere *n*. a gravitationally stable subcontinental zone above the mantle in which crustal or tectonic movements originate. The tectonosphere concept answers, at least indirectly, the geophysical question, based on heat-flow studies, of how a necessarily cooler subcontinental mantle can be in a condition of hydrostatic equilibrium with a hotter suboceanic mantle.

tectosilicates *n*. any of the three-dimensional framework SILICATES, such as QUARTZ, each tetrahedral group of which shares all its oxygen atoms with neighbouring groups, the ratio of silicon to oxygen being 1:2, (SiO_2).

teeth *n*. calcareous projections on the margin of one or both VALVES of a bivalved shell that interlock with depressions (sockets), allowing the valves to articulate. In BIVALVES, teeth occur on the HINGE PLATE of each valve. Those below the UMBO are known as *cardinal teeth*, those beyond are *lateral teeth*. See also DENTITION.

Only the articulate BRACHIOPODS have teeth: two on the HINGE LINE of the PEDICLE VALVE that fit into sockets in the BRACHIAL VALVE (Fig. 9).

tegmen *n*. (*pl*. **tegmina**) the domed, flexible cover on the oral surface of the CALYX of a CRINOID, forming the upper part of the THECA. The tegmen carries five primary food grooves leading to the central mouth and a lateral anus (see Fig. 21).

tektite *n*. a small rounded mass consisting almost entirely of GLASS. Prior to the Apollo missions, tektites were believed to be of lunar origin, but this was not supported by evidence from lunar studies. It is thought that tektites result from large meteoritic impacts on Earth; under the force of impact the heat generated instantaneously changes solid rock into drops of molten liquid that are widely sprayed for distances of hundreds or thousands of kilometres. The droplets solidify to glass prior to their deposition in STREWN FIELDS.

telluric current see EARTH CURRENT.

telluride *n*. a chemical compound that is a combination of tellurium with a metal. A mineral example is CALAVERITE.

teleost *n*. a ray-finned bony fish of the subclass Teleostei, a TAXON that includes all the modern bony fish. Teleosts evolved in the Late Jurassic and include such forms as eels, herrings, salmon and other common fishes.

telson *n*. the tail spine found in some ARTHROPODS, such as EURYPTERIDS.

temporary base level see BASE LEVEL.

tenacity *n*. a measure of how a mineral deforms. Terms used to describe this property include ELASTIC, where the mineral (or thin plates of it) can be bent but will straighten again, like MICA, and *flexible*, where the mineral can be bent but remains deformed, like TALC. A *malleable* mineral can be hammered into thin sheets; gold or copper show this behaviour. *Sectile* minerals can be cut with a knife but will break when hammered.

tennantite *n*. a dark, lead-grey mineral, $Cu_{12}As_4S_{13}$, the arsenic END MEMBER of the TETRAHEDRITE-tennantite series. It may contain some iron; the silver-bearing variety, freibergite, is a silver ore.

tenor *n*. the GRADE of an ore body.

tension *n*. the stress that tends to pull a solid body apart.

tension gashes *n*. small fractures, often filled with quartz or other minerals, formed during deformation of brittle rocks. They may be arranged EN ECHELON and related to SHEAR STRESS between beds during folding.

tension-saturated zone see CAPILLARY FRINGE.

tephra *n*. a collective term for all PYROCLASTIC MATERIAL ejected from a volcanic vent and deposited from pyroclastic flows, base surges or air fall. Compare TUFF.

tephrite *n*. a VOLCANIC ROCK, essentially composed of calcic PLAGIOCLASE, CLINOPYROXENE and NEPHELINE or another FELDSPATHOID, that forms more than 10 per cent of the FELSIC minerals. Unlike the BASANITES, they do not contain essential OLIVINE.

tephrochronology *n*. a stratigraphical technique that depends on the identification of TEPHRA layers produced by specific VOLCANIC ERUPTIONS, especially those that can be dated either historically or by radiometric DATING METHODS. By identifying and correlating tephra units, chronology can often be established over wide areas. *Tephrochronological* studies in Iceland, where the technique was developed, have proved effective for the correlation of geological and archaeological events.

terminal moraine see GLACIAL MORAINE.

terminus *n*. the extremity or outer margin of a GLACIER.

ternary system see PHASE DIAGRAM.

terra *n*. an upland region on the lunar surface characterised by a rough texture and a colour lighter than that of a MARE. These highland areas are heavily cratered and are thought to be part of the early crust. They are mainly composed of anorthosites and anorthositic gabbros.

terracotta *n*. a fired clay of a particular yellowish-red or brownish-red colour, used for ornamental objects and decorative work on the outside of buildings.

terrain see TERRANE.

terrain correction or **topographical correction** *n.* in geophysical surveys, a correction that accounts for the gravitational attraction of all material higher than the gravity station and also eliminates the effect of that material required to fill in below-station hollows in order to 'level up' the infinite slab hypothesised when applying the BOUGUER CORRECTION. See GRAVITY ANOMALY.

terrane or **terrain** *n.* a fault-bounded portion of the Earth's crust that has a distinct geology differing sharply from the areas adjacent to it. The term is particularly useful in regions of structural complexity where major Earth movements have brought together lithospheric plates or portions of them that were originally formed in remote parts of the Earth often in quite different circumstances. See ACCRETED TERRANE, CONTINENTAL ACCRETION, SUSPECT TERRANE.

terra rossa *n.* residual red clayey soil mantling LIMESTONE bedrock and extending down into open joints. It is a result of surface solution by descending GROUNDWATER.

terrestrial deposit or **continental deposit** *n.* **1.** a sedimentary deposit formed on land without the agency of water, e.g. sand DUNES.
2. any deposit formed on a land surface.

terrigenous deposits *n.* land-derived marine sediments that are carried down the CONTINENTAL SHELF by gravity and by geostrophic currents; their composition is influenced by the region of derivation. *Red mud* is reddish-brown SILT and CLAY that extends to ocean depths of about 2000 metres. The red colour, from the presence of ferric oxide, seems to be largely confined to the surface. Compare RED CLAY. *Blue* and *grey muds* range in depth from just below sea level to slightly greater than 5000 metres. At the surface, their colours are reddish to brownish, but beneath the surface the colours range from grey to blue. These muds are composed of about 15 per cent organic matter, a calcium oxide content ranging from 0 to 35 per cent and varying amounts of hydrogen sulphide; GLAUCONITE may or may not be present. *Green mud* is found at depths from less than 100 to more than 2500 metres, near the edge of the continental shelf. Apart from the presence of glauconite, from which it derives its colour, green mud is similar in composition to grey and blue muds. Its calcium carbonate content reaches about 50 per cent. Compare TERRESTRIAL DEPOSIT. See also OCEAN-BOTTOM DEPOSITS, BLACK MUD.

Tertiary *n.* a geological period that is part of the CENOZOIC.

Terzaghi's principle see STRESS.

teschenite *n.* an ALKALI GABBRO composed of OLIVINE, calcic PLAGIOCLASE (labrador-ite) CLINOPYROXENE, usually titaniferous, and ANALCIME. *Crinanite* is a variety of analcime gabbro or DOLERITE that contains less analcime than teschenite, although the name has sometimes been used as a synonym.

test *n.* **1.** the skeleton of an ECHINODERM, composed of a number of small, interlocking plates covered by skin. Each plate is a single CALCITE crystal.
2. the shell of a foraminiferid, see PROTOZOA.

test well see EXPLORATORY WELL.

Tethys *n.* the general name for the ocean that lay between the northern and southern continents of the Eastern Hemisphere from the Permian to the mid-Tertiary periods. Tethys began as an embayment of Panthalassa (see

PANGAEA). The Jurassic-Cretaceous ocean that opened along the northern margin of that embayment of Panthalassa is more correctly known as NEOTETHYS. It occupied the general region along the Alpine-Himalayan orogenic belt and separated LAURASIA and GONDWANA. The Alpine Himalayan continental collision obliterated all but the last vestige of Tethys – the Mediterranean Sea. At present, the continued movement of the African and Eurasian plates towards each other is shrinking the Mediterranean as well. See also PLATE TECTONICS. Compare PALAEOTETHYS.

tetracoralla *n.* an extinct CORAL with a four-fold asymmetry, i.e. septa arranged in quadrants. See RUGOSE CORALS.

tetragonal system see CRYSTAL SYSTEM.

tetrahedrite *n.* an isometric copper antimony sulphide mineral, $Cu_{12}Sb_4S_{13}$, isostructural with TENNANTITE in which arsenic is substituted for antimony; some iron and zinc may substitute for copper. It is usually found in veins associated with copper, silver, lead and zinc minerals.

textural maturity *n.* an end condition reached by detrital sediment, defined in terms of uniformity of particle size and degree of rounding. The textural maturity of a sediment is a result of certain modifying processes acting on it; it is a sequence of clay removal, sorting and rounding the rate of which depends on such factors as the current strength at the site of deposition and the mechanical resistance of the grains. Compare IMMATURE.

texture *n.* **1.** the general character or appearance of a rock as indicated by relationships between its component particles, specifically grain size and shape, degree of crystallinity and arrangement. Compare STRUCTURE.
2. the physical nature of a soil ex-

pressed in terms of relative proportions of sand, silt and clay.
3. TOPOGRAPHIC TEXTURE.

thalweg *n.* a line connecting the deepest parts of a river CHANNEL along its length.

thanatocoenosis or **death assemblage** *n.* an ASSEMBLAGE of fossils formed because they were brought together after death by sedimentary processes rather than because they had all shared the same habitat during life. Compare BIOCOENOSIS.

Thanetian *n.* the top STAGE of the PALAEOCENE.

theca *n.* (*pl* **thecae**) **1.** the cup that contained an individual GRAPTOLITE ZOOID. The shape of the thecae may be described as simple (a straight, short tube), hooked, lobate or sigmoidally curved (Fig. 41). Isolate thecae are widely separated. The aperture may have spines.
2. the skeleton of a CRINOID, which has two parts, the CALYX and the TEGMEN.

theodolite *n.* a precision surveying instrument equipped with a telescopic sight for determining horizontal and sometimes vertical angles.

theralite *n.* a NEPHELINE GABBRO composed of calcic PLAGIOCLASE (labradorite), titaniferous AUGITE and NEPHELINE, with variable amounts of OLIVINE.

thermal conductivity see CONDUCTIVITY.

thermal gradient see GEOTHERMAL GRADIENT.

thermal metamorphism *n.* another name for contact metamorphism, see METAMORPHISM.

thermal resistivity *n.* the reciprocal of thermal CONDUCTIVITY.

thermal spring *n.* a spring that discharges water heated by natural processes. If the water has a temperature higher than that of the human body (37°C), the spring is called a *hot* *spring*. When such a spring ejects steam and boiling water, it is called a GEYSER. Thermal springs are usually character-ised by the presence of travertine, silica and many other substances that are precipitated. Mineralised chimneys have been built up round submarine HYDROTHERMAL (hot-water) vents on the ocean crust along the MID-OCEANIC RIDGES by precipitation from solution when the hot mineral-laden water mixes with the cold ocean-bottom water. Many strange organisms, including giant tube worms, live in the warm waters around the vents. The superheated (350°C) fluids escaping from these vents are dark because of precipitated sulphides, giving rise to the name BLACK SMOKERS. The circulation of hydrothermal fluids through the oceanic crust in such environments is thought to be responsible for the formation of ore deposits such as those found in ophiolite complexes. See CYPRUS-TYPE DEPOSIT, also VOLCANIC-EXHALATIVE DEPOSITS, SEDIMENTARY-EXHALATIVE DEPOSITS.

Most hot springs occur in regions of recent VOLCANIC activity, where ground-water is heated by hot igneous rocks close to the surface. Thermal springs in other regions are thought to be caused by the circulation of meteoric water to considerable depths in the crust in areas of relatively high HEAT FLOW.

thermal stratification *n*. STRATIFICATION of lake waters because of temperature changes at different depths; this results in the formation of horizontal layers of different densities. See also DENSITY STRATIFICATION.

thermocline *n*. a layer of water with a sharper vertical gradient in tempera-ture than that of the layer below or above it; it originally referred to a LACUSTRINE environment. In the oceans, the boundaries of this sharp gradient layer usually form the division between the surface water and the deep and intermediate depth water below. These layers vary in depth, thickness, area and permanence. Thermoclines can be affected by almost all physical processes occurring in oceans and lakes and by above-surface meteorological processes.

thermohaline *n*. a slow vertical move-ment of sea water generated by density differences. Such movements are caused by temperature and salinity variations, which, in turn, induce convection and mixing.

thermoluminescence *n*. the property, exhibited by many minerals, of emitting light when heated.

thermoremanent magnetisation (TRM) see NATURAL REMANENT MAGNETISATION.

thin section *n*. a slice of rock or mineral that has been mechanically ground to a thickness of about 0.003 millimetres and mounted as a microscope slide. In such THIN SECTIONS most minerals are rendered translucent or transparent, thus allowing their optical properties to be studied.

thixotropy *n*. a property of fluids and plastic solids, characterised by high viscosity at low stress but decreased viscosity when the applied stress is increased. Certain colloidal substances, e.g. bentonitic clay, weaken when disturbed but increase in strength upon standing.

tholeiitic basalt or **tholeiite** *n*. a variety of BASALT that is poor in alkalis. It is characterised by the presence of orthopyroxene in addition to calcic plagioclase and clinopyroxene. It contains HYPERSTHENE in the NORM. Tholeiite is the commonest LAVA on the ocean floor (most MID-OCEANIC RIDGE basalts are tholeiitic), but they also occur in large volumes in some continental FLOOD BASALT provinces, for example the Columbia River province, North America. They are also found in ISLAND-ARC settings where they form part of the tholeiitic basalt–icelandite–DACITE–RHYOLITE series.

thorax *n*. the segmented body section of a TRILOBITE exoskeleton between the CEPHALON and the PYGIDIUM (Fig. 89). The segments are articulated, allowing the animal to roll up. There are three parts to each segment: the central, arched *axial ring* and the *pleurae* on either side.

threshold velocity *n*. the minimum velocity at which wind or water in a specified location, and under given conditions, will begin to move particles of sand, soil or other material.

throw *n*. the vertical component of the DIP SEPARATION measured in a vertical section at right angles to the FAULT surface.

thrust see FAULT.

tidal flat *n*. a broad, flat marshy or barren tract of land that is alternately uncovered and covered by the tide. It consists of unconsolidated sediment and may form the top surface of a deltaic deposit. See also MARSH, MUD FLAT.

tidal friction *n*. a tidal effect, particularly evident in shallow waters, that lengthens the tidal epoch and tends to retard the rotation of the Earth, thus very slowly increasing the length of the day.

tidal marsh see MARSH.

tidal wave see TSUNAMI.

tie-line *n*. a line on a PHASE DIAGRAM joining the compositions of phases that are in equilibrium with each other.

tiger's eye *n*. a yellowish-brown gem variety of quartz, pseudomorphous after CROCIDOLITE. The CHATOYANCY of the mineral is the result of the penetration of the quartz by asbestiform fibres of crocidolite.

tight fold see FOLD.

till or **boulder clay** *n*. generally non-stratified material deposited directly by glacial ice. Till is very poorly SORTED, with a wide range of grain sizes from CLAY to BOULDERS. CLASTS are usually ANGULAR because they have undergone little or no water transport. Till is of two types: *basal till*, which was carried in the GLACIER base and usually laid down under it, and *ablation till*, which was carried within or on the glacier surface and melted out at the snout of the glacier. Here, meltwater causes flowage and slumping within the till. Basal till is commonly placed by lodgement (plastered on to the glacier bed) and may be referred to as *lodgement till*. See also GLACIAL DRIFT, MORAINE.

tillite *n*. a sedimentary rock formed by the compaction and cementation of TILL.

time break *n*. the indication on a seismic record of the instant of detonation of a shot (see SEISMIC SHOOTING).

time-depth chart *n*. a chart, used in seismic work, by means of which time increments can be connected to corresponding depths. It is based on the relationship between the velocity and ARRIVAL TIME of vertically travelling seismic reflections that have been inflated by a series of shots.

time-depth curve see TIME-DEPTH CHART.

time-distance curve or **TD curve** *n*. (*seismology*) a relation between the travel times of various seismic phases and their epicentral distances. Data for establishing such relations can come from records of the same earthquake at a number of stations with different epicentral distances or from records noted at one station of several earthquakes at various distances. The paths of different phases, distribution of intra-earth seismic velocities and locations of discontinuities are deduced from these curves.

time-rock unit see STRATIGRAPHICAL UNIT.

time-stratigraphic *adj.* (of rock units) having boundaries that are based on their age or time of origin.

time tie *n.* the principle involved in continuous profiling, wherein seismic events on different records are identified by the coincidence of their travel times.

time-transgressive see DIACHRONOUS.

tin *n.* **1.** the native metallic element Sn. **2.** (in mining terminology) CASSITERITE and concentrates containing cassiterite and minor amounts of other minerals.

tinstone see CASSITERITE.

titanaugite see AUGITE.

titaniferous *adj.* containing titanium.

titanite *n.* a yellow, green or brown monoclinic mineral, $CaTiO(SiO_4)$, that occurs as an ACCESSORY MINERAL in granitic rocks and alkaline igneous rocks, especially nepheline syenite, and in gneisses, schists and contact metamorphosed limestones.

Tithonian *n.* the top STAGE of the JURASSIC.

Toarcian *n.* the top STAGE of the Lower JURASSIC.

tombolo *n.* a bar or barrier joining an island with a mainland or another island. Tombolos occur along SHORELINES of SUBMERGENCE, where islands are common. See also SPIT, BARRIER ISLAND.

Tommotian *n.* the basal STAGE of the CAMBRIAN.

tonalite *n.* a PLUTONIC igneous rock consisting essentially of quartz and sodic PLAGIOCLASE FELDSPAR with some MAFIC minerals, usually HORNBLENDE and BIOTITE. Tonalite grades into GRANODIORITE with increasing content of ALKALI FELDSPAR.

tongue *n.* **1.** any projection or offshoot of a larger body, such as a glacier tongue or a lava flow extending from a larger flow. **2.** a minor (informal) STRATIGRAPHICAL UNIT. See LENTIL.

tool mark *n.* a mark formed on the surface of a fine-grained SEDIMENT by an object carried by the current. Tool marks are typically preserved as MOULDS (counterparts) on the underside of a sandstone bed overlying mudrock, in which case they are types of SOLE MARK. Some tool marks are continuous, the result of some object being dragged along the sediment surface: these are GROOVE MARKS and *chevrons*. Chevrons appear on the base of a bed as a linear series of V-shaped ridges closing downstream, a result of sediment filling a wrinkled groove in the surface of the mud. Discontinuous marks such as bounce marks, prod marks or skip marks record the impact of objects on the sediment surface. *Prod marks* are asymmetrical, with one end better defined than the other: this is the downstream end where the object was buried more deeply. *Bounce marks* are more symmetrical, with both ends gently sloping. *Skip marks* are a linear series of bounce marks that are similar enough to have been formed by the same object 'skipping' along the surface. The 'tools' are rarely found but occasionally a shell fragment or something similar may be preserved at the end of a groove.

topaz *n.* an orthorhombic silicate mineral, $Al_2SiO_4(F,OH)_2$, sometimes found as enormous crystals in colours of yellow, blue, green or violet, or it may be colourless. Topaz is typical of pegmatitic pneumatolytic conditions found in gneisses, granites and microlitic cavities. When used as a gemstone it is sometimes sold as precious topaz to distinguish it from CITRINE.

topographic correction see TERRAIN CORRECTION.

topographic texture *n.* the disposition or general size of the topographic elements composing a particular topography. It usually refers to the spacing of drainage lines in stream-dissected areas.

topography *n.* the general configuration of a land surface, including size, relief and elevation.

topset bed see DELTA.

topsoil *n.* the fertile dark-coloured surface soil or a horizon. See SOIL PROFILE.

tor see CORESTONE.

torbanite *n.* a mineral substance intermediate between OIL SHALE and COAL. Coals produce aliphatic HYDRO-CARBONS, whereas torbanite, on destructive distillation, produces paraffinic and olefinic hydrocarbons. As its petroleum content increases, torbanite grades into CANNEL COAL. It takes its name from its type locality, Torbane Hill, in Scotland. Torbanite is often regarded as a type of BOGHEAD COAL.

torque *n.* the movement of a force or system of forces tending to cause rotation.

Torridonian *n.* thick deposits of TERRES-TRIAL sedimentary rocks of mid to late PROTEROZOIC age (1000 Ma–800 Ma) that occur in the northwest of Scotland and in the Inner Hebrides. The lower part of the succession, the Sleat Group, lies unconformably on the LEWISIAN GNEISS. The thicker Torridonian Group rests unconformably on the Sleat Group. The Torridonian is composed principally of red CLASTIC SEDIMENTARY ROCKS (CONGLOM-ERATES, SANDSTONES and MUDSTONES), mostly deposited in a FLUVIAL environment. The Torridonian sediments were derived largely from the west, apparently by the erosion of Precambrian metamorphic complexes of the Grenville mountain chains of south Greenland and Labrador.

torsion *n.* the twisting of a body by two equal and opposite TORQUES.

torsion balance see EÖTVÖS TORSION BALANCE.

Tortonian *n.* the lower STAGE of the Upper MIOCENE.

total alkali/silica diagram see IGNEOUS ROCK, Appendix 3.

total range biozone see RANGE ZONE.

total reflection *n.* REFLECTION in which all of the incident wave is returned.

tourmaline *n.* a group of hexagonal borosilicates minerals with variable chemical composition and the general formula $(Ca,Na,K)(Al,Fe^{2+},Fe^{3+},Li,Mg,Mn)_3(Al,Cr,Fe^{3+},V)_6(BO_3)_3Si_6O_{18}(O,OH,F)_4$, occurring as prismatic crystals or aggregates of parallel or radiating individuals. Tourmaline is a common ACCESSORY MINERAL in igneous and metamorphic rocks, and very common in pegmatites, where it sometimes occurs in crystals of enormous size. The black iron-bearing variety, SCHORL, is the most common. Many of the other varieties are attractively coloured and widely used as gemstones. Individual crystals sometimes show a variety of colours along their length. See DRAVITE, RUBELLITE.

Tournaisian *n.* the bottom SERIES of the CARBONIFEROUS.

T phase *n.* an occasionally recorded seismic phase for which the corresponding ray has most of its length in the ocean, where the wave velocity is about 1.5 kilometres per second. The T phase has been recorded in California from Hawaiian earthquakes.

trace *n.* the intersection of one geological surface with another, e.g. the trace of BEDDING on a FAULT PLANE.

trace element *n.* an element that is found in a mineral or rock in small quantities, significantly less than 1 per cent, and is not essential to the composition of the mineral.

trace fossil or **ichnofossil** *n*. a fossilised track, trail, boring or burrow formed by the movement of an animal in soft sediments.

trachyandesite *n*. an extrusive rock with a composition between trachyte and andesite. It contains sodic plagioclase, alkali feldspar and one or more mafic minerals (amphibole, biotite, pyroxene). Compare LATITE.

trachybasalt *n*. an extrusive rock with a composition intermediate between TRACHYTE and BASALT. Both calcic plagioclase and alkali feldspar are present, as well as augite and olivine. Leucite or analcime may be minor constituents. Compare LATITE.

trachyte *n*. a fine-grained extrusive alkaline rock, sometimes porphyritic, approximately silica-saturated and with a wide compositional range. Its main components are alkali feldspar and minor MAFIC minerals, sometimes with a small amount of quartz. Trachytes range from slightly over-saturated types with less than 10 per cent quartz to under-saturated types containing feldspathoids. With more or less silica, they grade into rhyolites and phonolites respectively. Trachytes are the extrusive equivalent of SYENITE. See TRACHYTIC.

trachytic *adj*. (of the texture of groundmasses of volcanic rock) in which feldspar MICROLITES are arranged in parallel or near-parallel manner, bending around PHENOCRYSTS (when these are present); the pattern corresponds to the flow lines of the LAVA from which the rock was formed. Such texture is common in TRACHYTES and is sometimes found in other types of lava, such as HAWAIITE. See also TRACHYTOID, PILOTAXITIC.

trachytoid *adj*. (of the texture of coarse-grained igneous rocks) in which the feldspars are disposed in parallel or sub-parallel fashion, thus resembling the TRACHYTIC texture of some volcanic rocks, e.g. varieties of nepheline syenite.

traction load see BED LOAD.

train *n*. **1.** a narrow glacial deposit, such as a boulder train, extending for a long distance. **2.** a series of oscillations on a SEISMOGRAPH record.

transcurrent fault see TRANSFORM FAULT, Fig. 88 below, FAULT.

transform fault *n*. a type of strike-slip FAULT that is typically found at plate boundaries. It derives its name from the fact that the fault ends at (and

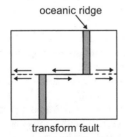

transform fault

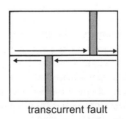

transcurrent fault

Fig. 88 **Transform fault**. Comparison of the direction of movement on a transform fault and a transcurrent fault

oceanic ridge

appears to 'transform' into) major features such as MID-OCEANIC RIDGES or SUBDUCTION ZONES, where the movement is different from that on the fault. The major transverse fracture zones that cross the mid-oceanic ridge systems are transform faults. Movement on these faults takes place mainly between the off-set sections of the ridge crest and is a consequence of the spreading of the OCEANIC CRUST on either side of the ridge (see Fig. 88 on the previous page). Spreading is parallel to the line of the transform fault and not necessarily at right angles to the ridge. The direction of movement on a transform fault is opposite that shown by a transcurrent fault with the same apparent offset (see Fig. 88).

Transform faults form one of the three types of boundary between lithospheric plates. They are known as *conservative boundaries* because along them, segments of the Earth's crust merely slide past each other and crust is neither created nor destroyed. See also PLATE TECTONICS.

transgression, invasion or **marine transgression** *n.* the incursion of the sea over land areas or a change that converts initially shallow-water conditions to deep-water conditions. Compare REGRESSION.

transit *n.* a THEODOLITE in which the telescope can be reversed in direction by rotating it 180° about its horizontal transverse axis.

transition temperature see DIFFERENTIAL THERMAL ANALYSIS.

transition zone *n.* any of several zones within the Earth in which there is a sharp increase in seismic velocity corresponding to phase or chemical changes.

translational movement *n.* apparent fault-block displacement in which

there has been no rotation of the blocks relative to each other; features that were parallel before movement are still parallel afterwards. Compare ROTATIONAL MOVEMENT.

translation gliding see GLIDING.

transpression *n.* a combination of STRIKE-SLIP motion and compression during crustal deformation. See also TRANSTENSION.

transtension *n.* a combination of STRIKE-SLIP motion and extension during crustal deformation. See RHOMBOCHASM. See also TRANSPRESSION.

transverse dune see DUNE.

transverse section see CROSS-SECTION.

transverse valley *n.* a valley the course of which cuts across the geological structure or the direction of which is at right angles to the general strike of the underlying strata. Compare LONGITUDINAL VALLEY.

transverse wave see SEISMIC WAVES.

trap *n.* 1. a dark, fine-grained IGNEOUS ROCK, such as BASALT or DOLERITE.
2. an OIL TRAP.

trapezohedron see ICOSITETRAHEDRON.

travertine *n.* a fine-grained crystalline CALCITE deposit of white, cream or tan colour, often banded, formed by chemical precipitation from solution by inorganic processes from cold or hot water. It occurs in caves as deposits of flowstone or dripstone and also forms in springs as TUFA, a coarsely crystalline, spongy limestone.

tree-ring dating see DATING METHODS.

trellis drainage pattern see DRAINAGE PATTERN.

Tremadoc *n.* the lowest SERIES of the Ordovician.

tremolite *n.* a white to pale greyish-green monoclinic mineral, $Ca_2Mg_5Si_8O_{22}(OH)_2$, of the AMPHIBOLE group; it forms a complete solid solution series with ACTINOLITE. It occurs as bladed crystals, frequently in radiating columnar aggregates, and sometime as silky asbestiform fibres. It is found in dolomitic marbles, serpentinites and talc schists associated with magnesite and calcite. See AMPHIBOLE.

trench see OCEAN TRENCH.

trend *n*. a term for the bearing (AZIMUTH) of a geological feature, as in the trend of a FAULT.

triangulation *n*. a method of surveying that is generally used where the area to be surveyed is large and requires the use of geodetic techniques. It establishes the distance between any two points by using such points as vertices of a triangle or series of triangles such that each triangle has a side of known length (the BASE LINE). This permits the angles of the triangle and length of its other two sides to be determined by observations taken from the two ends of the base line.

Triassic *n*. a period of geological time that began 245 Ma ago and ended 208 Ma; it is the earliest period of the Mesozoic Era. The term *Trias*, later modified to Triassic, was proposed in 1834 for a sequence of strata in Germany. It refers to a threefold division of the strata into a lower unit of continental red sediments (Bunter), a middle unit of marine limestone (Muschelkalk) and an upper, continental FACIES (Keuper). Zonal schemes for this period vary with location and are not well correlated. In the British Isles the Triassic is represented largely by continental deposits, principally sandstones and conglomerates in the Lower Triassic, and sandstones, red marls and evaporites in the Upper Triassic. Towards the end of the Triassic the sea gradually encroached upon the land, depositing marine shales, limestones and sandstones (the Rhaetic).

Large-scale EXTINCTIONS took place at the end of the Permian. AMMONITES are the dominant invertebrates of the Triassic. BIVALVES and BRACHIOPODS are much less abundant. RUGOSE CORALS are extinct, but the first Scleractinian corals appear in the Triassic. Nautiloids show no change in evolutionary pattern across the Permian-Triassic boundary. Tetrapod faunas of the Early Triassic are dominantly those surviving the Permian. New tetrapods that appear later are the proanurans (ancestors of frogs), chelonians (turtles), early dinosaurs of various kinds and the first mammals. Some late Permian sharks, as well as one family of coelacanth and one type of lungfish, lived into the Triassic. Gymnosperms typify Triassic flora; cycads and primitive conifers flourished in upland areas. Fossilised remains of their logs are found in the Petrified Forest of Arizona. Ferns and scouring rushes thrived in lower, moist areas, whereas seed ferns had vanished. CORDAITES are not conspicuous, and *Lepidodendron* is not represented.

The Triassic was a time of extensive continental emergence as the supercontinent of PANGAEA was finally assembled and immediately began to break up. Fossil and rock records suggest highly equable climatic conditions. There are no continental glacial deposits of this age. Triassic RED BEDS of sandstones and shales, found in places associated with EVAPORITE deposits, were formed as far north and south as 50° latitude.

tributary or **feeder** *n*. any stream that contributes water to another stream. Compare DISTRIBUTARY.

trichroism see PLEOCHROISM.

triclinic system see CRYSTAL SYSTEM.

tridymite *n*. a monoclinic silica mineral, SiO_2, a high-temperature polymorph of

QUARTZ. It occurs in siliceous volcanic rocks, such as obsidian and pitchstone, in small, white, blade-like crystals, spherulitic masses or rosettes in cavities. Tridymite is stable between 870 and 1470°C. Compare CRISTOBALITE.

trigonal system see CRYSTAL SYSTEM.

trilobite *n.* an extinct marine ARTHROPOD of the class Trilobita, characterised by a segmented body divided longitudinally into three lobes and transversely into three sections from head to tail: the CEPHALON, THORAX and PYGIDIUM (see Fig. 89 below). Each segment bore a pair of biramous (two-branched), jointed appendages. Some trilobites were eyeless, others had compound eyes (see HOLOCHROAL EYES and SCHIZO-CHROAL EYES). Trilobites had a chitinous exoskeleton that was shed as it grew (ecdysis). The earliest larval stage was the PROTASPIS.

Trilobites showed considerable diversity in shape and size. Their adult size ranged from 6 millimetres to 45 centimetres; some attained sizes of 70 to 75 centimetres. The intrinsic construction plan, however, seems to have remained the same. No single character such as facial sutures has proved satisfactory in the classification of these animals, since none demonstrates a progressive change within the group (other than adaptational change). The present classification is based on a range of morphological features.

Trilobites range from Lower CAMBRIAN to PERMIAN, and many kinds are used for STRATIGRAPHICAL CORRELATION. Because they appear fully developed in the Cambrian, it seems likely that ancestral forms originated in the PRECAMBRIAN as non-skeletal arthropods.

triple junction *n.* a point at which three tectonic (lithospheric) plates meet. Stable triple junctions retain their geometry as plates drift while unstable junctions change with drift. At present, six different triple junctions of spreading centres (ridges), SUBDUCTION ZONES (trenches) and transforms are found, although 16 types are theoretically possible. Magnetic patterns in sea-floor rocks provide clear evidence that old triple junctions have disappeared in subduction zones. See PLATE TECTONICS.

triple point *n.* a point on a PHASE DIAGRAM where the boundary curves for the fields of three phases meet and the three phases coexist.

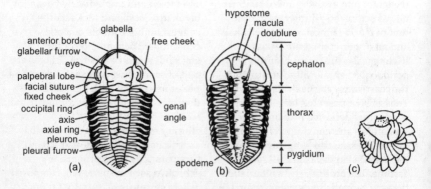

Fig. 89 **Trilobite**. General morphology: (a) features of the dorsal surface; (b) ventral surfaces; (c) enrolled trilobite

tripoli *n.* a light-coloured powdery or siliceous sedimentary rock produced by the WEATHERING of siliceous limestone or chert. Compare DIATOMACEOUS EARTH.

TRM *abbreviation for* thermoremanent magnetisation. See NATURAL REMANENT MAGNETISATION.

troctolite *n.* a coarse-grained IGNEOUS ROCK composed of OLIVINE and calcic PLAGIOCLASE with little or no PYROXENE. It is a variety of GABBRO.

trona *n.* a monoclinic mineral, $Na_3(CO)_3$ $(HCO_3)\cdot2H_2O$. It occurs in white or yellowish-white tabular crystals in saline lake deposits. Trona is an important source of sodium compounds.

tropical tree bog see BOG.

trough cross-bedding see CROSS-BEDDING.

true dip *n.* DIP, in comparison with APPARENT DIP.

true north *n.* geographic north, the North Pole of rotation of the Earth.

truncated spur *n.* a spur that projected into a valley and was completely or partially cut off by a moving GLACIER.

truncation *n.* the shortening or removal of a part of a landform or geological structure, as by EROSION. For example, a truncated SOIL PROFILE is one in which one or both upper horizons have been removed by erosion.

tsunami *n.* the gravity-wave system that follows large-scale disturbances of the sea floor. When the wave breaks over the coastline, water piles up, causing great destruction. Tsunamis seem to occur principally after earthquakes of magnitude greater than 6.5 (Richter scale) and focal depth of less than 50 kilometres. However, not all such earthquakes produce them. Once the tsunami is formed, the wave system closely resembles that which is produced by throwing a stone into a pond. The wave configuration is axisymmetric at the early stage and consists of concentric rings of crests and troughs. It is bounded at the outside by a kind of front. This front expands everywhere at the limiting velocity, $C = \sqrt{gh}$ for free waves in water of depth h (g = gravitational acceleration). The oscillation period of a tsunami is of the order of one hour, and the wavelength may measure several hundred kilometres. Wave velocities reach speeds of 900 kilometres per hour over deep ocean. It appears that a tsunami originates in the vicinity of an earthquake epicentre because of sudden changes in the level of the sea floor resulting from movements along tear faults or thrusts. Tsunamis might also be generated by submarine landslides caused by earthquakes or collapse of unstable slopes of volcanic islands. See also SEICHE.

tube feet see WATER-VASCULAR SYSTEM.

tubercle *n.* a projection on the skeleton or TEST of an ECHINOID to which a spine was attached. It consists of a gentle swelling, the BOSS, topped by a knob (the *mamelon*) and surrounded by a wide, smooth *areola*. See Fig. 27.

tufa *n.* a chemical sedimentary rock of calcium carbonate, precipitated by evaporation; it commonly occurs as an incrustation around the mouth of a spring or along a stream. The compact, dense variety is TRAVERTINE. It is not to be confused with TUFF. Compare SINTER.

tuff *n.* a PYROCLASTIC ROCK composed mainly of volcanic ash (fragments more than two millimetres in diameter). Tuffs may be classified as *crystal tuff* if they contain a large proportion of crystal fragments, *vitric tuff*, composed mainly of glass and pumice fragments, and *lithic tuff*, containing mainly rock fragments. A consolidated mixture of

LAPILLI and ash is a *lapilli tuff*. See also DUST TUFF, WELDED TUFF, PYROCLASTIC MATERIAL.

tuffite *n*. a consolidated mixture of pyroclastic and sedimentary detritus; it is usually well-bedded and sorted according to grain size.

tundra *n*. a generally level or undulating treeless region found mainly in the Arctic lowlands of Europe, Asia and North America. Vegetation is restricted to mosses, grasses, lichens and sedges; the subsoil is permanently frozen.

tungstate *n*. a chemical compound characterised by the WO_4 radical. An example of a tungstate mineral is wolframite, $(Fe,Mn)WO_4$. Tungsten and molybdenum may substitute for each other. Compare MOLYBDATE.

turbidites, turbidity currents or **density currents** *n*. currents caused by an excess density, the result of a sus-pended load of sediment, since horizontal differences in density within gaseous or fluid bodies can cause currents. These currents flow down-slope at very high speeds and spread along the horizontal. The distance covered is determined by the turbu-lence and velocity of the current. The sedimentary load is gradually dropped as the current slackens and the water comes to rest. Such deposits are called *turbidites* and are characterised by moderate sorting, graded bedding and well-developed primary structures. It is generally believed that SUBMARINE CANYONS are at least partially eroded by turbidity currents.

turbulent flow *n*. the motion of a liquid or gas is turbulent when the velocity at any point changes direction and magnitude in a random manner. Its diffusive action distinguishes turbu-lence from an irregular wave motion, i.e. turbulence tends to increase rates of transfer of momentum, heat and water vapour. When the velocity of a fixed-size flow is increased, a smooth or laminar motion will become unstable and turbulent. (Pipe flow is turbulent for REYNOLDS NUMBERS over about 2000.) Compare LAMINAR FLOW.

turnover *n*. **1.** an interval, especially in the autumn or spring, of uniform vertical temperature, when convective circulation occurs in a lake; the period of an OVERTURN. See also CIRCULATION. **2.** a process by which some species in a region become extinct and are replaced by other species.

Turonian *n*. a STAGE of the Upper CRETACEOUS.

turquoise *n*. a triclinic mineral, $CuAl_6(PO_4)_4(OH)_8 \cdot 5H_2O$, usually occurring in light-blue or green microcrystalline masses, nodules and veins. Turquoise is a SECONDARY MINERAL produced by the alteration (in arid regions) of alu-minium-bearing rocks. The phospho-rus may be derived from apatite and the copper from minerals such as chalcopyrite. It is a valuable ornamen-tal stone.

turrilite *n*. a Cretaceous cephalopod, about 13 centimetres in length, with a sharp-spired shell on which the whorls barely touch. It resembles a gastropod (compare TURRITELLA), but the presence of septa and its complex suture pattern differentiate it.

Turritella *n*. a genus of GASTROPODS ranging from Cretaceous to Recent time. Members of the genus have high-pointed shells, about 10 centimetres long, that are frequently incised with grooves, ridges or spiral lines. Com-pare TURRILITE.

twin gliding see GLIDING.

twinning *n*. the joining or intergrowth of two or more individuals of the same MINERAL species, with different orienta-

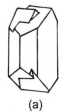 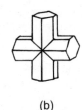 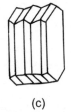  

(a)　　　　　(b)　　　　　(c)　　　　　(d)　　　　　(e)

Fig. 90 **Twinning**. Twinned crystals: (a) orthoclase; (b) staurolite:
(c) plagioclase feldspar; (d) rutile; (e) gypsum

tions of the CRYSTAL LATTICE. One part of a twinned crystal may appear to have been rotated about an axis (called the *twin axis*) relative to the other part or reflected across a *twin plane*. The geometrical relationships of the crystal lattices in the different parts of a twinned crystal are referred to as the *twin laws*. See Fig. 90 opposite.

A *contact twin*, such as occurs in GYPSUM, is made up of two crystals united along a plane known as the *composition plane*. In a *penetration twin*, two crystals interpenetrate one another, e.g. the 'Greek cross' of STAUROLITE.

FELDSPARS are commonly twinned according to the Carlsbad law. The twin axis is the C-CRYSTAL AXIS and the composition surface is irregular. *Multiple (polysynthetic) twinning* according to the ALBITE law is characteristic of PLAGIOCLASE FELDSPARS.

type concept *n.* a principle for regulating the application of scientific nomenclature by recognising and describing some unit that can then be used as a point of reference.

type locality *n.* **I.** the location in which a stratotype is situated and from which it usually takes its name. It is contained within the type area and contains the TYPE SECTION. Compare REFERENCE LOCALITY.

2. the place where some geological feature, e.g. a particular kind of metamorphic rock or the type specimen of a fossil species, was originally recognised and described.

type section *n.* **I.** the sequence of strata originally described as constituting a STRATIGRAPHICAL UNIT and serving as a standard of comparison when identifying geographically separated parts of a stratigraphical unit. A type section should be selected in an area where at least the top and bottom of the formation are exposed. Type section is better called a STRATOTYPE. Compare REFERENCE SECTION.

2. STRATOTYPE.

type specimens *n.* the specimens upon which the full description of a new species is based. There are various kinds of type specimen: the *holotype* is the reference specimen, selected because it shows the main features of the species. The *paratype* has additional characteristics and supplements the holotype. A *neotype* is a new type specimen selected when the original is lost or destroyed. A *lectotype* is a specimen selected later if there was no properly described original.

Tyrrhenian *n.* a STAGE of the PLEISTOCENE in southern Europe. See CENOZOIC, Fig 13.

437

Uu

ugrandite *acronym for* **u**varovite, grossular, *and* radite, END MEMBERS of a series of GARNET minerals showing SOLID SOLUTION. There appears, however, to be little solid solution between the ugrandite series and the PYRALSPITE series.

ulexite *n.* a triclinic mineral, $NaCaB_5O_6(OH)_6 \cdot 5H_2O$, occurring as light spongy rounded masses composed of aggregates of white silky hair-like fibres (cottonballs). Ulexite, an ore of boron, is produced by EVAPORITE precipitation in lake basins of arid regions and is usually associated with BORAX.

Ulsterian *n.* a series of the Lower Devonian of North America.

ultisol see DESILICATION.

ultrabasic see SILICA CONCENTRATION.

ultramafic *adj.* (of igneous rocks) containing more than 90 per cent of MAFIC minerals. See also COLOUR INDEX.

ultramylonite see MYLONITE.

ultraviolet (UV) *n.* that part of the electromagnetic spectrum corresponding to wavelengths in the range of 40 to 4000 angstroms.

umber *n.* a brown earth consisting of manganese oxides, hydrated ferric oxide, alumina, silica and lime. It is widely used as a pigment, either in the greenish-brown natural state, *raw umber*, or as *burnt umber*, the dark-brown calcined state. It is darker than either OCHRE or SIENNA.

umbilicus *n.* **1.** the depression at the centre of the coils on each side of a planispirally coiled shell, see AMMONOID Fig. 1, NAUTILOID Fig. 62. If the coiling is loose the umbilicus is wide, while tight coiling results in a small umbilicus. **2.** an opening at the base of the last WHORL in a helically coiled shell, caused by loose coiling. Compare COLUMELLA. See also GASTROPOD.

umbo *n.* (*p1* **umbones**) a rounded projection that is the earliest formed part of a VALVE in BRACHIOPODS (Fig. 9) and BIVALVES (Fig. 7). See also PROTEGULUM.

unaka *n.* a large residual form rising above a PENEPLAIN and occasionally showing in its summit or surface the remnants of an even older peneplain. It is greater in height and size than a MONADNOCK. The name is derived from the Unaka Mountains of the United States.

unconfined storativity see SPECIFIC YIELD.

unconformable or **discordant** *adj.* (of strata) not following the underlying rocks in immediate age sequence or not parallel with the rocks beneath. In particular, it applies to younger strata that do not have the same dip and strike as the underlying rocks. Compare CONFORMABLE. See also UNCONFORMITY.

unconformity *n.* a break in the sequence of strata in an area that represents a period of time during which no sediment was deposited. It indicates a change in the conditions prevailing in the area. An unconformity may be the result of uplift and erosion, an interruption in sedimentation or non-deposition of sedimentary material. The absence of rocks normally present in a sequence indicates a break in the geological record. There are four basic types of unconformity. In a *disconformity*, the buried erosion surface lies between two series of strata that are parallel on a large scale. Disconformity surfaces may be of low relief or may be highly irregular. Where the beds beneath an unconformity are not parallel with those above, an *angular unconformity* exists. This indicates folding or faulting of the lower layers before being levelled by erosion and covered by younger layers. An unconformity between overlying stratified sediments and underlying unbedded metamorphic or plutonic rocks is a *nonconformity* (or *heterolithic unconformity*). A *paraconformity* is similar to a disconformity in that the strata are parallel; however, there is little discernible evidence of erosion or prolonged non-deposition. See Fig. 91 below. Compare CONFORMITY. See also DIASTEM.

underclay see SEAT-EARTH.

underground water see GROUNDWATER.

underplating *n*. the addition of material of intermediate and felsic composition to the base of the over-riding crustal plate at a SUBDUCTION ZONE.

undersaturated or **silica undersaturated** *adj*. IGNEOUS ROCK composed of silica-unsaturated minerals such as

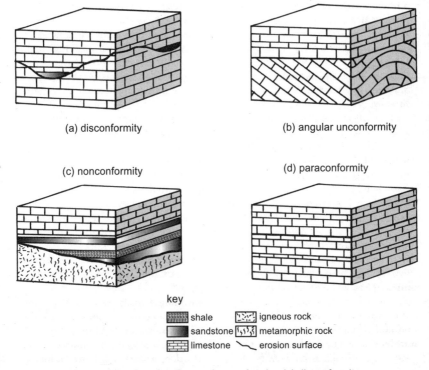

(a) disconformity

(b) angular unconformity

(c) nonconformity

(d) paraconformity

key

shale   igneous rock
sandstone   metamorphic rock
limestone   erosion surface

Fig. 91 **Unconformity**. Types of unconformity: (a) disconformity; (b) angular unconformity; (c) non-conformity; (d) paraconformity

olivine and feldspathoids or whose NORM contains olivine and feldspathoids. See SILICA SATURATION.

undulose extinction *n.* uneven or undulating extinction of a mineral viewed in THIN SECTION between CROSSED POLARISERS under the PETROLOGICAL MICROSCOPE. This is often an indication that a mineral has been subjected to stress. See EXTINCTION ANGLE.

uniaxial *adj.* (of crystals) having a single OPTIC AXIS, e.g. CRYSTALS of the tetragonal or hexagonal systems. Compare BIAXIAL.

uniclinal shifting *n.* the slow down-dip shift of a stream channel as a result of two factors: the tendency of streams in an area of inclined strata to follow the strike of less resistant strata, and the tendency for erosion to proceed more rapidly along the steeper slopes of a cuesta. Compare MIGRATION.

uniformitarianism *n.* the principle of James Hutton (1726–97), a leading figure of the Scottish Enlightenment and widely regarded as the founder of modern geology. In his *Theory of the Earth*, Hutton postulates that the laws of nature now prevailing have always prevailed and that, accordingly, the results of processes now active resemble the results of like processes of the past. The principle does not imply the invariance through time of conditions now prevailing; what is meant is that changes or modifications of the Earth's crust are produced only in accordance with the laws of physics and chemistry. Compare CATASTROPHISM.

uniserial *adj.* (of GRAPTOLITES) having a single row of THECAE (cups) along a STIPE. See Fig. 41.

unit cell see CRYSTAL LATTICE.

univalve, 1. *n.* a mollusc with a single valve, e.g. CEPHALOPOD, GASTROPOD. **2.** *adj.* having only one valve. Compare BIVALVE.

univariant *adj.* (of a chemical system) possessing one DEGREE OF FREEDOM. See also PHASE DIAGRAM.

universal stage *n.* an accessory device that can be attached to a PETROLOGICAL MICROSCOPE and allows the THIN SECTION under study to be rotated about two horizontal axes at right angles to one another, in addition to the vertical axis of rotation of the microscope stage. It is used in the optical study of ANISOTROPIC minerals.

unmixing see EXSOLUTION.

unsaturated or **silica unsaturated** see SILICA SATURATION.

unsaturated zone *n.* the zone at shallow depths in soil or rock where the pore space contains some air and is not completely filled with water. See WATER TABLE, VADOSE WATER.

uphole shooting *n.* the shooting of successive shots at several hole depths in SEISMIC EXPLORATION. The time required for the seismic impulse to travel from a given depth to the surface is the *uphole time*. This is one of the commonest methods for finding WEATHERING and subweathering velocity.

uphole time see UPHOLE SHOOTING.

upright fold see FOLD.

upthrown *adj.* (of the side of a fault) appearing to have moved upwards relative to the other side. The amount of upward vertical displacement is called the *upthrow*. Rocks on the upthrown side of a fault may be described as *upfaulted*. Compare DOWNTHROWN, HEAVE.

upwelling *n.* a process of vertical water motion in the sea whereby subsurface water is transported towards the surface. The converse downward displacement is called *downwelling* or *sinking*. The *upwelling area* is the geographical location of the vertical motion, but *upwelled water* and its influence on oceanographic conditions may extend for hundreds of kilometres. It may occur anywhere but is most common along the western coasts of continents.

upwelling area see UPWELLING.

Uralian orogeny *n.* an orogenic episode resulting from the collision of Siberia with Laurasia during the Permian.

uralitisation or **amphibolisation** *n.* the REPLACEMENT of PYROXENE by uralite (ACTINOLITE) as the result of late-magmatic processes in IGNEOUS ROCKS.

uraninite *n.* an isometric uranium mineral, UO_2, that is highly radioactive and is the primary ore of uranium. It usually occurs in granular masses or aggregates. Uraninite is found in pegmatites and high-temperature hydrothermal veins and in some placer deposits with minerals of lead, tin and copper. See PITCHBLENDE.

uranium-lead method see DATING METHODS.

Uriconian *n.* the Uriconian Group consists of a thick accumulation of volcanic material (lavas, breccias, agglomerates and tuffs) formed during the Late PRECAMBRIAN (*c.* 566 Ma) in Britain. It lies stratigraphically below the LONGMYNDIAN supergroup.

UV see ULTRAVIOLET.

uvala *n.* a depression in a LIMESTONE area resulting from the merging of two or more DOLINES. See KARST.

uvarovite *n.* the calcium-chromium END MEMBER of the GARNET series, $Ca_3Cr_2Si_3O_{12}$. It occurs as small brilliant green crystals in serpentinites that are rich in chromite; it is the rarest of all the garnets.

V𝑣

vadose water *n*. water that occurs between the ground surface and the WATER TABLE, i.e. in the zone of aeration. Compare PHREATIC WATER.

Valanginian *n*. a STAGE of the Lower CRETACEOUS.

valley *n*. a linear, low-lying tract of land bordered on both sides by higher land and frequently traversed by a stream or river. All valleys have been cut by running water over time; their cross-profiles change because of various factors. A V-shaped valley may be widened by sheetwash, WEATHERING and MASS WASTING. Glacial EROSION produces a valley that is U-shaped in cross-profile, e.g. a glacial trough.

valley train *n*. a long, relatively narrow train-like deposit of OUTWASH sand and gravel deposited by meltwater streams, usually in front of a terminal MORAINE. Compare SANDUR.

valve *n*. a single piece of skeletal material forming part or the whole of a shell, as in BRACHIOPODS, which are bivalved organisms, or GASTROPODS which are univalved.

vanadinite *n*. a hexagonal mineral, Pb_5 $(VO_4)_3Cl$, of the APATITE group, an ore of vanadium. It occurs in reddish-orange to yellow or brown hexagonal prisms or fibrous radiate masses or crusts. Vanadinite is found as a rare SECONDARY MINERAL in the oxidation zone of lead deposits.

Van der Waals bonds *n*. the weak electrical attraction existing between atoms or IONS in solids. They are much weaker than the other bond types and their effect is often masked by the presence of ionic or COVALENT BONDS. However, they can be important in some SILICATE minerals; for example, the silicate layers in TALC are linked by Van der Waals bonds so they are easily broken apart, giving talc its soapy feel. See also HYDROGEN BOND, IONIC BOND, COVALENT BONDS.

variation diagram *n*. a method used in the interpretation of petrochemical data in which compositional variations are plotted in terms of various chemical parameters. For example, a group of basalt lavas may exhibit compositional variations that can be related to the removal of different proportions of olivines by crystal-liquid fractionation. Two basic variation diagrams are used: a Cartesian graph of two variables or a triangular diagram of three variables. The variation of four variables can be plotted by projecting the points in a three-dimensional tetrahedron on to one of its sides. The *Harker diagram* plots the weight percentage of SiO_2 along the X-axis of a Cartesian graph and concentrations of other oxides along the ordinate axis.

variolitic *n*. a texture or structure that occurs only in basaltic rocks; it is equivalent to spherulitic texture in rhyolite. Variolitic structure is a common feature of the margins of certain PILLOW LAVAS. The *varioles*, which sometimes stand out as small knobs, are commonly composed of fanlike sprays of feldspar fibres. Some may be a devitrification texture of glassy rocks.

Variscan (Hercynian) orogeny *n*. the Upper Palaeozoic orogenic episode in

Europe, commencing in the Late Devonian and reaching its peak in the Late Carboniferous and ending in the Early Permian. It corresponds to the Allegheny OROGENY of North America (see APPALACHIAN FOLD BELT). Folded Hercynian mountains extended from England and Ireland through France and Germany. The more westerly folded areas are commonly called Armorican and the remainder Variscan. It is thought, therefore, that two separate deformational systems were involved in their formation. The folding and faulting that occurred during this disturbance were accompanied by large-scale igneous activity in England and on the continent. The Variscan deformation is the result of the closure of GONDWANA with LAURASIA, complicated by the presence of a number of terranes between the two major units. The Variscan OROGENIC BELT forms part of a larger orogenic belt extending from Georgia (USA) to the Czech Republic.

varve see RHYTHMITE, DATING METHODS.

vascular plant *n.* a plant characterised by a fully developed transport system for fluids and structural differentiation into roots, stems and leaves.

vein *n.* a tabular or sheet-like body of one or more minerals deposited in openings of fissures, joints or faults, sometimes with associated REPLACEMENT of the host rock. Compare LODE.

veld see SAVANNA.

velocity log see WELL LOGGING.

velocity profile *n.* a technique used to determine seismic velocity from the time-distance relationships of seismic reflections that are generated over a wide range of distances between a shot and a GEOPHONE.

Vendian *n.* the top division of the Russian PROTEROZOIC.

vent see VOLCANIC VENT.

venter *n.* the ventral wall of the shell in coiled CEPHALOPODS, which usually forms the outer margin. See AMMONOID Fig. 1. Compare DORSUM.

ventifact *n.* any stone shaped by the action of windblown sand. Ventifacts are generally pitted, polished and grooved. Abrasive wind action sometimes cuts facets on the windward side of large fragments. Two or more facets are often formed on smaller stones, and a ridge or edge is formed where facets intersect. See DREIKANTER, EINKANTER.

Venus hair *n.* needle-like crystals of yellow or reddish-brown RUTILE that form tangled masses of INCLUSIONS in QUARTZ. Crystals of rutilated dark smoky quartz are referred to as *Venus hairstone*. See also SAGENITE.

vergence *n.* a measure of the asymmetry shown by minor FOLDS developed on the limbs of larger folds. Changes in vergence of the minor folds can be used to locate the AXIAL TRACES of the major folds. See PARASITIC FOLDS and Fig. 67.

vermiculite *n.* a group of micaceous CLAY MINERALS related to chlorite and montmorillonite, with the same general formula, $(Mg,Fe,Al)_3,(Al,Si)_4O_{10}(OH)_2 \cdot 4H_2O$. It is frequently pseudomorphous after BIOTITE. Vermiculite is a hydrothermal alteration product of phlogopite and biotite with a TALC-like structure interlayered with water molecules. When heated to about 300°C, it loses water quickly and expands to about 18 to 25 times its original volume.

vertical accretion *n.* the vertical accumulation of a sedimentary deposit, such as the settling of sediment suspended in a stream that is prone to overflow. Compare LATERAL ACCRETION.

vesicle *n.* a small cavity in an IGNEOUS ROCK formed by the expansion of a gas bubble during solidification of the rock; they normally occur in LAVA FLOWS, particularly in their upper parts, or in dykes close to the surface. As MAGMA moves towards the surface, decompression immediately before and during extrusion causes the magmatic gases to come out of solution and form bubbles. Vesicles are generally less than two centimetres in diameter but may be larger where they coalesce. They tend to be spherical if formed in stationary lava or in lava of low viscosity, such as PAHOEHOE, because of surface tension. If formed in moving viscous lava, they may be elongated in the direction of flow because of stretching. *Pipe vesicles* are hollow, upward-rising tubes that form in some lava flows. The general opinion is that they are formed by steam from heated GROUNDWATER penetrating the base of the flow. *Inclined pipe vesicles*, formed while a flow is still moving, point in the direction of flow. Vesicles are generally larger in basaltic lava flows than in silicic to intermediate lavas. Their size seems to be a function of the lower-viscosity type of melt and higher temperature of basaltic extrusion, both of which permit the gas bubbles to expand before solidification. Rocks containing vesicles are said to have a *vesicular texture*. See LAVA, PUMICE.

vesuvianite or **idocrase** *n.* a tetragonal silicate mineral, $Ca_{19}Fe(Mg,Al)_8Al_4(SiO_4)_{10}(Si_2O_7)_4(OH)_{10}$, occurring as brown or yellow crystals or compact granular masses in brown or olive green. It is associated with contact METAMORPHISM of impure limestones.

vibration directions *n.* the mutually perpendicular directions of vibration of the two rays of light transmitted through ANISOTROPIC minerals. The positions of the vibration directions are fixed with respect to the mineral and can be located in THIN SECTION under CROSSED POLARS, as the crystal will be in extinction when its vibration directions are parallel to the directions of the POLARISER and ANALYSER. The rays have different refractive indices and travel with different velocities through the crystal. A quartz wedge or sensitive tint plate can be used to determine which is the faster ray. See also PLEOCHROISM.

virgella *n.* a spine projecting from the margin of the aperture of the SICULA of GRAPTOLITES, see Fig. 41.

Virgilian *n.* the uppermost series of the Pennsylvanian of North America.

virgula see NEMA.

viscosity *n.* the property of a substance that offers internal resistance to the force tending to cause the substance to flow.

viscous flow *n.* (*structural geology*) flow in which the rate of SHEAR STRAIN varies directly with the SHEAR STRESS. Compare SOLID STATE FLOW.

Viséan *n.* the upper SERIES of the DINANTIAN sub-SYSTEM (Lower CARBONIFEROUS).

vitrain see LITHOTYPE.

vitreous *adj.* having the appearance of GLASS.

vitric *adj.* of or pertaining to GLASS or containing glass, e.g. vitric TUFF.

vitriclastic *adj.* (of pyroclastic rock) having a structure characterised by fragmented GLASS.

vitrification *n.* the formation of a GLASS.

vitrinite see MACERAL.

vitrophyre *n.* a porphyritic igneous rock with a glassy matrix.

vogesite *n.* a CALC-ALKALINE LAMPROPHYRE essentially composed of PHENOCRYSTS of HORNBLENDE, diopsidic AUGITE and OLIVINE in a groundmass in which alkali FELDSPAR is dominant over PLAGIOCLASE.

voids *n.* empty spaces of various sizes within igneous and sedimentary rocks. If interconnected, they form pathways for fluid movement. See POROSITY.

volatile constituent *n.* a substance, dissolved in a MAGMA, that at surface temperatures and pressures would normally exist as a liquid or gas. As magma approaches the surface of the Earth the decrease in pressure causes the volatiles to exsolve, forming gas bubbles. During VOLCANIC ERUPTIONS a large proportion of the volatiles escape but some fluids are trapped as INCLUSIONS in CRYSTALS or, more commonly, in VESICLES as the rock solidifies. In volatile-rich LAVA the concentration of fluids, especially those rich in water, can lead to the chemical breakdown of early-formed minerals, such as OLIVINE, and their REPLACEMENT by late-stage SECONDARY MINERALS. See VOLCANIC GASES.

volatile matter or **volatiles** *n.* those substances, other than moisture, that are driven off as gases and vapours during the combustion of COAL.

volatiles *n.* **1.** VOLATILE MATTER.
2. VOLATILE CONSTITUENTS.

volcanic *adj.* **1.** relating to a VOLCANO.
2. see EXTRUSIVE VOLCANIC ROCKS.

volcanic ash *n.* fine-grained unconsolidated PYROCLASTIC MATERIAL under 2 millimetres in diameter.

volcanic bomb *n.* a clot of LAVA, usually of MAFIC to INTERMEDIATE composition, ejected from a VOLCANIC VENT while still molten; the less viscous varieties assume a rounded shape while solidifying in flight; if still molten when they hit the ground, they flatten to form *cowpat bombs*. Bombs composed of the more viscous types of lava are less well rounded and may form *breadcrust bombs*, which develop surface cracks because of the expansion of vesicles within them; occasionally the build-up of internal stress can cause such bombs to explode. Bombs range in size from 64 millimetres to several metres in diameter and show great variation in shape; they are normally vesicular, at least to some extent, and occasionally hollow. See also BLOCK, PYROCLASTIC MATERIAL, PYROCLASTIC ROCKS.

volcanic breccia *n.* a rock composed of abundant volcanic fragments greater than 64 millimetres in diameter. This term has no genetic significance. If the material was fragmented and deposited by PYROCLASTIC processes, it may be called a *pyroclastic breccia*. See AGGLOMERATE, PYROCLASTIC ROCKS.

volcanic conduit *n.* the channel through which volcanic material rises from depth. Compare VOLCANIC VENT.

   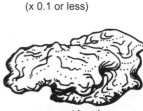

(x 0.1 or less)

spindle (fusiform) bomb   breadcrust bomb   ribbon bomb   cowpat bomb

Fig. 92 **Volcanic bomb**. Various shapes of volcanic bombs

volcanic cone *n*. a conical hill formed of LAVA and/or PYROCLASTIC MATERIALS deposited around a VOLCANIC VENT; it may be simple or composite. *Simple volcanic cones* are best typified by *cinder cones*. These are usually MONOGENETIC, sometimes symmetrical but more often open on one side, composed of cinders and other pyroclastics, rarely exceeding a few hundred metres in height. The exterior inclination is about 33°, the ANGLE OF REPOSE for loose cinder. *Composite cones* (*strato-volcanoes*) are much larger than the simple type, being built up by many eruptions, often of andesitic lava, over a long period (*polygenetic*). A significant proportion of pyroclastic material is erupted along with the lavas, resulting in a cone that is composed of layers of LAVA FLOWS alternating with pyroclastics. Compare SHIELD VOLCANO.

volcanic conglomerate *n*. a water-deposited CONGLOMERATE, generally derived from coarse PYROCLASTIC ROCKS and debris. Its consolidation is because of CEMENTATION.

volcanic crater *n*. a CRATER, normally steep-sided unless eroded, at the top or on the flanks of a VOLCANIC CONE, formed by eruption, explosion or subsidence. See MAAR, PSEUDOCRATERS, CALDERA.

volcanic dome *n*. an EXTRUSION of highly viscous MAGMA, normally of andesitic to rhyolitic composition, forming a thick bulbous dome over a VOLCANIC VENT. Endogenous domes grow by internal expansion, forcing blocks of already solidified LAVA out of and away from the vent in a chaotic tumbled mass. VOLCANIC SPINES sometimes develop during the growth of such domes if the rising magma column forces its way through the top of the dome. Collapse of a dome may result in the explosive

release of magma as a vertical or lateral blast of volcanic material, sometimes generating a NUÉE ARDENTE. Some effusions from the vent are almost solid and are thrust gradually upwards out of the vent with little lateral expansion; these are sometimes called *plug domes*. Slightly less viscous magma may extrude as a bulbous mass, growing with flow layering that is more concentric than radial. Much more rare are exogenous domes, which grow by the accumulation of viscous lava extruded from a summit vent building up layers of lava on the outside of the edifice.

volcanic dust see PYROCLASTIC MATERIAL.

volcanic earthquake *n*. SEISMIC activity associated with VOLCANIC activity. It often occurs before eruptions and in many cases can be related to the movement of hot MAGMA underground. See VOLCANIC TREMOR.

volcanic ejecta see PYROCLASTIC MATERIAL.

volcanic eruption *n*. the nature of a volcanic eruption is largely determined by the viscosity and gas content of its MAGMA. A low-viscosity magma will be easily extruded as LAVA. For a more viscous magma, the deciding factor in the degree of explosiveness is the gas content, because a viscous melt by itself cannot be explosive. Viscosity is a function of temperature and composition, namely SILICA content. A basaltic lava, which contains up to about 52 per cent silica, is much less viscous than a rhyolitic lava, which contains more than 65 per cent silica. The viscosity of andesitic lava, the silica content of which is about 60 per cent, is intermediate. BASALTS are produced at MID-OCEANIC RIDGES whereas RHYOLITES and ANDESITES are associated with SUBDUCTION ZONES. Volcanoes that occur along oceanic ridges are thus generally less explosive than subduction-zone volcanoes. See also VOLCANIC ERUPTION, TYPES OF; VOLCANO.

volcanic eruption, sites of *n.* VOLCANIC ERUPTIONS may occur at various sites relative to a pre-existing VOLCANO. In a *summit eruption* (sometimes referred to as a *central* or *terminal eruption*), activity is confined to the summit CRATER or craters. During *flank* or *lateral eruptions* some activity may take place at the summit crater, but all or most of the LAVA is erupted at a lower level from vents on the flank of the volcano. Flank eruptions that take place close to the summit are called *subterminal*. Small cones built on the flanks of volcanoes are called *parasitic* or *adventive cones*. See also FISSURE ERUPTION.

volcanic eruption, types of *n.* VOLCANIC ERUPTIONS have traditionally been arranged and classified into a sequence based on progressive explosiveness. The labels provided by this are accurate, however, only if they are applied to recognisable phases of an eruption, as the complete process may exhibit several eruptive patterns. (See VOLCANIC ERUPTION, SITES OF.) Although the grouping described here is still useful, it has been generally replaced by a more objective method based on the PYROCLASTIC MATERIAL associated with a particular eruption.

A *Hawaiian-type eruption*, the least explosive member of the sequence, is characterised by large quantities of very fluid basaltic MAGMA from which gases readily escape along with subordinate pyroclastic material. LAVA FOUNTAINS are commonly associated with this type of eruption. Although typical of Hawaiian volcanoes, similar activity has been noted elsewhere, often on oceanic-island volcanoes.

The *Strombolian-type eruption* is slightly more explosive. Less fluid, basaltic or andesitic lava permits spasmodic gas escape in the form of minor explosions that eject pyroclastic material.

The *Vulcanian-type eruption* is violent and characterised by much more viscous lava, large quantities of pyroclastic material (ash, BLOCKS) and ash clouds. During Vulcanian activity there are periods of intense explosive activity and gas-streaming interspersed with sporadic explosive activity. Violent explosive activity in the initial stages of renewed activity at a volcano is sometimes described as *ultra-vulcanian*.

The increased violence of the *Plinian-type eruption* is the result of extremely viscous gas-filled magma that is blasted out of the vent at nearly twice the velocity of sound, ejecting tremendous volumes of pyroclastic material. It is common for large parts of the volcano to blow away or collapse after such blasts. The eruption of Mount Vesuvius in AD 79 was of Plinian type.

volcanic-exhalative deposits or **sedimentary-exhalative deposits** (**SEDEX**) *n.* MASSIVE SULPHIDE deposits, usually lenticular to sheet-like in form (i.e. STRATIFORM), typically occurring at the interface between volcanic units or between volcanic and SEDIMENTARY units. There are three types of deposit: *zinc-lead-copper*, *zinc-copper* and *copper*, all containing iron in addition. The lead ORES are found associated only with RHYOLITE, which is the major host rock, while copper ores are commonly associated with MAFIC volcanics. For example, see CYPRUS-TYPE DEPOSIT. There is usually an underlying STOCKWORK, which apparently formed a conduit for the HYDROTHERMAL solutions that produced these deposits. The source of the fluids may be magmatic or may be a result of circulating sea water.

volcanic focus *n*. the apparent or assumed centre of activity beneath a VOLCANO or in a volcanic region.

volcanic gases *n*. compounds and elements that were previously dissolved in the MAGMA while under great pressure and are released as volatiles during a volcanic eruption. The gases are released when pressure is decreased as the magma reaches the surface. Most common constituents of the gases are water vapour and carbon dioxide; other constituents include sulphur dioxide, hydrogen sulphide, hydrogen chloride and nitrogen as a free element. In addition to its dependence on viscosity and composition of the lava, the explosiveness of a VOLCANIC ERUPTION is to a large degree determined by the proportion of gaseous material in the magma.

volcanic glass *n*. any GLASSY rock produced from molten LAVA or some liquid fraction of it. The cooling of such molten material is accompanied by an increase in viscosity. Because high viscosity inhibits CRYSTALLISATION, solidification at this stage by sudden cooling, such as after expulsion from a volcanic vent, tends to chill the material to a GLASS rather than to crystallise it. Volcanic glass is unstable and devitrifies in geologically short periods of time. A streaked or swirly structure is characteristic of many natural glasses. See also OBSIDIAN, TACHYLYTE.

volcanicity see VOLCANISM.

volcaniclastic *adj*. 1. (of rock texture) characteristic of rocks the clastic fabric of which is a result of any volcanic process. Compare PYROCLASTIC. 2. pertaining to fragmental rocks

containing volcanic material in any proportion whatever without regard to origin.

volcanic neck see VOLCANIC PLUG.

volcanic plug *n*. a roughly cylindrical mass of congealed MAGMA or TEPHRA that fills the conduit of an inactive VOLCANO. It may overlie a still-fluid magma column, and fragments of the solid plug are usually ejected when volcanic activity is resumed. Volcanic plugs are generally more erosion-resistant than the enclosing rock. The remnants that persist after much, or all, of the volcano has been destroyed are called *volcanic necks*. It should be noted, however, that the term 'neck' is used synonymously with 'plug', regardless of whether or not the formation stands up in relief. A large plug is often surrounded by smaller ones from *parasitic cones* (see VOLCANIC ERUPTION, SITES OF). Volcanic plugs are usually less than one kilometre in diameter but may be as high as 450 kilometres. Long, low DYKES may radiate from them; these are the erosion-resistant, exposed feeding fissures of the eroded volcano. Volcanic plugs are widely distributed, for example Edinburgh Castle Rock and the puys in the Auvergne area of central France.

volcanic products *n*. generally separated into three categories: VOLATILE CONSTITUENTS (gaseous products, including water vapour), LAVA (liquid products other than water) and PYROCLASTIC MATERIAL (solid products). Pyroclastic material includes matter solidified from fragments or drops of molten LAVA as well as pre-consolidated rock debris (lithics) derived from the walls of the conduit or the cone from explosions.

volcanic rocks *n*. 1. all extrusive IGNEOUS rocks and associated high-level intrusive ones.

2. fine-grained or glassy IGNEOUS ROCK produced by volcanic action at or near the Earth's surface, either extruded as LAVA (e.g. BASALT) or expelled explosively. Compare PYROCLASTIC ROCK.

volcanic spine *n.* an obelisk-like monolithic protrusion of extremely viscous LAVA squeezed up through the solidified surface crust of a VOLCANIC DOME or thick LAVA FLOW. Spines range from a few centimetres to hundreds of metres in height. A famous example is the spine that formed on Mount Pelée in Martinique (300 metres). *Megaspines*, which are even larger, include the PITONS of the West Indies.

volcanic-tectonic depression see CALDERA.

volcanic tremor *n.* a continuous low-frequency vibration often detected during VOLCANIC ERUPTIONS. It is thought to be caused by MAGMA flowing through underground channels.

volcanic vent *n.* an opening at the Earth's surface through which volcanic materials are extruded. Compare volcanic neck, see VOLCANIC PLUG.

volcanism, volcanicity or **vulcanism** *n.* any of various processes associated with the surficial discharge of MAGMA, VOLCANIC GASES, hot water and steam; included are volcanoes, geysers and fumaroles.

volcano *n.* a vent in the Earth's surface through which MAGMA and volatiles erupt; also the accumulated volcanic material around the vent. A volcano that is built up by a number of eruptive phases over a long period is *polygenetic*, while one built during a single eruptive phase is *monogenetic*.

Volcanoes may be either of the *fissure type*, where LAVA is erupted from linear cracks (see FISSURE ERUPTION and FLOOD BASALT), or the *central type*, where activity is concentrated around a point source. It is common for *parasitic cones*

to develop on the flanks of a volcano, and large polygenetic volcanoes may have a complex structure with a number of vents and CRATERS and in some cases a CALDERA.

Central volcanoes may build up distinctive shapes depending on the nature of the products. The eruption of very fluid basaltic lava at a volcanic centre forms broad, gently sloping SHIELD VOLCANOES, such as the Hawaiian volcanoes. Steeper VOLCANIC CONES may be built of interbedded PYROCLASTIC MATERIAL and lava (*strato-volcanoes*) or pyroclastic material alone (SCORIA cones, for example). The eruption of viscous silicic lava from a central vent forms VOLCANIC DOMES or VOLCANIC SPINES.

Most recently active volcanoes are found along zones associated with constructive and DESTRUCTIVE PLATE BOUNDARIES (see PLATE TECTONICS). They are also found in mid-plate regions, where they are thought to form over rising MANTLE PLUMES (see also HOT SPOT), and in continental RIFT VALLEYS.

Volcanoes formed along the MID-OCEANIC RIDGES and in the oceanic basins erupt mainly basaltic material from fissure eruptions and shield volcanoes.

Strato-volcanoes are characteristically associated with the SUBDUCTION ZONES of destructive plate boundaries. ANDESITE and BASALTIC ANDESITE are common products, but BASALT, DACITE or RHYOLITE may also occur. See Fig. 93 overleaf. See also VOLCANIC ERUPTION, TYPES OF.

volcanology *n.* the study of the structure, origin and petrology of volcanoes, and their role as contributors to the Earth's atmosphere, hydrosphere and the Earth's crust.

volume elasticity see BULK MODULUS, MODULUS OF ELASTICITY.

vug, vugh or **vugg** *n.* **1.** an unfilled cavity in a rock or mineral vein, commonly

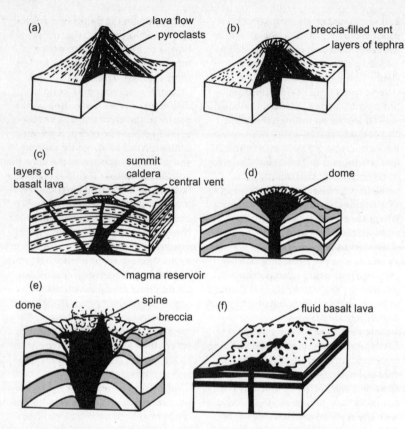

Fig. 93 **Volcano**. The shapes of the principal types of volcanic structure:
(a) strato-volcanic; (b) scoria or cinder cone; (c) shield volcano; (d) lava
dome within cone of pyroclasts; (e) dome with spine; (f) fissure eruption

not joined to other cavities and often lined with minerals deposited from hot fluids. The mineral linings of vugs are often in the form of spectacular EUHEDRAL crystals. Compare DRUSE, AMYGDALE.

2. an unfilled crystalline-encrusted cavity in a rock (see GEODE).

3. (*petroleum geology*) any opening from the size of a pea to that of a boulder. See also VUGGY.

vuggy *adj*. **1.** (of porosity) caused by VUGS

in calcareous rocks, e.g. the cavities formed by the dissolution of ooliths in limestone (see OOLITE).

2. (of a structure in volcanic rocks) resulting from the formation of irregularly shaped pockets of exsolved gas in an already crystalline matrix. Compare MIAROLITIC. See also VUG, DRUSE.

Vulcanian-type eruption see VOLCANIC ERUPTION, TYPES OF.

vulcanism see VOLCANISM.

W w

wacke *n.* a texturally and usually mineralogically immature SANDSTONE, consisting of poorly SORTED mineral and rock fragments in a matrix of fine silt and clay; in particular, an impure sandstone containing more than 10 per cent argillaceous matrix. It designates a category of sandstone, as distinguished from ARENITE. Compare GREYWACKE.

wackestone *n.* a mud-supported LIMESTONE containing more than 10 per cent carbonate grains (larger than 20 μm) in a finer-grained lime mud matrix. Wackestone is one of a number of terms in the DUNHAM CLASSIFICATION of limestones, which is based on depositional textures. See also MUDSTONE, PACKSTONE, GRAINSTONE and BOUNDSTONE.

wad *n.* manganese ore consisting of a black or dark brown impure mixture of manganese oxides.

Wadati-Benioff zone see BENIOFF ZONE.

wadi *n.* a valley with intermittent stream flow in a desert area.

wash load see STREAM LOAD.

washout *n.* a CHANNEL cut into an earlier deposit and infilled by later material. It marks the shifting of a river or stream channel (see AVULSION) and may contain blocks of the earlier deposit eroded from the banks or bed.

wastage *n.* **1.** the denudation of the Earth's surface.
2. ABLATION.

water cycle see HYDROLOGICAL CYCLE.

water gap *n.* a deep pass through a mountain ridge, occupied by a river or stream. Water gaps, in general, may result from two sets of conditions: in one case, the river forms its channel prior to uplift of the ridge and maintains its course as the ridge rises across the path. In the second case, the river forms its channel on rock lying above the structure but maintains its course while cutting down into the more resistant rocks below. Compare WIND GAP.

water of imbibition *n.* **1.** the quantity of water above the WATER TABLE that a rock can contain.
2. the quantity of water that a water-bearing substance can absorb without an increase in volume.

water of retention *n.* that portion of interstitial water in a sedimentary rock that remains in its pores under capillary pressure.

watershed or **divide** *n.* the high ground, separating adjacent DRAINAGE BASINS, that forms the boundary between them.

water table *n.* the upper surface of GROUNDWATER or the level below which an unconfined AQUIFER is permanently saturated with water. Above the water table (*zone of aeration/vadose zone*), the interstices in earth materials are partly filled with air. In the *zone of saturation*, below the water table, the interstices are completely filled with water. This zone extends from the water table downwards. The position of the water table below the ground surface is a function of the topography of an area and of local climatic conditions. See PHREATIC WATER, VADOSE WATER.

water-table well *n.* a well that taps into

unconfined GROUNDWATER. Its water level does not necessarily lie at the water-table level. Compare ARTESIAN.

water-vascular system *n*. a system of tubes and bladders inside the skeleton of an ECHINODERM that circulates fluid and operates the tube feet. These are retractable tentacles with suckers on the end and can be used for locomotion, feeding and respiration.

Waucoban *n*. the lowest of three Cambrian series in North America. A very long section of early Cambrian rocks, the Waucoban series, is located in California.

wave-cut platform or **abrasion platform** *n*. a gently sloping rock ledge that extends from high-tide level at a steep cliff base to below the low-tide level. It develops as a result of wave ABRASION at the sea-cliff base, which causes overhanging rocks to fall. Wave-cut platforms depend on rock structure and type.

wavefront *n*. **1.** a surface that is the locus of points representing the position of a progressing seismic disturbance at a specified time. **2.** the locus of all points reached by light that is emitted outwards in all directions from a point.

wavelength *n*. **1.** the distance between two successive crests or troughs in a series of harmonic waves. **2.** the distance between synformal or antiformal hinges in symmetrical periodic FOLD SYSTEMS.

wave ripple mark *n*. an oscillation ripple mark. See RIPPLE MARK.

way-up indicator *n*. a structure or feature that allows the original attitude of a rock mass to be determined and that can be used to show the direction of younging in a succession of STRATA (see YOUNG). FOSSILS in their life attitude and various types of tracks, burrows

and worm casts (TRACE FOSSILS) on the bedding surface or within the bed can help establish way-up.

Many SEDIMENTARY STRUCTURES have a characteristic relationship to the top or bottom of a bed and therefore make good way-up indicators. The most useful structures occurring within a bed are probably CROSS-BEDDING and GRADED BEDDING. The top bedding surface of a sedimentary horizon may contain RIPPLE MARKS, DESICCATION CRACKS or RAIN PRINTS, but these may also be found as CASTS on the bottom of a bed, so it is important to distinguish between the two. Some sedimentary structures, SOLE MARKS, are characteristically found on the bottom of a bed.

GEOPETAL STRUCTURES may be found in sedimentary or VOLCANIC ROCKS and allow the way-up to be determined. Pipe AMYGDALES (or pipe VESICLES) grow up from the base of lava flows, so can be used as way-up indicators. The top surface of a LAVA FLOW may be identified because it is usually more vesicular than the lower part or because of the reddish-brown WEATHERING products that develop. The upper surfaces of pillows in PILLOW LAVAS are convex upwards.

Wealden *n*. a FACIES of the British Lower CRETACEOUS. The Wealden facies consists of fluvial clays and sands deposited by meander and braided systems.

weathered layer *n*. (*seismology*) a zone of irregular thickness immediately below the surface. It is characterised by low seismic-wave velocities that are referred to as *weathering velocity*. Unless some correction is made, variations in the weathered zone near the surface might indicate fictitious 'formations' along the refraction horizon that is being mapped. The simplest method of correction determines the thickness and velocity of the weathered layer and then subtracts the delay time related to it from the observed intercept time; this is called a *weathering correction*.

weathering *n.* destructive natural processes by which rocks are altered with little or no transport of the fragmented or altered material. *Mechanical weathering* occurs with the freezing of confined water and the alternate expansion and contraction caused by temperature changes. *Chemical weathering* produces new minerals. The main chemical reactions are oxidation, hydration and solution. The main products of weathering are resistant minerals such as quartz, new minerals such as clay minerals produced through the weathering of feldspar and mica, and soluble substances carried away in solution.

weathering correction see WEATHERED LAYER.

weathering velocity see WEATHERED LAYER.

wehrlite *n.* a PERIDOTITE composed principally of OLIVINE and CLINOPYROXENE.

Weischelian see CENOZOIC, Fig. 13.

welded tuff *n.* a PYROCLASTIC ROCK formed by the sintering together of PUMICE, LAPILLI and GLASS shards. Welding can occur in air-fall TUFF and in IGNIMBRITE depending on the temperature and composition of glassy material and the weight of OVERBURDEN. It therefore tends to take place at the base of PYROCLASTIC FALL deposits where fragments of pumice collapse under the weight of overlying material and glassy material is flattened to produce a somewhat undulating planer FOLIATION or bedding fabric. The individual flattened particles often have an undulating shape in cross-section that gives them the appearance of tongues of flame,

hence the term FIAMME (from the Italian). In welded tuffs there are often three zones: a dark basal zone with complete welding, a lighter coloured zone of partial welding, and a pale, upper unwelded zone. The banded appearance of welded tuffs sometimes resembles FLOW LAYERING, thus they can easily be mistaken for RHYOLITES. See also EUTAXITIC TEXTURE.

well logging *n.* an operation wherein a continuous recording, or *log*, is made of the depth of some typical DATUM of a formation. Many types of well log are recorded by sondes, which are lowered into the borehole. Signals from the sonde are transmitted to the surface where a continuous log is made.

Electrical logging is basically the recording (made in uncased sections of a borehole) of the RESISTIVITIES or CONDUCTIVITIES of subsurface formations and of the SPONTANEOUS POTENTIALS (SP) generated in the borehole. In any means of electrical logging it is useful to classify RESERVOIR ROCKS according to porosity, composition and anisotropy. The SP log is a record of the naturally occurring potentials in the mud at different depths in the borehole. Uses of the SP log include the detection of permeable beds by obtaining values for the formation-water resistivity. *Induction logging* is a method wherein the conductivity or resistivity of formations is measured by means of induced currents without the aid of electrodes.

Radioactivity logging involves the use of gamma rays or neutrons. Since radiations from the rocks must penetrate fluids, and sometimes cement and casing, only GAMMA RAYS will do; the weak penetration capacity of alpha and beta particles renders them useless for this purpose. In *gamma-ray logging* differences in the radioactive content of the various rock layers surrounding a well are shown by corresponding differences in gamma radiation within the well bore. *Neutron logging* is an important technique for obtaining information relating to the fluid content of the rocks; the method, however, does not distinguish between water and oil. A neutron-emitting source bombards the rocks in a borehole with a constant flux of neutrons, the varying intensity of which is detected as gamma radiation. This is therefore properly described as a *neutron-gamma log*.

Unlike electrical logs, which are greatly affected by saltwater or by oil in the fluid column within the well, the radioactivity log is equally reliable with all ordinary well fluids. Other logging methods must be used to locate petroleum, however, since its natural radioactivity is low and accumulations of it cannot be observed on the gamma-ray curve. With rather few exceptions, similar rocks through-out the world provide similar res-ponses on a radioactivity log. On this basis, a particular formation can usually be identified on the log with a reasonable degree of certainty.

The purpose of a *directional log* is to ensure that a straight hole is being drilled. It is a record of the hole drift, i.e. deviation of the well bore from the vertical and the direction of that drift. It is usually obtained with a DIPMETER log. An *acoustic-velocity log* (*sonic log*) measures the time required for a sound wave to pass through a specified length of formation. The method is used extensively as a porosity and correlation log. It may be run in uncased boreholes containing oil, water or any type of mud; it cannot be

recorded in air- or gas-filled boreholes. The velocity of sound through a formation depends mainly on the velocity of sound through both the rock matrix and interpore fluid, and the amount of porosity. Sound sources include magnetostrictive metal alloys, piezoelectric quartz crystals, and the electromechanical hammer and anvil. Detectors include GEOPHONES, which use piezoelectric quartz crystals, and magnetostrictive metal alloys.

well shooting or **well velocity survey** *n*. a procedure used in SEISMIC EXPLORATION to determine velocity as a function of depth. Charges are exploded in a shallow drill hole alongside a deep exploratory borehole; an in-hole detector, placed at a number of depths from top to bottom, receives the waves, the ARRIVAL TIMES of which are then recorded.

well velocity survey see WELL SHOOTING.

Wenlock *n*. a SERIES of the SILURIAN.

Wentworth grade scale *n*. a scale to measure the grain size of sedimentary rocks. The scale ranges from less than $1/256$th of a millimetre (CLAY particles) to greater than 256 millimetres (BOULDERS). See Fig. 94 opposite. See also PHI GRADE SCALE.

Westphalian *n*. a SERIES of the SILESIAN sub-SYSTEM (CARBONIFEROUS). It is divided into four STAGES, A-D (bottom to top). The Westphalian is approximately equivalent to the COAL MEASURES.

white mica see MUSCOVITE.

whorl *n*. one complete turn of a coiled shell, see GASTROPOD, AMMONOID, NAUTILOID.

Widmanstätten pattern *n*. a pattern of intersecting lamellae seen on a cut, polished and acid-etched surface of an IRON METEORITE. The lamellae consist of a relatively nickel-poor phase, KAMACITE, which contains between 4 and 7 per

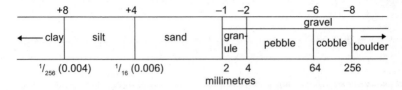

Fig. 94 **Wentworth grade scale**. Nomenclature of the Wentworth size classification for sedimentary grains and the corresponding φ scale units

cent Ni, bordered by a nickel-rich phase, TAENITE, which contains between about 27 and 65 per cent Ni. The structure is formed by EXSOLUTION of the phases from an originally homogeneous Ni-Fe alloy during very slow cooling. The lamellae form parallel to the octahedral faces of the original crystal, so meteorites showing this structure are called OCTAHEDRITES. The cooling rate has been calculated to be between 1° and 10°C per million years, suggesting formation within an ASTEROID-sized parent body.

wildcat well see EXPLORATORY WELL.

wind erosion *n*. the DEFLATION of surfaces and ABRASION of consolidated and unconsolidated material by wind action (see CONSOLIDATION). It is particularly important during dry periods in areas lacking vegetation cover. Deflation is very effective in removing the finer sands and silts from unconsolidated sediments, leaving coarser material to form a DESERT PAVEMENT. The sandblasting action of the particles carried by the wind abrades BOULDERS and PEBBLES to produce VENTIFACTS, while preferential erosion of softer or weaker rock surfaces may form YARDANGS or ZEUGEN.

wind gap *n*. a fairly shallow cut or notch in the upper part of a mountain ridge. Some wind gaps are the sites of former WATER GAPS that have been abandoned by the rivers or streams that eroded them.

window see NAPPE.

wind polish see DESERT POLISH.

Wisconsin *n*. the recognised classical fourth glacial stage of the Pleistocene in North America.

witherite *n*. a rare white or grey orthorhombic mineral, $BaCO_3$, used as a barium ore when found in large enough quantities.

wolframite *n*. a monoclinic, iron-manganese-tungstate mineral, $(Fe,Mn)Wo_4$, the principal ore of tungsten. Wolframite occurs as crystals or granular masses in pegmatites and high-temperature quartz veins.

wollastonite *n*. a triclinic silicate mineral, $CaSiO_3$, usually occurring as white or greyish radiating masses. It is normally formed as the result of contact metamorphism of limestones.

Wolstonian see CENOZOIC, Fig. 13.

wood opal see SILICIFIED WOOD.

wrench fault see FAULT.

wulfenite *n*. lead molybdate, $PbMoO_4$, a minor ore of molybdenum, found as massive granular or earthy aggregates, yellowish or reddish orange in colour. It is a secondary mineral occurring in the oxidation zone of lead deposits.

Würm see CENOZOIC, Fig. 13.

xeno- *prefix meaning* strange, different or foreign

xenoblast see CRYSTALLOBLAST.

xenocryst *n.* a CRYSTAL resembling a PHENOCRYST in an IGNEOUS ROCK but foreign to the rock body in which it occurs.

xenolith *n.* an INCLUSION in an IGNEOUS ROCK that has been introduced rather than crystallising from the same batch of MAGMA. If the xenolith is composed of material derived from a source genetically related to its host, it is called a *cognate xenolith*. If it is unrelated, such as a block of the COUNTRY ROCK, it is called an *accidental xenolith*. Xenoliths may react with the enclosing magma, resulting in chemical contamination and the formation of hybrid rocks. See ASSIMILATION, AUTOLITH, ENCLAVE.

xenomorphic *adj.* (of the texture of an igneous or metamorphic rock) characterised by minerals the crystals of which do not have their own faces. This is a result of interference of the individual minerals with each other during crystallisation. Xenomorphic crystals form when growth conditions favour continuous increase in crystal size but growth is ultimately limited by intercrystal contact. The texture is commonly shown by granitic rock. See also ANHEDRAL. Compare SUBHEDRAL.

X-ray *n.* electromagnetic radiation with a wavelength of less than 1×10^{-8}.

X-ray diffraction (XRD) *n.* an analytical method in which an X-ray beam of known wavelength is diffracted by a target crystal of an unknown mineral. The angular positions of the diffracted beams can be used to identify the mineral. See also Appendix 4.

X-ray fluorescence (XRF) *n.* an analytical method in which an X-ray beam excites atoms in elements within a rock sample, giving secondary (fluorescent) X-rays characteristic of the elements being excited. The secondary X-rays are then analysed to determine the concentrations of the elements surveyed. See also Appendix 4.

yardang *n.* an elongated ridge carved mostly from relatively weak earth materials by wind ABRASION in a desert region. Yardangs appear as sharp-edged topographical forms up to 15 metres in height but some in Iran stand over 200 metres. The ridges are separated by steep-sided troughs, both the ridges and troughs lying parallel to the prevailing winds. Yardangs are common in the deserts of Central Asia and on the Peruvian coast.

Yarmouth *n.* the classical second interglacial stage of the Pleistocene Epoch in North America; it follows the Kansan glacial stage and precedes the Illinoian.

yazoo stream *n.* a tributary stream that flows for a long distance parallel to a main stream before joining it. The type example, the Yazoo River in western Mississippi, is prevented from joining the Mississippi by the latter's elevated bed, making it higher than the back-swamp area where the Yazoo spills on to a FLOODPLAIN. There the Yazoo must continue alongside the main stream to the place where the Mississippi crosses over to the valley side.

Yeadonian *n.* the top STAGE of the NAMURIAN (CARBONIFEROUS) in Britain and western Europe.

yellow ground *n.* the soft weathered rock at the top of a KIMBERLITE pipe. See BLUE GROUND.

yellow ochre *n.* a LIMONITE mixture used as a pigment; it usually contains clay and silica.

yield-depression curve *n.* a plot for a confined AQUIFER of the DRAWDOWN versus the yield of a well or borehole. Energy dissipation as the water flows into the borehole leads to a non-darcian flow with the drawdown being greater than predicted, and the resulting graph is always a curve. See DARCY'S LAW.

yield point see ELASTIC LIMIT.

young *v.* to become younger. It is used when discussing the stratigraphical order of a succession of rocks, i.e. the direction in which the succession becomes younger. So, for example, rocks may be said to *young* towards the north. See also WAY-UP INDICATOR.

Young's modulus *n.* a MODULUS OF ELASTICITY expressing the relation between tension or compression and deformation in terms of change in length. In the equation, $E = \mu\sigma/\varepsilon$, E is Young's modulus, σ is the tensile or compressive stress and ε is the tensile or compressive strain.

Ypresian *n.* the lowest STAGE of the EOCENE.

Zz

Zanclian *n*. an AGE in the Lower PLIOCENE.

Zechstein *n*. a SERIES of the European Upper PERMIAN.

zeolite *n*. one of a large group of hydrous TECTOSILICATES. They are aluminosilicates of sodium, potassium and calcium with variable amounts of weakly bonded water molecules accommodated in large interconnected voids in the CRYSTAL LATTICE. Zeolites are characterised by their reversible loss of water of hydration and by their swelling and fission with a bubbling action (intumescence) when strongly heated. Zeolites commonly occur as well-formed crystals in cavities in BASALT and also as poorly crystallised deposits in beds of TUFF. The term 'zeolite' covers distinct mineral species including heulandite, $CaAl_2Si_7O_{18}$ $\cdot6H_2O$, natrolite, $Na_2Al_2Si_3O_{10}\cdot2H_2O$, stilbite, $NaCa_2Al_5Si_{13}O_{36}\cdot14H_2O$, and many others. As a result of their crystal configuration and 'open' structural pattern, zeolites manifest a strong capacity for BASE EXCHANGE, a property that makes them useful as water softeners. Natural and synthetic zeolites have a wide range of uses, including sewage treatment, odour control, heavy metal removal and in agricultural applications. They are useful as indicator minerals to determine the temperature and depth of burial. See ERIONITE.

zeolite facies see METAMORPHIC FACIES, Fig. 58

zeugen *n*. tabular erosional remnants, often 30 metres or more in height, formed in desert regions by the erosional effect of sand-laden winds. Zeugen are masses of resistant rock resting on supports of softer rock, having become separated by EROSION from the once continuous horizontal stratum. See also DUNE.

zinc blende see SPHALERITE.

zincite *n*. a hexagonal orange to dark-red mineral, $(Zn,Mn)O$, an ore of zinc; it is usually massive.

Zingg diagram *n*. a diagram used to plot the short, intermediate and long axes of rock particles or pebbles. The plot distinguishes four shape classes: bladed, oblate, equant and prolate.

zircon *n*. a tetragonal silicate mineral, $ZrSiO_4$, an important ore for zirconium, hafnium and thorium; some varieties are used as gemstones. Zircon is a typical ACCESSORY MINERAL of acidic igneous rocks and their metamorphic derivatives. It also occurs as a DETRITAL mineral. Hafnium substitutes for zircon in the CRYSTAL LATTICE. The trace amounts of uranium often found in zircon make it important for radiometric dating (see DATING METHODS).

zoisite *n*. an orthorhombic mineral of the EPIDOTE group, $Ca_2Al_3(SiO_4)_3(OH)$, a polymorph of CLINOZOISITE. Zoisite occurs in the form of prismatic crystals, rod-like aggregates and granular masses. It is found in IGNEOUS ROCKS where it is formed by the hydrothermal alteration of calcic plagioclase and also occurs in some medium-grade METAMORPHIC ROCKS.

zone *n*. **1.** (*crystallography*) a set of

crystal FACES the intersecting edges of which are all parallel; the direction of the edges is the *zone axis*. See also CRYSTAL SYMMETRY.

2. (*stratigraphy*) a minor STRATIGRAPHICAL UNIT, for example BIOZONE, CHRONOZONE. See also ACME ZONE, RANGE ZONE.

zone of aeration see WATER TABLE.

zone of capillarity see CAPILLARY FRINGE.

zone melting *n.* a concept borrowed from the metallurgical industry where it is applied to a process for the refining of metal by melting and recrystallisation of a mass of metal from the base upwards. The impurities are concentrated in the melt during its upward migration and can thus be removed when the melt reaches the top. It has been suggested that a similar process operates in the MANTLE to explain variations in the abundance of minor elements in basic MAGMAS.

zone of saturation see WATER TABLE.

zone of weathering *n.* the layer of the Earth's crust above the WATER TABLE that is subject to destructive atmospheric processes and in which soils develop.

zoning *n.* **1.** core-to-margin variation in the composition of a CRYSTAL. This is most easily detected in THIN SECTION between CROSSED POLARISERS under the PETROLOGICAL MICROSCOPE where the zones appear as concentric bands that mimic the outlines of EUHEDRAL crystals. In the commonest type, NORMAL ZONING, the internal zones have compositions that reflect the relatively high temperatures at the onset of crystallisation from a magma and successive zones reflect changes in composition as the magma cools. Thus, in normally zoned OLIVINE, the centre of the crystal will be rich in magnesium but the outer zones are successively enriched in iron. Zoning is an indication of non-equilibrium conditions of crystallisation. See also INVERSE ZONING.

2. in metamorphosed rocks, the development of regions in which particular minerals predominate, thus reflecting spatial variation in temperature and pressure at the time of METAMORPHISM. See METAMORPHIC ZONE.

3. the distribution pattern of minerals or elements in the region of ORE deposits.

zooid *n.* any individual animal in a colony in which all the organisms are connected, for example GRAPTOLITES, BRYOZOANS, some CNIDARIANS.

zooplankton *n.* all animals unable to swim effectively against horizontal

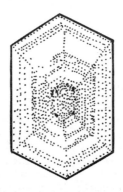

Fig. 95 **Zoning**. An augite crystal showing compositional zoning

ocean currents. The most abundant members are copepods, which are minute CRUSTACEANS. Some very large organisms, such as the Portuguese man-of-war, also belong to this group because they are poor swimmers. Zooplankton are part of the PELAGIC realm. See also PHYTOPLANKTON.

zweikanter *n.* a VENTIFACT with two edges.

Appendix I
SI units

Quantity	Unit	Symbol
length	metre	m
mass	kilogram	kg
time	second	s
electric current	ampere	A
temperature	kelvin	K
luminous intensity	candela	cd
amount of substance	mole	mol

Supplementary and derived SI units

plane angle	radian	rad
area	square metre	m^2
volume	cubic metre	m^3
velocity	metre per second	$m\ s^{-1}$
acceleration	metre per second^2	$m\ s^{-2}$
density	kilogram per metre^3	$kg\ m^{-3}$
momentum	kilogram metre per second	$kg\ m\ s^{-1}$
force	newton	$N = kg\ m\ s^{-2}$
surface tension	newton per metre	$N\ m^{-1}$
pressure, stress	newton per metre^2	$N\ m^{-2}$
	= pascal	Pa
viscosity, dynamic	newton second per metre^2	$N\ s\ m^{-2}$
work, energy, quantity of heat	joule	$J = N\ m$
power, heat flow rate	watt	$W = J\ s^{-1}$
electric charge	coulomb	C
potential difference	volt	V
resistance	ohm	Ω
intensity of magnetisation	ampere per metre	$A\ m^{-1}$
magnetic flux	weber	Wb
magnetic induction	tesla	T

Commonly used SI prefixes and magnitudes

Prefix	Magnitude	Symbol
giga-	10^9	G
mega-	10^6	M
kilo-	10^3	k
milli-	10^{-3}	m
micro-	10^{-6}	μ
nano-	10^{-9}	n
pico-	10^{-12}	p

Appendix 2
Geological Time Scale

Eon	Era	Period/sub-period	Epoch	Age	Approx. age (Ma)
P H A N E R O Z O I C	C E N O Z O I C		QUATERNARY	Holocene Pleistocene	
					1.8
		T E R T I A R Y	NEOGENE	Pliocene	Piacenzian Zanclian
				Miocene	Messinian Tortonian Serravallian Langhian Burdigalian Aquitanian
					4.0
			PALAEOGENE	Oligocene	Chattian Rupelian
				Eocene	Priabonian Bartonian Lutetian Ypresian
				Palaeocene	Thanetian Danian
					65

Eon	Era	Period/sub-period	Epoch	Age	Approx. age (Ma)
P H A N E R O Z O I C	M E S O Z O I C	CRETACEOUS	Senonian	Maastrichtian Campanian Santonian Coniacian Turonian Cenomanian	
			Neocomian	Albian Aptian Barremian Hauterivian Valanginian Ryazanian	
					144
		JURASSIC	Malm	Tithonian Kimmeridgian Oxfordian	
			Dogger	Callovian Bathonian Bajocian Aalenian	
			Lias	Toarcian Pliensbachian Sinemurian Hettangian	
					208
		TRIASSIC	Late	Rhaetian Norian Carnian	
			Middle	Ladinian Anisian	
			Early	Olenekian Induan	
					245

Eon	Era		Period/sub-period	Epoch	Age	Approx. age (Ma)
P H A N E R O Z O I C	P A L A E O Z O I C		PERMIAN	Lopingian	Changhsingian Wuchiapingian	
				Guadalupian	Capitanian Wordian Roadian	
				Cisuralian	Kungurian Artinskian Sakmarian Asselian	286
		C A R B O N I F E R O U S		Stephanian	Stephanian C Stephanian B Barruelian Cantabrian	
				Westphalian	Westphalian D Bolsovian Duckmantian Langsettian	
				Namurian	Yeadonian Marsdenian Kinderscoutian Alportian Chokierian Arnsbergian Pendleian	
			DINANTIAN	Viséan	Brigantian Asbian Holkerian Arundian Chadian (upper part)	
				Tournasian	Chadian (lower part) Courceyan	
						360

Eon	Era	Period/sub-period	Epoch	Age	Approx. age (Ma)
P H A N E R O Z O I C	P A L A E O Z O I C	DEVONIAN	Late	Famennian Frasnian	
			Middle	Givetian Eifelian	
			Early	Emsian Pragian Lochkovian	408
		SILURIAN	Přidoli		
			Ludlow	Ludfordian Gorstian	
			Wenlock	Homerian Sheinwoodian	
			Llandovery	Telychian Aeronian Rhuddanian	440
		ORDOVICIAN	Ashgill Caradoc Llanvirn Arenig Tremadoc		510
		CAMBRIAN	Merioneth	Dolgellian Maentwrogian	
			St David's	Menevian Solvan Lenan	
			Caerfai	Atdabanian Tommotian	544

Eon	Era	Epoch	Approx. age (Ma)
P R E C A M B R I A N	PROTEROZOIC	Neoproterozoic	
			900
		Mesoproterozoic	
			1600
		Palaeoproterozoic	
			2500
	ARCHAEAN	Late Archean	
			3000
		Middle Archean	
			3500
		Early Archean	
			3900
	HADEAN		
		origin of the Earth	4530

Appendix 3
Classification and Nomenclature of Igneous Rocks

Based on the QAPF and TAS Diagrams

The QAPF system is used for plutonic and volcanic igneous rocks for which mineral modes are available. It is not suitable for ultramafic rocks (M > 90), carbonatites, lamprophyres, kimberlites, charnockites or rocks containing more than 10 per cent modal melilite.

Q = quartz and other polymorphs of SiO_2

A = alkali feldspars, including orthoclase, microcline, perthite, anorthoclase, sanidine and albite (An_0 to An_5)

P = plagioclase (An_5 to An_{100}) and scapolite

F = feldspathoids (foids)

M = mafic and related minerals, including amphiboltes, olivine, pyroxene, mica, opaque minerals, accessory minerals such as apatite, zircon, sphene, etc, epidote, garnet, melilite and primary carbonate

Q + A + P + F + M = 100%.

To use the diagrams, Q + A + P + F must be recalculated to 100 per cent. See Fig. 45 for plutonic rocks and Fig. 46 for volcanic rocks.

Volcanic rocks that are too fine-grained for modes to be determined are classified on the basis of chemical composition using the TAS diagram (total alkali/silica); see Fig. 47.

The full recommendations of the International Union of Geological Sciences Subcommission on the Systematics of Igneous Rocks are given in Le Maitre, R. W. (editor), *A Classification of Igneous Rocks and Glossary of Terms*, Blackwell Scientific Publications, 1989.

Appendix 4
Apparatus and Techniques

Atomic absorbtion spectroscopy
This is a technique used in the analysis of rocks, minerals or aqueous solutions for a variety of elements. It depends on the fact that atoms of a substance in vapour form will absorb particular wavelengths of light. Light characteristic of the element to be analysed is passed through a flame and then to a detector. When the specimen in liquid form is sprayed through the flame, it absorbs light of a particular wavelength, so reducing the intensity recorded. Comparison of measurements made before and during the time the speciment was present in the flame gives the absorbance of the specimen, which is proportional to the concentration of the atoms in the vapour. Solutions of known composition are used as standards to calibrate the machine.

Cathodoluminescence (CL)
Cathodoluminescence is the process whereby visible light is emitted when electrons strike a specimen. Cathodoluminescence observations may be made using a scanning electron microscope, but specialised CL equipment is available for use with standard petrological microscopes. Specimens may be uncovered thin sections, rock slices or loose mineral grains. The colour and brightness of the CL depends not only on the composition of the specimen but on a number of factors relating to the operation of the equipment, such as the beam voltage and current, nature of the specimen surface, etc. However, fine textures and structures such as vein fillings, authigenic overgrowths and complex mineral intergrowths are well displayed in CL, as is compositional zoning that is not obvious in transmitted light microscopy.

Diamond anvil cell
In this apparatus small samples are subjected to high temperatures and pressures by compressing them between two opposing diamond crystals. Temperatures of thousands of degrees Celsius and pressures of hundreds of gigapascals can be attained, so the technique is used to simulate conditions that occur deep within the interior of the Earth and other planets. The physical properties of diamond permit X-ray experiments to be carried out on the samples *in situ* at high temperatures and pressures. This enables identification of the crystalline phases present to be made. Important

findings have been made regarding assemblages of the high-pressure mineral perovskite, which forms the bulk of the Earth's lower mantle.

Electron microprobe

A fine beam of electrons generated at several kilovolts is focused on the polished carbon-coated surface of a rock or mineral specimen. This generates an X-ray spectrum characteristic of the elements present in the area of impact. The intensity of the characteristic spectrum for each element is proportional to the concentration of that element in the sample. Generally, analysis is limited to the heavier elements with atomic numbers > 11, and detection limits are of the order of 50 to 100 ppm.

The specimen is mounted on a moveable stage, and an optical microscope is incorporated into the instrument to allow inspection and selection of the areas to be analysed. The electron beam can be focused into an area of about 1 μm diameter on the surface of the specimen, allowing analysis of selected areas within a mineral for comparison of the differences in composition between core and rim, or across a zoned specimen, for example. It is also possible to scan the beam over the surface and display the distribution of selected elements as an image on a cathode-ray screen.

Electron microscope

An instrument that uses a beam of electrons focused on the surface of a sample to produce enlarged images of objects that are much less than 0.5 μm across. Electrons have much shorter effective wavelengths than light, so electron microscopes are capable of resolving much smaller objects. The images may be produced on a fluorescent screen or photographic plate.

In the scanning electron microscope (SEM), the electron beam is scanned across the surface of the specimen to build up a picture of the whole area. It has the advantage of greater depth of focus than the conventional electron microscope and is useful for producing images of the surface texture of mineral grains or delicate structures such as microfossils, pore fillings, etc, where the relief is too great for conventional microscopes. Many machines also have an analytical facility that detects the characteristic secondary electrons or X-rays produced from the specimen by the action of the electron, in much the same way as for the electron microprobe. For quantitative analysis, however, a polished specimen is necessary since rough surfaces affect the production of X-rays.

In the transmission electron microscope (TEM) and scanning transmission electron microscope (STEM), a beam of high-energy electrons is transmitted through a very thin specimen (less than 500 nm) to produce a magnified image on a fluorescent screen or photographic plate. Information on crystal structure may be obtained from electron diffraction patterns of selected areas of the mineral grains analysed.

Ion microprobe

The ion microprobe is an instrument used in the analysis of minerals. A focused beam of ions directed at the polished surface of a rock of mineral specimen produces secondary ions from the surface, which are characteristic of the elements present. The type and amount of the different ions present are measured by means of a mass spectrometer, which sorts the ions according to their mass/charge ratio. The small size of the incident beam (a few μm in diameter) makes it possible to analyse a very small area of a mineral. The area analysed is larger than than with the electron microprobe (of the order of 10 μm compared with 1 μm) but the depth of material analysed is less.

The ion probe is in general more sensitive than the electron microprobe and is capable of detecting and measuring the concentration of trace elements in amounts less than ppm. It can also detect light elements with atomic numbers < 10, which cannot be analysed by electron microprobe. The ion probe may be used to measure the relative abundance of isotopes of an element, although the accuracy is less than for conventional mass spectroscopy. However, it has been used to measure $^{207}Pb/^{206}Pb$ ratios for radiometric dating, to study stable isotopes (for example, isotopes of O, C and S) and to investigate isotope anomalies in meteorites.

It is also possible to produce images displaying the distribution of a selected element over an area.

Laser Raman spectrometry

A laser is used to generate a beam of photons that are aimed at a sample. When the electrons strike a molecule they are scattered elastically (Rayleigh scattering) and inelastically (Raman scattering), generating Stoke's and anti-Stoke's lines. Solid or aqueous samples of any size or shape can be studied, including microscopic material down to 10 microns. Applications include the detection of different crystalline forms of elements or compounds, e.g. distinguishing between rutile and anatase (polymorphs of TiO_2) or between diamond, graphite and amorphous carbon.

Mass spectrometer

An instrument used to measure differences in the abundance of atoms and molecules by separating them on the basis of their mass/charge ratio. The specimen may be introduced into the instrument as a gas and ionised by bombardment with electrons, or as as solid deposited on a filament and ionised by heating. The ions are separated into different beams according to their mass and charge as they pass through a magnetic field before reaching the collectors.

In the Earth sciences, mass spectrometers are mainly used to measure the relative abundance of the different isotopes of elements produced by radioactive decay or the difference in the ratios of stable isotopes (such as $^{18}O/^{16}O$ or H/D) resulting from mass fractionation. They are also used in trace element analysis by isotope dilution, where a 'spike' of known isotopic ratio is added to a specimen in which the ratio of the isotopes is unknown. The concentration of the isotopes in the mixture depends upon the ratios of both 'spike' and specimen, allowing the concentration in the unknown specimen to be calculated.

Mössbauer spectrometry

A technique used in geology mainly for characterising the local environment of iron in iron-bearing minerals. A source of γ radiation is used as a probe and the γ-ray absorption spectrum of the sample determined. Mössbauer spectrometry is useful in studying a wide range of applications arising out of the prescence of iron in the Fe^{2+} and Fe^{3+} states in many minerals. These include valence states, oxidation states in relation to different crystallographic sites, electron delocalisation, site occupancy and distortion, magnetic properties and coordination of iron in silicate glasses.

Neutron activation analysis

A technique used to analyse rock specimens for rare-earth elements and trace elements such as Ba, Hf, Sc, U and Th. It is a very sensitive technique, allowing detection and measurement of elements present in very small amounts (of the order of tens of ppb). It depends on the bombardment of a specimen with nuclear particles in a reactor, resulting in the formation of unstable isotopes that decay with the emission of a characteristic energy.

Optical emission spectroscopy

This was formerly a common method of chemical analysis, particularly for trace elements. The specimen, in powder or liquid form, is volatilised by a high-energy source and emits a polychromatic beam of light that contains the characteristic emission spectra of all the elements present. The beam is refracted and dispersed according to wavelength on passing through a prism, producing a series of lines corresponding to the wavelengths of the characteristic radiations for each element. The lines corresponding to the emission spectrum for a particular element can be identified, and the intensity of the lines is a measure of the amount of that element present in the specimen.

X-ray diffraction (XRD)

This is an important method for identifying minerals. X-rays have very short wavelengths, about the dimensions of the spacing between atoms or ions in a mineral lattice. A beam of X-rays directed at a mineral is diffracted from successive parallel planes of atoms in the mineral according to Bragg's law: $n\lambda = 2d_{hkl} \sin \theta$ (see BRAGG EQUATION), where n is an integer, λ is the wavelength of the X-rays, d_{hkl} is the spacing between the planes of atoms (in angstroms) and θ is the angle of diffraction. As all crystalline materials have characteristic diffraction patterns, minerals may be identified by comparing the diffraction pattern from an unknown specimen with a set of standard patterns.

X-ray diffraction may be used to identify the minerals present in a whole rock specimen as well as identifying separated mineral specimens. The commonest method uses a finely ground powder (mineral or whole rock) packed in a special holder or smeared on a glass slide. The X-ray beam is usually monochromatic (consists of a single wavelength), although some techniques use 'white radiation', with a range of wavelengths.

Diffraction patterns are commonly recorded on a chart as a succession of peaks corresponding to the diffracting planes within the mineral(s). The x-axis of the chart is calibrated in degrees 2θ, and the d spacing between diffracting planes may be calculated using Bragg's law, since the wavelength of the X-rays is known. For identification of individual mineral specimens, the ten tallest peaks on the chart are used, as these correspond to the most strongly diffracting planes. In whole rock specimens at least three matching peaks are necessary to identify the presence of particular minerals.

Another technique, now less frequently used, records the diffraction pattern by using a special camera, producing a 'powder photograph' that is characteristic of the atomic structure of the mineral.

X-ray fluorescence spectroscopy (XRF)

This is a standard technique for determining the chemical composition of rocks. The specimen to be analysed may be in the form of finely ground loose powder, compressed powder brickettes or fused discs.

The specimen is bombarded with high-energy X-rays and emits secondary radiation, which is characteristic of the elemens present. For each element, the intensity of its characteristic radiation is proportional to the amount of that element in the specimen. The radiation emitted is measured and compared with that emitted by standards of known composition to identify the elements present.

XRF analysis is suitable for determining the concentration of major and minor elements, such as Si, Al, Mg, Ca, Fe, K, Na, Ti, S and P, and trace elements, such as Pb, Zn, Cd, Cr and Mn.

X-ray fluorescence may be used qualitatively to identify the elements present in an unknown specimen.

Appendix 5
Selected Abbreviations
and Symbols

AAS atomic absorption spectrometer
AMS accelerator mass spectrometry
ASTM American Society for Testing and Materials
ATEM analytical TEM
BS British Standard
CCD carbonate compensation depth
CEC cation-exchange-capacity
CL cathode luminescence
CMAS a projection into $CaO–MgO–Al_2O_3–SiO_2$ space
DCP direct current plasma (emission spectrometry)
DOC dissolved organic carbon
DNA deoxyribonucleic acid
DTA differential thermal analysis
EDA energy dispersive analysis
EDS energy dispersive X-ray spectroscopy
EELS electron energy-loss spectroscopy
EMPA electron probe micro-analysis
EXAFS extended X-ray absorption fine structure
GC gas chromatography
GPTS geomagnetic polarity timescale
GSSP global stratotype section and point
HFSE high field strength element
HIMU high μ mantle source region (μ is the ratio $^{238}U/^{204}Pb$
HREE heavy rare earth elements
HRSC high-resolution stepped combustion (mass spectrometry)
HRTEM high-resolution TEM
ICP inductively coupled plasma emission spectrometry

ICP–MS inductively coupled plasma emission mass spectrometry
IFS initial flooding surface
IR infrared
LA-ICP-MS laser ablation – inductively coupled plasma – mass spectrometer
LIL (LILE) large ion lithophile elements
LFS (LFSE) low field strength element
LREE light rare earth elements
LST low stand systems tract
MAR mid-Atlantic Ridge
MAS mass absorption spectrometry
MOM metabolisable organic matter
MC-ICP-MS multiple collector – inductively coupled plasma – mass spectrometer
MORB mid-ocean ridge basalt
MS mass spectrometer
NAA neutron activation analysis
NMR nuclear magnetic resonance
O.A.P. optic axial plane
ODP ocean drilling programme
OIB ocean island basalt
PGE platinum group elements
PGM platinum group minerals
PIXE proton micro-beam induced X-ray emission
POM particulate organic matter
ppb parts per billion (1 in 10^9)
ppm parts per million (1 in 10^6)
REE rare earth elements
RNA ribonucleic acid
SEM scanning electron microscope
SHRIMP sensitive high-resolution ion microprobe
SMOW standard mean ocean water

TEM transmission electron microscope
TGA thermogravimetric analysis
TL thermo-luminescence
TOC total organic carbon
UHP ultra-high pressure
UHT ultra-high temperature
XRD X-ray diffraction
XRF X-ray fluorescence

Symbols for geological time scale

T	Tertiary	D	Devonian
K	Cretaceous	S	Silurian
J	Jurassic	Ɵ	Ordovician
Ʀ	Triassic	Ɛ	Cambrian
P	Permian	V	Precambrian
C	Carboniferous		

δ (*delta*) stable isotope ratio expressed relative to a standard
Δ (*cap. delta*) difference in stable isotopic ratio (δ) for two coexisting minerals
ε (*epsilon*) neodymium (Nd) isotopic composition relative to a mantle reservoir
κ (*kappa*) the isotopic ratio $^{232}Th/^{238}U$
λ (*lambda*) radioactive decay constant
μ (*mew*) the isotopic ratio $^{238}U/^{204}Pb$

Appendix 6
Average Abundance of the
Elements (weight %)

The Earth's crust, the only part of the Earth that can be sampled directly, is composed mainly of silicate minerals, the feldspars being the commonest. The oceanic crust is essentially basaltic in composition so the abundances given are based on the average composition of ocean ridge basalts. The continental crust is much more varied but is assumed here to have a composition equivalent to approximately equal proportions of granite and basalt. The very different abundance of the elements in the Earth as a whole is a reflection of the contrasting mineralogical and geochemical nature of the subcrustal regions. Although the upper mantle is composed of peridotitic rocks rich in olivine and pyroxenes, more than 70 per cent of the deeper zones are thought to be made up of a high-pressure form of the mineral perovskite $(Mg,Fe)SiO_3$. Geophysical and cosmological evidence suggests that the Earth's core is composed mainly of iron with lesser amounts of nickel and sulphur.

Oceanic crust of the Earth

Element	symbol	%
oxygen	O	44
silicon	Si	24
aluminium	Al	8.1
iron	Fe	8.1
calcium	Ca	8.0
magnesium	Mg	4.6
sodium	Na	2.0
potassium	K	0.17
titanium	Ti	0.95
manganese	Mn	0.15
phosphorus	P	0.11
hydrogen	H	0.060
other elements		<0.03

Continental crust of the Earth

Element	symbol	%
oxygen	O	47
silicon	Si	28
aluminium	Al	8.1
iron	Fe	5.0
magnesium	Mg	2.2
calcium	Ca	3.6
sodium	Na	2.8
potassium	K	2.6
titanium	Ti	0.44
hydrogen	H	0.14
phosphorus	P	0.01
other elements		<0.01

Bulk composition of the Earth

Element	symbol	%
iron	Fe	35
oxygen	O	30
silicon	Si	15
magnesium	Mg	13
nickel	Ni	2.4
sulphur	S	1.9
calcium	Ca	1.1
aluminium	Al	1.1
other elements		<1

Appendix 7
Further Reading

General geology

Catermole, P. 2000. *Building Planet Earth*. Cambridge University Press, Cambridge.

Duff, P. McL. D., and Smith, A.J. 1992. *Geology of England and Wales*. The Geological Society Publishing House, Bath.

Marshak, S. 2001. *Earth: Portrait of Planet*. Norton, London

Press, F. and Siever, R. 2000. *Understanding Earth* (3rd edition). Freeman, New York.

Trewin, N. H. (editor). 2003. *Geology of Scotland*. (4th edition). The Geological Society Publishing House, Bath.

Woodcock, N. H., Strachan, R. A. 2000. *The Geological History of Britain and Ireland*. Blackwell Science, Oxford.

Geomorphology

Allen, P.A. 2003. *Earth Surface Processes*. Blackwell, Oxford.

Ehlers, J. 1996. *Quaternary and Glacial Geology*, John Wiley, Chichester

Geochemistry, geophysics and radiometric age dating

Bolt, B. A. 2000. *Earthquakes*. W. H. Freeman, New York.

Emsley, J. 1991. *The Elements*. Oxford University Press, Oxford.

Faure, G. 1986. *Principles of Isotope Geology* (2nd edition). John Wiley, New York.

History of geology

McIntyre, D. B. and McKirdy, A. 1997. *James Hutton: The Founder of Modern Geology*. Scottish Natural Heritage, HMSO, Edinburgh.

Sigurdsson, H. 1999. *Melting the Earth*. Oxford University Press, Oxford and New York

Igneous and metamorphic petrology

Best, M. G. 2002. *Igneous and Metamorphic Petrology*. Blackwell, Oxford.

Bowes, D. R. (editor). 1989. *The Encyclopedia of Igneous and Metamorphic Petrology*. Van Nostrand Reinhold, New York.

Cas, R. A. F. and Wright, J. V. 1987. *Volcanic Successions*. Allen & Unwin, London.

Hall, A. 1996. *Igneous Petrology*. Longman, Harlow, Essex.

MacDonald, G.A. 1972. *Volcanoes*. Prentice-Hall, New York.

Major Earth structures and plate tectonics

Kearey, P. A. and Vine, F. J. 1996. *Global Tectonics* (2nd edition). Blackwell Science, Oxford.

Redfern, R. 2000. *Origins: The Evolution of Continents, Oceans and Life*. Cassell, London.

Mineralogy and crystallography

Deer, W. A., Howie, R. A. and Zussman, J. 1992. *An Introduction to the Rock-forming Minerals*. (2nd edition) Longman Scientific and Technical, Harlow.

Johnsen, O. 2002. *Photographic Guide to Minerals of the World*. Oxford University Press, Oxford.

Klein, C. 2002. *Mineral Science*. John Wiley, New York.

Palaeontology

Bell, P. R., Helmsley, A. R. 2000. *Green Plants: Their Origin and Diversity* (2nd edition). Cambridge University Press. Cambridge.

Benton, M. J. 2000. *Vertebrate Palaeontology*. Blackwell, Oxford.

Clarkson, E. N. K. 1999. *Invertebrate Palaeontology and Evolution* (4th edition). Blackwell Science, Oxford.

Sedimentology and stratigraphy

Doyle, P., Bennett, M. R. (editors). 1998. *Unlocking the Stratigraphical Record: Advances in Modern Stratigraphy*. Wiley, Chichester.

Fritz, W. J. and Moore, J. N. 1988. *Basics of Physical Stratigraphy and Sedimentology*. John Wiley, New York.

Rawson, P. F. *et al.* 2002. *Stratigraphical Procedure*. The Geological Society Publishing House, Bath.

Finding geology resources on the internet

The internet has a vast amount of information on geology, the earth sciences and related subjects available to students, academics and enthusiastic amateurs alike. Finding reliable and free data should not be difficult although a few points need to be borne in mind. The material should be up to date. Small organizations and departments within academic institutions sometimes encounter funding difficulties and are unable to continue with their researches. Make sure to look at the 'Last updated' section of the main website page before using any data. Try clicking on the links to make sure that they have been maintained properly and do not result in error messages. Ideally information should be obtained from websites run by universities, research institutes, and other reputable organizations. Websites maintained by individuals may not be up to date and comprehensive. It is also possible that the prejudices of those maintaining the websites will be reflected in the content and list of links.

There is a very useful free 'tutorial' to finding and using information on geology and earth sciences, which can be found at:

www.vts.rdn.ac.uk/tutorial/earth

General geology and earth sciences sites

BUBL Link

http://bubl.ac.uk/link/linkbrowse.cfm?menuid=6408

A major gateway site maintained by the University of Strathclyde. The earth sciences section, within its science directory, covers many topics including geology, properties and structure of the earth, volcanoes, earthquakes, geomorphology, oceanography, global change, petrology, and economic geology.

The Physical Sciences Information Gateway (PSIgate)

www.psigate.ac.uk/newsite/earth-gateway.html

The earth sciences section of this major academic gateway site contains over 3,700 links in such areas as seismology, geochemistry, remote sensing, geomorphology, geophysics, hydrology, minerals and gemstones, petrology, planetary geology, sedimentology, stratigraphy, structural geology and tectonics, and volcanoes.

Geo-Guide

www.geo-guide.de/

The Geo-Guide is a subject gateway to over 3,000 scholarly relevant websites in earth sciences, geography and mining maintained by the: Goettingen State and University Library and the University Library 'Georgius Agricola' Freiberg.

Ocean Portal

http://oceanportal.org/

Ocean Portal is a high-level directory of ocean data and information related web sites. Its objective is to help scientists and other ocean experts in locating such data and information.

Geology Web Links

www.geologynet.com/indexa.htm

A site provided by MinServ in Australia, with a very wide range of links.

CIESIN

www.ciesin.org/

The Center for International Earth Science Information Network (CIESIN) is a centre within the Earth Institute at Columbia University. Its site provides information and many links.

Organizations

American Geological Institute
www.agiweb.org/

American Geophysical Union
www.agu.org/

British Geological Survey
www.bgs.ac.uk/

Geological Society of America
www.geosociety.org/

Geological Society of Canada
http://sparky2.esd.mun.ca/~gac/
index.html

Geologist Association
www.geologist.demon.co.uk/index.html

Geoscience Australia
http://www.ga.gov.au/

The Geological Society
www.geolsoc.org.uk/

Geological Society of New Zealand
www.gsnz.org.nz/

Geological Survey of Canada
http://gsc.nrcan.gc.ca/index_e.php

International Union of Geological
Sciences
www.iugs.org/

National Geophysical Data Center (US)
http://www.ngdc.noaa.gov/

US Geological Survey
http://www.usgs.gov/

Interesting sites

Volcanoes

USGS Volcano Information
http://volcanoes.usgs.gov

Volcano World
http://volcano.und.nodak.edu

Smithsonian Institute Global Volcanism
Program
www.volcano.si.edu/gvp

Earthquakes

USGS National Earthquake Information
Center
http://neic.usgs.gov

The Earth

NASA Natural Hazards
http://earthobservatory.nasa.gov/
NaturalHazards

NASA Observing the Earth
http://www.visibleearth.nasa.gov

Minerals

Mineralogy database
http://webmineral.com/

Mineralogy websites and information
www.geologyshop.co.uk/minera~1.htm

Department of Mineral Sciences,
Smithsonian National Museum of
Natural History
www.minerals.si.edu/

Miscellaneous

Ask an Earth Scientist (University
of Hawaii
www.soest.hawaii.edu/GG/ASK/
askanerd.html

Frequently Asked Questions
(US Geological Survey)
www.usgs.gov/search/faq.html

Geoenvironmental search engine
www.geoindex.com/geoindex/
index.html